Brief Contents

Problem-Solving Strategies and Tactics

Volume 1 (pp. 1–328) contains Chapters 1–14
Volume 2 (pp. 329–634) contains Chapters 15–26

Essential COLLEGE PHYSICS

FIRST EDITION

Volume 2
Chapters 15–26

Andrew F. Rex
University of Puget Sound

Richard Wolfson
Middlebury College

Addison-Wesley

Boston Columbus Indianapolis New York San Francisco Upper Saddle River
Amsterdam Cape Town Dubai London Madrid Milan Munich Paris Montréal Toronto
Delhi Mexico City São Paulo Sydney Hong Kong Seoul Singapore Taipei Tokyo

Publisher: Jim Smith
Executive Editor: Nancy Whilton
Development Director: Michael Gillespie
Development Editor: Gabriele Rennie
Senior Development Editor: Margot Otway
Editorial Manager: Laura Kenney
Senior Project Editor: Martha Steele
Editorial Assistant: Dyan Menezes
Editorial Assistant: Claudia Trotch
Media Producer: David Huth
Director of Marketing: Christy Lawrence
Executive Marketing Manager: Scott Dustan
Executive Market Development Manager: Scott Frost
Market Development Coordinator: Jessica Lyons

Managing Editor: Corinne Benson
Senior Production Supervisors: Nancy Tabor and Shannon Tozier
Production Service: Pre-Press PMG
Illustrations: Rolin Graphics
Text Design: Elm Street Publishing Services
Cover Design: Derek Bacchus
Manufacturing Manager: Jeff Sargent
Director, Image Resource Center: Melinda Patelli
Manager, Rights and Permissions: Zina Arabia
Image Permission Coordinator: Elaine Soares
Photo Research: Kristin Piljay
Text and Cover Printer and Binder: Courier, Kendallville
Cover Image: Mark Madeo Photography. "Formation", four men jumping over a wall at the beach in San Francisco

Credits and acknowledgments borrowed from other sources and reproduced, with permission, in this textbook appear on p. C-1.

Many of the designations used by manufacturers and sellers to distinguish their products are claimed as trademarks. Where those designations appear in this book, and the publisher was aware of a trademark claim, the designations have been printed in initial caps or all caps.

MasteringPhysics™ is a trademark, in the U.S. and/or other countries, of Pearson Education, Inc. or its affiliates.

Library of Congress Cataloging-in-Publication Data
Rex, Andrew F., 1956-
   Essential college physics / Andrew F. Rex, Richard Wolfson.—1st ed.
      p. cm.
   Includes index.
   ISBN-13: 978-0-321-61116-1 (v. 1)
   ISBN-10: 0-321-61116-0 (v. 1)
   ISBN-13: 978-0-321-61117-8 (v. 2)
   ISBN-10: 0-321-61117-9 (v. 2)
      1. Physics—Textbooks.   I. Wolfson, Richard.   II. Title.

QC23.2.R49 2009
530—dc22   2009024991

ISBN 10: 0-321-61117-9; ISBN 13: 978-0-321-61117-8 (Student edition Volume 2)

ISBN 10: 0-321-69470-8; ISBN 13: 978-0-3321-69470-6 (Professional copy Volume 2)

Addison-Wesley
is an imprint of

1 2 3 4 5 6 7 8 9 10—CRK—13 12 11 10 09
Manfactured in the United States of America.

About the Authors

Andrew F. Rex

Andrew F. Rex has been professor of physics at the University of Puget Sound since 1982. He frequently teaches the College Physics course, so he has a deep sense of student and instructor challenges. He is the author of several textbooks, including *Modern Physics for Scientists and Engineers* and *Integrated Physics and Calculus.* In addition to textbook writing, he studies foundations of the second law of thermodynamics, which has led to the publication of several papers and the widely acclaimed book *Maxwell's Demon: Entropy, Information, Computing.*

Richard Wolfson

Richard Wolfson has been professor of physics at Middlebury College for more than 25 years. In addition to his textbooks, *Essential University Physics, Physics for Scientists and Engineers,* and *Energy, Environment, and Climate,* he has written two science books for general audiences: *Nuclear Choices: A Citizen's Guide to Nuclear Technology,* and *Simply Einstein: Relativity Demystified.* His video courses for the Teaching Company include *Physics in Your Life* and *Einstein's Relativity and the Quantum Revolution: Modern Physics for Non-Scientists.*

Preface to the Instructor

During the three decades we have been teaching physics, algebra-based physics textbooks have grown in length, complexity, and price. We've reached the point where textbooks can be overwhelming to students, many of whom are taking physics as a requirement for another major or profession and will never take another physics class. And yet, we've also seen many students in the algebra-based course who are eager to learn how physics explains what they see in their everyday lives, how it connects to other disciplines, and how exciting new ideas in physics can be.

A Concise and Focused Book

The first thing you'll notice about this book is that it's more concise than most algebra-based textbooks. We believe it is possible to provide a shorter, more focused text that better addresses the learning needs of today's students while more effectively guiding them through the mastery of physics. The language is concise and engaging without sacrificing depth. Brevity needn't come at the expense of student learning! We've designed our text from the ground up to be concise and focused, rather than cutting down a longer book. Students will find the resulting book less intimidating and easier to use, with well coordinated narrative, instructional art program, and worked examples.

A Connected Approach

In addition to making the volume of the book less overwhelming, we've stressed connections, to reinforce students' understanding and to combat the preconception that physics is just a long list of facts and formulas.

Connecting ideas The organization of topics and the narrative itself stress the connections between ideas. Whenever possible, the narrative points directly to a worked example or to the next section. A worked example can serve as a bridge, not only to the preceding material it is being used to illustrate, but also forward by introducing a new idea that is then explicated in the following section. These bridges work both ways; the text is always looking forward and back to exploit the rich trail of connections that exist throughout physics.

Connecting physics with the real world Instead of simply stating the facts of physics and backing them up with examples, the book develops some key concepts from observations of real-world phenomena. This approach helps students to understand what physics is and how it relates to their lives. In addition, numerous examples and applications help students explore the ideas of physics as they relate to the real world. Connections are made to phenomena that will engage the students—applications from everyday life (heating a home, the physics of flight, DVDs, hybrid vehicles, and many more), from biomedicine (pacemakers, blood flow, cell membranes, medical imaging), and from cutting-edge research in science and technology (superconductivity, nanotechnology, ultracapacitors). These applications can be used to motivate interest in particular topics in physics, or they might emerge from learning a new physics topic. One thing leads to another. What results is a continuous story of physics, seen as a seamless whole rather than an encyclopedia of facts to be memorized.

Connecting words and math In the same way, we stress the connections between the ideas of physics and their mathematical expression. Equations are statements about physics—sentences, really—not magical formulae. In algebra-based physics, it's important to stress the basics but not the myriad details that cloud the issues for those new to the subject. We've reduced the number of enumerated equations, to make the essentials clearer.

- Complete edition Volumes 1–2 (shrinkwrapped) (ISBN 978-0-321-59854-7): Chapters 1–26
- Volume 1 (ISBN 978-0-321-61116-1): Chapters 1–14
- Volume 2 (ISBN 978-0-321-61117-8): Chapters 15–26
- Complete edition Volumes 1–2 (shrinkwrapped) with MasteringPhysics™ (ISBN 978-0-321-59856-1): Chapters 1–26
- Volume 1 with MasteringPhysics™ (ISBN 978-0-321-61118-5): Chapters 1–14
- Volume 2 with MasteringPhysics™ (ISBN 978-0-321-61119-2): Chapters 15–26

Connecting with how students learn Conceptual worked examples and end-of-chapter problems are designed to help students explore and master the qualitative ideas developed in the text. Some conceptual examples are linked with numerical examples that precede or follow them, linking qualitative and quantitative reasoning skills. Follow-up exercises to worked examples ("Making the Connection") prompt students to explore further, while "Got It?" questions (short concept-check questions found at the end of text sections) help ensure a key idea is grasped before the student moves on.

Students benefit from a structured learning path—clear goals set out at the start, reinforcement of new ideas throughout, and a strategic summary to wrap up. With these aids in place, students build a solid foundation of understanding. We therefore carefully structure the chapters with learning goals, "Reviewing new concepts" reminders, and visual chapter summaries.

Connecting with how students use their textbook Many students find using a textbook to be a chore, either because English is not their first language or because their reading skills are weak or their time limited. Even students who read with ease prefer their explanations lucid and brief, and they expect key information to be easy to find. Our goal, therefore, is a text that is clear, concise, and focused, with easy-to-find reference material, tips, and examples. The manageable size of the book makes it less intimidating to open and easier to take to class.

To complement verbal explanations in the text, the art program puts considerable information directly on the art in the form of explanatory labels and "author's voice" commentary. Thus, students can use the text and art as parallel, complementary ways to understand the material. The text tells them more, but often the illustrations will prove more memorable and will serve as keys for recalling information. In addition, a student who has difficulty with the text can turn to the art for help.

Connecting the chapters with homework After reading a chapter, students need to be able to reason their way through homework problems with some confidence that they will succeed. A textbook can help by consistently demonstrating and modeling how an expert goes about solving a problem, by giving clear tips and tactics, and by providing opportunities for practice. Given how important it is for students to become proficient at solving problems, a detailed explanation of how our textbook will help them is provided below.

Problem-Solving Strategies

Worked examples are presented consistently in a three-step approach that provides a model for students:

Organize and Plan The first step is to gain a clear picture of what the problem is asking. Then students gather information they need to address the problem, based on information presented in the text and considering similarities with earlier problems, both conceptual and numerical. If a student sketch is needed to help understand the physical situation, this is the place for it. Any known quantities that will be needed to calculate the answer or answers are gathered at the end of this step.

Solve The plan is put into action, and the required steps carried out to reach a final answer. Computations are presented in enough detail for the student to see a clear path from start to finish.

Reflect There are many things that a student might consider here. Most important is whether the answer is reasonable, in the context of either the problem or a similar known situation. This is the place to see whether units are correct or to check that symbolic answers reduce to sensible results in obvious special cases. The student may reflect on connections to other solved problems or real-life situations. Sometimes solving a problem raises a new question, which can lead naturally to another example, the next section, or the next chapter.

Conceptual examples follow a simpler two-step approach: Solve and Reflect. As with the worked examples, the Reflect step is often used to point out important connections.

Worked examples are followed by "Making the Connection," a new problem related to the one just solved, which serves as a further bridge to earlier material or the next section of text. Answers to Making the Connection are provided immediately, and thus they also serve as good practice problems—getting a second example for the price of one.

Strategy boxes follow the three-step approach that parallels the approach in worked examples. These give students additional hints about what to do in each of the three steps. "Tactic" boxes give additional problem-solving tools, outside the three-step system.

End-of-Chapter Problems

There are three types of problems:

1. *Conceptual questions,* like the conceptual worked examples, ask the students to think about the physics and reason without using numbers.

2. *Multiple-choice problems* serve three functions. First, they prepare students for their exams, in cases where instructors use that format. Second, those students who take this course in preparation for the MCAT exam or other standardized exam will get some needed practice. Third, they offer more problem-solving practice for all students.

3. *Problems* include a diversity of problem types, as well as a range of difficulty, with difficulty levels marked by one, two, or three "boxes." Problems are numerous enough to span an appropriate range of difficulty, from "confidence builders" to challenge problems. Most problems are listed under a particular section number in the chapter. General problems at the end are not tied to any section. These problem sets include multi-concept problems that require using concepts and techniques from more than one section or from an earlier chapter.

Organization of Topics

The organization of topics should be familiar to anyone who has taught College Physics. The combined Volumes 1 and 2 cover a full-year course in algebra-based physics, divided into either two semesters or three quarters.

Volume 1: Following the introductory Chapter 1, the remainder of Volume 1 is devoted to mechanics of particles and systems, including one chapter each on gravitation, fluids, and waves (including sound). Volume 1 concludes with a three-chapter sequence on thermodynamics.

Volume 2: Volume 2 begins with six chapters on electricity and magnetism, culminating and concluding with a chapter on electromagnetic waves and relativity. Following this are two chapters on optics—one on geometrical optics and one on wave optics. The final four chapters cover modern physics, including quanta, atoms, nuclei, and elementary particles.

Instructor Supplements

NOTE: For convenience, all of the following instructor supplements can also be downloaded from the "Instructor Area," accessed via the left-hand navigation bar of Mastering-Physics™ (www.masteringphysics.com).

The **Instructor Solutions Manual**, written by Brett Kraabel, Freddy Hansen, Michael Schirber, Larry Stookey, Dirk Stueber, and Robert White, provides *complete* solutions to all the end-of-chapter questions and problems. All solutions follow the Organize and Plan/Solve/Reflect problem-solving strategy used in the textbook for quantitative problems and the Solve/Reflect strategy for qualitative ones. The solutions are available by chapter in Word and PDF format and can be downloaded from the *Instructor Resource Center* (www.pearsonhighered.com/educator).

The cross-platform **Instructor Resource DVD** (ISBN 978-0-321-61126-0) provides invaluable and easy-to-use resources for your class. The contents include a comprehensive library of more than 220 applets from **ActivPhysics OnLine™**, as well as all figures, photos, tables, and summaries from the textbook in JPEG format. In addition, all the Problem-Solving Strategies, Tactics Boxes, and Key Equations are provided in editable Word as well as JPEG format. PowerPoint slides containing all the figures from the text are also included, as well as Classroom Response "Clicker" questions.

MasteringPhysics™ (www.masteringphysics.com) is a homework, tutorial, and assessment system designed to assign, assess, and track each student's progress. In addition to the textbook's end-of-chapter problems, MasteringPhysics for *Essential College Physics* also includes prebuilt assignments and tutorials.

MasteringPhysics provides instructors with a fast and effective way to assign uncompromising, wide-ranging online homework assignments of just the right difficulty and duration. The tutorials coach 90% of students to the correct answer with specific wrong-answer feedback. The powerful post-assignment diagnostics allow instructors to assess the progress of their class as a whole or to quickly identify individual students' areas of difficulty.

ActivPhysics OnLine™ (accessed through the Self Study area within www.masteringphysics.com) provides a comprehensive library of more than 420 tried and tested *ActivPhysics* applets updated for web delivery using the latest online technologies. In addition, it provides a suite of highly regarded applet-based tutorials developed by education pioneers Professors Alan Van Heuvelen and Paul D'Alessandris. The *ActivPhysics* margin icon directs students to specific exercises that complement the textbook discussion.

The online exercises are designed to encourage students to confront misconceptions, reason qualitatively about physical processes, experiment quantitatively, and learn to think critically. They cover all topics from mechanics to electricity and magnetism and from optics to modern physics. The highly acclaimed *ActivPhysics OnLine* companion workbooks help students work through complex concepts and understand them more clearly. More than 220 applets from the *ActivPhysics OnLine* library are also available on the *Instructor Resource DVD*.

The **Test Bank** contains more than 2000 high-quality problems, with a range of multiple-choice, true/false, short-answer, and regular homework-type questions. Test files are provided in both TestGen® (an easy-to-use, fully networkable program for creating and editing quizzes and exams) and Word format, and can be downloaded from www.pearson-highered.com/educator.

Student Supplements

The **Student Solutions Manuals Volume 1 (Chapters 1–14)** (ISBN 978-0-321-61120-8) and **Volume 2 (Chapters 15–26)** (ISBN 978-0-321-61128-4), written by Brett Kraabel, Freddy Hansen, Michael Schirber, Larry Stookey, Dirk Stueber, and Robert White, provide *detailed* solutions to half of the odd-numbered end-of-chapter problems. Following the problem-solving strategy presented in the text, thorough solutions are provided to carefully illustrate both the qualitative (Solve/Reflect) and quantitative (Organize and Plan/Solve/Reflect) steps in the problem-solving process.

MasteringPhysics™ (www.masteringphysics.com) is a homework, tutorial, and assessment system based on years of research into how students work physics problems and precisely where they need help. Studies show that students who use MasteringPhysics significantly increase their final scores compared to those using handwritten homework. MasteringPhysics achieves this improvement by providing students with instantaneous feedback specific to their wrong answers, simpler sub-problems upon request when they get stuck, and partial credit for their method(s) used. This individualized, 24/7 Socratic tutoring is recommended by nine out of ten students to their peers as the most effective and time-efficient way to study.

Pearson eText is available through MasteringPhysics, either automatically when MasteringPhysics is packaged with new books or as a purchased upgrade online. Allowing students access to the text wherever they have access to the Internet, Pearson eText comprises the full text, including figures that can be enlarged for better viewing. Within Pearson eText, students are also able to pop up definitions and terms to help with vocabulary and the reading of the material. Students can also take notes in Pearson eText, using the annotation feature at the top of each page.

Pearson Tutor Services (www.pearsontutorservices.com) Each student's subscription to MasteringPhysics also contains complimentary access to Pearson Tutor Services, powered by Smarthinking, Inc. By logging in with their MasteringPhysics ID and password, students will be connected to highly qualified e-instructors™ who provide additional, interactive online tutoring on the major concepts of physics. Some restrictions apply; offer subject to change.

ActivPhysics OnLine™ (accessed via www.masteringphysics.com) provides students with a suite of highly regarded applet-based tutorials (see above). The following workbooks help students work though complex concepts and understand them more clearly. The *ActivPhysics* margin icons throughout the book direct students to specific exercises that complement the textbook discussion.

ActivPhysics OnLine Workbook Volume 1: Mechanics • Thermal Physics • Oscillations & Waves (ISBN 978-0-805-39060-5)

ActivPhysics OnLine Workbook Volume 2: Electricity & Magnetism • Optics • Modern Physics (ISBN 978-0-805-39061-2)

Acknowledgments

A new full-year textbook in introductory physics doesn't just happen overnight or by accident. We begin by thanking the entire editorial and production staff at Pearson Education. The idea for this textbook grew out of discussions with Pearson editors, particularly Adam Black, whose initial encouragement and vision helped launch the project; and Nancy Whilton, who helped hone and guide this text to its current essentials state. Other Pearson staff who have rendered invaluable service to the project include Ben Roberts, Michael Gillespie, Development Manager; Margot Otway, Senior Development Editor; Gabriele Rennie, Development Editor; Mary Catherine Hagar, Development Editor; Martha Steele; Senior Project Editor; and Claudia Trotch, Editorial Assistant. In the project's early days, we were bolstered by many stimulating discussions with Jon Ogborn, whose introductory textbooks have helped improve physics education in Great Britain. In addition to the reviewers mentioned below, we are grateful to Charlie Hibbard, accuracy checker, for his close scrutiny of every word, symbol, number, and figure; to Sen-Ben Liao for meticulously solving every question and problem and providing the answer list; and to Brett Kraabel, Freddy Hansen, Michael Schirber, Larry Stookey, Dirk Stueber, and Robert White for the difficult task of writing the *Instructor Solutions Manual*. We also want to thank production supervisors Nancy Tabor and Shannon Tozier for their enthusiasm and hard work on the project; Jared Sterzer and his colleagues at Pre-Press PMG for handling the composition of the text; and Kristin Piljay, photo researcher.

Andrew Rex: I wish to thank my colleagues at the University of Puget Sound, whose support and stimulating collegiality I have enjoyed for almost 30 years. The university's staff, in particular Neva Topolski, has provided many hours of technical support throughout this textbook's development. Thanks also to student staff member Dana Maijala for her technical assistance. I acknowledge all the students I have taught over the years, especially those in College Physics classes. Seeing how they learn has helped me generate much of what you see in this book. And last but foremost, I thank my wife Sharon for her continued support, encouragement, and amazing patience throughout the length of this project.

Richard Wolfson: First among those to be acknowledged for their contributions to this project are the thousands of students in my introductory physics courses over three decades at Middlebury College. You've taught me how to convey physics ideas in many different ways appropriate to your diverse learning styles, and your enthusiasm has convinced me that physics really can appeal to a wide range of students for whom it's not their primary interest. Thanks also to my Middlebury faculty colleagues and to instructors around the world who have made suggestions that I've incorporated into my textbooks and my classrooms. It has been a pleasure to work with Andy Rex in merging our ideas and styles into a coherent final product that builds on the best of what we've both learned in our years of teaching physics. Finally, I thank my family, colleagues, and students for their patience during the intensive period when I was working on this project.

Reviewers

Chris Berven, *University of Idaho*
Benjamin C. Bromley, *University of Utah*
Michelle Chabot, *University of South Florida–Tampa*
Orion Ciftja, *Prairie View A & M University*
Joseph Dodoo, *University of Maryland–Eastern Shore*
Florence Egbe Etop, *Virginia State University*
Davene Eyres, *North Seattle Community College*
Delena Bell Gatch, *Georgia Southern University*
Barry Gilbert, *Rhode Island College*
Idan Ginsburg, *Harvard University*
Timothy T. Grove, *Indiana University–Purdue University, Fort Wayne*
Mark Hollabaugh, *Normandale Community College*
Kevin Hope, *University of Montevallo*
Joey Houston, *Michigan State University*
David Iadevaia, *Pima County Community College*
Ramanathan Jambunathan, *University of Wisconsin–Oshkosh*
Monty Mola, *Humboldt State University*
Gregor Novak, *United States Air Force Academy*
Stephen Robinson, *Belmont University*
Michael Rulison, *Ogelthorpe University*
Douglas Sherman, *San Jose State University*
James Stephens, *University of Southern Mississippi*
Rajive Tiwari, *Belmont Abbey College*
Lisa Will, *San Diego City College*
Chadwick Young, *Nicholls State University*
Sharon T. Zane, *University of Miami*
Fredy Zypman, *Yeshiva University*

Preface to the Student

Welcome to physics! Whether you're taking this course as a requirement for a pre-professional program, as a cognate for your college major, or just because you're curious, we want you to enjoy your physics experience and we hope you'll find that it's enriching and stimulating, and that it connects you with both nature and technology.

Physics is fundamental. To understand physics is to understand how the world works, both in everyday life and on scales of time and space unimaginably large and small. For that reason we hope you'll find physics fascinating. But you'll also find it challenging. Physics demands precision in thought and language, subtle interpretation of universal laws, and the skillful application of mathematics. Yet physics is also simple, because there are really only a very few basic principles to learn. Once you know those principles, you can apply them in a vast range of natural and technological applications.

We've written this book to make it engaging and readable. So read it! And read it thoroughly—*before* you begin your homework assignments. The book isn't a reference work, to be consulted only when you need to solve a particular problem or answer a particular question. Rather, it's an unfolding story of physics, emphasizing connections among different physics principles and applications, and connections to many other fields of study—including your academic major, whatever it is.

Physics is more about big ideas than it is about the nitty-gritty of equations, algebra, and numerical answers. Those details are important, but you'll appreciate them more and approach them more successfully if you see how they flow from the relatively few big ideas of physics. So look for those big ideas, and keep them in mind even as you burrow down into details.

Even though you'll need algebra to solve your physics problems, don't confuse physics with math. Math is a tool for doing physics, and the equations of physics aren't just math but statements about how the world works. Get used to understanding and appreciating physics equations as succinct and powerful statements about physical reality—not just places to "plug in" numbers.

We've written this book to give you our help in learning physics. But you can also learn a lot from your fellow students. We urge you to work together to advance your understanding, and to practice a vigorous give-and-take that will help you sharpen your intuition about physics concepts and develop your analytical skills.

Most of all, we hope you'll enjoy physics and appreciate the vast scope of this fundamental science that underlies the physical universe that we all inhabit.

Detailed Contents

15 Electric Charges, Forces, and Fields

■ This aircraft was struck by lightning, but the flight continued with the passengers unhurt and the plane undamaged. Why?

This is the first of three chapters on electric charge. We'll begin by describing the basic properties of electric charges and the electric force between charges, which, unlike gravitation, can be attractive or repulsive. Like gravity, the electric force can be described in terms of a field that surrounds charges. The electric field helps us, both visually and quantitatively, understand how charges interact, so the field will play a key role throughout Chapters 15–17.

15.1 Electric Charges

It's no exaggeration to say that electricity both runs the modern world and ties it together. From where you sit, you can surely see a machine, a communication tool, or entertainment device that uses electricity. You probably have an assortment of electronic gadgets. Your computer may be the most obvious. Portable electronic devices are everywhere: cell phones, laptops, audio and video players, games and toys. Many electronic systems are less obvious but vital, including those on cars and airplanes and in business and manufacturing.

Charge through the Ages

The ancient Greeks knew of electric charge. The Greek ηλεκτρον and the Latin *electrum* describe *amber*, a fossilized resin that becomes charged when rubbed with cloth or fur. Such a charged material exhibits **static electricity**, so called because charges on the material don't move easily. Early observers noted a slight shock or spark when the charged material was touched, and they also recognized that two charged materials attract or repel each other.

By the 1700s there were two competing electrical theories. In the **single-fluid** theory advanced by Benjamin Franklin (1706–1790), the electrical forces result from the excess or deficiency of an electrical fluid that flows from one material to another. Materials then attract or repel based on the fluid's tendency to equalize: Two objects with an excess (plus) and deficiency (minus) are attracted to one another, while a plus is repelled by a plus (and a minus by a minus), so as to avoid a further imbalance. This theory is not correct, but we've retained the *plus* and *minus* (or *positive* and *negative*) designations.

The competing theory, which eventually proved more accurate, was advanced by the French scientist Charles Du Fay (1698–1739) and his colleagues. This theory holds that there are two distinct kinds of electrical charge, with like charges repelling and opposites attracting.

In England, Stephen Gray (1696–1736) distinguished two types of materials: **conductors**, which carry electricity effectively, and **insulators**, which do not. Most materials fall into one of these categories. Two other categories emerged more recently. **Semiconductors**, the heart of modern electronics, conduct electricity much better than insulators but not as well as conductors. **Superconductors** are materials that become perfect conductors at low temperatures. We'll explore semiconductors and superconductors in Chapter 17.

Charge and Matter

Electric charge is everywhere, but it's usually not obvious. That's because atoms contain equal amounts of positive and negative charge, so they're electrically neutral. The positive charge resides on protons in the atomic nucleus. (The nucleus is described in more detail in Chapter 25.) The protons attract an equal number of negative electrons, which swarm around the nucleus to complete the neutral atom.

In 1913, Niels Bohr presented a model of the atom much like a miniature solar system, with the nucleus as the Sun and the lighter electrons orbiting like planets. (More on the Bohr model appears in Chapter 24.) As you'll learn in Section 15.2, the force between protons and electrons has the same $1/r^2$ distance dependence as in Newton's law of gravitation, so electrons in the Bohr model describe orbits similar to those of the planets. Although Bohr's model has been superseded by quantum mechanics, it's a reasonable first step in describing atoms.

Because electrons are so light, they're highly mobile. An atom can lose one or more electrons to become a **positive ion**, or it can gain one or more electrons to become a **negative ion**. Figure 15.1 shows how attraction binds such oppositely charged ions to form crystals of table salt.

The attractive force between opposite charges binds all molecules, not just ionic ones like NaCl. **Molecular bonds** ultimately result from the arrangement of positive and negative charges in neighboring atoms. That's true not only of individual molecules, but also of solids, where the electrical attraction between neighboring atoms gives the solid its structure. When you stand on the floor, electrical forces from atoms in the floor support you and keep the floor from collapsing. This is the source of the normal force you studied in Chapter 4.

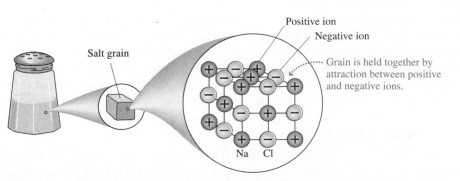

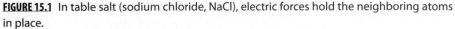

FIGURE 15.1 In table salt (sodium chloride, NaCl), electric forces hold the neighboring atoms in place.

Experiment 1

❶ Rub plastic rod with fur

❷ Touch rod to plastic ball

❸ Release ball

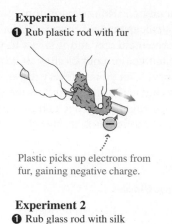

Plastic picks up electrons from fur, gaining negative charge.

Rod transfers some negative charge to ball.

Rod and ball repel each other because both are negatively charged.

Experiment 2

❶ Rub glass rod with silk

❷ Hold rod near charged ball from Experiment 1

Silk picks up electrons from glass, leaving rod with positive charge.

Rod and ball attract each other because they have opposite charges.

FIGURE 15.2 Experiments demonstrating that there are two kinds of charge.

Moving Charge Around—Do It Yourself

It's easy to convince yourself that there are two kinds of charge—for instance, follow the experiments shown in Figure 15.2. Rub a plastic rod with fur, and electrons are transferred from fur to plastic, giving the rod a negative charge. Because plastic is an insulator, the charge remains where you rubbed. Touch the rod to a light plastic ball hanging from a thread, and charge moves onto the ball. This is **charging by conduction**. Now both rod and ball are negatively charged. Bring the rod near the ball, and you'll observe the repulsive force between the two like charges. Now rub a glass rod with silk. This transfers electrons from glass to silk, leaving the rod positive. Bring this rod near your negatively charged ball, and you'll observe the attractive force between the opposite charges.

Try other experiments, such as giving an uncharged ball a positive charge and approaching with either charged rod; again you'll confirm that there are two types of charges, with opposites attracting and likes repelling.

Polarization and Electric Dipoles

Observe carefully and you'll see that an uncharged ball is slightly attracted to a charged rod *before* any charge transfer takes place. The reason is **charge polarization**. Although individual charges in the insulating ball are stuck within neutral molecules, those molecules become **polarized** by the presence of the charged rod, having more positive charge on one side and a compensating negative charge on the other. They act as **electric dipoles** that aren't free to move but can change orientation. (An electric dipole is any pair of oppositely charged point particles, and the polarized molecule approximates this ideal.) When the positively charged rod is brought nearby, the negative portion of the dipole is closer to the rod, as shown in Figure 15.3. The net effect of the positive rod on all the dipoles is a weak attraction.

✓**TIP**

Polarized molecules still have zero net charge, but the charge isn't distributed uniformly.

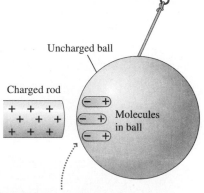

Uncharged ball

Charged rod

Molecules in ball

Presence of charged rod causes nearby molecules in ball to become polarized dipoles.

FIGURE 15.3 Attraction caused by charge polarization.

Charge polarization is a common phenomenon. Run a comb through dry hair, and you'll find it attracts small pieces of paper. Break a chunk of Styrofoam, and bits cling annoyingly to you. Rub a balloon to give it excess charge, and it sticks to your hand or a wall. In these examples the paper, the wall, and you or your clothes all become polarized and attract charged objects. A polarized molecule that you use every day is water, which is sometimes called the "universal solvent." Its chemical structure makes the water molecule a permanent electric dipole, with the hydrogen slightly positive and the oxygen negative. The molecular dipoles attract ions in an ionic solid, such as salt (NaCl), which helps dissolve the solid.

CONCEPTUAL EXAMPLE 15.1 **Attraction of Electric Dipoles**

You're holding a plastic rod near a plastic ball that's suspended by a string. (a) Suppose the ball is uncharged and the rod carries a negative charge. Explain what happens when you move the rod near the ball. (b) Now suppose the ball carries a net negative charge, but the rod is uncharged. What happens when you bring the rod near the ball?

SOLVE (a) This is just like the situation in the text, but with the charges reversed. Electric dipoles in the ball respond to the negative rod as shown in Figure 15.4a, with the positive ends of the dipoles toward the rod. This results in an attractive force between rod and ball.

(b) Now the ball carries a negative charge. Since the plastic rod is also an insulator that contains electric dipoles, the positive ends of those dipoles are attracted to the negative ball, as shown in Figure 15.4b. Again the result is an attractive force between ball and rod.

REFLECT Newton's third law is still in effect here. In each of the two cases (a) and (b), the ball exerts a force on the rod with the same magnitude and opposite direction of the force exerted on the ball by the rod. You don't notice that force because the rod and your hand are so much more massive than the ball.

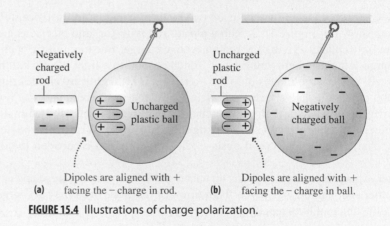

(a) Dipoles are aligned with + facing the − charge in rod.

(b) Dipoles are aligned with + facing the − charge in ball.

FIGURE 15.4 Illustrations of charge polarization.

Charging Yourself

You frequently acquire excess charge, especially in dry weather. The classic case is walking across a carpet. Your shoes pick up electrons from the insulating carpet, and if you remain insulated there's nowhere for this charge to go. Because no insulator is perfect, the charge will slowly drain off. But if you touch a metal doorknob—a conductor—much of the excess charge flows quickly into the neutral conductor, and you feel a shock.

It's a lot harder to retain charge when the air is humid. Water isn't an excellent conductor, but it's good enough to draw away excess charge. The weak conducting properties of water also make it dangerous to handle any electrical equipment when you're wet.

The shock you get from static discharge may be uncomfortable, but it's usually not dangerous. However, some situations require care. You may acquire a charge when sliding across the seat as you get out of your car; then you'll get a shock when you touch the metal of the car. If you're filling your gas tank and discharge yourself around the pump nozzle, the spark can ignite the gasoline—hence the warning shown in Figure 15.5.

Grounding—literally, connecting conductors to the Earth—can reduce the risk from electric discharge. Tanker trucks carrying flammable fuels sometimes drag a chain, which keeps the truck grounded. Before fueling an airplane, the fuel technician connects a wire between the plane and the fuel tank to prevent an electric spark. When you fill a metal gas can at the gas station, you should put the can on the ground—never in the back of your car—so it remains firmly grounded. Technicians working on computers ground themselves to prevent static discharges from damaging sensitive electronic components.

Charge Units, Conservation, and Quantization

Electric charge is a distinct physical quantity, like length, mass, and time. The SI unit of charge is the **coulomb** (C), named for the French physicist Charles Augustin de Coulomb (1736–1806), who discovered the force law for electric charges (Section 15.2). The coulomb is defined in terms of the ampere (A), the SI unit of electric current, as you'll see in Chapter 18. However, it's also convenient to think of the coulomb as being about 6.25×10^{18} **elementary charges**, where the elementary charge, e, is the magnitude of the charge on the electron and the proton and is equal to 1.602×10^{-19} C. It's remarkable that the proton and electron, although very different in mass, carry charges of exactly the same magnitude: $+e$ for the proton and $-e$ for the electron (Table 15.1). Notice that the proton and neutron have nearly the same mass, but the electron is much lighter.

In a 1911 experiment, Robert Millikan first measured the electron charge and determined that charge is **quantized**, coming in discrete multiples of the elementary charge. Today we know that there are smaller units of charge, $\pm\frac{2}{3}e$ and $\pm\frac{1}{3}e$, carried on the **quarks** that make up protons, neutrons, and many other particles.

Experiments show that charge is a **conserved** quantity. That is, net charge—the algebraic sum of all charges including their + or − signs—within a closed system remains constant. You can move charge around and transfer it from one body to another, as in charging by conduction. But the net charge in any closed system doesn't change.

FIGURE 15.5 Warning on a gas pump.

TABLE 15.1 Properties of Some Subatomic Particles

Particle name	Charge	Mass
Proton	$+e = +1.602 \times 10^{-19}$ C	$m_p = 1.673 \times 10^{-27}$ kg
Neutron	0	$m_n = 1.675 \times 10^{-27}$ kg
Electron	$-e = -1.602 \times 10^{-19}$ C	$m_e = 9.109 \times 10^{-31}$ kg

15.2 Coulomb's Law

In 1785, Coulomb used a torsion balance, a schematic of which is shown in Figure 15.6, to make quantitative measurements of the forces between different charges. The device is similar to the one Cavendish used to measure gravitational forces in 1798 (Chapter 9). Coulomb found that the magnitude of the force between two charges q_1 and q_2 is proportional to the magnitude of each charge and inversely proportional to the square of the distance between them. As you learned in Section 15.1, the direction of the force is attractive if the charges have unlike signs and repulsive if they're opposite. Coulomb's experiments showed that the magnitude of this **electrostatic force** is

$$F = \frac{k|q_1||q_2|}{r^2} \quad \text{(Coulomb's law; SI unit: N)} \quad (15.1)$$

where k is a constant and r is the distance between the two charges.

Equation 15.1 is the scalar portion of **Coulomb's law**. Because force is a vector, it's necessary also to specify the direction of the force, based on whether it's attractive or repulsive. In Section 15.3 we'll show you how to combine both magnitude and direction into a single vector equation. Notice that Equation 15.1 uses the absolute values (magnitudes) of the charges, so that all the quantities that go into this equation are positive. This guarantees that the result is positive, as a vector magnitude must be.

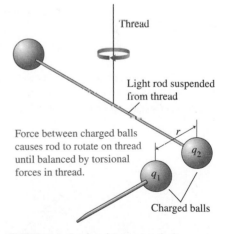

Force between charged balls causes rod to rotate on thread until balanced by torsional forces in thread.

FIGURE 15.6 Coulomb's torsion balance.

✓**TIP**

Remember that force is a vector, with a magnitude and direction.

Strictly speaking, Coulomb's force law is valid only for **point charges**—that is, charged particles of negligible size. But as you learned for gravity in Chapter 9, there's something special about an inverse-square force: Any spherically symmetric distribution can be treated as a point particle at the sphere's center. Thus the force between spherical charges—like the balls in Coulomb's experiment—is also given by Equation 15.1, provided that charge distribution is spherically symmetric and the distance r is measured from the spheres' centers.

The Constants k and ϵ_0

The quantity k in Equation 15.1 governs the strength of the electrostatic force and is analogous to the constant G in Newton's law of gravitation. In SI units,

$$k = 8.988 \times 10^9 \, \text{N} \cdot \text{m}^2/\text{C}^2$$

With charge in coulombs and distance in meters, the force in Equation 15.1 is in newtons. For reasons that will become apparent later, physicists often write k in terms of another constant ϵ_0, called the **permittivity of free space**:

$$k = \frac{1}{4\pi\epsilon_0}$$

where numerically

$$\epsilon_0 = 8.854 \times 10^{-12} \, \text{C}^2/(\text{N} \cdot \text{m}^2)$$

Many good scientific calculators "know" k and ϵ_0, the elementary charge e, and the electron, proton, and neutron masses.

✓ TIP

Find the constants k, ϵ_0, and e in your calculator. If these aren't built in, you can store them in the calculator's memory.

EXAMPLE 15.2 **Electrostatic Force in the Hydrogen Atom**

In a model of the hydrogen atom, the electron orbits a proton in a circle of radius 5.29×10^{-11} m. (a) Find the magnitudes of the electrostatic and gravitational forces between the proton and electron at that distance. (b) Find the ratio of the two magnitudes you computed in part (a).

ORGANIZE AND PLAN We've sketched the orbit in Figure 15.7. The force laws are Coulomb's law and Newton's law of gravitation. The distance between the proton and electron is known, and the numerical values of charge and mass are given in Table 15.1.

The electrostatic force is given by Equation 15.1:

$$F = \frac{k|q_1||q_2|}{r^2}$$

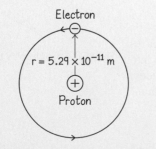

Electron

$r = 5.29 \times 10^{-11}$ m

Proton

FIGURE 15.7 Our diagram for Example 15.2.

The absolute value of the charge is the same for both the electron and proton: $|q_1| = |q_2| = e = 1.60 \times 10^{-19}$ C. The gravitational force comes from Newton's law of gravitation:

$$F = \frac{Gm_p m_e}{r^2}$$

with $m_p = 1.67 \times 10^{-27}$ kg and $m_e = 9.11 \times 10^{-31}$ kg.

Known: $r = 5.29 \times 10^{-11}$ m.

SOLVE (a) Finding the numerical values, the electrostatic force has magnitude

$$F_e = \frac{k|q_1||q_2|}{r^2} = \frac{(8.99 \times 10^9 \, \text{N} \cdot \text{m}^2/\text{C}^2)(1.60 \times 10^{-19} \, \text{C})^2}{(5.29 \times 10^{-11} \, \text{m})^2}$$

$$= 8.22 \times 10^{-8} \, \text{N}$$

Similarly, for the gravitational force:

$$F_g = \frac{Gm_p m_e}{r^2}$$

$$= \frac{(6.67 \times 10^{-11} \, \text{N} \cdot \text{m}^2/\text{kg}^2)(1.67 \times 10^{-27} \, \text{kg})(9.11 \times 10^{-31} \, \text{kg})}{(5.29 \times 10^{-11} \, \text{m})^2}$$

$$= 3.63 \times 10^{-47} \, \text{N}$$

cont'd.

(b) The ratio of the two force magnitudes is

$$\frac{F_e}{F_g} = \frac{8.22 \times 10^{-8}\,\text{N}}{3.63 \times 10^{-47}\,\text{N}} = 2.26 \times 10^{39}$$

REFLECT The electrostatic and gravitational forces differ by roughly 40 orders of magnitude! That's true not just in the hydrogen atom; because both forces have the same $1/r^2$ dependence, their ratio is the same regardless of distance. This suggests that the large bodies in the solar system must be very nearly electrically neutral; otherwise, the electrostatic force would affect their orbits, which are in fact determined almost entirely by gravity.

EXAMPLE 15.3 **Dust Levitation**

Through friction, a dust particle picks up an excess charge $q = 3.4 \times 10^{-10}$ C. Assume a spherical particle with radius $r = 5.0 \times 10^{-5}$ m and uniform density $\rho = 3500\,\text{kg/m}^3$. The dust particle is directly above an identical one with the same charge. For what distance between the particles does the electrostatic repulsion just balance the weight of the upper particle?

ORGANIZE AND PLAN Figure 15.8a shows one dust particle centered a distance d above the other. The force diagram in Figure 15.8b shows that the two forces should have equal magnitudes in order to balance.

Coulomb's law, Equation 15.1, gives the electrostatic force:

$$F = \frac{k|q_1||q_2|}{r^2} = \frac{kq^2}{d^2}$$

where d is the unknown distance. The gravitational force equals the weight of the dust particle, which near Earth is just mg. The mass m is

density times volume ($m = \rho V$), and the volume of a sphere of radius r is $V = \frac{4}{3}\pi r^3$.

Known: $r = 5.0 \times 10^{-5}$ m, $q = 3.4 \times 10^{-10}$ C, $\rho = 3500\,\text{kg/m}^3$.

SOLVE Setting the electrostatic force equal to the gravitational force (weight):

$$\frac{kq^2}{d^2} = mg = (\rho V)g = \rho\left(\frac{4}{3}\pi r^3\right)g$$

Rearranging,

$$d^2 = \frac{kq^2}{\frac{4}{3}\pi\rho g r^3} \text{ or } d = \sqrt{\frac{kq^2}{\frac{4}{3}\pi\rho g r^3}}$$

Inserting numerical values,

$$d = \sqrt{\frac{kq^2}{\frac{4}{3}\pi\rho g r^3}}$$

$$= \sqrt{\frac{(8.99 \times 10^9\,\text{N}\cdot\text{m}^2/\text{C}^2)(3.4 \times 10^{-10}\,\text{C})^2}{\frac{4}{3}\pi(3500\,\text{kg/m}^3)(9.8\,\text{m/s}^2)(5.0 \times 10^{-5}\,\text{m})^3}} = 0.24\,\text{m}$$

REFLECT In this example, a very small charge is sufficient to balance the force of gravity, again showing the relative strengths of the electrostatic and gravitational forces. Dust particles are moved easily by electrostatic forces. Look at how they stick to walls or clothing.

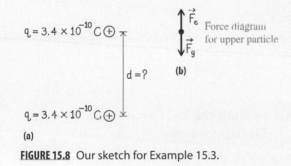

FIGURE 15.8 Our sketch for Example 15.3.

So what's a significant amount of charge? Unlike mass, length, and time, you don't yet have an intuitive feel for charge. The excess charge you develop by rubbing a plastic rod or walking across a dry carpet is typically on the order of nanocoulombs (10^{-9} C). Electronic devices called capacitors (see Chapter 16) routinely store charges ranging from picocoulombs to coulombs. Remember that 1 C contains a huge number of elementary charges (almost 10^{19}).

✓**TIP**

Review the common SI prefixes, often used with charge units.

Electrostatic attraction is essential to the operation of photocopiers and laser printers. These devices contain a cylindrical drum coated with photoconductive material that conducts only when it's illuminated. The drum is initially charged, and an image of the page to be copied is focused on the drum—either optically or via a laser that scans to "write" the image. The dark parts of the drum retain their charge, but charge leaves the illuminated, conductive parts. Then tiny plastic particles called "toner" are sprayed onto the drum, where they stick only to the charged regions. Finally, the copy paper is rolled over the drum, where it picks up the toner particles, and then through hot rollers that melt the plastic into the paper, forming a permanent copy.

GOT IT? Section 15.2   Which of the following is true of the forces on two nearby point charges, $q_1 = 2$ nC and $q_2 = 4$ nC? (a) The forces are attractive, and the force on q_1 is larger than the force on q_2. (b) The forces are attractive, and the force on q_1 is smaller than the force on q_2. (c) The forces are repulsive, and the force on q_1 is larger than the force on q_2. (d) The forces are repulsive, and the force on q_1 is smaller than the force on q_2. (e) The forces are repulsive, and the magnitudes of the two forces are the same.

15.3 Coulomb's Law for Multiple Charges

In this section, we'll show you how to handle situations involving more than two charges, as in Figure 15.9. What's the force acting on charge q_3? You might guess that it's the sum of the forces due to the other two charges:

$$\vec{F}_{3net} = \vec{F}_{13} + \vec{F}_{23} \qquad \text{(Principle of superposition)} \qquad (15.2)$$

Experiments prove that this guess is right. The fact that electric forces combine in this simple way is called the principle of **superposition**.

Force is a vector, and Equation 15.2 shows that you have to add the forces as vectors; it won't do simply to add their magnitudes. We use the notation $\vec{F}_{13}$ to mean "the force that the charge q_1 exerts on q_3," consistent with the notation introduced in Chapter 4.

Computing the Net Force

You've learned that it's usually easiest to add vectors in component form. Here's how that works in an electrostatic situation:

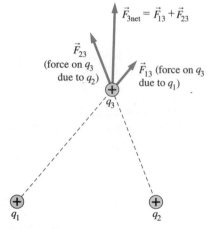

FIGURE 15.9 The net force due to multiple charges.

PROBLEM-SOLVING STRATEGY 15.1   **Finding the Net Force Due to Multiple Electric Charges**

ORGANIZE AND PLAN
- To visualize the situation (see Problem-Solving Strategy 4.1), make a schematic diagram.
- Locate the charges and note their signs.
- Identify the charge on which you want to calculate the force.
- Sketch vectors suggesting the forces on this charge from each of the others, assigning directions based on the rules of attraction and repulsion. Don't worry about magnitudes at this point.

SOLVE
- Use Coulomb's law to find the magnitudes of the individual forces.
- Set up a coordinate system and find components of the individual forces.
- Find the net force by adding vector components.

REFLECT
- Check the order of magnitude of the answer (the net force). Is it reasonable, given the individual forces you summed?
- Look at the direction of the net force. See if it makes sense, relative to the directions of the individual forces you identified.

EXAMPLE 15.4   **Three Charges in a Plane**

Three point charges lie in the x-y plane: $q_1 = 86\ \mu C$ at the origin $(0,0)$, $q_2 = 32\ \mu C$ at the point $(2.5\ m, 0)$, and $q_3 = -53\ \mu C$ at $(1.5\ m, 2.2\ m)$. Find the net force on q_1.

ORGANIZE AND PLAN Coulomb's law gives the force on q_1 due to each of the other two charges. Figure 15.10a shows q_1 repelled by the positive q_2, so $\vec{F}_{21}$ points to the left, the $-x$-direction. But q_1 is attracted to the negative q_2, so $\vec{F}_{31}$ points up and to the right, and its components have to be found using trigonometry.

The magnitudes of the two forces are

$$F_{21} = \frac{k|q_1||q_2|}{r_{21}^2}$$

where r_{21} is the distance between q_1 and q_2, and

$$F_{31} = \frac{k|q_1||q_3|}{r_{31}^2}$$

Because $\vec{F}_{31}$ points at an angle θ above the x-axis, its components are

$$F_{31,x} = F_{31}\cos\theta \quad \text{and} \quad F_{31,y} = F_{31}\sin\theta$$

SOLVE Computing the first force magnitude,

$$F_{21} = \frac{k|q_1||q_2|}{r_{21}^2}$$
$$= \frac{(8.99\times10^9\ N\cdot m^2/C^2)(86\times10^{-6}\ C)(32\times10^{-6}\ C)}{(2.5\ m - 0\ m)^2}$$
$$= 4.0\ N$$

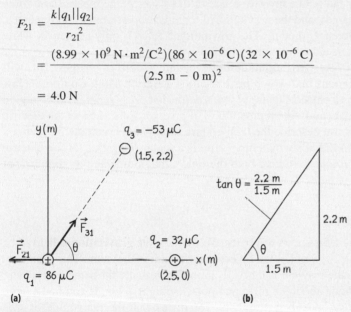

FIGURE 15.10 Our sketch for Example 15.4.

Given its direction, the force is $\vec{F}_{21} = -4.0\ N\ \hat{\imath}$.

To compute F_{31}, we use the Pythagorean theorem: $r_{31}^2 = (\Delta x)^2 + (\Delta y)^2$. Therefore,

$$F_{31} = \frac{k|q_2||q_2|}{r_{31}^2}$$
$$= \frac{(8.99\times10^9\ N\cdot m^2/C^2)(86\times10^{-6}\ C)(53\times10^{-6}\ C)}{(1.5\ m - 0\ m)^2 + (2.2\ m - 0\ m)^2}$$
$$= 5.8\ N$$

Figure 15.10b shows that $\tan\theta = \dfrac{2.2\ m}{1.5\ m} = 1.47$, so $\theta = \tan^{-1}(1.47) = 55.8°$. Then the components of $\vec{F}_{31}$ are

$$F_{31,x} = F_{31}\cos\theta = (5.8\ N)\cos(55.8°) = 3.3\ N$$
$$F_{31,y} = F_{31}\sin\theta = (5.8\ N)\sin(55.8°) = 4.8\ N$$

Using unit vectors, $\vec{F}_{31} = 3.3\ N\ \hat{\imath} + 4.8\ N\ \hat{\jmath}$. By the principle of superposition, the net force on q_1 is

$$\vec{F}_{1net} = \vec{F}_{21} + \vec{F}_{31}$$
$$= -4.0\ N\ \hat{\imath} + (3.3\ N\ \hat{\imath} + 4.8\ N\ \hat{\jmath})$$
$$= -0.7\ N\ \hat{\imath} + 4.8\ N\ \hat{\jmath}$$

REFLECT After computing vector components, finding the final vector sum is straightforward. The procedure here works with any number of charges—there are just more forces to add. It's also more difficult to tell whether the final answer is sensible. In this case, the attraction that q_1 feels toward q_3 (which lies above the x-axis) is consistent with the final answer's positive y-component.

MAKING THE CONNECTION Suppose q_1 in Example 15.4 is the net charge on a dust particle with mass 10^{-6} kg. What's the particle's acceleration?

ANSWER Force and acceleration are related by $\vec{F} = m\vec{a}$. Therefore, $\vec{a} = \vec{F}/m = -7.0\times10^5\ m/s^2\ \hat{\imath} + 4.8\times10^6\ m/s^2\ \hat{\jmath}$. As this example shows, it's no wonder dust particles are so easily accelerated. In Section 15.5 you'll study the accelerated motion of charged particles in more detail.

You've now seen how Coulomb's law lets you compute the net force on a charge due to any number of other charges. Knowing the net force on a charge is important because net force leads to acceleration (Newton's second law). Moving charges around is what makes electronic devices work and is also at the heart of chemical reactions.

Charged particles are generally quite small, and often there are huge numbers of them. Does this mean that your calculations will be tedious beyond belief, involving millions of force vectors? No! Fortunately, the idea of electric force leads quite naturally to another tool—the *electric field*—that helps with both the visualization and computations of how charges move under the influence of others. The remainder of this chapter will be devoted to electric fields.

Reviewing New Concepts

- Electric charges exert forces on one another; like charges repel, and unlike charges attract.
- The forces between electric charges are governed by Coulomb's law.
- Like any other force, the electrostatic force is a vector, with a magnitude and a direction.
- When multiple forces are present, the net force on any charge is the sum of the individual forces due to the other charges; this is the superposition principle.

GOT IT? Section 15.3 Three point charges are located on the *x*-axis as shown. The net force on the middle charge is (a) zero; (b) directed toward the left; (c) directed toward the right.

$q_1 = 1\ \mu C \qquad q_2 = 1\ \mu C \qquad\qquad q_3 = 2\ \mu C$

⊕   ⊕   ⊕ —— *x* (m)

$x = 0 \qquad x = 1\ m \qquad\qquad x = 3\ m$

15.4 Electric Fields

You've seen how the force laws for electrostatics (Coulomb's law) and gravity (Newton's law) are similar. Both depend on the inverse square of the distance; the electrostatic force is proportional to both charges, and the gravitational force is proportional to both masses. There's an important difference, though: The gravitational force is only attractive, while the electrostatic force can be either attractive or repulsive.

We'll use the concept of **electric field** to describe the interactions of charges. This will help you understand situations that would be difficult to describe using Coulomb's law alone. In physics, a field is a physical quantity that's defined at every point in some region of space. It may be a scalar field, like temperature, or a vector field such as the electric or gravitational field. You can describe the temperature throughout a volume by giving a single number at each point; that's why temperature is a scalar field. But for the electric and gravitational fields, you need to give both the field's magnitude and its direction at each point.

The Gravitational Field

The room you're sitting in defines a set of points. We'll define the **gravitational field** to be the gravitational acceleration ($\vec{g}$) of a small object at each point in the room. This is an easy field to visualize, because the gravitational acceleration $\vec{g}$ is virtually the same everywhere in the room: It has magnitude about 9.8 m/s² and points straight down. If you had a large enough room, and sensitive enough instruments, you'd find that the magnitude of $\vec{g}$ is slightly smaller the higher you go, but for most purposes you'd say that $\vec{g}$ is the same everywhere in the room. Figure 15.11a shows vectors of the gravitational field through a room's door. This is a **uniform field**, because all the vectors have the same magnitude and direction.

There's an alternate way to draw vector fields, using continuous **field lines**. Start at any point and move a short distance in the direction of the field. Then evaluate the field direction again—it may have changed—and go a short distance in the new direction. The result is a continuous field line. Repeat with different starting points, and you've got a representation of the entire field. Figure 15.11b shows the result for the uniform gravitational field.

Field represented by   **Field represented by**
field vectors   **field lines**

Draw vectors to represent the field at selected points.

Connect the vectors to form field lines.

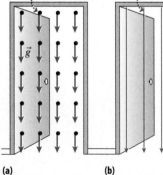

(a)   (b)

FIGURE 15.11 Two ways to represent the gravitational field in a doorway.

✓ **TIP**

A field can be represented graphically two ways, either with individual arrows (field vectors) or with continuous field lines.

If you look at a larger region around Earth, the gravitational field isn't uniform. How does it look? Remember the definition of the field $\vec{g}$: the gravitational acceleration at each point. This means each field vector points toward Earth's center. The magnitude g of this field was derived in Section 9.1 using Newton's law of gravitation:

$$g = G\frac{M}{r^2}$$ (Gravitational field of Earth [magnitude]; SI unit: m/s^2)   (15.3)

Now you can draw the $\vec{g}$ vector field. It points toward Earth's center and has a magnitude that decreases with the inverse square of distance. Figure 15.12a shows some of the field vectors, and Figure 15.12b gives the equivalent field-line picture.

We'll generally use field-line diagrams, because they show clearly both the direction and magnitude of the field. In general:

- At any point on a field line, the direction of the field vector is tangent to the field line.
- The field magnitude depends on the density of field lines. That is, where the lines are closer together, the field is stronger; where the lines are farther apart, the field is weaker.

You should convince yourself that these two facts make sense for the gravitational field in Figure 15.12. The gravitational field is a property of Earth, independent of any mass m we place in the field. Notice that only Earth's mass M, and not the "test mass" m, appears in our expression for g. The gravitational field surrounds Earth and "waits" for you to introduce the test particle of mass m; the particle then experiences a force $\vec{F} = m\vec{g}$. You can "map" the field experimentally by moving the test mass m to different locations, measuring the gravitational force $\vec{F}$, and then computing the field vector

$$\vec{g} = \frac{\vec{F}}{m}$$ (Gravitational field vector, test particle; SI unit: N/kg)   (15.4)

for each location. The direction of the force tells you the direction of the field at that location. Finally, Equation 15.4 provides another way to conceptualize the gravitational field: **The gravitational field is the force per unit mass (N/kg) on a test particle placed in the field.**

Defining the Electric Field

We've used a familiar force—gravity—to introduce vector fields. The similarity between the gravitational force and the electrostatic force makes it reasonable to define the electric field using an analogous approach. This time, we'll use a charged test particle, with positive charge q. At each point in space, **the electric field $\vec{E}$ is the force per unit charge on the test particle**. That is,

$$\vec{E} = \frac{\vec{F}}{q}$$ (Electric field; SI unit: N/C)   (15.5)

The unit, N/C, is evident in the structure of Equation 15.5.

Why should you care about electric fields? Again, consider the gravitational analog: If you know the gravitational field $\vec{g}$ at some point, you can compute the force $\vec{F} = m\vec{g}$ on a particle of any mass m at that point. From force you get acceleration, so you can determine how the particle will move. It's the same with electric fields. From Equation 15.5, a particle with charge q experiences a force $\vec{F} = q\vec{E}$ in the field. Once again, this force determines the particle's motion.

Magnitudes of field vectors decrease as the inverse square of the distance from the Earth's center.

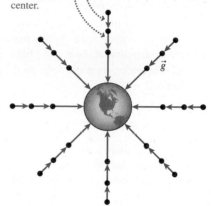

(a) Represented by field vectors

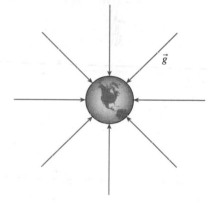

(b) Represented by field lines

FIGURE 15.12 Two ways to represent Earth's gravitational field.

EXAMPLE 15.5 **Electron in a Thunderstorm**

Below a thundercloud is a uniform electric field of magnitude 15.0 kN/C, pointing downward. Find (a) the force on an electron in this field and (b) the electron's acceleration.

ORGANIZE AND PLAN Figure 15.13 shows the uniform field and the electron located in the field. The electric field results in a force on the charged electron. Once you know the force, the acceleration follows from Newton's second law.

In general the force on a charge q follows from rearranging Equation 15.5: $\vec{F} = q\vec{E}$. In this case the particle is an electron, with charge $q = -e$, so $\vec{F} = q\vec{E} = -e\vec{E}$. With a negative charge, the force is in the opposite direction of the electric field. Newton's second law says $\vec{F} = m\vec{a}$. Therefore, once the force on the electron is known, its acceleration is $\vec{a} = \vec{F}/m$.

Known: $E = 15.0$ kN/C, elementary charge $e = 1.60 \times 10^{-19}$ C, electron mass $m_e = 9.11 \times 10^{-31}$ kg.

SOLVE (a) We'll adopt a coordinate system with the positive y-axis upward. Because the electric field points downward, it's expressed in terms of the unit vector $\hat{j}$ as $\vec{E} = -15.0$ kN/C $\hat{j} = -1.50 \times 10^4$ N/C $\hat{j}$. Then the force is

$$\vec{F} = -e\vec{E} = -(1.60 \times 10^{-19} \text{ C})(-1.50 \times 10^4 \text{ N/C } \hat{j})$$
$$= 2.40 \times 10^{-15} \text{ N } \hat{j}$$

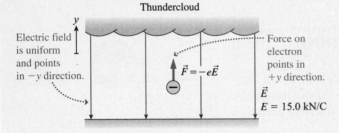

Thundercloud

Electric field is uniform and points in $-y$ direction.

$\vec{F} = -e\vec{E}$

Force on electron points in $+y$ direction.

$\vec{E}$
$E = 15.0$ kN/C

FIGURE 15.13 An electron in an electric field.

(b) This results in acceleration

$$\vec{a} = \frac{\vec{F}}{m_e} = \frac{2.4 \times 10^{-15} \text{ N } \hat{j}}{9.11 \times 10^{-31} \text{ kg}} = 2.63 \times 10^{15} \text{ m/s}^2 \, \hat{j}$$

REFLECT Although the field points *downward*, the force and thus the acceleration are *upward*. That's because of the electron's *negative* charge. The equation $\vec{F} = q\vec{E}$ shows that the force on a negatively charged particle is in the opposite direction of the electric field, while a positively charged particle experiences a force in the same direction as the field, as shown in Figure 15.14. Our electron's acceleration is large—over 10^{15} m/s². This may seem unrealistic, but their small masses mean subatomic particles can easily have large accelerations.

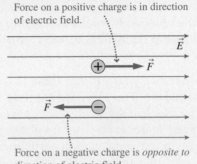

Force on a positive charge is in direction of electric field.

$\vec{E}$

$\vec{F}$

$\vec{F}$

Force on a negative charge is *opposite to* direction of electric field.

FIGURE 15.14 Force exerted by an electric field on positively and negatively charged particles.

MAKING THE CONNECTION What's the acceleration of a proton in the same electric field?

ANSWER The proton's charge is $+e$, so the force on it has the same magnitude but opposite direction as that on the electron: $\vec{F} = +e\vec{E} = -2.40 \times 10^{-15}$ N $\hat{j}$. Dividing this force by the (larger) proton mass gives the proton's acceleration: -1.44×10^{12} m/s² $\hat{j}$.

Electric Field Due to Point Charges

Electric fields influence electric charges. Conversely, electric charges create electric fields. We'll consider some common charge distributions, starting with the simplest: a single point charge.

Let our point charge be q. From Coulomb's law, the magnitude of the force q exerts on a positive test charge q_0 is

$$F = k\frac{|q|q_0}{r^2}$$

But the definition of the electric field (Equation 15.5) shows that the field magnitude is

$$E = \frac{F}{q_0} = k\frac{|q|}{r^2} \quad \text{(Electric field due to point charge; SI unit: N/C)} \quad (15.6)$$

What about the field direction? Because the test charge is positive, it's in the same direction as the force on the test charge—namely, directly away from a positive charge q shown in Figure 15.15a. Moving the test charge around to "map" the electric field gives the field shown in Figure 15.15b. It points radially away from the charge q and has a magnitude given by Equation 15.6.

Compare this result with the gravitational field of the point mass, given by Equation 15.3 and illustrated in Figure 15.12b. They look the same, except that the gravitational field points radially inward, while the electric field points radially outward. The difference is that gravity is an attractive force, but the electrostatic force is repulsive for these like charges. If we make q negative, then the attractive force on a positive test charge means the field vectors point inward, as in Figure 15.16.

Table 15.2 provides a quick summary and comparison of all the point-particle fields: gravitational fields of point masses and electric fields of positive and negative point charges. Notice the similarities and differences in the magnitudes and the directions of these fields.

Because the superposition principle applies to the electric force, it also applies to the electric field. That is, to find the electric field of a number of point charges, sum the fields of the individual charges.

To map the electric field of this charge q…

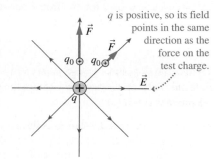
… we explore the space around it with a test charge q_0, noting the force on q_0 at each point.

(a) Using a test charge to map the field around q

q is positive, so its field points in the same direction as the force on the test charge.

(b) The resulting field $\vec{F}$

FIGURE 15.15 Mapping the electric field around a positively charged point particle.

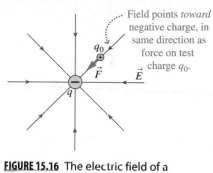
Field points *toward* negative charge, in same direction as force on test charge q_0.

FIGURE 15.16 The electric field of a negatively charged point particle.

TABLE 15.2 Summary of Point-Particle Fields

Object generating field	Type of field	Field magnitude	Field direction
Mass m	Gravitational	$g = G\dfrac{m}{r^2}$	Inward
Positive charge q	Electric	$E = k\dfrac{\lvert q\rvert}{r^2}$	Outward
Negative charge q	Electric	$E = k\dfrac{\lvert q\rvert}{r^2}$	Inward

EXAMPLE 15.6 **Electric Field of a Proton**

A proton is so small that for most purposes it can be considered a point charge. Find the magnitude of the electric field at the following distances from the proton: (a) 5.29×10^{-11} m (the distance from the proton to the electron in a hydrogen atom) and (b) 1.0 m.

ORGANIZE AND PLAN The proton's electric field points radially outward, as shown for the positive point charge in Figure 15.15b, and its magnitude is given by Equation 15.6. Its charge is $e = 1.6 \times 10^{-19}$ C.

In both cases, the electric field is given by Equation 15.6: $E = k\lvert q\rvert/r^2$, where r is the distance from the proton.

SOLVE (a) With $r = 5.29 \times 10^{-11}$ m,

$$E = k\frac{\lvert q\rvert}{r^2} = (8.99 \times 10^9 \text{ N} \cdot \text{m}^2/\text{C}^2)\frac{1.60 \times 10^{-19} \text{ C}}{(5.29 \times 10^{-11} \text{ m})^2}$$

$$= 5.14 \times 10^{11} \text{ N/C}$$

(b) At $r = 1.0$ m,

$$E = k\frac{\lvert q\rvert}{r^2} = (8.99 \times 10^9 \text{ N} \cdot \text{m}^2/\text{C}^2)\frac{1.60 \times 10^{-19} \text{ C}}{(1.0 \text{ m})^2}$$

$$= 1.44 \times 10^{-9} \text{ N/C}$$

REFLECT The electric field inside the atom is large—on the order of 10^{11} N/C. By contrast, the proton's electric field shrinks to an extremely low value at a distance of 1.0 m. Like the electrostatic force, the electric field has an inverse-square dependence, so it falls off rapidly with distance.

MAKING THE CONNECTION Find the electric field at the surface of a uranium nucleus, which you can assume is a sphere with charge $+92e$ and a radius of about 7.4×10^{-15} m. Assume that the charge is spread uniformly throughout the sphere. By analogy with what you learned about gravity, the electric field outside a uniform sphere of charge is the same as if all the charge were concentrated at the sphere's center.

ANSWER The result is an amazing 2.4×10^{21} N/C. The field is larger than either of the answers in Example 15.6. This stands to reason, because the charge creating the field is larger ($92e$ instead of e), and because you are computing the field right next to the nucleus, not out where the electrons orbit.

The Electric Dipole

An electric dipole lies on the x-axis, consisting of a -2.5-nC charge at $x = -10.0$ cm, and a $+2.5$-nC charge at $x = +10.0$ cm. Find the electric field (a) on the x-axis, at the point (20.0 cm, 0) and (b) on the y-axis, at the point (0, 10.0 cm).

ORGANIZE AND PLAN Here we need to find the fields of both point charges, and then form their vector sum. For each charge, the magnitude of the electric field follows from Equation 15.6:

$$E = k\frac{|q|}{r^2}$$

The field of the positive charge points away from the positive charge, while the field of the negative charge points toward the negative charge. The net field is their vector sum.

SOLVE (a) For convenience, call $E_{(+)}$ the magnitude of the electric field due to the positive charge and $E_{(-)}$ the magnitude of the field due to the negative charge. The distance from the positive charge to the point (20 cm, 0) is $r_{(+)} = 10$ cm $= 0.10$ m. Similarly, the distance from the negative charge to the point (20.0 cm, 0) is $r_{(-)} = 0.30$ m. Therefore,

$$E_{(+)} = k\frac{|q|}{r^2_{(+)}} = (8.99 \times 10^9\,\text{N}\cdot\text{m}^2/\text{C}^2)\frac{2.5 \times 10^{-9}\,\text{C}}{(0.10\,\text{m})^2}$$
$$= 2250\,\text{N/C}$$

and

$$E_{(-)} = k\frac{|q|}{r^2_{(-)}} = (8.99 \times 10^9\,\text{N}\cdot\text{m}^2/\text{C}^2)\frac{2.5 \times 10^{-9}\,\text{C}}{(0.30\,\text{m})^2}$$
$$= 250\,\text{N/C}$$

$\vec{E}_{(+)}$ points to the right, away from the positive charge, but $\vec{E}_{(-)}$ points to the left—toward the negative charge, as shown in Figure 15.17. In terms of the unit vector $\hat{\imath}$

$$\vec{E}_{(+)} = 2250\,\text{N/C}\,\hat{\imath} \text{ and } \vec{E}_{(-)} = -250\,\text{N/C}\,\hat{\imath}$$

Adding these vectors gives the net electric field at the point (20 cm, 0):

$$\vec{E}_{\text{net}} = \vec{E}_{(+)} + \vec{E}_{(-)} = 2250\,\text{N/C}\,\hat{\imath} + (-250\,\text{N/C}\,\hat{\imath})$$
$$= 2000\,\text{N/C}\,\hat{\imath}$$

(b) This computation is more difficult, because the vectors don't point along the same line. Using the two-dimensional

distance formula for $r_{(+)}$ and $r_{(-)}$, the magnitudes of the electric fields are

$$E_{(+)} = k\frac{|q|}{r^2_{(+)}} = (8.99 \times 10^9\,\text{N}\cdot\text{m}^2/\text{C}^2)\frac{2.5 \times 10^{-9}\,\text{C}}{(0.10\,\text{m})^2 + (0.10\,\text{m})^2}$$
$$= 1120\,\text{N/C}$$

$$E_{(-)} = k\frac{|q|}{r^2_{(-)}} = (8.99 \times 10^9\,\text{N}\cdot\text{m}^2/\text{C}^2)\frac{2.5 \times 10^{-9}\,\text{C}}{(0.10\,\text{m})^2 + (0.10\,\text{m})^2}$$
$$= 1120\,\text{N/C}$$

The direction of $\vec{E}_{(+)}$ is up and to the left, as shown in Figure 15.17, at an angle θ above the $-x$-axis. Looking at the same angle in the triangle below,

$$\tan \theta = \frac{0.10\,\text{m}}{0.10\,\text{m}} = 1.00$$

and therefore $\theta = 45°$. This means that the x- and y-components of $\vec{E}_{(+)}$ are

$$E_{(+)x} = -(1120\,\text{N/C})\cos 45° = -792\,\text{N/C} \text{ and}$$
$$E_{(+)y} = (1120\,\text{N/C})\sin 45° = 792\,\text{N/C}$$

Combining into a single vector,

$$\vec{E}_{(+)} = -792\,\text{N/C}\,\hat{\imath} + 792\,\text{N/C}\,\hat{\jmath}$$

The field due to the negative point charge has the same magnitude but points down and to the left at the same 45-degree angle, so

$$\vec{E}_{(-)} = -792\,\text{N/C}\,\hat{\imath} - 792\,\text{N/C}\,\hat{\jmath}$$

Adding the two point-charge fields gives the net electric field at (0, 10.0 cm):

$$\vec{E}_{\text{net}} = \vec{E}_{(+)} + \vec{E}_{(-)} = (-792\,\text{N/C}\,\hat{\imath} + 792\,\text{N/C}\,\hat{\jmath})$$
$$+ (-792\,\text{N/C}\,\hat{\imath} - 792\,\text{N/C}\,\hat{\jmath})$$
$$\vec{E}_{\text{net}} = -1.6 \times 10^3\,\text{N/C}\,\hat{\imath}$$

The net electric field on the $+y$-axis points straight left, as you might expect from the symmetry of the problem.

REFLECT You should convince yourself that these results are fairly general. For example, points on the $+x$-axis to the right of the positive charge are always closer to the positive charge, so its field dominates and therefore the net field is to the right. Points on the y-axis are equidistant from both charges, so the magnitudes of the field contributions from both are equal, but the different directions to the two charges make for a net field that points to the left—in the general direction from positive to negative. The fields decrease as you move farther away from the charges.

MAKING THE CONNECTION For this electric dipole, find the electric field at the origin (0, 0).

ANSWER The field points to the left with magnitude $E = 4.5$ kN/C.

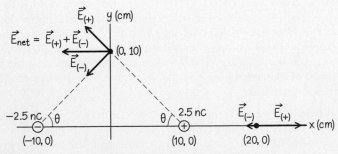

FIGURE 15.17 Our sketch for the two electric fields.

In principle, you can compute the field of an electric dipole anywhere, although for points on either axis, symmetry makes the computation easier. For points off the main axes, the electric field is most easily found using a computer to compute the field at representative points. Doing these computations leads to an electric field map, as shown in Figure 15.18. Note the symmetry of this field.

Electric Fields of Other Point-Charge Distributions

The electric field of the dipole (Figure 15.18) suggests some principles for sketching the fields of other charge distributions. First, notice that close to each charge, the field lines are approximately radial, pointing straight away from the positive charge and straight toward the negative charge. This makes sense: If you're close to one charge, its electric field dominates due to the inverse-square dependence of the field on distance. Second, note that electric field lines always emerge from positive charges and terminate on negative charges, consistent with the electric fields of single positive and negative charges (Figure 15.15b and Figure 15.16). Also notice that the number of field lines emerging from the positive charge is the same as the number of lines terminating on the negative charge. This is required by the principle that the density of field lines is proportional to the magnitude of the electric field. In an electric dipole, the charge magnitudes are equal, so the field has the same magnitude around each of the two charges.

Figure 15.19a shows the electric field around two equal positive point charges. The field is still symmetric and shows the combined effects of the two individual fields. In contrast, consider the electric field shown in Figure 15.19b, in the region around a positive charge $+Q$ and a negative charge $-Q/2$. Here, there are twice as many lines leaving the positive charge as there are ending on the negative, indicating the magnitudes of the charges and their individual fields. The "extra" field lines from the positive charge extend indefinitely, and far from the charge distribution its field begins to resemble that of a point charge $+Q/2$—which is the net charge of the entire distribution.

Electric Field of a Uniformly Charged Plane

In principle, superposition allows you to find the electric field of any charge distribution. The calculation may be difficult, however, with a large number of charges. As an example, consider the distribution of charge over an infinite plane (Figure 15.20a). The charge is spread uniformly over the surface, with the symbol σ (Greek lowercase sigma) used to represent the surface charge density—that is, the charge per unit area, in coulombs per meter squared. On a very fine scale, charge is, of course, quantized as individual electrons and protons. But if there's enough excess charge on a surface, the charge distribution appears uniform on a macroscopic scale, and the "graininess" of the charge isn't noticeable.

What's the electric field of this uniformly charged plane? Figure 15.20b shows a method for finding the electric field at a point P above the plane. Look at the electric field at P generated by the areas of charge at P' and P''. When you add the electric field from those two charges, the field's components parallel to the plane cancel, but components perpendicular to the plane add. If you repeat this process until you've accounted for all the charge in the plane, the result is an electric field vector perpendicular to the plane, and pointing away from it.

Summing the field contributions from the entire plane isn't easy, but the result gives a simple expression for magnitude of the electric field at P:

$$E = \frac{\sigma}{2\epsilon_0} \quad \text{(Electric field of a uniformly charged plane; SI unit: N/C)} \quad (15.7)$$

where ϵ_0 is the permittivity of free space. (The constant ϵ_0 in Equation 15.7 can be traced back to the electric field of a point charge, which involves the constant $k = 1/4\pi\epsilon_0$.)

It may surprise you that the magnitude of the field doesn't depend on the distance between P and the charged plane. However, this is exactly true only for an infinite plane.

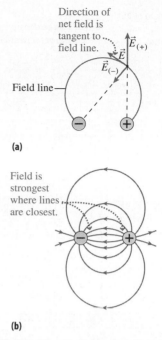

FIGURE 15.18 Electric field of an electric dipole. (a) At each point, the field-line direction is that of the *net* electric field $\vec{E}_{net} = \vec{E}_{(+)} + \vec{E}_{(-)}$. (b) Tracing several field lines shows the overall dipole field.

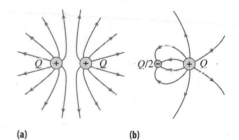

(a)   (b)

FIGURE 15.19 Electric field around (a) two equal positive charges and (b) a positive and a negative charge of unequal magnitude.

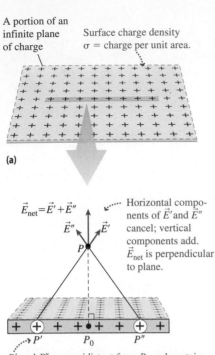

(a)

(b)

P' and P'' are equidistant from P_0 and contain equal amounts of charge. P_0 is directly below P.

FIGURE 15.20 Electric field of a uniformly charged plane.

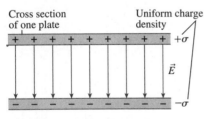

FIGURE 15.21 The electric field between oppositely charged parallel plates.

Therefore, you should take Equation 15.7 as an *approximation*, valid for points P that are close to a flat, uniformly charged surface. "Close" means at a distance that is small compared with the linear dimensions of the surface. For the symmetry argument of Figure 15.20 to be valid, the point must also be far from the edges.

Charged Parallel Planes

We've picked the charged plane as an example not just because there's a simple expression for its field, but because it's also of practical importance. That's because, as we've just seen, the field of the charged plane is essentially uniform over a wide region of space—similar to the gravitational field near Earth.

You can make an even more uniform electric field using two parallel planes, carrying charge densities $\pm\sigma$ (Figure 15.21). The electric field due to the negative plane has the same magnitude ($\sigma/2\epsilon_0$) as the positive plane, but its field points toward the plane rather than away.

Between the plates, the field contributions from the two are in the same direction, so the magnitude of the net field is just their sum:

$$E_{net} = E_{(+)} + E_{(-)} = \frac{\sigma}{2\epsilon_0} + \frac{\sigma}{2\epsilon_0} = \frac{\sigma}{\epsilon_0}$$

or

$$E_{net} = \frac{\sigma}{\epsilon_0} \quad \text{(Electric field, between oppositely charged parallel planes; SI unit: N/C)} \quad (15.8)$$

For example, if the numerical value of σ is 2.10×10^{-9} C/m^2, the electric field between the plates is

$$E_{net} = \frac{\sigma}{\epsilon_0} = \frac{2.10 \times 10^{-9}\,\text{C/m}^2}{8.854 \times 10^{-12}\,\text{C}^2/(\text{N}\cdot\text{m}^2)} = 237\,\text{N/C}$$

Notice how the units combine to give the correct units for electric field (N/C).

So a pair of parallel, charged plates is an excellent device for producing a uniform electric field. A second important application of this geometry is in an electronic component called a *capacitor*. A capacitor consists of a pair of oppositely charged conducting plates. Capacitors store charge and energy in electric circuits and are discussed further in Section 16.4.

Reviewing New Concepts: Electric Fields

- A field is a quantity defined for every point in a region of space.
- The electric field is the force per unit charge on a positive test charge at that point.
- Electric field vectors and field lines allow you to visualize the field throughout space.
- Electric field lines emerge from positive charges and end on negative charges.

Electric Fields and Conductors

Metallic conductors contain many free electrons—one or more per atom. This gives conductors some interesting properties with regard to electric fields. When a conductor is given excess charge, positive or negative, the charges reach equilibrium and stop moving almost instantly. That means the net electric field inside a conductor in equilibrium is zero. If that weren't true, free electrons in the conductor would continue to experience a force $\vec{F} = q\vec{E}$ and would therefore accelerate.

Any excess charge placed on a conductor quickly goes to the conductor's surface. You might expect this, because of the mutual repulsion of the like charges, although it also depends on the precise nature of the inverse-square electric force.

APPLICATION **Electric Fish**

A number of fish species routinely generate electric fields that, depending on the species, range from minimal to about 1000 N/C. Fish that produce large electric fields use them to stun their prey. In others, such as the glass knifefish shown here, the field helps with navigation, communication, and prey detection. Sharks don't produce electric fields, but they have a highly refined capacity to detect fields as small as 10^{-6} N/C. They use this detection system to sense nearby prey, which generate weak electric fields via their swim muscles. Stronger fields actually disturb sharks and make them leave the area. Electric field generators for swimmers have proved effective in discouraging shark attacks.

The net electric field is zero inside a conductor, even if you attempt to impose a field from the outside. In this case, the free charges in the conductor rearrange themselves to keep the electric field zero inside (Figure 15.22). Experimenters who make sensitive electronic measurements often do so inside a so-called "Faraday cage"—a metal box or metal screen enclosure—to cancel any stray electric fields from outside.

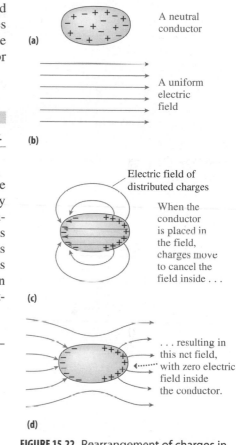

(a) A neutral conductor

(b) A uniform electric field

(c) Electric field of distributed charges
When the conductor is placed in the field, charges move to cancel the field inside . . .

(d) . . . resulting in this net field, with zero electric field inside the conductor.

FIGURE 15.22 Rearrangement of charges in a conductor placed in a uniform electric field.

TIP

To insulate yourself from the effects of electric fields, surround yourself with a conductor.

Mutual repulsion is what sends charge to the surface of a conductor. If that surface includes narrow projections or sharp points, charge accumulates there because that way it can get farther from charge elsewhere on the surface (Figure 15.23), and such a distribution is required to keep the electric field zero inside the conductor. This is the physics behind the **lightning rod**, a metal rod placed vertically on top of a building. Charge builds readily on the rod's end, and this charge leaks gradually into the atmosphere and thus helps prevent a sudden lightning discharge. If lightning does strike, the discharge is drawn to the charged lightning rod. A heavy wire connects the rod to ground, delivering the lightning's charge to the ground without harming the building.

GOT IT? Section 15.4 Four point charges are located on the x-axis as shown. The electric field at $x = 0$ is (a) zero; (b) directed toward the left; (c) directed toward the right.

$$-2\ \mu C \qquad +1\ \mu C \qquad\qquad +1\ \mu C \qquad +2\ \mu C$$

$x = -2\ \text{m} \qquad x = -1\ \text{m} \qquad 0 \qquad x = 1\ \text{m} \qquad x = 2\ \text{m}$

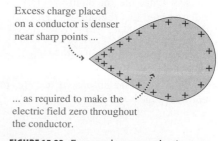

Excess charge placed on a conductor is denser near sharp points . . .

. . . as required to make the electric field zero throughout the conductor.

FIGURE 15.23 Excess charge gathering on a conductor concentrates on sharp points.

15.5 Charged Particles in Electric Fields

Ultimately, it's useful to understand electric fields because they exert forces on charged particles and thus determine their motion. From the definition of electric field, the force on a particle with charge q in an electric field $\vec{E}$ is

$$\vec{F} = q\vec{E} \qquad \text{(Force on a charged particle in an electric field; SI unit: N)} \qquad (15.9)$$

Again, Equation 15.9 shows that the force on a positively charged particle is in the same direction as the electric field, while the force on a negatively charged particle is opposite the field (Figure 15.14). Here we'll consider some examples of charged particles responding to electric fields.

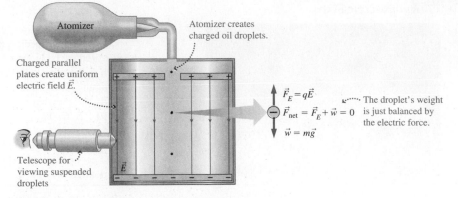

FIGURE 15.24 How Millikan measured the elementary charge.

Measuring the Elementary Charge

In 1911, the American physicist Robert A. Millikan made the first direct measurement of the elementary charge (e). (Prior to Millikan, e was inferred from chemical reactions.) Millikan's idea was simple: Balance the downward gravitational force on a charged particle with an upward electric force resulting from a vertical electric field (Figure 15.24). In balance, the electric force has the same magnitude as the particle's weight. Knowing the electric field and the weight lets you calculate the charge. By Equation 15.9, it's just

$$q = \frac{F}{E} = \frac{mg}{E}$$

where mg is the particle's weight.

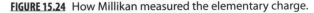

✓ TIP

To balance a charged particle against the force of gravity, the force from the electric field must point upward.

Millikan sprayed small droplets of oil into the space between a pair of horizontal conducting plates. In the process, some droplets acquire electric charge. Measuring the charges on many drops reveals a pattern: they're found with charge $\pm e$, $\pm 2e$, $\pm 3e$, and so on, so the value of e can be computed from all the data.

Millikan experiments in most undergraduate physics laboratories use small plastic spheres with the same mass, instead of oil. Millikan couldn't control the sizes of his oil droplets. Instead, he switched off the electric field and let a droplet fall until it reached a terminal speed due to its interaction with the air. He used fluid dynamics to calculate the droplet's size and mass from its terminal speed.

EXAMPLE 15.8 **Levitating an Oil Droplet**

An oil droplet of density $\rho = 927$ kg/m^3 is suspended motionless in an electric field of 9.66 kN/C directed vertically upward. After the field is switched off, the oil droplet falls, and using fluid dynamics, it's determined that the droplet's radius is 4.37×10^{-7} m. Find the charge on this droplet.

ORGANIZE AND PLAN When the droplet is suspended (Figure 15.25), the electric force pointing upward just balances the droplet's downward weight; that is, the magnitudes qE and mg are equal: $qE = mg$, so the unknown charge is

$$q = \frac{mg}{E}$$

The mass is the density ρ times the volume V, and the volume of a spherical droplet with radius r is $V = \frac{4}{3}\pi r^3$.

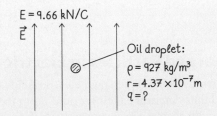

FIGURE 15.25 Our sketch for Example 15.8.

Known: $\rho = 927$ kg/m^3, $E = 9.66$ kN/C, radius $r = 4.37 \times 10^{-7}$ m.

SOLVE Combining the results to express the charge in terms of the droplet density and radius,

$$q = \frac{mg}{E} = \frac{\rho V g}{E} = \frac{4\rho \pi r^3 g}{3E}$$

cont'd.

Inserting the numerical values gives

$$q = \frac{4\pi\rho r^3 g}{3E} = \frac{4\pi(927 \text{ kg/m}^3)(4.37 \times 10^{-7} \text{ m})^3(9.80 \text{ m/s}^2)}{3(9.66 \times 10^3 \text{ N/C})}$$

$$q = 3.29 \times 10^{-19} \text{ C}$$

In the last step, you should verify that the charge units reduce to coulombs (C). Remember that $1 \text{ N} = 1 \text{ kg} \cdot \text{m/s}^2$.

REFLECT The result of this calculation reveals a charge close to $2e$. The numerical value differs only slightly from the exact value of $2e \ (= 3.20 \times 10^{-19} \text{C})$, with the difference due to experimental errors and rounding errors in the calculation. The droplet in this case probably carries an excess of two elementary charges.

MAKING THE CONNECTION What's the sign on the charge in this example? How should the experiment be modified for a charge of the opposite sign?

ANSWER The electric field is upward, and the force is in the same direction. Therefore, the charge is positive. If the charge were negative, the electric field would have to be switched to point downward, so that the electric force would still be upward to balance gravity.

The particles in a Millikan experiment can be suspended because they have so little charge that the electric and gravitational forces are comparable. Physicists use the **charge-to-mass ratio** as an indicator of how susceptible a particle is to electric fields. The charge-to-mass ratio is the absolute value of a particle's net charge divided by its mass, or, symbolically, $|q/m|$. For the oil drop in the preceding example, the charge-to-mass ratio is

$$\left|\frac{q}{m}\right| = \frac{2e}{\frac{4}{3}\pi\rho r^3} = \frac{2(1.60 \times 10^{-19} \text{ C})}{\frac{4}{3}\pi(927 \text{ kg/m}^3)(4.37 \times 10^{-7} \text{ m})^3} = 9.88 \times 10^{-4} \text{ C/kg}$$

By comparison, the charge-to-mass ratio of subatomic particles is much higher. Specifically, for protons

$$\left|\frac{q}{m}\right| = \frac{e}{m_\text{p}} = \frac{1.60 \times 10^{-19} \text{ C}}{1.67 \times 10^{-27} \text{ kg}} = 9.58 \times 10^7 \text{ C/kg}$$

and for electrons

$$\left|\frac{q}{m}\right| = \frac{e}{m_\text{e}} = \frac{1.60 \times 10^{-19} \text{ C}}{9.11 \times 10^{-31} \text{ kg}} = 1.76 \times 10^{11} \text{ C/kg}$$

The larger the charge-to-mass ratio, the greater the particle's acceleration in a given electric field. This has some important practical applications, as the following examples illustrate.

EXAMPLE 15.9 **An Electron Microscope**

A scanning electron microscope uses a uniform 15.0-kN/C electric field to accelerate electrons horizontally toward the subject to be imaged. (a) Determine the magnitude of an electron's acceleration. (b) Assuming that the electron begins at rest, what is its speed after traveling 5.0 cm?

ORGANIZE AND PLAN The electrons experience a force $\vec{F} = q\vec{E}$. Because they are negatively charged ($q = -e$), the force is opposite the electric field, as shown in Figure 15.26. Acceleration and force are related by Newton's second law, $F = ma$. This is a case of constant acceleration, for which the kinematic equations (Chapter 2) can be used to find the speed of an electron with known acceleration. Using the electric force for F in Newton's second law gives $eE = ma$, so the acceleration is $a = eE/m$. For constant acceleration, the final speed after a horizontal distance $x - x_0$ is given by Equation 2.10:

$$v^2 = v_0^2 + 2a(x - x_0)$$

SOLVE (a) Inserting the numerical values,

$$a = \frac{eE}{m} = \frac{(1.60 \times 10^{-19} \text{ C})(15.0 \text{ kN/C})}{9.11 \times 10^{-31} \text{ kg}} = 2.63 \times 10^{15} \text{ N/kg}$$

or $a = 2.63 \times 10^{15} \text{ m/s}^2$.

(b) The electron's final speed v is found using the numerical value of acceleration found in part (a):

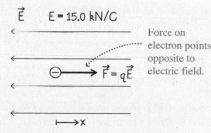

FIGURE 15.26 Our sketch for Example 15.9.

$$v^2 = v_0^2 + 2a(x - x_0)$$
$$= (0 \text{ m/s})^2 + 2(2.63 \times 10^{15} \text{ m/s}^2)(0.050 \text{ m})$$
$$= 2.63 \times 10^{14} \text{ m}^2/\text{s}^2$$

Taking the (positive) square root to get the speed, $v = 1.62 \times 10^7 \text{ m/s}$.

REFLECT The acceleration is extremely high, but this is acceptable for a subatomic particle. Note too that its speed is more than 5% of the speed of light. If it were much faster, you'd need to take into account Einstein's theory of special relativity.

MAKING THE CONNECTION How much time does it take for the electron to travel that 5 cm?

ANSWER Using the kinematic equations from Chapter 2, the time is only about 6 ns.

Is Gravity Important?

In the preceding example, how important is gravity in determining the electron's speed?

SOLVE The acceleration due to gravity is $9.8 \, \text{m/s}^2$—lower than our answer by a factor greater than 10^{14}. Another way to look at this is to consider the time of flight—about 6 ns in this example. That's enough

to tell us that gravity is totally negligible. In doesn't matter whether the electron beam is horizontal or vertical; either way, gravity's effect is immeasurably small.

REFLECT Gravity is typically unimportant when electric forces accelerate small (especially subatomic) particles.

EXAMPLE 15.11 **Deflecting Electrons in a Scanning Electron Microscope**

Consider again the electron microscope described in Example 15.9, in which the electrons are accelerated to a speed of $1.62 \times 10^7 \, \text{m/s}$. The next step is to deflect the electrons so that they can scan across the sample—hence the *scanning* electron microscope. To accomplish this, the electrons are directed between a pair of oppositely charged parallel plates, which produce a uniform electric field of $6.42 \times 10^3 \, \text{N/C}$, perpendicular to the electron beam. This field extends for 3.5 cm along the electron's path. (a) What direction is the force due to this field? (b) What is the electron's new velocity when it emerges from the electric field? (c) In what direction is the electron traveling, relative to its initial path?

ORGANIZE AND PLAN With a force perpendicular to its initial direction, the electron will undergo two-dimensional motion. Choose a coordinate system in which the electron is originally traveling in the $+x$-direction (Figure 15.27), and let the field point in the $-y$-direction. Given the field and charge, we can find the acceleration. Kinematic equations will give the two velocity components, from which the deflection angle can be found.

The electron is originally traveling in the $+x$-direction, with a velocity component in that direction $v_x = 1.62 \times 10^7 \, \text{m/s}$. Because the force is in the $+y$-direction, v_x won't change. But v_y, initially zero, will change because the force produces acceleration in the y-direction: $F_y = ma_y = qE_y$.

Known: $q = -e$, $v_x = 1.62 \times 10^7 \, \text{m/s}$, and $E_y = -6.42 \, \text{kN/C}$.

SOLVE (a) The electric field is in the $-y$-direction, giving a force in the $+y$-direction on the *negative* electron.

(b) Writing Newton's law, $F_y = ma_y = qE_y$, and solving gives the acceleration:

$$a_y = \frac{qE_y}{m_e} = \frac{(-1.60 \times 10^{-19} \, \text{C})(-6.42 \times 10^3 \, \text{N/C})}{9.11 \times 10^{-31} \, \text{kg}}$$
$$= 1.13 \times 10^{15} \, \text{m/s}^2$$

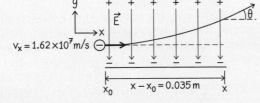

FIGURE 15.27 Our sketch for Example 15.11.

The velocity component v_y then follows from the kinematic equation $v_y = v_{y0} + a_x t$. Because the electron's initial velocity was in the x-direction, $v_{y0} = 0$. In the x-direction, the velocity component is a constant $v_x = 1.62 \times 10^7 \, \text{m/s}$ over a distance $x - x_0 = 3.5 \, \text{cm} = 0.035 \, \text{m}$. Therefore, the time is

$$t = \frac{x - x_0}{v_x} = \frac{0.035 \, \text{m}}{1.62 \times 10^7 \, \text{m/s}} = 2.16 \times 10^{-9} \, \text{s}$$

Returning to the equation for v_y,

$$v_y = v_{y0} + a_y t$$
$$= 0 \, \text{m/s} + (1.13 \times 10^{15} \, \text{m/s}^2)(2.16 \times 10^{-9} \, \text{s})$$
$$= 2.44 \times 10^6 \, \text{m/s}$$

(c) The final velocity vector has components $v_x = 1.62 \times 10^7 \, \text{m/s}$ and $v_y = 2.44 \times 10^6 \, \text{m/s}$, so the angle the velocity vector makes with the x-axis (its original direction) is given by

$$\tan \theta = \frac{v_y}{v_x} = \frac{2.44 \times 10^6 \, \text{m/s}}{1.62 \times 10^7 \, \text{m/s}} = 0.151$$

and so $\theta = \tan^{-1}(0.151) = 8.59°$.

REFLECT The deflection angle is small, but over the length of the microscope it's sufficient for a substantial deflection. Continually changing the deflecting field allows the beam to scan over the object being imaged.

MAKING THE CONNECTION What is the electron's trajectory under these conditions—that is, what path does it trace?

ANSWER In analogy with the projectiles you studied in Chapter 3, the x motion here is at constant velocity while the y motion is one of constant acceleration. Therefore, just like the projectile, the trajectory is a parabola.

CONCEPTUAL EXAMPLE 15.12   **Electric Dipole in an Electric Field**

Explain how an electric dipole reacts when placed at some random orientation in a uniform electric field.

SOLVE A dipole consists of charges $+Q$ and $-Q$ separated by a fixed distance. The forces on the two charges are

$$\text{Positive: } \vec{F}_{(+)} = +Q\vec{E}$$

and

$$\text{Negative: } \vec{F}_{(-)} = -Q\vec{E}$$

The net force on the electric dipole is

$$\vec{F}_{\text{net}} = \vec{F}_{(+)} + \vec{F}_{(-)} = +Q\vec{E} + (-Q\vec{E}) = 0$$

The net force on the dipole is zero, so its center of mass doesn't move. However, the fact that the two forces are being applied away from the center of mass creates a torque (Chapter 8) on the dipole. Figure 15.28 shows how this torque will rotate the dipole into alignment with the electric field. That is precisely what happens when an insulator becomes polarized in an electric field, as described in Section 15.1.

REFLECT In real insulating materials, the two-point-charge model here is an extreme simplification. In reality, some insulating materials contain polar molecules that have slightly more positive charge toward one end and slightly more negative charge on the other end, and the relative rigidity of the molecular bonds is consistent with the fixed separation distance in the dipole model. Other types of molecules are not normally polarized but may become so in the presence of an applied field, and in this case the applied field determines the charge separation.

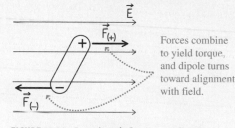

Forces combine to yield torque, and dipole turns toward alignment with field.

FIGURE 15.28 Our sketch for Example 15.12.

GOT IT? Section 15.5   Suppose a proton and an electron are placed in identical electric fields. Relative to the acceleration of the electron, the proton's acceleration is (a) zero; (b) in the same direction but smaller; (c) in the same direction but larger; (d) in the opposite direction but smaller; (e) in the opposite direction but larger.

Chapter 15 in Context

This chapter is the first in a set of three chapters (15–17) devoted to the properties and uses of electric charge. The new ideas here—electric charge, force, and field—have many connections with the physics introduced earlier. Because electric charges are subject to forces and are often mobile, the kinematics and dynamics that you learned earlier help you understand the behavior of electric charges. Because Coulomb's law is structurally similar to Newton's law of gravitation, the electric fields of point charges resemble the gravitational fields of point and spherical masses.

Looking Ahead In Chapters 16 and 17 we'll continue with electric charges and their applications. Much of Chapter 16 is about electrical energy, and again we'll use energy concepts that you learned in the context of mechanics and motion. In Chapter 17 you'll learn how electric current flows through circuits, and you'll see how to design and analyze simple circuits. In later chapters (18–20) we'll move on to consider magnetism and its relation to electricity. Throughout these chapters the basic properties of electric charge that you learned here will be useful.

APPLICATION

Gel Electrophoresis

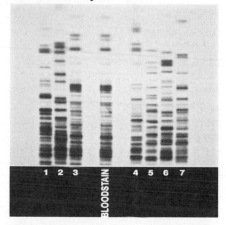

The analysis of DNA in humans and other animals is made possible by a technique called *gel electrophoresis*, which uses electric fields to separate different kinds of macromolecules (proteins or nucleic acids). When placed in a solution at a certain pH, these molecules become charged, making them susceptible to an electric field. The sample molecules are in a gel, a thick substance through which they drift very slowly under the influence of the field. Molecules of different shapes and sizes are affected differently by the electric field and gel, and so they are separated as shown. The result for a sample extracted from human DNA gives a characteristic "fingerprint" of that person. The DNA pattern can be used to identify and understand genetic abnormalities. Criminal investigators use these DNA patterns to match the DNA of a suspect to that obtained from cells left at a crime scene.

CHAPTER 15 SUMMARY

Electric Charges

(Section 15.1) Like charges repel; unlike charges attract.

Charge is **conserved:** The sum of all positive and negative charge within a closed system is a constant.

Charge is **quantized:** Free charges always appear in integer multiples of the elementary charge e.

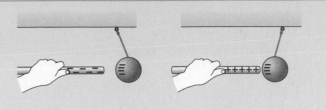

Charges of subatomic particles:
proton: $+e = +1.602 \times 10^{-19}$ C   neutron: charge $= 0$
electron: $-e = -1.602 \times 10^{-19}$ C

Coulomb's Law

(Section 15.2) **Coulomb's law** governs the electrostatic force between point charges.

Coulomb's law: $F = \dfrac{k|q_1||q_2|}{r^2}$

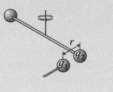

Coulomb's Law for Multiple Charges

(Section 15.3) By the principle of **superposition**, the net electric force on a charge is the sum of the forces due to all the other charges.

Principle of superposition: $\vec{F}_{3net} = \vec{F}_{13} + \vec{F}_{23}$ for the three charges shown.

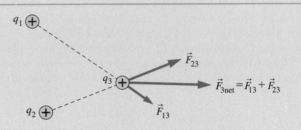

Electric Fields

(Section 15.4) The **electric field** is a vector field that represents the force per unit charge acting on a test charge.

Electric field lines originate from positive charges and terminate on negative charges.

Electric field defined: $\vec{E} = \dfrac{\vec{F}}{q_0}$

Electric field of a point charge: $E = \dfrac{F}{q_0} = k\dfrac{|q|}{r^2}$

Electric field between a pair of oppositely charged planes: $E = \dfrac{\sigma}{\epsilon_0}$

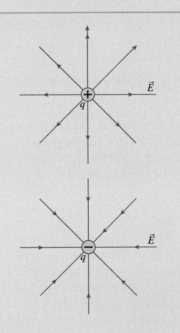

Charged Particles in Electric Fields

(Section 15.5) Charged particles are accelerated in electric fields. The force on a positively charged particle is in the same direction as the electric field; the force on a negatively charged particle is in the opposite direction of the electric field.

Force on a charged particle in an electric field: $\vec{F} = q\vec{E}$

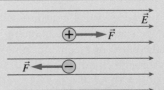

NOTE: Problem difficulty is labeled as ■ straightforward to ■ ■ ■ challenging. Problems labeled BIO are of biological or medical interest.

Conceptual Questions

1. Suppose you bring a plastic rod near a plastic ball that is suspended by a thread. If the ball is repelled by the rod, what can you conclude about the net charge on the ball? Answer the same question if the ball is attracted to the rod.
2. Pull off a long strip of transparent tape and hold it vertically with one hand. Bring your other hand near the hanging tape, and notice that the tape is attracted to your (electrically neutral) hand. Explain.
3. Why is charge usually transferred by electrons rather than protons?
4. List the similarities and differences between Coulomb's law and Newton's law of gravitation.
5. An electron and proton are released from rest some distance apart. Will they meet halfway between their original positions? Explain.
6. A small ball with charge Q is held fixed. A second identical ball (also with charge Q) is released some distance directly above the fixed ball. (a) Argue that the second ball will oscillate along a vertical line. (b) Is the oscillation simple harmonic motion? Explain.
7. How is action at a distance (the force between two distant particles) explained using the exchange of other particles?
8. Justify the rule that the electric field strength is proportional to the local density of electric field lines. *Hint:* Consider the electric field around a point charge, and look at the electric field lines passing through two concentric spheres of different radii. Then compare the density of field lines along the surface of each sphere.
9. Can two electric field lines cross each other? Explain.
10. Sketch the electric field lines generated by a pair of $-Q$ point charges.
11. Sketch the electric field diagram in the region around the point charges $+Q$ and $-3Q$.
12. Describe how the electric field appears if you are far away from the point charges $+Q$ and $-Q/2$ shown in Figure 15.19b.
13. Argue why an electric dipole placed in a uniform electric field should oscillate with simple harmonic motion.
14. Why doesn't the electric field strength near an infinite, uniformly charged plane vary depending on the distance from the plane?

Multiple-Choice Problems

15. What is the net charge of a collection of 2.5×10^{18} electrons? (a) 0.025 C; (b) 0.40 C; (c) 2.5 C; (d) 4.0 C.
16. The net charge of one mole of protons is (a) 1.60×10^{-19} C; (b) 1.00 C; (c) 1.60×10^4 C; (d) 9.65×10^4 C.
17. What is the magnitude of the force acting on a proton at a distance 0.15 m from another proton? (a) 1.1×10^{-36} N; (b) 1.0×10^{-26} N; (c) 6.4×10^{-8} N; (d) 1.5×10^{-27} N.
18. Two charges lie on the $+x$-axis: 12 nC at $x = 0.24$ m and 16 nC at $x = 0.64$ m. Where on the x-axis could a third charge be placed so that the net force on it is zero? (a) There is no such place; (b) $x = 0.47$ m; (c) $x = 0.43$ m; (d) $x = 0.29$ m.
19. Two charged particles each experience a force of 20 N when they are a distance d apart. The force will be 40 N when the separation distance is changed to (a) $d/4$; (b) $d/2$; (c) $d/\sqrt{2}$; (d) $2d$.

20. Which one of the following is true regarding a proton and an electron in identical electric fields? (a) The force on the electron is larger. (b) The force on the proton is larger. (c) The acceleration of the electron is larger. (d) The acceleration of the proton is larger.
21. Three identical charges of $10.0\,\mu\text{C}$ are placed at the vertices of an equilateral triangle, that is 0.25 m on a side. The net force acting on each charge has a magnitude (a) 57.6 N; (b) 28.8 N; (c) 24.9 N; (d) 14.4 N.
22. A point charge $-1.2\,\mu\text{C}$ is 0.50 m away from a second point charge $+1.0\,\mu\text{C}$. The force on a third charge, $+1.4\,\mu\text{C}$, placed exactly halfway between the other two is (a) 0.44 N, directed toward the negative charge; (b) 0.11 N, directed toward the negative charge; (c) 1.76 N, directed toward the negative charge (d) 0.11 N, directed toward the positive charge.
23. The electric field halfway between two point charges, -5.0 nC and $+5.0$ nC, separated by a distance of 0.30 m is (a) zero; (b) 1000 N/C, directed toward the negative charge; (c) 2000 N/C, directed toward the negative charge; (d) 4000 N/C, directed toward the negative charge.
24. A proton produces an electric field of exactly 500 N/C at a distance of (a) 2.9×10^{-12} m; (b) 1.7×10^{-6} m; (c) 1.3×10^{-3} m; (d) 1.0 m.
25. The electric field and gravitational field (a) have the same units; (b) both depend on mass; (c) are both attractive and repulsive forces; (d) both have inverse-square distance dependence.
26. An alpha particle (produced in many nuclear reactions) has a mass of 6.64×10^{-27} kg and a charge of $+2e$. In a uniform electric field inside a linear accelerator, an alpha particle is observed to have an acceleration of 6.0×10^{13} m/s^2. The magnitude of the electric field is (a) 1.25×10^6 N/C; (b) 2.50×10^6 N/C; (c) 5.00×10^6 N/C; (d) 2.50×10^7 N/C.

Problems

Section 15.1 Electric Charges

27. ■ (a) How much charge is contained in 1 kg of electrons? (b) How much charge is contained in 1 kg of protons?
28. ■ The Faraday is a unit of charge used in electrolysis and is defined as the charge of one mole of protons. Find the conversion factor between coulombs and Faradays.
29. ■ ■ Estimate the number of electrons present in a (neutral) 1.0-kg block of carbon.
30. ■ ■ Suppose Earth contains roughly an equal number of protons and neutrons. Estimate (a) the number of protons in Earth and (b) the charge Earth would have if all its electrons were removed.
31. ■ ■ ■ Suppose that the charge on the electron and proton did not have the same magnitude but instead differed by one part in 10^{10}. What would be the net charge on a 1.0-kg sample of helium gas?
32. ■ ■ A plastic rod is rubbed with a cloth, and in the process it acquires a charge of -2.3×10^{-8} C. Did the mass of the rod increase or decrease, and by how much?

Section 15.2 Coulomb's Law

33. ■ Two small charged balls have a repulsive force of 0.12 N when they are separated by a distance of 0.85 m. The balls are moved closer together, until the repulsive force is 0.60 N. How far apart are they now?

34. ■ Treat as point charges two quarks of charge separated by a distance of in a proton, and find the electrostatic force between them.

35. ■■■ Suppose all the electrons could be removed from Earth. (a) Find the force on a proton just above Earth's surface under these conditions. (b) Compare your answer in part (a) with the weight of the proton.

36. ■ A free electron and a free proton are exactly 1.0 cm apart. Find the magnitude and direction of (a) the acceleration of the proton and (b) the acceleration of the electron.

37. ■■ Two Ping-Pong balls each have a mass of 1.4 g and carry a net charge of 0.450 μC. One ball is held fixed. At what height should the second ball be placed directly above the fixed ball if it is to remain at rest there?

38. BIO ■ **Forces in a DNA molecule.** In a chemical reaction, the two ends of a 2.45-μm-long DNA molecule acquire charges of $+e$ and $-e$, respectively. What is the force between the two ends of the molecule?

39. ■ How far should an electron be from a proton so that the electrostatic force on the electron is just equal to the electron's weight? Comment on the significance of your result.

40. ■■ (a) Use the astronomical data in Appendix E to compute the gravitational force between Sun and Earth. (b) Suppose gravity were turned off, and you wanted to replace the gravitational force with an electrostatic force of the same magnitude you computed in part (a). If the Sun had charge $+Q$ and Earth $-Q$, determine the value of Q.

41. ■■■ Two identical pendulums of length L hang side by side from a common support point, with their bobs (each having mass m) just touching. An equal charge Q is then placed on each bob. Find an equation relating θ (the angle each pendulum string makes with the vertical, as shown in Figure P15.41) to the other parameters given in the problem.

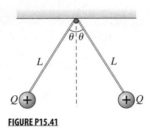

FIGURE P15.41

Section 15.3 Coulomb's Law for Multiple Charges

42. ■■ Three point charges are placed on the x-axis as follows: 20 μC at $x = 0$; 30 μC at $x = 0.50$ m; and $-10\,\mu$C at $x = 1.0$ m. Find the net force on each point charge.

43. ■■ Two point charges, $-25\,\mu$C and $-75\,\mu$C, are held fixed exactly 1.0 m apart. (a) Where should a third charge be placed so that the net force on it is zero? (b) Explain why the magnitude and sign of the third charge do not affect your answer to part (a).

44. ■■■ Two point charges, -10 nC and $+15$ nC, are held fixed a distance 0.50 m apart. (a) Where can a $+20$-nC charge be placed so that the net force on it is zero? (b) Repeat part (a) if the third charge is -20 nC.

45. ■■■ Three identical charges of 2.0 μC are placed at the vertices of an equilateral triangle, 0.15 m on a side. Find the net force acting on each charge.

46. ■■■ Three charges are fixed in the x-y plane as follows: 1.5 nC at the origin $(0, 0)$; 2.4 nC at $(0.75$ m, $0)$; -1.9 nC at $(0, 1.25$ m). Find the force acting on the charge at the origin.

47. ■■ Four identical point charges of 5.6 nC are placed at the corners of a square, 0.25 m on a side. Find the force acting on each charge.

48. ■■■ Four point charges are placed on the corners of a square of side d as shown in Figure P15.48. (a) Find the net force (magnitude and direction) on each charge. (b) If a fifth charge of size $+Q$ is brought into the picture and placed at the center of the square, what is the net force acting on it?

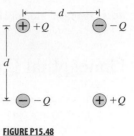

FIGURE P15.48

49. ■■■ A point charge $+6.5\,\mu C$ is placed at the origin, and a second charge $-4.2\,\mu$C is placed in the x-y plane at $(0.35$ m, 0.45 m). Where should a third charge be placed in the plane so that the net force acting on it is zero?

50. ■■ One electron is at the origin of a coordinate system. A second electron, somewhere in the x-y plane, feels a force 2.4×10^{-22} N $\hat{\imath} + 3.6 \times 10^{-22}$ N $\hat{\jmath}$. Where is the second electron located?

51. ■■■ Three charges are fixed in the x-y plane as follows: 1.5 nC at the origin $(0, 0)$; 2.4 nC at $(0.75$ m, $0)$; -1.9 nC at $(0, 1.25$ m). Find the force acting on the negative charge.

Section 15.4 Electric Fields

52. ■ Compute the electric field at the surface of (a) a proton, assuming the proton is a sphere of radius 1.2×10^{-15} m; (b) a lead nucleus, which has charge $82e$, assuming the nucleus is a sphere of radius 7.1×10^{-15} m.

53. ■ What is the electric field a distance 1.0 mm from an electron?

54. ■■ Two positive point charges, each 15 μC, lie along the x-axis at $x = -0.15$ m and $x = +0.15$ m. Find the electric field at (a) the origin $(0, 0)$ and (b) the point $(0, 0.20$ m$)$ on the y-axis.

55. ■■ For the point charges described in the preceding problem, rank in increasing order the magnitude of the electric field at the following positions: (a) $(0, 0)$; (b) $(0.25$ m, $0)$; (c) $(0.10$ m, $0)$; (d) $(0, 0.10$ m$)$.

56. ■■ Two point charges, $-25\,\mu$C and $+50\,\mu$C, are separated by a distance of 20.0 cm. What is the electric field (magnitude and direction) at a point halfway between the two charges?

57. ■■ For the electric dipole in Example 15.7, compute the electric field at the following locations: (a) $(-15$ cm, $0)$; (b) $(0, -5$ cm$)$.

58. ■■■ An electric dipole on the x-axis consists of a charge $-Q$ at $x = -d$ and a charge $+Q$ at $x = +d$. (a) Find a general expression (in terms of Q and d) for the electric field at points on the x-axis with $x > d$. (b) Show that for $x \gg d$ the magnitude of the field you computed in part (a) is *approximately* equal to $E = 4kQd/x^3$ and therefore approximates an *inverse-cube* distance law.

59. ■■■ A pendulum consists of a 0.155-kg ball hanging vertically from a light thread. The ball has a net charge of 235 μC. If the pendulum is placed into a uniform, horizontal electric field of 1250 N/C (Figure P15.59), find the angle the pendulum string makes with the vertical when the system is in equilibrium.

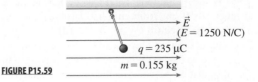

FIGURE P15.59

60. ■ A large plane having uniform charge density has an electric field just outside the plane that points directly toward the plane with magnitude 1.25×10^4 N/C. What is the surface charge density?

61. ■ If a charged plane has a uniform surface charge density of -1.5×10^{-7} C/m^2, how many electrons will be on the surface in a square that is 1.0 mm on a side?

62. ■ For the charged plane described in the preceding problem, what is the electric field just outside the plane's surface?

63. ■■ A pair of charged, parallel planes are used to create a uniform electric field between the planes. The charge densities on the two planes are equal in magnitude but have opposite sign. Find the charge density on each plane if the electric field between them has a magnitude of 10,000 N/C.

64. ■■ A 23.6-μC point charge lies at the origin. Find the electric field at the following points in the x-y plane: (a) (0.95 m, 0); (b) (1.25 m, 0.75 m).

65. ■ Assume that the hydrogen atom consists of an electron in a circular orbit around a proton, with an orbital radius of 5.29×10^{-11} m. (a) What is the electric field acting on the electron? (b) Use your answer in part (a) to find the force acting on the electron.

66. ■■ Four point charges lie in the x-y plane: $+Q$ at (a, a); $+2Q$ at $(-a, -a)$ $-Q$ at $(-a, a)$; and $-2Q$ at $(a, -a)$. Find the electric field at the origin.

67. ■■ Two point charges lie on the x-axis: 0.25 μC at $x = 0.25$ m and 0.16 μC at $x = 0.75$ m. Find the place(s) where the electric field is zero.

68. ■■■ Two point charges lie on the x-axis: 1.65 μC at $x = 1.00$ m and -2.30 μC at $x = 2.50$ m. Find the place(s) where the electric field is zero.

69. ■■ In a particular region of space, there exists a uniform electric field of magnitude 12.4 kN/C in the $+x$-direction. Now a -1.0-μC charge is placed in this field, at a position we will call the origin $(0, 0)$. (a) Find the net electric field at the point (0.50 m, 0). (b) Find the net electric field at the point (0, 0.50 m).

Section 15.5 Charged Particles in Electric Fields

70. ■ What is the acceleration of an electron that is placed in a uniform electric field of 6500 N/C pointing in the $+y$-direction?

71. ■ A small bit of photocopier toner having mass of 3.5×10^{-15} kg carries a net charge of $2e$ and has an acceleration of 0.50 m/s^2. What is the magnitude of the electric field causing the acceleration?

72. ■ Two parallel plates carry uniform charge densities -0.50 nC/m^2 and $+0.50$ nC/m^2. (a) Find the electric field between the plates. (b) Find the acceleration of an electron between these plates.

73. ■ An electron moving in the x-y plane has an acceleration of 3.25×10^{13} m/s^2 directed at an angle $30°$ above the $+x$-axis. Find the electric field responsible for that acceleration.

74. ■■ Two large parallel plates are separated by 0.75 cm and carry uniform charge densities equal in magnitude and opposite in sign. A proton is between the plates and has an acceleration of magnitude 1.39×10^{11} m/s^2. Find the charge density on the plates.

75. ■■ In a Millikan oil-drop experiment, using oil with a density of 940 kg/m^3, a droplet with a charge of $-3e$ is just balanced against gravity in an electric field of 11.5 kN/C. (a) What is the direction of the electric field? (b) What is the size of the (spherical) droplet?

76. ■■ A pair of parallel plates are charged with uniform charge densities of -35 pC/m^2 and $+35$ pC/m^2. The distance between the plates is 2.3 mm. If a free electron is released at rest from the negative plate, find (a) its speed when it reaches the positive plate and (b) the time it takes to travel across the 2.3-mm gap.

77. ■■■ An electron in an oscilloscope tube is traveling horizontally with a speed of 3.9×10^6 m/s. It passes through a 2.0-cm-long region with an electric field pointing upward with a magnitude of 1500 N/C. What is the electron's velocity when it emerges from the electric field region?

78. ■■ In the preceding problem, how far downward (from its original trajectory) is the electron deflected during the time it's in the electric field?

79. ■■ In an x-ray tube, electrons are accelerated in a uniform electric field and then strike a metal target. Suppose an electron starting from rest is accelerated in a uniform electric field directed horizontally and having a magnitude of 2300 N/C. The electric field covers a region of space 10.5 cm wide. (a) What is the speed of the electron when it strikes the target? (b) How far does it fall under the influence of gravity during its flight?

80. ■ In an optics experiment, a discharge tube is filled with singly ionized argon atoms, each having a mass of 6.64×10^{-26} kg. If the electric field inside the tube is 7500 N/C, find the acceleration of each argon ion.

81. ■■ An electron is traveling with a speed of 2.1×10^6 m/s when it enters a uniform electric field that is parallel to its direction of travel. After moving for another 9.2 cm, the electron comes to rest. (a) What is the magnitude of the electric field? (b) What is the direction of the electric field, relative to the electron's original motion?

82. ■■ A proton, initially traveling in the $+x$-direction with a speed of 5.45×10^5 m/s, enters a uniform electric field directed vertically upward. After traveling in this field for 3.92×10^{-7} s, the proton's velocity is directed $45°$ above the $+x$-axis. What is the strength of the electric field?

General Problems

83. ■■ A 2.1-g Ping-Pong ball is rubbed with a piece of wool, resulting in a net positive charge of 7.4 μC on the ball. How many electrons were transferred from the ball to the wool in this process?

84. ■■ You have three charges, two of which have a charge of $+4$ μC and the other a charge of -1 μC. Show how you can place the three charges along a line so that there is no net force on any of them.

85. ■■ A charge of 3 μC is at the origin and a charge of -2 μC is on the y-axis at $y = 50$ cm. (a) Where can you place a third charge so that the force acting on it is zero? (b) What is the electric field at the location you found in part (a)?

86. ■■ Two charges, one twice as large as the other, are located 15 cm apart, and each experiences a repulsive force of 95 N from the other. (a) What is the magnitude of each charge? (b) Does your answer to part (a) depend on whether the charges are positive or negative? Why or why not? (c) At what location is the electric field zero?

87. ■■ In his famous experiment that demonstrated quantization of electric charge, Robert A. Millikan suspended small oil drops in an electric field. (a) With a field strength of 20 MN/C, what is the mass of a drop containing 10 elementary charges that can just be suspended in that field? (b) Indicate the direction of the electric field if the elementary charges are (i) positive; (ii) negative.

88. ■■■ A proton moving at 3.8×10^5 m/s to the right enters a region where a uniform electric field of magnitude 56 kN/C points

to the left. (a) How far will the proton travel before it stops moving? (b) What will happen to the proton after it stops moving? (c) Graph the proton's position versus time, velocity versus time, and acceleration versus time after it enters the field.

89. BIO ■ ■ **Electric bees.** As they fly through the air, honeybees can pick up a small amount of electric charge. Suppose a 0.12-g bee acquires a charge of -55 pC. (a) How many excess electrons has the bee acquired? (b) Find the electric force on the bee in the electric field just above Earth's surface, 120 N/C directed vertically downward. (c) Compare your answer in part (b) to the bee's weight.

90. ■ ■ ■ An electric dipole lies on the y-axis and consists of an electron at $y = 0.60$ nm and a proton at $y = -0.60$ nm. Find the electric field (a) at a point halfway between the two charges; (b) at the point $x = 2.0$ nm, $y = 0$; and (c) at the point $x = -2.0$ nm, $y = 0$.

91. ■ ■ ■ Two identical small metal spheres carry charges q_1 and q_2. When they're 1.0 m apart, they experience a 2.5-N attractive force. Then they are brought together so that charge moves from one sphere to the other until the net charges are equal. They are again placed 1.0 m apart and now they repel with a 2.5-N force. What are the original charges q_1 and q_2?

92. ■ ■ ■ Two 2.3-g Ping-Pong balls hang suspended, with the balls just touching when they hang from a common support. Each ball's center is initially 60 cm below the support. The balls are then each given an identical amount of charge q, and each string makes an angle of 15° with the vertical as shown in Figure GP15.92. Find the magnitude of the charge q.

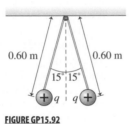

0.60 m   0.60 m

15° 15°

$+$ q q $+$

FIGURE GP15.92

93. ■ ■ (a) For the scanning electron microscope described in Example 15.11, find the electron's vertical displacement. (b) What would be the electron's vertical displacement if the scanning electric field were reduced to 3.21×10^3 N/C (half its original value)?

94. ■ ■ ■ Suppose you have a friend who has a part-time job translating computer maintenance training manuals from Japanese to English. A section of the manual for an inkjet printer is missing, and you need to replace the missing information. The printer works by steering ink drops with charge q and mass m between two parallel surfaces where there is a uniform electric field of magnitude E that deflects them an appropriate amount. The field extends over a length L and width d, and each ink drop enters halfway between the charged plates as shown in Figure GP15.94. Find the minimum speed v the drop must have in order to make it through the region without striking one of the plates. *Note*: You may assume that the electric field is large enough that gravity can be neglected.

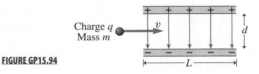

Charge q
Mass m v d

FIGURE GP15.94 L

95. BIO ■ ■ ■ **DNA compression.** Due to its double-helix structure, a DNA molecule acts like a spring that can be compressed. Biophysicists have estimated the force constant for one type of DNA molecule to be 0.60 nN/nm. Find the compression of a 75-μm-long DNA molecule that acquires a charge of $+1.2$ pC at one end and -1.2 pC at the other.

Answers to Chapter Questions

Answer to Chapter-Opening Question
The conducting metal skin of the aircraft shields the interior. Free electric charges within the metal rearrange themselves quickly in the presence of external charge, in order to prevent penetration of the electric field.

Answers to GOT IT? Questions
Section 15.2   (e) The forces are repulsive, and the magnitudes of the two forces are the same.

Section 15.3   (c) Directed toward the right.

Section 15.4   (b) Directed toward the left.

Section 15.5   (d) In the opposite direction but smaller.

16 Electric Energy, Potential, and Capacitors

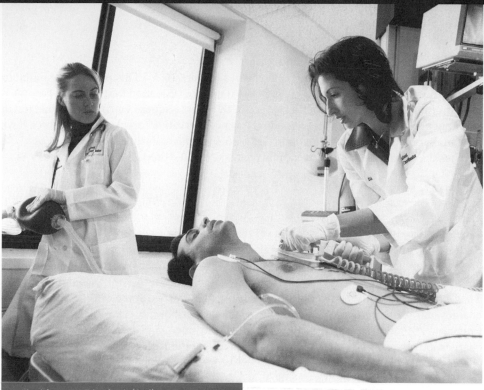

■ The lifesaving jolt of a defibrillator requires a large amount of energy delivered in a short time. What's the source of that energy?

This chapter focuses on the **energy** associated with electric forces and fields. Electric energy is of fundamental and practical significance, so we'll explore basic concepts as well as applications.

Electric energy is often described in terms of electric potential. Here we'll explore applications of potential—especially capacitors, which store energy and charge. Potential will prove essential in understanding the circuits of Chapter 17.

16.1 Electric Potential Energy

Potential Energy: A Review

Chapter 5 introduced energy, both kinetic (K) and potential (U). You saw that with conservative forces, **potential energy** (Section 5.4) provides a "shortcut" for solving motion problems without explicitly using Newton's law.

You may want to review Chapter 5 briefly, along with Section 9.3 on gravitational potential energy. Here are some highlights:

1. The total mechanical energy of a system, $E = K + U$, is constant when forces are conservative (Equation 5.18).
2. Potential energy is solely a function of the position(s) of particle(s). For example, a particle with mass m near Earth has potential energy $U = mgy$ (Equation 5.15), with y the height above some reference point.

To Learn

By the end of this chapter you should be able to

■ Understand the concepts of electric potential energy and electric potential.

■ Apply conservation of energy in electrostatics.

■ Relate potential and electric field, both numerically and graphically.

■ Understand capacitors as storing electric charge and energy.

■ Describe dielectrics and explain how they affect capacitance.

3. The change in potential energy of a particle moving between two points is independent of the path taken.

4. The gravitational potential energy of two masses a distance r apart is $U = -\dfrac{Gm_1m_2}{r}$ (Equation 9.5).

Calculating Potential Energy

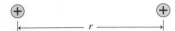

FIGURE 16.1 Two point charges, separated by a distance r.

What's the potential energy of two point charges (Figure 16.1)? This problem is similar to finding the gravitational potential energy of two masses (Equation 9.5), because Newton's law of gravitation and Coulomb's law are both inverse-square force laws. Changes in potential energy result from the work done by a conservative force, so similar force laws give similar potential energy functions. Table 16.1 summarizes this analogy.

Thus the potential energy of two point charges q_1 and q_2 separated by a distance r is

$$U = \frac{kq_1q_2}{r} \qquad \text{(Potential energy of two point charges; SI unit: J)} \qquad (16.1)$$

where the zero of potential energy is at very large separation r $(r \to \infty)$.

TABLE 16.1 Force and Potential Energy for Gravity and Electricity

	Force law	Potential energy
Gravitational	$F = \dfrac{GMm}{r^2}$ (attractive)	$U = -\dfrac{Gm_1m_2}{r}$
Electric	$F = \dfrac{k\lvert q_1\rvert\lvert q_2\rvert}{r^2}$ (attractive or repulsive)	$U = \dfrac{kq_1q_2}{r}$

The expressions in Table 16.1 are similar: Each involves the appropriate constant (G for gravity, k for electricity). The forces are both inverse-square, and both potential energies are proportional to $1/r$. The one difference is the negative sign for gravitational energy. That's because gravity is always attractive. But the electric force is repulsive for like charges, attractive for opposites. Although Figure 16.1 shows two positive point charges, Equation 16.1 covers both possibilities: Potential energy U is negative for opposite charges, positive for like charges. These signs reflect the fact that you do work bringing charges of like sign together, while the attractive force between opposite charges, like the force of gravity acting on masses, does work as the charges approach. Example 16.1 shows how this works.

✓ **TIP**

Potential energy can be positive or negative. It's positive for two like charges and negative for two unlike charges.

CONCEPTUAL EXAMPLE 16.1 **Attraction or Repulsion?**

Charge q_1 is fixed at the origin, and q_2 is free to move. (a) Suppose both charges are positive. If charge q_2 is released from rest on the $+x$-axis, describe its subsequent motion. Tell what happens to the kinetic energy K, potential energy U, and total energy E. (b) Repeat part (a) assuming the charges q_1 and q_2 have *opposite* signs.

SOLVE Coulomb's law gives the force, and Equation 16.1 the potential energy. Coulomb's law says that like charges repel, and unlike charges attract. Equation 16.1, $U = kq_1q_2/r$, holds whether the charges are like or unlike. In a conservative system, the total energy $E = K + U$ is constant.

(a) The charges repel, so q_2 feels a force to the right, and it accelerates in that direction (Figure 16.2a). Initially, the kinetic energy K is zero. The potential energy U is initially positive, because the product q_1q_2 is positive. So the total energy E is positive, and it remains constant even after q_1 starts moving. But as q_2 accelerates, its kinetic energy becomes increasingly positive, and its potential energy drops so that E remains constant. This makes sense, because $U = kq_1q_2/r$, and r is increasing as q_2 moves away from q_1.

cont'd.

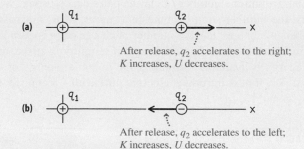

FIGURE 16.2 (a) Two like charges. (b) Opposite charges.

(b) The opposite charges attract. With q_1 at the origin, q_2 feels a force to the left, so it accelerates leftward (Figure 16.2b). Again, q_2's kinetic energy K is initially zero, but now its potential energy U is negative, because the product q_1q_2 is negative. So the total energy E is negative, and it remains constant even after q_2 starts moving. But again q_2's K increases as it accelerates, so U must become even more negative to keep E constant. Is this consistent with the separation r decreasing? Yes: As r decreases, $U = kq_1q_2/r$ becomes *more negative*. So U decreases as K increases, and the total energy E remains constant.

REFLECT Conservation of energy holds for the electric force—a conservative force. Potential energy can be positive or negative, depending on whether the charges are like or unlike.

Equation 16.1 gives the potential energy associated with two charges. The result should be in joules (J), as with other forms of energy. Is it? Remember that the Coulomb constant k has units $\text{N} \cdot \text{m}^2/\text{C}^2$. With charges in C and distance in m, the units in Equation 16.1 are joules, as expected:

$$\frac{\text{N} \cdot \text{m}^2}{\text{C}^2} \frac{\text{C} \cdot \text{C}}{\text{m}} = \text{N} \cdot \text{m} = \text{J}$$

For example, in a hydrogen atom, the proton and electron (charge $= \pm e - \pm 1.60 \times 10^{-19}$ C) are separated by 5.29×10^{-11} m. Equation 16.1 then gives their potential energy:

$$U = \frac{kq_1q_2}{r} = \frac{(8.99 \times 10^9 \, \text{N} \cdot \text{m}^2/\text{C}^2)(1.60 \times 10^{-19} \, \text{C})(-1.60 \times 10^{-19} \, \text{C})}{5.29 \times 10^{-11} \, \text{m}}$$

$$= -4.36 \times 10^{-18} \, \text{N} \cdot \text{m} = -4.36 \times 10^{-18} \, \text{J}$$

This tiny energy is typical of atomic electrons. In Chapter 24 you'll see how we determine such energies using light emitted when electrons jump between atomic energy levels. Small though it is, this potential energy is much larger than the gravitational potential energy of the proton and electron, which is $U = -Gm_1m_2/r = 1.9 \times 10^{-57}$ J. This is consistent with a fact you saw throughout Chapter 15: The electric force between elementary particles is far stronger than the gravitational force between them.

Multiple Charges

What's the potential energy of three or more charges? Figure 16.3 shows three charges q_1, q_2, and q_3. Each pair has potential energy given by Equation 16.1. The total potential energy of the distribution is then the sum over all three pairs:

$$U = \frac{kq_1q_2}{r_{12}} + \frac{kq_1q_3}{r_{13}} + \frac{kq_2q_3}{r_{23}} \qquad \begin{array}{l}\text{(Potential energy of multiple}\\\text{charges; SI unit: J)}\end{array} \qquad (16.2)$$

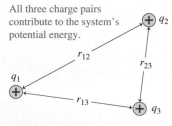

FIGURE 16.3 Three charges.

Add a fourth charge, and the number of charge pairs becomes six. With five charges, there are ten pairs; the number soon becomes enormous. Most macroscopic situations involve huge numbers of charges, and the number of charge pairs is even larger. For example, charge in a thundercloud provides the energy of a lightning strike, and you'd like to know how much energy is available. Summing the potential energy of all those charge pairs is impractical! We'll next introduce **electric potential**—a concept that can help us avoid this counting game.

GOT IT? Section 16.1 Three 1.0-μC charges lie at the corners of an equilateral triangle 0.10 m on a side. What's the potential energy of this configuration: (a) 0.09 J; (b) 0.18 J; (c) 0.27 J; (d) 0.63 J?

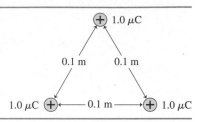

16.2 Electric Potential

In Chapter 15 we defined electric field as the electric force per unit charge. We'll use an analogous approach here, defining **electric potential**, V, as the potential energy per unit charge (Table 16.2). Thus, the electric potential a distance r from a point charge q is

$$V = \frac{kq}{r} \quad \text{(Electric potential of a point charge; SI unit: V)} \tag{16.3}$$

Electric potential has a value at every point in space. Knowing its value at some point tells how much potential energy a given charge would have if placed at that point. That's because electric potential V is potential energy *per unit charge*, so the potential energy U of a charge q_0 is

$$U = q_0 V \quad \text{(Potential energy from electric potential; SI unit: J)} \tag{16.4a}$$

This is analogous to the force $(\vec{F} = q\vec{E})$ on a charged particle in an electric field.

TABLE 16.2 Electric potential defined by analogy with electric field

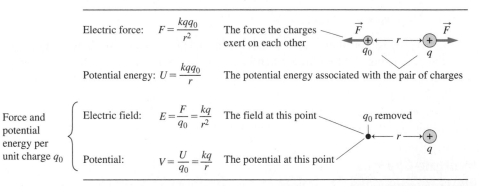

Electric force:	$F = \dfrac{kqq_0}{r^2}$	The force the charges exert on each other
Potential energy:	$U = \dfrac{kqq_0}{r}$	The potential energy associated with the pair of charges
Electric field:	$E = \dfrac{F}{q_0} = \dfrac{kq}{r^2}$	The field at this point
Potential:	$V = \dfrac{U}{q_0} = \dfrac{kq}{r}$	The potential at this point

Force and potential energy per unit charge q_0

There's one important difference between the electric potential and electric field: The electric field is a vector, while potential is a scalar. That means it's often easier to calculate potential than field.

✓TIP

Electric potential can be positive or negative; near a positive charge, potential is positive; near a negative charge, it's negative.

Potential Difference

In Chapter 5 you saw that only *differences* in potential energy matter physically. Therefore, only electric **potential difference** is meaningful. When, as in Equation 16.3, we talk about the potential at some point, we really mean the potential difference between that point and another where we've taken $V = 0$. Equation 16.3 came from Equation 16.1, with the zero of potential energy taken when the charges have infinite separation. Therefore, Equation 16.3 describes the potential difference between infinity and a point a distance r from the point charge q. In other situations we may choose other points for the zero of potential; for example, in electric power systems, the ground is often chosen as the zero point. In terms of potential difference, we can write Equation 16.4a as

$$\Delta U = q_0 \Delta V \quad \text{(Potential energy difference; SI unit: J)} \qquad (16.4b)$$

The Volt

The SI unit of potential is the **volt** (V), defined as one joule per coulomb: $1\ V = 1\ J/C$. The term "voltage" is often used, but it's unnecessary, just as it would be to use "meterage" for displacement or "joulage" for energy.

EXAMPLE 16.2   **Energy of a Lightning Strike**

Turbulent motions in thunderstorms cause charge separations that result in a potential difference of around 10 MV between cloud and ground. When lightning occurs, it transfers some 50 C of charge between cloud and ground. How much energy is released in the lightning strike?

ORGANIZE AND PLAN Potential difference is the potential energy difference per unit charge. Equation 16.4b relates potential energy difference and potential difference: $\Delta U = q_0 \Delta V$. Here q_0 is the 50-C charge and ΔV is the 10-MV potential difference between cloud and ground (Figure 16.4).

SOLVE The charge's potential energy drops as it "falls" through the 10-MV potential difference, so ΔU is the energy released:

Energy released $= \Delta U = q_0 \Delta V = (50\ \text{C})(10\ \text{MV}) = 500\ \text{MJ}$

REFLECT That's a lot of energy. You don't want to be on the receiving end of a lightning strike! In what forms do you notice this energy being released?

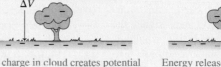

Net charge in cloud creates potential difference between cloud and ground.

Energy released by lightning strike is proportional to the charge transferred and the potential difference:

$$\Delta U = q \Delta V$$

FIGURE 16.4 Energy released in a lightning strike.

Electron-Volts

Equation 16.4 suggests an energy unit, the **electron-volt**, which physicists often use in atomic systems. One electron-volt (eV) is the energy associated with an elementary charge $e = 1.60 \times 10^{-19}$ C crossing a potential difference of 1 volt. Thus the conversion

Cathode:
negative charge,
lower potential

Anode:
positive charge,
higher potential

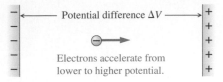

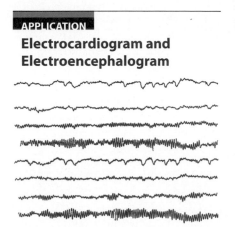

FIGURE 16.5 Accelerating electrons from cathode to anode in an x-ray tube.

Electrocardiogram and Electroencephalogram

Cardiologists study heart function using the electrocardiogram (EKG), a record of the heart's electric potential throughout its beating cycle. Deviations from a healthy EKG pattern help diagnose abnormal heart conditions such as a blocked coronary artery. An irregular EKG may signal the body's failure to provide the potentials needed to stimulate regular heartbeat. Artificial pacemakers introduce the correct potentials and ensure a regular beat.

The brain, too, produces electric potentials, with patterns depending on the patient's activity (as shown here). Electroencephalograms (EEGs) record these minute potential differences—on the order of 1 microvolt (10^{-6} V), versus 1 millivolt (10^{-3} V) in the EKG. Subtle differences in EEG patterns link to electrical activity in different areas of the brain, because many functions of body and mind can be traced to control centers in specific brain regions. Lack of function in one area may indicate a tumor there. EEGs also help diagnose nervous system diseases such as epilepsy and meningitis.

between joules and electron-volts is $1 \text{ eV} = 1.60 \times 10^{-19} \text{ J}$. In the example of the hydrogen atom (Section 16.1), the potential energy $-4.36 \times 10^{-18} \text{ J}$ becomes

$$-4.36 \times 10^{-18} \text{ J} \left(\frac{1 \text{ eV}}{1.60 \times 10^{-19} \text{ J}} \right) = -27.3 \text{ eV}$$

This more manageable number is why electron-volts are used at the atomic and subatomic level, where charges often have the values $\pm e$.

EXAMPLE 16.3 **An Electron's Energy**

Find the potential energy change for a single electron moving from cloud to ground in the thunderstorm of Example 16.2. Express your answer in electron-volts.

ORGANIZE AND PLAN Again Equation 16.4b applies, but now q_0 is 1 elementary charge e.

Known: $\Delta V = 10 \text{ MV}$.

SOLVE One electron-volt is the energy change when 1 elementary charge moves through a potential difference of 1 volt. Here we have a difference of 10 MV. Therefore, the electron's energy change is

$$(10 \text{ MV})(1 \text{ elementary charge}) = 10 \text{ MeV}$$

REFLECT It's easy to work with charge in elementary charges and energy in electron-volts! Had we wanted joules, we would have had to use the conversion $1 \text{ eV} = 1.60 \times 10^{-19} \text{ J}$.

Medical X Rays

X rays used in medicine are produced when electrons with energies of tens of kiloelectron-volts (keV) slam into matter. An x-ray tube (Figure 16.5) is an evacuated structure containing a hot **cathode** that "boils off" electrons and an **anode** typically of tungsten. A potential difference of tens of kilovolts is applied between anode and cathode, with the anode positive. Electrons start at the cathode with negligible kinetic energy but a large potential energy. As they "fall" through the potential difference, they gain kinetic energy and lose potential energy to keep their total energy $E = K + U$ constant. So what's their kinetic energy when they reach the anode? Example 16.3 shows how to calculate this. With a potential difference of, say, 50 kV across the x-ray tube, the electrons gain kinetic energy of 50 keV. Next time you have a medical or dental x ray, you may see the radiologist adjust a control labeled "kilovolts" to set the electron energy. That, in turn, determines the x rays' penetrating power.

Electric Potential and Life

The Italian physician Luigi Galvani (1737–1798) noticed that a frog's muscles twitched when he applied an electric potential to its spinal cord. Research since Galvani has shown just how important electric potentials are to life. Nervous systems of animals, including humans, use changes in electric potential to transmit information throughout the body. As Galvani observed, electrical signals trigger muscle actions. Proper muscle function requires maintaining the correct electric potentials.

GOT IT? **Section 16.2** An electron accelerated from rest through a potential difference of 100 V acquires kinetic energy (a) $1.6 \times 10^{-19} \text{ J}$; (b) $3.2 \times 10^{-19} \text{ J}$; (c) $1.6 \times 10^{-18} \text{ J}$; (d) $1.6 \times 10^{-17} \text{ J}$.

16.3 Electric Potential and Electric Field

Electric potential provides a "shortcut" for calculating energy changes without having to consider details of force and acceleration. Behind those energy changes are forces resulting from electric fields. Here we'll explore the relation between potential and field.

An Example: Parallel Conducting Plates

In Chapter 15 you saw that a pair of oppositely charged, parallel conducting plates produces a nearly uniform field (Figure 16.6). If you move a charge q along this uniform field, the work done on the charge is just the electric force, whose magnitude is $F = qE$ (Equation 15.5), times the displacement Δx:

$$W = qE\Delta x \quad \text{(Work done on charge in a uniform electric field; SI unit: J)} \tag{16.5}$$

By the work-energy theorem (Chapter 5), this changes the charge's kinetic energy by $\Delta K = W$. Because the electric force is conservative, there's an opposite change in potential energy: $W = \Delta K = -\Delta U$. But the change in a charge's potential energy is the charge multiplied by the potential difference ΔV. Therefore,

$$W = -\Delta U = -q\Delta V \quad \text{(Work from potential difference; SI unit: J)} \tag{16.6}$$

Equating the expressions for work in Equations 16.5 and 16.6,

$$\Delta V = -E\Delta x \quad \text{(Potential difference from electric field; SI unit: V)} \tag{16.7}$$

Equation 16.7 is what we're after: the relationship between potential difference and electric field. The equation shows that if you move in the direction of a uniform electric field, the electric potential *decreases* by $E\Delta x$. Move opposite the field, and the potential *increases* by this amount.

Suppose the electric field in Figure 16.6 is 500 N/C, and that the plate separation is 4.0 cm. Then by Equation 16.7, the potential difference between the plates is $\Delta V = E\Delta x = (500 \text{ N/C})(0.040 \text{ m}) = 20 \text{ N} \cdot \text{m/C} = 20 \text{ V}$. The units work out because $1 \text{ N} \cdot \text{m} = 1 \text{ J}$, and $1 \text{ V} = 1 \text{ J/C}$.

Finding Field from Potential

Divide Equation 16.7 through by Δx and you get

$$E = -\frac{\Delta V}{\Delta x} \quad \text{(Electric field from potential difference; SI unit: V/m)} \tag{16.8}$$

This equation makes two points. First, it shows that the units of electric field (which Chapter 15 introduced as N/C) can also be expressed as V/m. N/C is preferable when you're relating force and field, and V/m when you're relating field and potential.

More importantly, Equation 16.8 shows that the electric field depends on the rate at which potential changes with position. Where potential changes rapidly, the field is strong; where potential changes slowly, the field is weak. The minus sign in Equation 16.8 shows that potential *increases* when you move *against* the field—just as gravitational potential energy increases when you move uphill.

Equipotentials

An **equipotential** is a surface over which the potential has the same value. Figure 16.7 shows some equipotential surfaces for the uniform field of Figure 16.6. The equipotentials are planes parallel to the charged plates. Because the field is uniform, the potential increases linearly with position, so the equipotentials are evenly spaced. If the field were stronger in some region, the equipotentials would be closer together there.

Constant force exerted by electric field does work on charged particle.

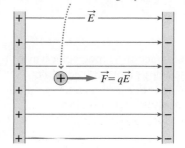
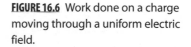

FIGURE 16.6 Work done on a charge moving through a uniform electric field.

$\Delta V = -E\Delta x$, so all points equidistant from a plate are at the same potential. To show this . . .

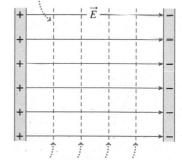

. . . we draw *equipotential lines* connecting points with the same potential. These lines represent planes of equal potential between the charged plates; they are perpendicular to the electric field.

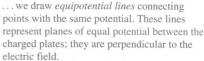

FIGURE 16.7 Equipotentials and electric field lines.

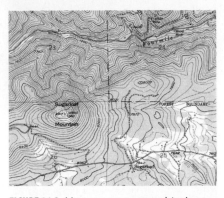

FIGURE 16.8 Lines on a topographical map are lines joining points with equal altitude. Where lines are closer together, the slope is steeper.

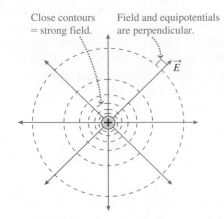

Close contours = strong field.

Field and equipotentials are perpendicular.

FIGURE 16.9 Electric field (solid lines) and equipotentials (dashed lines) around a positive point charge.

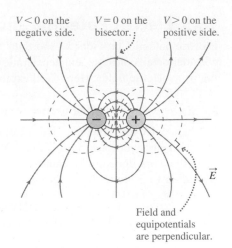

$V < 0$ on the negative side.

$V = 0$ on the bisector.

$V > 0$ on the positive side.

Field and equipotentials are perpendicular.

FIGURE 16.10 Electric field (solid lines) and equipotentials (dashed lines) around an electric dipole.

Force on positive charge is directed toward lower potential.

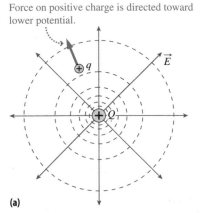

(a)

Force on negative charge is directed toward higher potential.

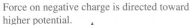

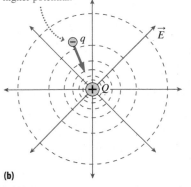

(b)

FIGURE 16.11 Relating force to electric potential.

Equipotentials are like contours on a topographical map (Figure 16.8), where closely spaced lines imply a steep slope. On a topographical map, the direction of the steepest slope is perpendicular to the contour lines. Similarly, on an equipotential diagram, **the electric field is perpendicular to the equipotential surfaces, with the direction of the field pointing from higher to lower potential**. That's clearly true for the uniform field shown in Figure 16.7, but in fact it's true for all electric fields.

The electric field of a positive point charge extends radially outward from the charge. Equipotentials are perpendicular to the radial field lines; therefore, the equipotential surfaces are spheres. Figure 16.9 shows a two-dimensional cut in a plane containing the point charge; here the equipotentials appear as circles. The uneven spacing of the equipotentials in Figure 16.9 reflects the nonuniform field of the point charge: The equipotentials are closely spaced where the field is stronger—namely, near the charge.

You can also draw the equipotentials in Figure 16.9 using the point-charge potential, $V = kq/r$. Surfaces with constant potential therefore have constant r. The equipotentials for an electric dipole (Figure 16.10) aren't as easy to construct. Here the electric potential is the sum of two point-charge potentials, with charges $\pm q$:

$$V = \frac{kq}{r_{(+)}} + \frac{k(-q)}{r_{(-)}} = \frac{kq}{r_{(+)}} - \frac{kq}{r_{(-)}}$$

where $r_{(+)}$ and $r_{(-)}$ are the distances of a given point from the positive and negative charge, respectively. The equipotential $V = 0$ runs down the center of the diagram, because all points on that line are equidistant from the two charges. Figure 16.10 shows the dipole equipotentials and the dipole field. As always, the two are everywhere perpendicular.

The electric field points from higher to lower potential, while the direction of the electric force, $\vec{F} = q\vec{E}$, depends on the sign of the charge. Thus the force on a positive charge points from higher to lower potential, and the force on a negative charge points from lower to higher potential. Figure 16.11a shows equipotentials around a positive point charge $+Q$. The potential a distance r from the point charge is $V = kQ/r$, so points closer to the charge have a higher potential. As shown in Figure 16.11a, a second positive charge q experiences a force that accelerates it from higher to lower potential—that is, away from $+Q$. And so it does, because like charges repel. Figure 16.11b shows a negative charge in the same situation. It moves from lower to higher potential, consistent with the attraction between opposite charges.

Consider our map analogy, and you can see that a positive charge moves "downhill" on the equipotential map, just like a ball rolling downhill. Negative charge, in contrast, moves "uphill." Although this may seem strange, it's just the result of there being two kinds of charge but only one kind of mass.

GOT IT? Section 16.3 Two parallel plates 0.50 mm apart have a potential difference of 1500 V. What's the electric field in the region between the plates? (a) 1500 V/m; (b) 7500 V/m; (c) 1.5 MV/m; (d) 3.0 MV/m.

```
+ + + + + +
↕ 0.50 mm   ΔV = 1500 V
- - - - - -
```

Reviewing New Concepts: Electric Potential Energy, Potential, and Field

■ Any distribution of electric charges has potential energy, with the potential energy of a pair of point charges given by $U = \dfrac{kq_1q_2}{r}$.

■ Electric potential is the potential energy per unit charge, so the potential of a point charge is $V = \dfrac{kq}{r}$.

■ Electric field lines are perpendicular to equipotential lines, with a uniform electric field given by $E = -\Delta V/\Delta x$.

16.4 Capacitors

Capacitors (Figure 16.12a) are important components in electric circuits, where they store charge and energy. Here we'll explore capacitors and their applications. A capacitor consists of a pair of conductors that are insulated from each other and carry charges of equal magnitude Q but opposite signs (Figure 16.12b). It's important that the conductors be insulated; otherwise, charge would flow from one to the other, and the capacitor would become **discharged**.

✓ **TIP**

Capacitors are neutral! When we say "the capacitor charge is Q," we really mean that one conductor carries $+Q$ and the other $-Q$. Overall, the capacitor remains neutral even when it's "charged."

Capacitance

The charge on each conductor of a capacitor almost instantaneously reaches an equilibrium in which there's no bulk charge movement. Then the entire conductor must be at a single potential; if it weren't, there would be an electric field that would accelerate individual charges. But the oppositely charged conductors are at different potentials, so there's a potential difference ΔV between them. Here we'll switch to commonly used but slightly sloppy notation and designate this potential difference by V, without the "Δ" for "difference."

The **capacitance** C of a capacitor is then

$$C = \frac{Q}{V} \quad \text{(Capacitance defined; SI unit: C/V = F)} \qquad (16.9)$$

Thus capacitance is the ratio of the magnitude of the charge on each conductor to the magnitude of the potential difference. Capacitance C is a constant that depends only on the configuration of the conductors. If the stored charge increases, then Equation 16.9 shows that the potential difference increases proportionally. Equation 16.9 also shows that the SI unit for capacitance is the coulomb per volt (C/V), which defines the **farad** (**F**): 1 F = 1 C/V. The farad is named for the English physicist Michael Faraday (1791–1867).

(a)

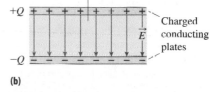

Insulation (e.g., air, plastic, or vacuum)

$+Q$ · · · Charged conducting plates

$\vec{E}$

$-Q$

(b)

FIGURE 16.12 (a) The assortment of capacitors shown here includes some you might see in your physics lab or in electronic instruments. They represent a wide range of capacitance values, and the largest capacitor does not necessarily have the highest capacitance. (b) A parallel-plate capacitor carries charges of equal magnitude and opposite sign.

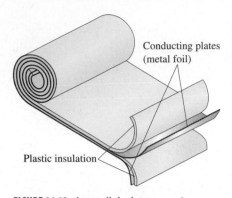

FIGURE 16.13 A parallel-plate capacitor rolled to fit into a cylinder.

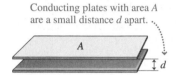

Conducting plates with area A are a small distance d apart.

FIGURE 16.14 Dimensions of a parallel-plate capacitor.

Parallel-Plate Capacitors

The simplest capacitor consists of two parallel conducting plates. Although the capacitors you see in the lab or in personal electronics don't look like they consist of parallel plates, many still use parallel-plate geometry. A common trick is to sandwich an insulator between two foil sheets, and roll the whole thing into a cylinder (Figure 16.13).

The parallel-plate capacitor in Figure 16.14 consists of two identical conducting plates with area A, separated by a distance d. We'll now use the definition $C = Q/V$ to find its capacitance. In Chapter 15 you saw that the electric field between charged parallel planes is $E = \sigma/\varepsilon_0$, where σ is the surface charge density. Here $\sigma = Q/A$, so the capacitor charge Q becomes $Q = \sigma A = \varepsilon_0 EA$.

Equation 16.7, $\Delta V = -E\Delta x$, relates potential difference and field. Here we're only interested in the magnitude V, and for Δx we have the spacing d. So $V = Ed$, and our capacitance becomes $C = Q/V = \varepsilon_0 EA/Ed$, or

$$C = \frac{\varepsilon_0 A}{d} \quad \text{(Capacitance of parallel-plate capacitor; SI unit: F)} \quad (16.10)$$

Note that the capacitance depends only on the constant ε_0 and the geometrical factors A and d. This confirms our earlier statement that capacitance depends only on the configuration of the two conductors.

✓**TIP**

You can show in Conceptual Question 12 that the units of ε_0 can be expressed as farads per meter (F/m)—convenient when ε_0 is needed in a capacitance calculation.

EXAMPLE 16.4 **Computing Capacitance**

A lecture-demonstration capacitor consists of two parallel circular metal plates, each with radius 12 cm. The plate separation is adjustable. (a) What's the capacitance when the plates are separated by 0.1 m? (b) How close would the plates have to be in order to make a 1.0-μF capacitor?

ORGANIZE AND PLAN The capacitance $C = \epsilon_0 A/d$ of a parallel-plate capacitor depends only on the plate area A and separation d. In part (a) you know these quantities. In part (b) the same relationship gives d, because you know A and the capacitance C.

Known: $r = 12$ cm; in part (a), $d = 0.1$ m; in part (b), $C = 1.0\ \mu$F.

SOLVE (a) The circular plates have area $A = \pi r^2$. Then the capacitance is

$$C = \frac{\varepsilon_0 A}{d} = \frac{(8.85 \times 10^{-12}\ \text{F/m}) \cdot \pi (0.12\ \text{m})^2}{1.0 \times 10^{-4}\ \text{m}}$$

$$= 4.0 \times 10^{-9}\ \text{F} = 4.0\ \text{nF}$$

(b) Solving for d gives

$$d = \frac{\varepsilon_0 A}{C} = \frac{(8.85 \times 10^{-12}\ \text{F/m}) \cdot \pi (0.12\ \text{m})^2}{1.0 \times 10^{-6}\ \text{F}}$$

$$= 4.0 \times 10^{-7}\ \text{m}$$

That's extremely close, and even a modest potential difference will result in sparks jumping and discharging the capacitor.

REFLECT Small values of capacitance are typical, often measured in pF, nF, or F. That's a good reason to review your SI prefixes. And in practical capacitors you might expect a smaller A than in this example, giving even smaller capacitance. Yet capacitors of up to several F are available in small packages. In Section 16.5 we'll explain how this is possible.

MAKING THE CONNECTION If the plate separation is kept at 0.1 mm, what plate diameter would be needed to make a 1-μF capacitor?

ANSWER The area works out to 11.3 m^2, or a diameter of 3.8 m. That's obviously impractical!

Circuit Diagrams

How do you charge a capacitor? Simple: Just connect the two conductors to the terminals of a battery or other source of potential difference, as drawn realistically in Figure 16.15a and schematically, as a circuit diagram, in Figure 16.15b. We'll explore batteries in Chapter 17; for now, all you need to know is that they produce an essentially constant potential difference. With a battery's potential difference V between the plates of a capacitor, those plates carry charges $\pm Q$, where by Equation 16.9, $Q = CV$. Connect a 0.22-μF capacitor across a 1.5-volt battery, for example, and the stored charge is

$$Q = CV = (0.22 \, \mu F)(1.5 \, V) = 0.33 \, \mu C$$

Again, this is the magnitude of the charge on either plate; with $+Q$ on one plate and $-Q$ on the other, the capacitor remains overall neutral.

Recall from Chapter 15 that metallic conductors contain electrons that are free to move. So in Figure 16.15, the positive battery terminal attracts negative electrons from the upper capacitor plate, leaving a net positive charge on that plate. Meanwhile, electrons move from the negative battery terminal to the lower plate, depositing a net negative charge there. Both plate charges have the same magnitude, $Q = CV$, where in this case V is the battery's potential difference.

Capacitors in Parallel

Capacitors and other electronic components are often connected. Figure 16.16a shows a **parallel combination**, in which one plate of each capacitor is connected to one plate of the other. What's the **equivalent capacitance** of this parallel combination? By equivalent capacitance we mean the single capacitor (Figure 16.16b) that will store the same charge as the parallel combination.

We've connected our parallel capacitors across a battery, as shown. This puts the battery's potential difference V across each capacitor. But $C = Q/V$ for each capacitor, so the charges on the two capacitors are $Q_1 = C_1 V$ and $Q_2 = C_2 V$, respectively. If you think of the parallel combination as a single capacitor, its total charge is $Q_1 + Q_2$. With the battery's potential difference V across it, the equivalent capacitance is

$$C_p = \frac{Q}{V} = \frac{Q_1 + Q_2}{V} = \frac{Q_1}{V} + \frac{Q_2}{V}$$

The last two terms are just the individual capacitances, so

$$C_p = C_1 + C_2 \qquad (16.11a)$$

Thus **the equivalent capacitance of parallel capacitors is the sum of the individual capacitances**. For example, suppose the capacitors in Figure 16.16a have capacitances $C_1 = 2.0 \, \mu F$ and $C_2 = 5.0 \, \mu F$. Then the equivalent capacitance of the combination is $C_p = C_1 + C_2 = 2.0 \, \mu F + 5.0 \, \mu F = 7.0 \, \mu F$. The charges on the two capacitors are

$$Q_1 = C_1 V = (2.0 \, \mu F)(9.0 \, V) = 18 \, \mu F \cdot V = 18 \, \mu C$$

and

$$Q_2 = C_2 V = (5.0 \, \mu F)(9.0 \, V) = 45 \, \mu F \cdot V = 45 \, \mu C$$

The total charge, 63 C, follows from summing these values, or directly from $Q = C_p V$. Note that the equivalent capacitance is larger than either individual capacitance. Physically, you can think of the parallel combination as a single capacitor with larger plate area. As you've already seen, that results in a larger capacitance. Our derivation readily generalizes to three or more parallel capacitors, so

$$C_p = C_1 + C_2 + C_3 + \dots \qquad \text{(Parallel capacitors; SI unit: F)} \qquad (16.11b)$$

Capacitors in Series

Figure 16.17 shows two capacitors connected in **series**. Two components are in series when current going through the first component has nowhere to go but through the second; you can see from the figure that this is the case for our two capacitors. What's the equivalent capacitance of the series combination?

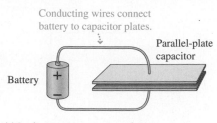
Conducting wires connect battery to capacitor plates.

Parallel-plate capacitor

Battery

(a) Realistic representation of circuit

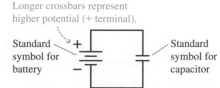
Longer crossbars represent higher potential (+ terminal).

Standard symbol for battery

Standard symbol for capacitor

(b) Schematic circuit diagram

FIGURE 16.15 (a) Capacitor connected across a battery. (b) Circuit diagram for a battery-capacitor circuit.

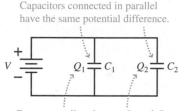
Capacitors connected in parallel have the same potential difference.

V $Q_1 \! = \! C_1$ $Q_2 \! = \! C_2$

Battery supplies charges Q_1 and Q_2.

(a) Capacitors connected in parallel to a battery

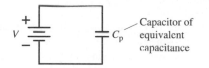
V C_p — Capacitor of equivalent capacitance

(b) The two capacitors are replaced by one of equivalent capacitance C_p.

FIGURE 16.16 Two parallel capacitors.

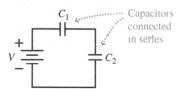
C_1 — Capacitors connected in series

V C_2

(a) Capacitors connected in series to a battery

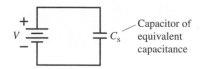
V C_s — Capacitor of equivalent capacitance

(b) The two capacitors are replaced by one of equivalent capacitance C_s.

FIGURE 16.17 Two capacitors in series.

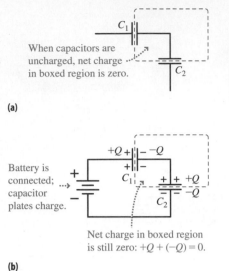

When capacitors are uncharged, net charge in boxed region is zero.

(a)

Battery is connected; capacitor plates charge.

Net charge in boxed region is still zero: $+Q + (-Q) = 0$.

(b)

FIGURE 16.18 (a) Two capacitors in series, initially uncharged. (b) When the battery is connected, the capacitors carry the same charge.

The key here is to realize that both capacitors in a series combination carry the same charge Q. To see why, focus on the dashed box in Figure 16.18a. Before the battery is connected, there's no charge on either capacitor, so the net charge in the boxed region is zero. With the battery connected, charge can't jump across the insulating gaps between capacitor plates, so the net charge in the boxed region remains zero. This leads to the rearrangement of charge shown in Figure 16.18b.

Although the charge on each capacitor is the same, the potential differences across them are generally *not* the same, because the definition $C = Q/V$ leads to $V = Q/C$—and thus different values of C imply different values of V. Here the battery's potential difference V appears across the series combination, so $V = V_1 + V_2$, and therefore

$$\frac{Q}{C_s} = \frac{Q}{C_1} + \frac{Q}{C_2}$$

Canceling the common factor Q,

$$\frac{1}{C_s} = \frac{1}{C_1} + \frac{1}{C_2} \qquad (16.12a)$$

There's the rule: **For capacitors in series, the reciprocal of the equivalent capacitance is the sum of the reciprocals of the individual capacitances.** As with the parallel rule, this one can be extended to n capacitors in series:

$$\frac{1}{C_s} = \frac{1}{C_1} + \frac{1}{C_2} + \frac{1}{C_3} + \ldots + \frac{1}{C_n} \quad \text{(Series capacitors)} \qquad (16.12b)$$

✓**TIP**

To make a larger capacitance, put capacitors in parallel. To make a smaller capacitance, put capacitors in series.

EXAMPLE 16.5 **Series Capacitors**

Capacitors $C_1 = 6.0\,\text{nF}$ and $C_2 = 3.0\,\text{nF}$ are connected in series. (a) What's the equivalent capacitance of this combination? (b) Find the charge on each capacitor and the potential difference across each when the series pair is connected across a 9.0-V battery.

ORGANIZE AND PLAN We start by drawing a circuit diagram (Figure 16.19). For the two capacitors in series, the equivalent capacitance follows from the reciprocal rule, Equation 16.12a. Each capacitor carries the same charge, which follows from $Q = C_s V$. Knowing Q, we can get the individual potential differences from $V = Q/C$.

Known: $C_1 = 6.0\,\text{nF}$; $C_2 = 3.0\,\text{nF}$.

SOLVE (a) The equivalent capacitance follows from Equation 16.12a:

$$\frac{1}{C_s} = \frac{1}{C_1} + \frac{1}{C_2} = \frac{1}{6.0\,\text{nF}} + \frac{1}{3.0\,\text{nF}} = \frac{1}{2.0\,\text{nF}}$$

Therefore, $C_s = 2.0\,\text{nF}$.

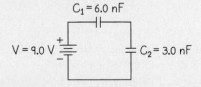

$C_1 = 6.0\,\text{nF}$

$V = 9.0\,\text{V}$

$C_2 = 3.0\,\text{nF}$

FIGURE 16.19 Our sketch for Example 16.5.

(b) The charges on the two capacitors are the same:

$$Q = C_s V = (2.0\,\text{nF})(9.0\,\text{V}) = 18\,\text{nC}$$

Then the potential differences are

$$V_1 = \frac{Q}{C_1} = \frac{18\,\text{nC}}{6.0\,\text{nF}} = 3.0\,\text{C/F} = 3.0\,\text{V}$$

$$V_2 = \frac{Q}{C_2} = \frac{18\,\text{nC}}{3.0\,\text{nF}} = 6.0\,\text{C/F} = 6.0\,\text{V}$$

REFLECT Here's a check to see if the last result makes sense: The potential differences V_1 and V_2 should add up to the potential difference across the battery terminals, and they do—it's 9.0 V.

Note that now the equivalent capacitance is *smaller* than the capacitance of any individual capacitor—which follows because their *reciprocals* add. Notice throughout this example how we repeatedly applied the capacitance definition $C = Q/V$, first solving for one quantity, then for another.

MAKING THE CONNECTION Find a third capacitance to be placed in series with this combination to make the equivalent capacitance 1.0 nF.

ANSWER Adding a term $1/C_3$ to the sum with $C_s = 1.0\,\text{nF}$ and solving for C_3 gives $C_3 = 2.0\,\text{nF}$.

Armed with the series and parallel rules, you can tackle circuits containing both series and parallel combinations. It's a good idea to break the problem into smaller pieces, finding the equivalent capacitance of each piece and then joining those, again with the parallel/series rules.

PROBLEM-SOLVING STRATEGY 16.1 **Networks of Capacitors**

ORGANIZE AND PLAN

■ Draw a circuit diagram.
■ Study the connections to determine which capacitors are in series and which are in parallel. If necessary, redraw the diagram to make this clear.
■ Review the information you have, and plan how to use that information to find the equivalent capacitance.

SOLVE

■ Follow the rules for parallel and series combinations to simplify the network, one step at a time, until you've found the equivalent capacitance.
■ If you need to find the charge on a capacitor, use $Q = CV$ for that capacitor. You may want to apply this equation repeatedly, solving sometimes for Q and sometimes for V.

REFLECT

■ Is your equivalent capacitance reasonable, given the sizes of the individual capacitors? Remember that adding capacitors in parallel increases capacitance, and adding capacitors in series decreases capacitance.

EXAMPLE 16.6 **Equivalent Capacitance**

Three capacitors are combined as in Figure 16.20a. (a) What's the equivalent capacitance of this combination? (b) What's the charge on each capacitor when the combination is connected across a 14-V battery?

ORGANIZE AND PLAN This problem involves a network of capacitors, so it's necessary to apply both parallel and series rules. The parallel rule will give the equivalent capacitance of the combination of C_1 and C_2. Call this $C_4 = C_1 + C_2$. Next, redraw the circuit, with C_4 replacing C_1 and C_2 (Figure 16.20b).

Now C_4 and C_3 are in series, so the equivalent capacitance follows from

$$\frac{1}{C_s} = \frac{1}{C_4} + \frac{1}{C_3}$$

To find the charge on each capacitor, we'll then use $Q = CV$.

SOLVE (a) For the parallel combination,

$$C_4 = C_1 + C_2 = 1.0\ \mu F + 4.0\ \mu F = 5.0\ \mu F$$

Now for the series combination of $C_4 = 5.0\ \mu F$ and $C_3 = 2.0\ \mu F$:

$$\frac{1}{C_s} = \frac{1}{C_4} + \frac{1}{C_3} = \frac{1}{5.0\ \mu F} + \frac{1}{2.0\ \mu F} = \frac{7}{10\ \mu F}$$

Thus the equivalent capacitance for the three-capacitor network is $C = 10/7\ \mu F \approx 1.43\ \mu F$.

(b) Consider first the charge on the equivalent capacitance. With the 14-V battery across the combination, $Q = CV = (10/7\ \mu F)(14\ V) = 20\ \mu F \cdot V = 20\ \mu C$. This C actually consists of the 2-μF capacitor C_3 in series with the parallel combination C_4. Since the charge on series capacitors is the same, $Q_3 = 20\ \mu C$ and $Q_4 = 20\ \mu C$. So we've got our answer for the charge on C_3, but for C_1 and C_2 we're first going to need the potential difference. For the parallel combination C_4,

$$V_4 = \frac{Q_4}{C_4} = \frac{20\ \mu C}{5.0\ \mu F} = 4.0\ V$$

The potential difference across each of the individual parallel capacitors is the same as across the combination, so the charges are

$$Q_1 = C_1 V_1 = (1.0\ \mu F)(4.0\ V) = 4\ \mu C$$
$$Q_2 = C_2 V_2 = (4.0\ \mu F)(4.0\ V) = 16\ \mu C$$

REFLECT Do these answers seem about right? It's not as easy to tell as in the case of a single series or parallel combination. The equivalent capacitance is less than the 2.0-μF series capacitor, so that much is as it should be. Note again how we used $C = Q/V$, alternately solving for Q or V, to "burrow down" into the circuit and end up with the charges on the individual capacitors.

MAKING THE CONNECTION If you switched the 1-μF and 2-μF capacitors, does the equivalent capacitance of the combination increase, decrease, or remain the same?

ANSWER The equivalent capacitance decreases to 6/7 μF.

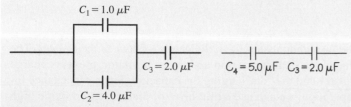

(a) Network of three capacitors   **(b)** C_4 replaces C_1 and C_2

FIGURE 16.20 Finding the equivalent capacitance.

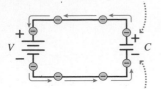

Battery pulls electrons from this side of capacitor, leaving positive charge.

Battery sends electrons to this side of capacitor, producing negative charge.

(a) Why charging a capacitor requires work

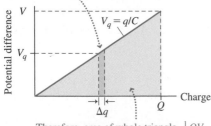

Area of strip, $\approx V_q \Delta q$, is work done in moving charge Δq between the plates when potential difference is V_q.

Therefore, area of whole triangle, $\frac{1}{2}QV$, is total work needed to charge capacitor.

(b) Finding the amount of work required

FIGURE 16.21 Work required to charge a capacitor.

Energy Storage in Capacitors

As capacitors store charge, they also store energy. Why? Imagine charging a capacitor by moving electrons, sequentially, from one plate to the other. That makes one plate increasingly negative, the other positive (Figure 16.21a). Once there's charge on the plates, there's a potential difference between them, so it takes work to move additional charge. The work to move a small additional charge ΔQ is $V_q \Delta q$, where V_q is the potential difference when the charge is q. As the charge increases, so does the potential difference. Figure 16.21b shows that the total work needed to charge the capacitor is therefore $W = \frac{1}{2}QV$. This work ends up as stored potential energy U. Since $V = Q/C$, the stored energy can be written

$$U = \frac{Q^2}{2C} \quad \text{(Energy stored in a capacitor; SI unit: J)} \tag{16.13}$$

or

$$U = \frac{1}{2}CV^2 \quad \text{(Capacitor energy in terms of V; SI unit: J)} \tag{16.14}$$

✓**TIP**

Capacitors store energy, with the stored energy proportional to the *square* of the charge or the potential difference.

Capacitors can release their stored energy quickly to supply large amounts of power. A **defibrillator** (as shown in the photo at the beginning of this chapter) delivers a jolt of electrical energy to restart a heart that's either stopped or undergoing ventricular fi brillation, a rapid irregular beating. To supply this energy, a capacitor of several hundred farads is charged to about 1 kV before being quickly discharged through the patient. For typical values $C = 500\ \mu\text{F}$ and $V = 1.2$ kV, Equation 16.14 shows that the stored energy is 360 J.

Short bursts from high-power lasers are used in a number of cutting-edge applications, including experiments with nuclear fusion. The highest-power lasers, at the National Ignition Facility (NIF) in Livermore, California, deliver some 2 MJ in about 1 ns. That gives a power of 2×10^6 J/10^{-9} s $= 2 \times 10^{15}$ W—more than 100 times humanity's entire energy consumption rate! How is that possible? By slowly charging huge capacitor banks, the overall power consumption is low. But discharging the capacitors quickly gives the phenomenal power of the short laser burst.

GOT IT? Section 16.4 Each of the capacitors shown is identical. Rank in decreasing order the equivalent capacitances.

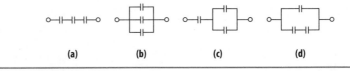

(a)   (b)   (c)   (d)

16.5 Dielectrics

Most capacitors have a solid insulating material—a **dielectric**—between their plates. The dielectric maintains the plates' physical separation and, as an insulator, prevents charge from moving between the plates and thus discharging the capacitor. Since capacitance increases with decreasing separation, we can use a thin layer of dielectric to provide high capacitance. Dielectrics also raise capacitance in another way, by lowering the electric field between the plates. You'll now see how that works.

APPLICATION **Ultracapacitors in Transportation**

Ride San Francisco's BART rapid transit system (shown here), and capacitors help speed you on your way, saving huge amounts of energy in the process. As a BART train approaches a station, on-board generators slow the train by converting its kinetic energy into electrical energy. The energy is stored in so-called ultracapacitors. When the train leaves the station, the stored energy drives the electric motors that accelerate the train. Energy storage in capacitors saves the BART system some 300 megawatt-hours of energy each year.

Dielectrics: A Microscopic View

The molecules in a dielectric are electric dipoles. As you saw in Chapter 15, dipoles align with an applied electric field. The dipoles of a capacitor's dielectric therefore align as shown in Figure 16.22, with their negative ends toward the positive plate and vice versa. The electric field produced by the dipoles themselves points opposite the field of the charged capacitor plates. Thus the dipoles' field reduces the electric field between the plates. The amount of reduction depends on the dielectric material.

✓ TIP

A dielectric is an insulator containing electric dipoles; the dipoles align with an applied electric field, reducing the field strength within the dielectric.

The factor by which a dielectric reduces the electric field is called the **dielectric constant** (symbol κ). Table 16.3 lists dielectric properties of a few materials.

CONCEPTUAL EXAMPLE 16.7 **Dielectric Constants for Vacuum and Perfect Conductor**

What's the dielectric constant κ for a vacuum? How about for a perfect conductor? Explain, based on the response of each when it's between the plates of a capacitor. Assume the conductor fills the space between the capacitor plates almost completely, with its faces adjacent to the capacitor plates but not quite touching.

SOLVE (a) A vacuum is literally nothing (Figure 16.23a), so the electric field is unaffected and $\kappa = 1$, exactly.

(b) Charges in the conductor are free to move, and, as you've seen, they move so as to cancel completely any electric field inside the conductor, as shown in Figure 16.23b. So the electric field between the plates becomes essentially zero. In practice there's still a very small region of nonzero field between each plate and the conductor, but that's negligible if the gap is small. We can therefore conclude that the dielectric constant of a conductor is infinite!

REFLECT The vacuum and the perfect conductor represent the two extremes of dielectric constant. Normal materials have dielectric constants in the range $1 < \kappa < \infty$.

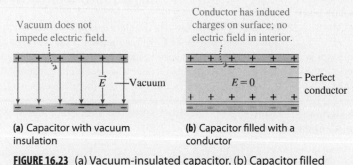

(a) Capacitor with vacuum insulation

(b) Capacitor filled with a conductor

FIGURE 16.23 (a) Vacuum-insulated capacitor. (b) Capacitor filled with a conductor.

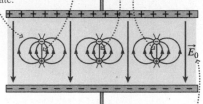

Molecular dipoles align with negative ends toward the positive plate.

The dipoles' electric fields superpose with the original field $\vec{E}_0$, reducing the net field...

...charge Q stays the same, so the reduced field $\vec{E} = \vec{E}_0/\kappa$ results in a lower potential $V = V_0/\kappa$ and therefore larger capacitance $C = \kappa C_0$.

FIGURE 16.22 Molecular view of a dielectric in a capacitor.

TABLE 16.3 Dielectric Properties of Selected Materials (Measured at 20°C)

Material	Dielectric constant κ	Dielectric strength E_{max} (MV/m)
Vacuum	1 (exact)	——
Air	1.00058	3.0
Teflon	2.1	60
Polystyrene	2.6	25
Nylon	3.4	14
Paper	3.7	16
Glass (Pyrex)	5.6	14
Neoprene	6.7	12
Tantalum oxide	26	500
Water	80	Depends on purity
Strontium titanate	256	8.0

So how does a dielectric affect capacitance? Recall that the potential difference between capacitor plates is $V = Ed$. With fixed plate spacing d, a reduction in E by a factor κ results in a reduction of V by the same factor. If V_0 is the potential difference between the plates without a dielectric present, then with the dielectric it becomes $V = V_0/\kappa$. If Q is the capacitor charge, then the capacitance without the dielectric is $C_0 = Q/V_0$. The charge Q doesn't change when the dielectric is inserted, so with the dielectric the capacitance becomes

$$C = \frac{Q}{V} = \frac{Q}{V_0/\kappa} = \kappa \frac{Q}{V_0}$$

or

$$C = \kappa C_0 \qquad \text{(Capacitance with dielectric; SI unit: F)} \qquad (16.15)$$

Therefore: **Inserting a material with dielectric constant κ between capacitor plates increases the capacitance by a factor κ.** By choosing the right dielectric, it's possible to increase capacitance significantly.

Dielectric Strength

No insulator is perfect, and a strong enough electric field will tear electrons out of their atoms and make the material conduct. If that happens in a capacitor, it discharges and permanent damage can result. The limiting value of electric field E_{max} within a dielectric is the **dielectric strength** of the material, as listed in Table 16.3.

Note in Table 16.3 that the dielectric strength of air is low, compared with that of common insulators. Lightning occurs when the electric field exceeds air's dielectric strength. The strong electric field near the sharp end of a conductor (for example, aircraft wing tips or a boat's mast) can result in **corona discharge**—known to sailors as St. Elmo's Fire. This effect is put to good use in laser printers and pollution control devices known as electrostatic precipitators.

The finite dielectric strength results in a maximum potential difference that can be applied between capacitor plates. In using capacitors it's as important to know this quantity as it is to know the capacitance!

EXAMPLE 16.8 **Capacitance and Dielectrics**

A parallel-plate capacitor has square plates 1.1 cm on a side separated by 0.15 mm of Teflon. Find its capacitance and the maximum potential difference that it can sustain.

ORGANIZE AND PLAN Capacitance depends on two things: the capacitor's dimensions (Section 16.4) and the dielectric. The maximum potential difference depends on plate spacing and the maximum electric field; the latter follows from the dielectric strength.

Without a dielectric, the capacitance of a parallel-plate capacitor is $C = \varepsilon_0 A/d$, with A the plate area and d the separation. Inserting the dielectric (Figure 16.23) increases the capacitance by a factor of κ, so $C = \kappa C_0 = \kappa \varepsilon_0 A/d$. To find the maximum potential difference, recall that $V = Ed$ for the uniform field in a parallel-plate capacitor. So the maximum V is $V_{max} = E_{max}d$, where E_{max} is the dielectric strength.

Known: $A = (1.1 \text{ cm})^2$; $d = 0.15 \text{ mm}$.

SOLVE Table 16.3 shows that $\kappa = 2.1$ for Teflon, so

$$C = \kappa \frac{\varepsilon_0 A}{d} = (2.1)\frac{(8.85 \times 10^{-12} \text{ F/m})(0.011 \text{ m})^2}{1.5 \times 10^{-4} \text{ m}}$$

$$= 1.5 \times 10^{-11} \text{ F} = 15 \text{ pF}$$

Table 16.3 gives 60 MV/m for the dielectric strength of Teflon, so the maximum potential difference for our capacitor is

$$V_{max} = E_{max}d = (60 \times 10^6 \text{ V/m})(1.5 \times 10^{-4} \text{ m})$$

$$= 9.0 \times 10^3 \text{ V} = 9.0 \text{ kV}$$

REFLECT That 0.15-mm plate spacing is actually fairly large, and results in a small capacitance C. But at the same time it gives a large V_{max}. As in this example, it's easy to make a small C with a large V_{max}. And it's easy to make a large C by decreasing the plate spacing—at the cost of reducing V_{max}. What's difficult, and therefore expensive, is to make a capacitor with both a large capacitance and a large maximum potential difference.

MAKING THE CONNECTION Would an air-insulated capacitor of the same size have a larger or smaller capacitance? A larger or smaller V_{max}?

ANSWER The capacitance is 2.1 times smaller, and the lower dielectric strength of air means that for the same plate spacing the maximum potential difference is also smaller.

Energy Storage

How does a dielectric affect energy storage in a capacitor? Suppose a vacuum-insulated capacitor with capacitance C_0 is given charge Q and then disconnected from the charging battery. Then by Equation 16.13 its energy is $U_0 = Q^2/2C_0$. Inserting a material with dielectric constant κ doesn't alter the charge, but increases the capacitance by a factor κ. The new capacitance is $C = \kappa C_0$, so the energy becomes

$$U = \frac{Q^2}{2C} = \frac{Q^2}{2\kappa C_0}$$

or $U = U_0/\kappa$. Thus the stored energy *decreases* by a factor κ.

The Cell Membrane: A Natural Capacitor

Most living cells are contained within a thin membrane whose exterior surface normally carries a positive electric charge, while the inside surface is negative. The membrane is therefore a capacitor, with capacitance given, as usual, by the ratio of charge to potential difference. This difference is on the order of 0.1 V, and the capacitance is typically around 1 μF per square centimeter of membrane area. Of course, cell areas are far less than 1 cm^2, so the actual membrane capacitance is correspondingly smaller. The membrane is thin enough that, even though it's not flat, it can be treated as a parallel-plate capacitor. A cell's area is readily computed from its size, and the membrane capacitance can be determined with electrical measurements. Then the parallel-plate formula yields the membrane thickness. In fact, the first determination of cell membrane thickness involved measuring the capacitance per unit area in a suspension of cells, and then computing the thickness.

Chapter 16 in Context

This chapter is the second of three covering electric charge. Here you learned about *electrical energy*, building on energy concepts from Chapter 5 and the electric field from Chapter 15. You explored *electric potential* and its relation to potential energy. All these concepts proved useful in understanding *capacitors*—systems of charged conductors we use to store charge and energy.

Looking Ahead Electric potential and potential difference will prove crucial for understanding electric circuits in Chapter 17. Especially important will be the relation of potential difference and electric current. At the end of Chapter 17, we'll look at circuits that combine capacitors with other electrical components. Electric potential will continue to be important in later chapters—for example, in the AC circuits of Chapter 19 and particle accelerators of Chapter 26.

Electric Potential Energy

(Section 16.1) A system of two or more electric charges has **potential energy**. Potential energy is solely a function of the position(s) of the particle(s).

The change in potential energy of a particle that moves from one point to another in an electric field is independent of the path taken.

Potential energy of two point charges: $U = \dfrac{kq_1q_2}{r}$

Electric Potential

(Section 16.2) **Electric potential** is potential energy per unit charge.

Electric potential of a point charge: $V = \dfrac{kq}{r}$ (with zero of potential taken at an infinite distance)

Potential energy and potential: $U = q_0V$

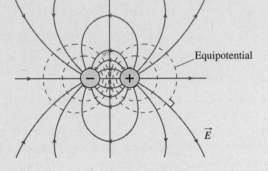

Electric Potential and Electric Field

(Section 16.3) Electric field lines are perpendicular to equipotential surfaces and are directed from higher to lower potential.

An electric field pushes positive charges from higher to lower potential and negative charges from lower to higher potential.

Electric field and potential difference: $E = -\Delta V/\Delta x$ (for uniform electric field)

Capacitors

(Section 16.4) A **capacitor** consists of two insulated conductors. Capacitors store charge and energy.

Capacitance is the ratio of charge to potential difference.

Capacitance: $C = \dfrac{Q}{V}$

Capacitors in parallel: $C_p = C_1 + C_2$

Capacitors in series: $\dfrac{1}{C_s} = \dfrac{1}{C_1} + \dfrac{1}{C_2}$

Dielectrics

(Section 16.5) A **dielectric** between capacitor plates increases capacitance and allows the capacitor to store more charge.

Capacitor with a dielectric: $C = \kappa C_0$

NOTE: Problem difficulty is labeled as ■ straightforward to ■ ■ ■ challenging. Problems labeled BIO are of biological or medical interest.

Conceptual Questions

1. Suppose you have three negative point charges at different locations. Is the electric potential energy greater than or less than zero, or could it be either?
2. Repeat the preceding question for two negative charges and one positive charge.
3. If you have two charges of the same magnitude but opposite sign, separated by some distance d, is there any place where the electric potential is zero? Explain.
4. Repeat the preceding question for the case of two charges that have the same magnitude and same sign.
5. Lightning carries a tremendous amount of electrical energy. Into what forms do you see this energy transformed?
6. If an electron and a proton are each accelerated from rest to a kinetic energy of 300 eV, which one is traveling faster?
7. Is it possible for two equipotential surfaces to intersect? Why or why not?
8. At a point where the electric potential is zero, is the electric field necessarily zero too? Why or why not?
9. At a point where the electric field is zero, is the electric potential necessarily zero too? Why or why not?
10. Two identical positive charges $+Q$ are separated by a distance d. Sketch some electric field lines and equipotentials.
11. Using the electric potential map shown in Figure CQ16.11, indicate the approximate directions and relative magnitudes of the electric field vectors at the points labeled A, B, C, and D.

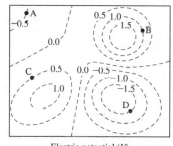

FIGURE CQ16.11   Electric potential (V)

12. Show that the units $C^2/(N \cdot m^2)$ for the constant ε_0 reduce to F/m.
13. Suppose you have a parallel-plate capacitor attached to a battery with fixed potential difference V. If you increase the plate spacing, explain whether each of the following increases, decreases, or remains the same: (a) the capacitance; (b) the charge on the plates; (c) the stored energy.
14. Suppose you disconnect the capacitor of the preceding problem from the battery before you move the plates, ensuring that the charge Q is fixed. As you increase the plate separation, explain whether each of the following increases, decreases, or remains the same: (a) the capacitance; (b) the potential difference between the plates; (c) the stored energy.
15. An air-filled parallel-plate capacitor with plate separation d has capacitance C_0. A dielectric with dielectric constant κ and thickness $d/2$ is inserted between the plates as shown in

Figure CQ16.15. What's the new capacitance? Does it matter whether the dielectric is closer to one plate or the other? Explain.

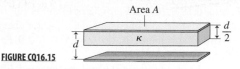

FIGURE CQ16.15

16. An air-filled parallel-plate capacitor with plate area A and separation d has capacitance C_0. A dielectric with cross-sectional area $A/2$, dielectric constant κ, and thickness d is inserted between the plates as shown in Figure CQ16.16. What's the new capacitance?

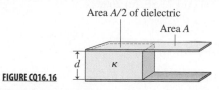

FIGURE CQ16.16

17. An air-filled parallel-plate capacitor is charged to some fixed value Q. A dielectric slab is then inserted between the plates. Is the work done by an external agent to insert the slab positive, negative, or zero? Explain.
18. For the situation with two like charges described in Conceptual Example 16.1, make a plot showing the speed of charge q_2 as a function of its position on the x-axis.

Multiple-Choice Problems

19. Electric potential is (a) force per unit charge; (b) potential energy per unit charge; (c) force per unit distance; (d) electric field per unit charge.
20. Two charges, one positive and one negative, are separated by a fixed distance. The sign of the potential energy of this pair (a) is always positive; (b) is always negative; (c) is sometimes positive and sometimes negative; (d) depends on the distance.
21. A helium nucleus consists of two protons (charge $e = 1.60 \times 10^{-19}$ C) and two neutrons (zero charge). Assuming the protons are separated by 1.9×10^{-15} m, the electric potential energy is (a) 4.8×10^{-13} J; (b) 2.4×10^{-19} J; (c) 3.4×10^{-11} J; (d) 1.2×10^{-13} J.
22. The distance from a proton at which the electric potential is 1 V is (a) 1.4×10^{-9} m; (b) 3.8×10^{-5} m; (c) 2.0×10^{-18} m; (d) 7.2×10^{-5} m.
23. A square 2.5 cm on a side has identical charges -7.1×10^{-9} C on three of its corners. The electric potential at the fourth (empty) corner is (a) 4400 V; (b) -4400 V; (c) -6900 V; (d) -3500 V.
24. When a proton at rest is accelerated through a potential difference of -40 V, its final speed is (a) c (the speed of light); (b) 4000 m/s; (c) 880 m/s; (d) 8.8×10^4 m/s.
25. Two parallel conducting plates carry uniform charge densities $\pm 3.7 \times 10^{-6}$ C/m². If the plate separation is 3.0 mm, the potential difference between the plates is (a) 830 V; (b) 1250 V; (c) 66 V; (d) 130 V.

26. A particle with charge $-3e$ is accelerated from rest through a potential difference of 150 V. This gives the particle a kinetic energy of (a) 150 eV; (b) 50 eV; (c) 450 eV; (d) -150 eV.

27. What's the potential difference across a 380-nF capacitor carrying a charge of 15 μC? (a) 200 V; (b) 26 mV; (c) 26 V; (d) 39 V.

28. A μF · mV is equivalent to a (a) mC; (b) μC; (c) nC; (d) pC.

29. The equivalent capacitance of 10-F, 20-F, and 30-F capacitors in series is (a) 0.18 F; (b) 5.5 F; (c) 32 F; (d) 60 F.

30. If you want to store 25 J of energy in a capacitor charged to 12 V, the required capacitance is (a) 0.35 F; (b) 2.9 F; (c) 2.1 F; (d) 0.48 F.

31. In terms of the charge stored (Q) and the potential difference (V), the energy stored in a capacitor is (a) $2QV$; (b) QV; (c) $\frac{1}{2}QV$; (d) Q/V.

32. The capacitance of a water-filled parallel-plate capacitor with plate area 2.4×10^{-3} m^2 and separation 0.065 mm is (a) 520 nF; (b) 0.52 nF; (c) 72 μF; (d) 26 nF.

Problems

Section 16.1 Electric Potential Energy

33. ■ Two protons in a nucleus are separated by 6.5×10^{-15} m. What's the potential energy of the pair?

34. ■■ In a classical model of the hydrogen atom, the proton and electron are 5.29×10^{-11} m apart. Compute their *gravitational* potential energy and compare it with the electric potential energy computed in the text. That is, find the ratio of the electric to the gravitational potential energy.

35. ■■ Suppose both Earth and Moon carried the same net charge Q spread uniformly over their surfaces. (a) How large would Q need to be so that the system's electric and gravitational potential energies had equal magnitude? (b) Under these conditions, what would be the average charge per square meter on Earth's surface? (c) Comment on the result of part (b).

36. ■■ Two identical 25-g particles each carry $+1.0$ μC of charge. One is held fixed, and the other is placed 2.5 cm away and released. (a) Describe the subsequent motion of the second charge. (b) Find the speed of the moving charge when it's (i) 10 cm from the fixed charge and (ii) 1.0 km from the fixed charge.

37. ■■ Find the potential energy of charges q_1 and q_2 located in the x-y plane: $q_1 = -3.6 \times 10^{-9}$ C at (0.12 m, 0.45 m) and $q_2 = 1.6 \times 10^{-9}$ C at (0.36 m, 0.55 m).

38. ■■■ Find the potential energy of the charges q_1 and q_2 at the following positions: $q_1 = -3.6 \times 10^{-6}$ C at (1.1 m, 1.4 m, 2.4 m) and $q_2 = -1.6 \times 10^{-6}$ C at (0.85 m, 1.9 m, 1.4 m).

39. ■■ Three point charges of 2.0 μC, 4.0 μC, and 6.0 μC form an equilateral triangle with each side 2.5 cm. Find the electric potential energy of this distribution.

40. ■■ A proton is held at a fixed position. An electron is placed 0.75 m away and released. Find the electron's speed when it's 1.0 mm from the proton.

41. ■■ Two protons are released from rest 5.0 mm apart. (a) Describe the subsequent motion after they're released. (b) How fast are the protons moving when they're a great distance apart?

Section 16.2 Electric Potential

42. ■ If the potential difference between a cloud and the ground is about 12 MV, how much charge is transferred in a lightning bolt that releases 1.6 GJ of energy?

43. ■ At 1.0 m from a point charge, the electric potential is 280 V. (a) What's the potential 2.0 m from the same charge? (b) What are the magnitude and sign of the charge?

44. ■■ Two point charges, $+2.0$ μC and -4.0 μC, lie 0.45 m apart on the x-axis. Are there any points on the x-axis where the electric potential is zero? If so, find the point(s). If not, explain.

45. ■ After walking across a dry carpet, there's a potential difference of 4500 V between your hand and the doorknob. As you reach toward the knob, a spark jumps, releasing energy 1.6 μJ. (a) How much charge was transferred? (b) Assuming all the charges were electrons, how many were there?

46. ■■ It's generally okay to ignore relativistic effects when a particle's speed is less than 1% of the speed of light. Find the accelerating potential difference that will bring each of the following particles to that speed, starting from rest: (a) electrons; (b) protons; (c) alpha particles (helium nuclei, consisting of two protons and two neutrons).

47. ■■ A 900-kg electric car uses a 270-V battery pack. As the car accelerates from rest, 400 C flows from the battery. What's the maximum speed the car can attain during this time? (The actual speed will be less, due to friction and other losses.)

48. ■■ In a mass spectrometer, ions accelerate through a fixed potential difference of 150 V. Find the kinetic energy and speed of the following ions after they've accelerated from rest: (a) He^{2+}; (b) ^{107}Ag$^+$; (c) ^{109}Ag$^+$; (d) ^{235}U$^+$.

49. ■■ The "charmed lambda" particle (symbol Λ^+) is a subatomic particle with mass 4.1×10^{-27} kg and charge $+e$. Find the potential difference needed to accelerate the Λ^+ from rest to 3.0×10^6 m/s (1% the speed of light).

50. ■■ X rays are produced in a process called "Bremsstrahlung," in which electrons slow and convert their kinetic energy into electromagnetic energy. Find the energy of an x ray produced when an electron moving at 2.8 Mm/s is suddenly brought to rest. Express your answer in both J and eV.

51. ■■ (a) An ion that passes through a 300-V potential difference gains 9.6×10^{-17} J of kinetic energy. What's its charge? (b) Repeat for an ion that loses the same amount of kinetic energy.

52. ■■ An electron moves through an electric field, and its speed drops from 2500 m/s to 1500 m/s. What's the potential difference between the two points at which the speed was measured?

Section 16.3 Electric Potential and Electric Field

53. ■ The potential difference between a pair of parallel conducting plates is 65 V. The plates are 3.2 cm apart. What's the magnitude of the electric field between the plates?

54. ■■ You're in a uniform electric field of magnitude 260 N/C. Compute the potential difference if you travel 1 cm in each of the following directions: (a) in the direction of the field; (b) opposite to the field; (c) perpendicular to the field; (d) at a 45° angle to the field.

55. ■■ A pair of parallel plates produces a uniform electric field of 850 V/m. If an electron is released at rest from the negative plate, how fast is it moving when it hits the positive plate, 7.5 mm away?

56. ■■ A uniform electric field of about 100 V/m, pointing straight down, exists just above Earth's surface. (a) Find the potential difference between the ground and the top of a 75-m-tall building. (b) What uniform surface charge density would be required on Earth's surface to create such a field? What would be the sign of the surface charge?

57. ■ Two parallel plates separated by 6.5 mm have a potential difference of 250 V between them. (a) Find the electric field between the plates. (b) What's the force on an electron in this field?

58. ■■ Consider a 1.5-C point charge, fixed at the origin. (a) Compute the electric potential 5.00 m away on the *x*-axis. (b) Compute the electric potential a distance 5.01 m away on the *x*-axis. (c) Compute the electric field a distance 5.00 m away on the *x*-axis. (d) Using your answer from parts (a)–(c), show that the relationship $E = -\Delta V/\Delta x$ is approximately true.

59. ■ Spherical equipotential surfaces surround a point charge. If the electric potential is -175 V on a sphere of radius 0.65 m, what is the charge?

60. ■■ A uniform electric field exists between two conducting plates separated by 1.5 mm. A dust grain with charge 65 nC between the plates feels a 25-N electric force. Find (a) the electric field and (b) the potential difference between the plates.

61. ■■ In a Millikan oil-drop experiment used to measure the elementary charge (Chapter 15), two horizontal parallel plates are 1.0 cm apart. With a 180-V potential difference between the plates, a charged droplet of mass 8.8×10^{-16} kg is suspended motionless. (a) If the upper plate is at the higher potential, is the charge on the droplet positive or negative? (b) What's the electric field between the plates under these conditions? (c) Compute the charge on the droplet. To how many elementary charges (*e*) does this correspond?

Section 16.4 Capacitors

62. ■ Suppose you want to make a parallel-plate capacitor with square plates 1 cm on a side. (a) How far apart should the plates be for a capacitance of 3.0 pF? (b) Repeat for $C = 3.0\ \mu$F. (c) Are both designs practical? Comment.

63. ■ A 250-pF parallel-plate capacitor consists of two identical circular plates 0.0875 mm apart. Find their radius.

64. ■ Find the equivalent capacitance of three capacitors $C_1 = 4.0\ \mu$F, $C_2 = 9.0\ \mu$F, and $C_3 = 12\ \mu$F, when they're (a) in parallel and (b) in series.

65. ■■ A 0.25-F capacitor and a 0.45-F capacitor are in parallel. (a) Find the equivalent capacitance. (b) Find the charge stored on each capacitor and the total charge stored when the parallel combination is connected across a 24-V battery.

66. ■■ Two capacitors, $C_1 = 4.0$ mF and $C_2 = 6.0$ mF, are in series. (a) Find the equivalent capacitance. (b) If the combination is connected across a 16-V battery, find the charge stored on each capacitor and the potential difference across each.

67. ■■■ Suppose the three-capacitor network in Example 16.6 (Figure 16.20a) is connected across a 12-V battery. Follow these steps to find the charge on each capacitor: (a) Use the equivalent capacitance to find the charge supplied by the battery. (b) Note that the charge supplied by the battery equals the charge Q_3 on the capacitor C_3. (Why?) Use this fact to compute the potential difference V_3 across C_3. (c) Compute the potential difference across the other two capacitors. (d) Use your answer from part (c) to find the charge stored on each of C_1 and C_2. *Note: Q_1* and Q_2 aren't the same.

68. ■■ For the capacitor network shown in Figure P16.68, (a) find the equivalent capacitance; (b) find the charge on each capacitor when the entire combination is connected across a 6.0-V battery.

In Problems 69–74, you have three 3.0-μF capacitors.

69. ■ Find the equivalent capacitance if the three are in parallel.

70. ■■ The three capacitors in parallel are connected across a 30-V battery. Find the charge on each capacitor and potential difference across each.

71. ■ Find the equivalent capacitance if the three capacitors are in series.

72. ■■ The three capacitors in series are connected across a 30-V battery. Find the charge on each capacitor and the potential difference across each.

73. ■■ Two of the capacitors are in parallel, and this parallel combination is in series with the third capacitor. Find the equivalent capacitance.

74. ■■ The capacitor combination in the preceding problem is connected across a 30-V battery. Find the charge on each capacitor and the potential difference across each.

75. ■■ Three capacitors are in series: $C_1 = 45\ \mu$F, $C_2 = 65\ \mu$F, and $C_3 = 80\ \mu$F. (a) Compute the equivalent capacitance. The series combination is connected across a 48-V battery. (b) Find the charge on each capacitor. (c) Compute the energy stored in each capacitor. (d) Show that the total energy you found in part (c) is the same as would be stored in a single capacitor with the equivalent capacitance you found in part (a).

76. ■■ Repeat the preceding problem if the three capacitors are in parallel.

77. **BIO** ■■ **Defibrillator: energy and power.** A defibrillator uses a 225-μF capacitor charged to 2400 V. (a) How much energy is stored in the capacitor? (b) What's the power delivered to the patient if the capacitor discharges completely in 2.5 ms?

78. **BIO** ■■ **Defibrillator: capacitance and energy.** (a) If a 2200-V defibrillator is designed to carry a maximum charge of 0.40 C, what should its capacitance be? (b) How much energy will it store?

79. ■■ A parallel-plate capacitor has plate area 2.5×10^{-3} m^2 and plate spacing 0.050 mm. (a) What's its capacitance? (b) Find the stored charge and energy when this capacitor is connected across a 50-V battery.

80. ■■ (a) For the preceding problem, find the energy per unit volume between the plates; this is the *electric energy density*. (b) Show that the energy density is equal to $\epsilon_0 E^2/2$, where E is the uniform electric field between the plates. (The expression $\epsilon_0 E^2/2$ is actually the energy density in any electric field, not just this configuration.)

81. ■■ You have a 250-nF capacitor and wish to combine it with one other to make a combined capacitance of 1.50 μF. How large a capacitor do you need, and should it be combined in series or parallel with the first one?

82. ■■■ You have three capacitors: 10 μF, 15 μF, and 20 μF. Find the values of all the possible capacitances you can create with different combinations using one, two, or all three capacitors.

Section 16.5 Dielectrics

83. ■■ A parallel-plate capacitor has plates with area $A = 3.25 \times 10^{-5}$ m^2 separated by 0.12 mm. With paper insulation between the plates, find (a) the capacitance; (b) the charge on each plate when a potential difference of 12.0 V is across the capacitor; and (c) the maximum possible potential difference across the plates.

84. ■■ An oil-filled capacitor has square parallel plates 25 cm on a side, separated by 0.50 mm. When the capacitor is charged to 90 V, the stored charge is 7.7 μC. Find the dielectric constant of the oil.

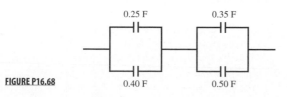

FIGURE P16.68

0.25 F   0.35 F

0.40 F   0.50 F

85. ■■ A 0.45-nF parallel-plate capacitor has plate area $A = 1.7 \times 10^{-4}\,\text{m}^2$. If its dielectric has $\kappa = 3.8$, what's the plate separation?

86. ■ The electric field near Earth's surface on a clear day is typically about 100 V/m. How does this compare to the electric field needed for an electrical discharge in air?

87. ■■ A polystyrene-insulated capacitor has $C = 220$ nF. If the separation between its parallel plates is 0.45 mm, what's the maximum charge it can hold?

88. ■■ The parallel-plate capacitor used on each key of a computer keypad is a square 0.8 cm on a side. Between the plates is a dielectric with $\kappa = 4.2$. When the key is in the "up" position, the plate separation is 0.75 mm. (a) Find the initial capacitance. (b) How far must the key be depressed for a keystroke to be registered, which happens when the capacitance increases by 4.0 pF?

General Problems

89. ■■ Consider a *spherical capacitor*, in which a spherical conductor of radius a lies inside a concentric spherical conducting shell of radius b. Suppose the inner and outer shells carry charges $\pm Q$, respectively. (a) Compute the potential difference between the two shells. (Note that the electric potential outside of a uniform sphere of charge is the same as if all the charge were concentrated at the sphere's center. Your answer should be a function of Q and the two radii.) (b) Use your answer from part (a) and the definition of capacitance to show that the capacitance of the spherical capacitor is $C = 4\pi\varepsilon_0 ab/(b - a)$. (c) Compute the capacitance of a spherical capacitor with inner radius 2.0 cm and outer radius 10.0 cm.

90. ■■■ Consider the spherical capacitor of the preceding problem. Show that in the limit as b approaches a, the capacitance formula is equivalent to that for a parallel-plate capacitor.

91. ■■ A 0.01-μF, 300-V capacitor costs \$0.25; a 0.10-μF, 100-V capacitor costs \$0.35; and 30-μF, 5.0-V capacitor costs \$0.88. (a) Which capacitor can store the most charge? (b) Which can store the most energy? (c) Which is the most effective as an energy storage device, in terms of energy stored per unit cost?

92. ■■ A camera flashtube requires 5.0 J of energy per flash. The flash duration is 1.0 ms. (a) What power does the flashtube use while it's flashing? (b) If the flashtube operates at 200 V, what capacitance is needed to supply the flash energy? (c) If the flashtube is fired once every 10 s, what's the average power consumption?

93. ■■ The NOVA laser fusion experiment at Lawrence Livermore Laboratory in California uses energy at the rate of 10^{14} W (roughly 100 times the output of all the world's power plants) while its lasers are on. However, one laser pulse lasts for only 10^{-9} s. (a) Only about 0.17% of the capacitor energy actually appears as light. How much light energy is delivered in one pulse? (b) The capacitor bank supplying this energy has a total capacitance of 0.26 F. What's the potential difference across the capacitor bank when it's fully charged?

94. ■■ (a) Find the equivalent capacitance of the four identical capacitors shown in Figure GP16.94, measured between points A and B, if $C = 10\ \mu\text{F}$. (b) If a 50-V battery is connected across A and B, what's the charge on each capacitor?

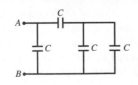

FIGURE GP16.94

95. ■■■ Figure GP16.95 shows something we've heretofore ignored: that there's a nonuniform electric field *outside* a capacitor. Consider the dipole shown in the figure, and show that it experiences an attractive force toward the capacitor. That's the origin of the work done *on* a dielectric slab while it's being inserted into the capacitor, and that's why the stored potential energy is lower in the presence of a dielectric.

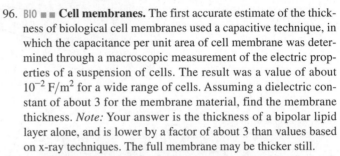

FIGURE GP16.95

96. BIO ■■ **Cell membranes.** The first accurate estimate of the thickness of biological cell membranes used a capacitive technique, in which the capacitance per unit area of cell membrane was determined through a macroscopic measurement of the electric properties of a suspension of cells. The result was a value of about $10^{-2}\,\text{F/m}^2$ for a wide range of cells. Assuming a dielectric constant of about 3 for the membrane material, find the membrane thickness. *Note:* Your answer is the thickness of a bipolar lipid layer alone, and is lower by a factor of about 3 than values based on x-ray techniques. The full membrane may be thicker still.

97. ■■■ An unknown capacitance C is in series with a 3.0-μF capacitor. This pair is in parallel with a 1.0-μF capacitor, and the resulting combination is in series with a 2.0-μF capacitor. (a) Make a circuit diagram of this network. (b) When a potential difference of 100 V is applied across the ends of the network, the total energy stored in all capacitors is 5.8 mJ. Find C.

98. BIO ■■ **Neuromuscular microstimulator.** A *neuromuscular microstimulator* is a device designed to be implanted into paralyzed muscles or other tissue suffering neuromuscular dysfunction. The device supplies an electric current that stimulates the dysfunctional tissue. A particular device uses a capacitor built into a microchip; the capacitor can supply 10 mA of current for 200 μs before it's fully discharged. (a) Find the initial charge on the capacitor and (b) the capacitance if it's initially charged to 4.5 V.

Answers to Chapter Questions

Answer to Chapter-Opening Question
The energy is stored in the electric field of a pair of conductors—a capacitor—and dumped quickly through the defibrillator into the patient's chest when needed.

Answers to GOT IT? Questions
Section 16.1   (c) 0.27 J
Section 16.2   (d) 1.6×10^{-17} J
Section 16.3   (d) 3.0 MV/m
Section 16.4   (b) > (d) > (c) > (a)

17 Electric Current, Resistance, and Circuits

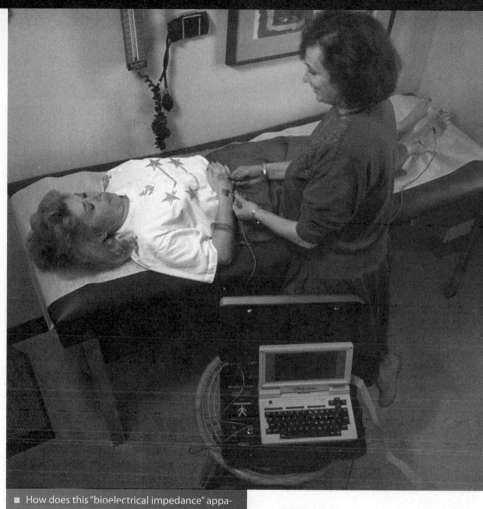

■ How does this "bioelectrical impedance" apparatus help determine the body's fat content?

Electric current makes things happen in technology, from personal electronics to the interconnected power grid, and in natural phenomena from lightning to cell membranes. Here you'll learn about electric current and the resistance that limits its flow, and you'll see how batteries work. You'll then analyze networks of resistors and circuits with resistors and capacitors. You'll also learn about semiconductors and superconductors.

17.1 Current and Resistance

Introduction to Current

Electric current is a flow of electric charge; quantitatively, it's the rate at which charge flows past any point. Symbolically, if charge Δq passes a point in time Δt, then the current at that point is

$$I = \frac{\Delta q}{\Delta t} \quad \text{(Electric current; SI unit: A)} \quad (17.1)$$

To Learn

By the end of this chapter you should be able to

- Understand the relationships among electric current, resistance, and potential difference.
- Apply Kirchhoff's rules to determine currents in circuits.
- Analyze circuits containing series and parallel resistor combinations.
- Compute power in electric circuits.
- Analyze circuits containing capacitors and resistors.
- Recognize the basic properties of semiconductors and superconductors.

The free electrons in a wire move from lower to higher potential...

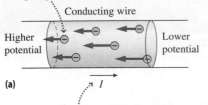

(a)

...but the *current I* is defined as moving from higher to lower potential. Thus, current leaves the positive terminal of a battery...

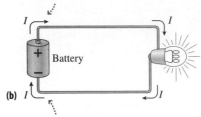

(b)

...flows through the conducting circuit, and returns to the negative terminal.

FIGURE 17.1 (a) Electron flow is opposite the direction of the current. (b) Current flow in a circuit.

Strictly speaking, Equation 17.1 should be taken in the limit as Δt approaches zero, making I the instantaneous current. If the current is steady, this distinction is unimportant. But when current varies, charge flow must be sampled over a short time for an accurate measure of the instantaneous current. An example of time-varying current is alternating current (AC), described in Chapter 19.

The SI unit for current is the **ampere** (A), defined as 1 C/s and named for the French physicist André Marie Ampère (1775–1836). A 100-W lightbulb uses a current of about 1 A, while the motors of power tools, vacuum cleaners, and air conditioners draw several amperes. Small electronic devices use much less—milliamperes (mA) or even microamperes (μA).

Electron Flow and Positive Current

Equation 17.1 implies that current flows in the direction of positive charge motion. However, in most electric circuits negative electrons in conducting wires carry the current. By convention, we still speak of the current direction as if positive charges were flowing—but in the direction opposite the negative electrons (Figure 17.1a). This avoids extra negative signs in circuit analysis.

In Chapter 16 you saw that positive charges move from higher to lower potential. In a circuit, that means you can think of current flowing from the positive battery terminal around the circuit to the negative terminal (Figure 17.1b). The current is the same—from positive to negative—whether it's actually positive charges going from the positive to the negative battery terminal, or negative electrons flowing from negative to positive. If you find this confusing, blame Ben Franklin for his choice of "negative" for the charge we now know is associated with electrons! When current flows in only one direction, it's called **direct current**, or **DC**. In Chapter 19, you'll learn about **alternating current (AC)**, which reverses periodically.

EXAMPLE 17.1 **Through the Cell Membrane**

The cell membrane separates the interior of a living cell from its surroundings. So-called ion channels penetrate the membrane, allowing passage of materials into and out of the cell (Figure 17.2). A particular channel opens for 1.0 ms and allows passage of 1.1×10^4 singly ionized potassium ions during this time. What's the current in the channel?

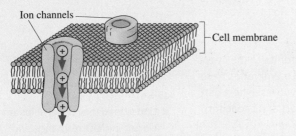

FIGURE 17.2 A cell membrane, showing ions passing through an ion channel.

ORGANIZE AND PLAN Current is the charge per time passing a given point: $I = \Delta q / \Delta t$. Here we know the charge on each ion and the number of ions, so we can find Δq, and we're given the time.

Known: $\Delta t = 1.0$ ms, 1.1×10^4 ions passing, with 1 elementary charge $e = 1.6 \times 10^{-19}$ C per ion.

SOLVE The total charge Δq on all those ions is

$$\Delta q = (1.1 \times 10^4 \text{ ions})(1.6 \times 10^{-19} \text{ C/ion}) = 1.76 \times 10^{-15} \text{ C}$$

Then

$$I = \frac{\Delta q}{\Delta t} = \frac{1.76 \times 10^{-15} \text{ C}}{1.0 \times 10^{-3} \text{ s}} = 1.8 \times 10^{-12} \text{ A} = 1.8 \text{ pA}$$

REFLECT That's a tiny current, but it's typical of microscopic systems like these.

MAKING THE CONNECTION How would the current change if the ions were doubly ionized?

ANSWER Each ion carries twice as much charge, so with the other parameters unchanged, the current would double.

Current: A Microscopic View

Metals are good conductors because they contain free electrons that aren't bound to individual atoms. Apply a potential difference across a metal, and the resulting electric field gives these **conduction electrons** an average **drift speed**, v_{drift}, which results in an electric current. How is the current related to the microscopic properties of the metal?

Figure 17.3 shows electrons moving through a metal wire. The charge Δq within a volume is

$$\Delta q = \frac{\text{charge}}{\text{volume}} \times \text{volume}$$

Consider the electrons crossing the area A in a time Δt. Moving at speed v_{drift}, they occupy a length $v_{drift}\Delta t$, and thus a volume $Av_{drift}\Delta t$. The charge per volume is the product of the **number density**, or number of electrons per unit volume, with the electron charge:

$$\text{number density } n = \frac{\text{charge}}{\text{volume}} = \frac{\text{number of electrons}}{\text{volume}} \times \text{charge per electron}$$

The number density, n, is characteristic of each metallic conductor; see Table 17.1. Each electron carries charge e. (Here we're just concerned with the magnitude, not the sign.) Putting our results together,

$$\Delta q = \frac{\text{charge}}{\text{volume}} \times \text{volume} = (ne)(Av_{drift}\Delta t)$$

Finally, from the definition of current

$$I = \frac{\Delta q}{\Delta t} = neAv_{drift} \quad \text{(Current and drift speed; SI unit: A)} \quad (17.2)$$

Equation 17.2 relates the macroscopic current to the density of free electrons, the electron charge, and the drift speed. This linear relationship between current and drift speed makes sense, because faster electrons mean more current.

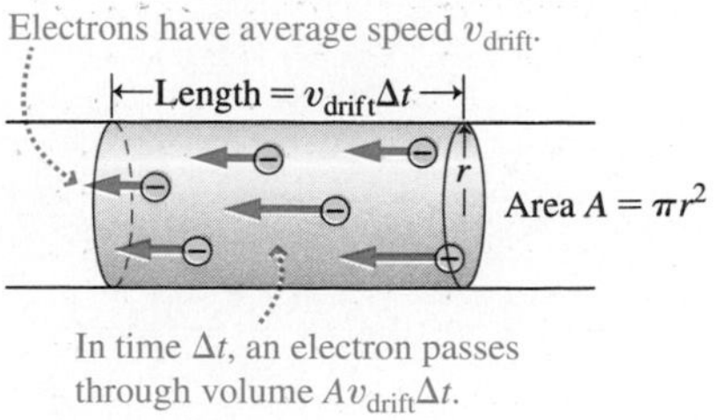

In time Δt, an electron passes through volume $Av_{drift}\Delta t$.

FIGURE 17.3 Current and drift speed.

TABLE 17.1 Number Densities n for Selected Conductors at $T = 27°C$

Conductor	Number density n (m^{-3})
Silver	5.86×10^{28}
Copper	8.47×10^{28}
Gold	5.86×10^{28}
Iron	1.70×10^{29}
Niobium	5.56×10^{28}
Aluminum	1.81×10^{29}
Tin	1.48×10^{29}
Gallium	1.54×10^{29}
Zinc	1.32×10^{29}
Lead	1.32×10^{29}

EXAMPLE 17.2 Drift Speed in Copper

An 8-gauge copper wire is 3.26 mm in diameter and can carry a maximum current of 24 A. Find the drift speed of the electrons at this maximum current.

ORGANIZE AND PLAN The relationship between drift speed and current is developed in the text. It's important to know the type of wire, because the number density depends on that. The relationship between current and drift speed is given by Equation 17.2:

$$I = \frac{\Delta q}{\Delta t} = neAv_{drift}$$

You'll want to solve this equation for the drift speed. From Table 17.1, the number density for copper is $n = 8.47 \times 10^{28}$ m^{-3}. The cross-sectional area (as in Figure 17.3) is $A = \pi r^2$, with $r = d/2 = 1.63$ mm.

Known: $I = 24$ A, $r = 1.63$ mm $= 0.00163$ m.

SOLVE Solving Equation 17.2 for drift speed gives

$$v_{drift} = \frac{I}{neA} = \frac{24 \text{ A}}{(8.47 \times 10^{28} \text{ m}^{-3})(1.60 \times 10^{-19} \text{ C})\pi(0.00163 \text{ m})^2}$$
$$= 2.1 \times 10^{-4} \text{ m/s}$$

REFLECT The units for drift speed do in fact reduce to m/s, because amperes (A) in the numerator are C/s. The numerical result may seem low—it's truly a snail's pace! But this is a typical electron drift speed. It can be so small because the electron density in a good conductor is so high that electrons don't have to move very fast to carry a large current.

MAKING THE CONNECTION Suppose you have a wire of the same size carrying the same current as in this example, but of a material with a larger number density n. Is the electron drift speed higher, lower, or the same?

ANSWER With the other parameters unchanged, larger n implies a lower drift speed. Conceptually, more conduction electrons means they don't have to move as fast to carry the same current.

With drift speeds well below 1 m/s, you might wonder why your electric light comes on almost instantly when you flip the wall switch. A good analogy is drinking a milk shake through a straw. The milk doesn't move very fast through the straw—perhaps a few millimeters per second, comparable to the electron drift speed. However, if the straw is already full of milk shake, you get a taste right away. A conducting wire is like the full straw. Free electrons are everywhere, ready to spill out into the lightbulb. You don't have to wait for electrons to travel from the wall switch.

✓ **TIP**

Drift speeds in conductors are typically on the order of millimeters per second or smaller.

Thermal Motion and Collisions

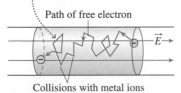

Free electrons exhibit a slight drift superimposed on random motion.

Path of free electron

$\vec{E} \rightarrow$

Collisions with metal ions

FIGURE 17.4 An electron's drift velocity is superimposed on its random motion.

Our discussion of drift speed suggests a slow, steady progression of electrons through a current-carrying wire. But actually, the electrons are whizzing around with thermal speeds on the order of 10^5 m/s. They're moving in random directions, so this thermal motion doesn't constitute a current. Apply a potential difference across the wire, though, and the associated electric field accelerates the electrons in the same direction, producing a net current. As Figure 17.4 shows, the electrons collide frequently with the ions of the metal, bouncing off in random directions. But they're again accelerated by the electric field, with the result that the electrons acquire a small average drift speed—just as a car in city traffic is always accelerating and braking, but nevertheless progresses with a constant average speed.

Resistance, Conductivity, and Ohm's Law

Those collisions between electrons and metal ions are inelastic, meaning the electrons give up some of the energy they've gained from the electric field. The result is heating of the wire—which is what limits the wire's safe current. Collisions also reduce the average electron speed—that is, the drift speed—and therefore limit the current. Energy loss from collisions is the origin of **electrical resistance**.

The electrical resistance R of a conductor is the ratio of the potential difference across the conductor to the current through it, or

$$R = \frac{V}{I} \quad \text{(Electrical resistance; SI unit: ohm, } \Omega \text{)} \qquad (17.3)$$

Rearrange Equation 17.3 to get $I = V/R$, and it's clear that for a fixed potential difference V, high resistance results in little current, while low resistance results in a large current.

The SI unit of resistance is the **ohm** (Ω), defined as 1 Ω = 1 V/A, and named for the German physicist Georg Ohm (1787–1854). For many materials, the resistance R is approximately constant, independent of the potential difference. Such materials are called **ohmic**, and for them Equation 17.3 is called **Ohm's law**. Ohm's law, unlike Coulomb's law, is not fundamental, but is rather an empirical law that holds approximately for some materials. It's a good thing that there are non-ohmic materials; electronic devices made from semiconductors owe their useful properties to non-ohmic behavior.

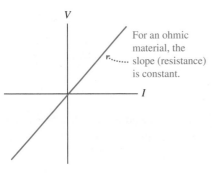

V

I

For an ohmic material, the slope (resistance) is constant.

(a) V versus I for an ohmic material

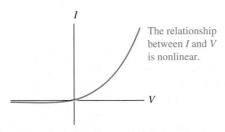

I

V

The relationship between I and V is nonlinear.

(b) I versus V for a non-ohmic material, here a semiconductor diode

FIGURE 17.5 The relationship between current and potential difference for different kinds of materials.

✓ **TIP**

The resistance of an ohmic material is constant—but not all materials are ohmic!

Figure 17.5a graphs potential difference versus current for an ohmic material. Because $I = V/R$, the graph is a straight line. Note that the slope is the same for both positive and negative V and I, showing that ohmic materials carry current equally well in both directions. Figure 17.5b shows a non-ohmic device, for which this isn't true.

Resistors

Resistors are devices with specific resistances for use in electric circuits. Commercial resistors have resistances ranging from a fraction of an ohm to millions of ohms. Typical resistors have a tolerance of 5% or 10%, meaning their actual resistances fall within ±5% or ±10% of their nominal value. Precision resistors with tolerances of 1% or better are available at higher cost. Resistors are also rated by the maximum power they can handle without overheating.

Resistors help establish currents and voltages in circuits. Figure 17.6 shows a single resistor R connected across a battery. In a steady state the current must be the same everywhere; otherwise, charge would build up and lead to huge potential differences. The conducting wires normally have such a low resistance that it can be ignored, so R is the only resistance that matters. Therefore, the potential difference across the resistor is the battery's potential difference V, so the current in the circuit is $I = V/R$. For example, with a 1.5-V battery and 20-Ω resistor, the current is

$$I = \frac{V}{R} = \frac{1.5 \text{ V}}{20 \text{ }\Omega} = 0.075 \text{ V}/\Omega = 0.075 \text{ A} = 75 \text{ mA}$$

For a given battery or other source of potential difference, the current is inversely proportional to the resistance: The higher the resistance, the lower the current. Thus resistors can be used to limit and control the current in a circuit.

Resistivity and Conductivity

Resistance depends on a resistor's material and on its shape. A typical cylindrical resistor of length L and cross-sectional area A (Figure 17.7) has resistance

$$R = \rho\frac{L}{A} \quad \text{(Resistance of a resistor; SI unit: } \Omega\text{)} \qquad (17.4)$$

where ρ, the **resistivity**, is a characteristic of the material; higher ρ means the material is a poorer conductor. From Equation 17.4, it's clear that the SI unit for resistivity is $\Omega \cdot$ m. Table 17.2 shows resistivity for selected materials.

Equation 17.4 makes intuitive sense: Lengthen a resistor and there's more opportunity for electrons to collide with ions. That's why resistance R is proportional to length L. On the other hand, increasing the cross-sectional area A provides more paths for an electron to get through the resistor—just as widening a highway lowers its "resistance" to traffic flow. That's why resistance R is inversely proportional to cross sectional area A.

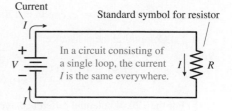

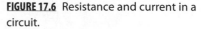

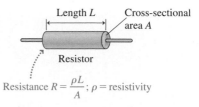

FIGURE 17.6 Resistance and current in a circuit.

FIGURE 17.7 A cylindrical resistor.

TABLE 17.2 Resistivity and Temperature Coefficients for Selected Materials Measured at 20°C

Material	Resistivity $\rho\,(\Omega \cdot \text{m})$	Temperature coefficient $\alpha\,(°C^{-1})$
Silver	1.59×10^{-8}	3.8×10^{-3}
Copper	1.69×10^{-8}	3.9×10^{-3}
Gold	2.44×10^{-8}	3.4×10^{-3}
Aluminum	2.75×10^{-8}	3.9×10^{-3}
Tungsten	5.61×10^{-8}	3.9×10^{-3}
Platinum	1.06×10^{-7}	3.9×10^{-3}
Lead	2.23×10^{-7}	3.9×10^{-3}
Nichrome	1.50×10^{-6}	4.1×10^{-4}
Carbon	3.52×10^{-5}	-5.0×10^{-4}
Germanium	0.46	-4.8×10^{-2}
Silicon	640	-7.5×10^{-3}
Glass	10^{10} to 10^{14}	
Rubber	10^{13}	
Teflon	10^{14}	

EXAMPLE 17.3   **Resistivity, Resistance, and Current**

Nichrome is a nickel-chromium alloy used in heating applications like electric toasters, because it has a relatively high resistivity and heats up when current passes through it. Suppose you have a nichrome wire 0.20 mm in diameter and 75 cm long. (a) What's its resistance? (b) Find the current when a potential difference of 120 V is connected across the wire's ends.

ORGANIZE AND PLAN The resistance of nichrome depends on resistivity (Table 17.2) and the dimensions of the wire. Equation 17.4 gives the resistance in terms of the other parameters in the problem: $R = \rho L/A$. Table 17.2 gives $\rho = 1.50 \times 10^{-6}$ $\Omega \cdot$ m for nichrome. You're given the length L and diameter d; then the area is $A = \pi r^2$, with $r = d/2 = 0.10$ mm. Current follows from resistance and potential difference: $I = V/R$.

cont'd.

Known: $r = 0.10$ mm; $L = 75$ cm; $V = 120$ V.

SOLVE (a) Computing the resistance:

$$R = \rho \frac{L}{A} = (1.50 \times 10^{-6}\ \Omega \cdot m)\frac{0.75\ m}{\pi(0.00010\ m)^2} = 36\ \Omega$$

(b) For the 120-V potential difference, the current is

$$I = \frac{V}{R} = \frac{120\ V}{36\ \Omega} = 3.3\ A$$

REFLECT This is reasonable for a household circuit, whose circuit breakers limit current to typically 15–20 A.

MAKING THE CONNECTION What's the resistance of a comparably sized copper wire?

ANSWER Using the resistivity of copper from Table 17.2, $\rho = 1.69 \times 10^{-8}\ \Omega \cdot m$, the resistance is $R = 0.40\ \Omega$. Copper wires have low resistance.

Resistance and Temperature

Resistivity of metallic conductors increases with temperature. You might expect this, because increasing temperature means faster thermal motion and therefore more frequent electron-ion collisions. Over a wide temperature range the relationship between resistivity and temperature is linear, described by

$$\rho = \rho_0[1 + \alpha(T - T_0)] \qquad \text{(Resistivity versus temperature; SI unit: } \Omega \cdot m) \quad (17.5)$$

where ρ_0 is the resistivity at temperature T_0, and ρ is the resistivity at temperature T. The parameter α is the **temperature coefficient of resistivity**, and it's listed in Table 17.2. For good conductors, α is around $4 \times 10^{-3}\ °C^{-1}$. This means that a temperature increase of only about 2.5°C raises the resistivity by 1%.

Carbon, germanium, and silicon are semiconductors, and Table 17.2 shows that they have negative values of α. Thus their resistivity *decreases* with increasing temperature. This isn't consistent with our model of electron-ion collisions producing resistance, so there must be another mechanism at work in semiconductors—as you'll see in Section 17.6.

EXAMPLE 17.4 **Temperature and Resistance**

A platinum wire has 25-Ω resistance at room temperature (20°C). The wire's temperature increases to 240°C when it carries a high current. Find its new resistance.

ORGANIZE AND PLAN Resistivity increases with temperature according to Equation 17.5. Resistance is proportional to resistivity (Equation 17.4), so resistance R follows a similar equation: $R = R_0[1 + \alpha(T - T_0)]$.

Known: For platinum, $\alpha = 3.9 \times 10^{-3}\ °C^{-1}$.

SOLVE Computing the new resistance:

$$R = R_0[1 + \alpha(T - T_0)]$$
$$R = (25\ \Omega)[1 + (3.9 \times 10^{-3}\ °C^{-1})(240°C - 20°C)] = 46\ \Omega$$

REFLECT The resistance has nearly doubled. Platinum's resistance varies enough with temperature that it's often used as a **thermistor**, a device in which resistance serves to measure temperature. The fact that platinum is corrosion resistant makes it especially good for this application.

MAKING THE CONNECTION Determine what happens to the resistance when this same platinum wire is immersed in liquid nitrogen at −196°C.

ANSWER Resistance decreases when temperature falls. In this case, R drops to 3.9 Ω.

A practical application of electrical resistance is to determine the fat content of the human body. Electrodes send a small current (less than 1 mA) through the body, as shown in the photo on the first page of this chapter. From the potential difference V measured between the electrodes, the resistance $R = V/I$ is computed (in this context, resistance is called **bioelectrical impedance**). Fat tissue has much higher resistance than lean muscle, so the resistance provides a fairly accurate measure of the body's fat content—usually good to within 3%. The method is quick and painless, and the equipment is portable, unlike the alternate method of weighing a person underwater to determine body density.

Reviewing New Concepts: Electric Current and Resistance

- Electric current is the rate of charge flow: $I = \dfrac{\Delta q}{\Delta t}$.

- Current depends on drift speed v_{drift} and charge carrier density n: $I = \dfrac{\Delta q}{\Delta t} = neAv_{\text{drift}}$.

- Resistance is potential difference divided by current: $R = \dfrac{V}{I}$.

- Resistance depends on resistivity ρ and geometrical factors: $R = \rho\dfrac{L}{A}$.

- Resistivity and resistance are temperature dependent: $\rho = \rho_0[1 + \alpha(T - T_0)]$.

GOT IT? Section 17.1 The cylindrical (round) wires shown are all made of the same material. Rank in decreasing order the electrical resistance.

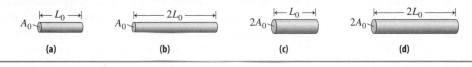

(a) (b) (c) (d)

17.2 Batteries: Real and Ideal

Batteries have been around since the Italian physicist Alessandro Volta (1745–1827) developed a crude battery called a "Voltaic pile," consisting of layers of zinc and silver. With acid-soaked paper between the layers, the pile produced a potential difference that could drive a current. The SI unit for potential difference, the volt, honors Volta's work.

Conceptually, today's batteries are similar to Volta's pile. The common alkaline battery (Figure 17.8) uses zinc (Zn) and manganese dioxide (MnO_2) in place of Volta's metals. An aqueous paste of potassium hydroxide (KOH) separates the two. OH^- ions react at the zinc electrode to produce zinc hydroxide and free electrons. The electrons move through an external circuit and return to the manganese dioxide electrode, where they react to produce manganese trioxide (MnO_3) and OH^- ions. Together, these reactions result in a potential difference of about 1.5 V between the two electrodes. The battery wears down as its supplies of Zn and MnO_2 are exhausted.

There are many other kinds of batteries, but the basic principles are similar. Zinc-carbon batteries are cheaper than alkalines but don't last as long. Your car's battery uses lead and lead-oxide, in a strong acid. This arrangement produces potential difference of 2.0 V, and a 12-V car battery houses six cells in series.

Electromotive Force and Internal Resistance

In Chapter 16 we described a battery as providing a constant potential difference. It's common to refer to a battery as a source of **electromotive force**, or emf. This is obsolete and misleading language because it relates only indirectly to the physics concept of force. A battery's emf is the potential difference of its cell or cells, such as 1.5 V for a flashlight battery. The emf is given the symbol $\mathcal{E}$ to distinguish it from potential differences that aren't associated with sources of electrical energy. While those potential differences may vary with circuit conditions, a battery's emf $\mathcal{E}$ is, ideally, fixed by its chemistry.

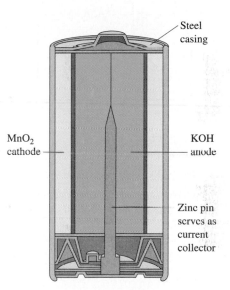

FIGURE 17.8 Structure of an alkaline battery.

✓**TIP**

emf is a potential difference, not a force.

When a battery is connected in a complete circuit, current flows through the circuit. The circuit external to the battery is the **load**—for example, a lightbulb or whatever you're trying to supply with electrical energy. In an ideal battery, the potential difference across the battery terminals is simply the battery's emf and is independent of the load. This isn't true in practice, however, because any real battery contains **internal resistance**. We'll use r for the internal resistance, to distinguish it from the load resistance R.

Kirchhoff's Loop Rule

The circuit in Figure 17.9a consists of a battery and resistor R. Figure 17.9b shows a model that accounts for the battery's internal resistance, with the battery replaced by a fixed emf $\mathcal{E}$ and internal resistance r in series. We analyze this circuit by considering the potential difference across each of the three elements: the fixed emf, the internal resistance, and the load resistance. We'll assume there's no potential difference across the connecting wires—valid if the wires' resistances are negligible compared with r and R.

The important thing to know in this or any circuit—whether it's a washing machine or computer—is the current in each part of the circuit. It's current—the flow of charge—that makes things happen. Figure 17.9b introduces a method you can use to find the current in any circuit containing batteries and resistors. Later you'll see the method extended to circuits with other components.

Figure 17.9b shows that the sum of all three potential differences is zero. That's because the potential at any point in a steady-state circuit doesn't change, so you have to return to the same potential after moving around a complete loop. Therefore,

> The sum of the potential differences around a closed loop is zero.

This is **Kirchhoff's loop rule**, named for the German physicist Gustav Kirchhoff (1824–1887). It may be helpful to recall the similarity between electric potential energy and gravitational potential energy (Section 16.1). The gravitational equivalent of Kirchhoff's loop rule is to walk up and downhill, returning to your starting point. Once you're back, you're at the same altitude, where you have the same gravitational potential energy you started with.

Kirchhoff's loop rule applies to the battery-resistor circuit in Figure 17.9b. Starting at P and moving clockwise, the first element encountered is the emf $\mathcal{E}$. Because we're going from negative to positive, that's a potential gain of $+\mathcal{E}$. Next, moving through the internal resistance r, the potential drops by Ir, because current flows from higher to lower potential through a resistor. Thus, the change in potential is $-Ir$. Similarly, there's a change $-IR$ in passing through the load resistor (see Equation 17.3). Then we're back at P, so we set the sum of the potential changes to zero: $+\mathcal{E} - Ir - IR = 0$. Solving for the current I gives

$$I = \frac{\mathcal{E}}{r + R} \quad \text{(Current from Kirchhoff's loop rule; SI unit: A)} \quad (17.6)$$

Notice that including the battery's internal resistance makes the current less than $\mathcal{E}/R$, which is what the current would be if there were no internal resistance.

Internal resistance depends on the physical size of the battery and the chemistry of its cells. Equation 17.6 suggests we should keep the load resistance R much larger than the internal resistance r if we want the battery's behavior to approximate the ideal. Battery manufacturers don't normally report the internal resistance, which increases as the battery ages. You can measure r, however, by connecting a known load resistance across the battery and measuring the current. If you know the battery's emf, you can then solve Equation 17.6 for r.

Equation 17.6 shows that the potential difference across a real battery is a function of the load. For an ideal battery, the potential difference across the terminals would remain constant at the battery's emf $\mathcal{E}$. For a real battery, our loop analysis shows that the potential difference across the terminals is $\mathcal{E} - Ir$. The following example shows that the difference between ideal and real batteries can be significant.

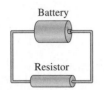

Battery

Resistor

(a) Circuit consisting of battery and resistor

Starting at an arbitrary point such as P, go around the circuit, noting the potential difference across each component.

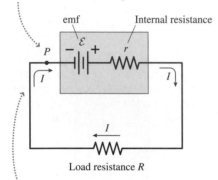

emf   Internal resistance

$\mathcal{E}$ r

P $-$ $+$

I I

I

Load resistance R

When you get back to P, the sum of all the potential differences must be zero.

(b) Applying Kirchhoff's loop rule

FIGURE 17.9 (a) Resistor connected across a battery. (b) Applying Kirchhoff's loop rule.

EXAMPLE 17.5 **Hard Starting!**

A 12-V car battery can deliver high current for the short period when it's cranking the starter motor. A particular starter's resistance is 0.058 Ω, and it draws 161 A. Find (a) the battery's internal resistance, and (b) the potential difference across the battery terminals while starting the car.

ORGANIZE AND PLAN We've started by sketching the circuit (Figure 17.10), which makes it easier to analyze using Kirchhoff's loop rule. The internal resistance follows from the potential difference, load resistance, and current. Then the potential difference across the battery terminals is the sum of the potential differences within the battery. Kirchhoff's loop rule for this circuit gives $\mathcal{E} - Ir - IR = 0$, where $\mathcal{E}$ is the battery emf, r the internal resistance, and R the starter's resistance. The potential difference across the battery terminals is the sum of the emf and the potential difference across the internal resistance: $V_{\text{terminals}} = \mathcal{E} - Ir$.

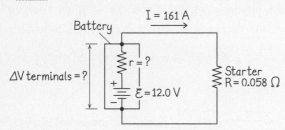

FIGURE 17.10 Our sketch of the starter circuit.

Known: $r = 0.058\ \Omega$, $I = 161$ A, $\mathcal{E} = 12.0$ V.

SOLVE (a) Solving the loop equation for r gives

$$r = \frac{\mathcal{E} - IR}{I} = \frac{12.0\ \text{V} - (161\ \text{A})(0.058\ \Omega)}{161\ \text{A}} = 0.0165\ \Omega$$

(b) With the internal resistance known, the potential difference across the terminals becomes

$$V_{\text{terminals}} = \mathcal{E} - Ir = 12.0\ \text{V} - (161\ \text{A})(0.0165\ \Omega) = 9.34\ \text{V}$$

REFLECT The starter's low resistance and the associated high current reduce the terminal potential difference significantly. The battery can't supply such a high current for long without running down. Fortunately, the lead-acid battery is rechargeable. When the engine runs, your car's *alternator* produces current that flows backward through the battery to recharge it.

MAKING THE CONNECTION Your car hasn't started after several tries. How has the high current affected the starter's resistance? What effect does this have on the starter current if you make another attempt to start the car right away?

ANSWER The high current raises the starter's temperature, increasing its resistance R. With a higher resistance, the battery supplies less current, making the starter less effective. You need to give it time to cool!

APPLICATION **Hybrid Cars**

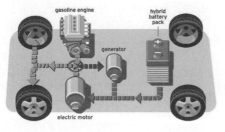

Hybrid cars use both a gasoline engine and an electric motor. A generator run by the gas engine charges a large battery, which allows the electric motor alone to accelerate the car from rest to about 15 mph, at which point the gas engine comes on and gradually takes a bigger share of the load at higher speeds. Hybrids are particularly energy efficient, because instead of friction they use "regenerative braking," in which an electric generator puts the kinetic energy of the slowing car back into the battery. The Toyota Prius (shown here) contains 28 nickel-metal-hydride batteries, with a total potential difference of 201.6 volts (168 1.2-V cells). Its electric and gasoline motors are rated at 67 hp and 76 hp, respectively. A newer development is the "plug-in" hybrid, whose battery can also be charged directly from the power grid—although the environmental soundness of this approach depends on the source of electrical energy.

GOT IT? Section 17.2 A 9-V battery has an internal resistance of 0.5 Ω. What is the load that will result in the battery having an 8.0-V potential difference across its terminals? (a) 1 Ω; (b) 2 Ω; (c) 3 Ω; (d) 4 Ω.

17.3 Combining Resistors

Resistors in Series

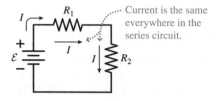

(a) Resistors connected in series

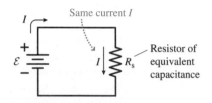

(b) The two resistors are replaced by one of equivalent resistance R_s.

FIGURE 17.11 Combining resistors in series.

Figure 17.11a shows two resistors in series, connected across a battery. We'll assume that the internal resistance of the battery is small enough compared with R_1 and R_2 that it can be ignored. What single resistor R_s could replace the two series resistors R_1 and R_2 and give the same current in the circuit, as in Figure 17.11b?

In a series circuit, the current is the same everywhere, because there's no buildup or loss of charge. Applying Kirchhoff's loop rule beginning at the battery and going clockwise around the loop in Figure 17.11a, the sum of the potential differences is $\mathcal{E} - IR_1 - IR_2 = 0$. Solving for the current gives $I = \mathcal{E}/(R_1 + R_2)$. Now in the equivalent circuit of Figure 17.11b, the loop rule gives $\mathcal{E} - IR_s = 0$, so $I = \mathcal{E}/R_s$. The equivalent resistance must give the same current as R_1 and R_2 in series, which will happen if

$$R_s = R_1 + R_2 \quad \text{(Equivalent resistance of series resistors; SI unit: } \Omega\text{)} \quad (17.7)$$

That's the rule for resistors in series:

> The equivalent resistance of resistors in series is the sum of the individual resistances.

This example readily extends to n series resistors:

$$R_s = R_1 + R_2 + \cdots + R_n$$

For example, connect three resistors in series: $R_1 = 200\ \Omega$, $R_2 = 250\ \Omega$, and $R_3 = 350\ \Omega$; the equivalent resistance is $R_s = R_1 + R_2 + R_3 = 800\ \Omega$. Connect this combination across a 12-V battery, and the current throughout the circuit is

$$I = \frac{\mathcal{E}}{R_s} = \frac{12\ \text{V}}{800\ \Omega} = 0.015\ \text{A} = 15\ \text{mA}$$

Resistors in Parallel

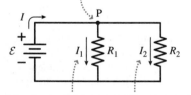

(a) Resistors connected in parallel

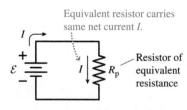

(b) The two resistors are replaced by one of equivalent resistance R_p.

FIGURE 17.12 Combining resistors in parallel.

What's the equivalent resistance of the parallel resistor combination in Figure 17.12? What's the current through each resistor? Again, "equivalent resistance" means a single resistor that draws the same current as the parallel combination.

The key here is that parallel resistors share connections with each other and with the battery. Therefore, the same potential difference $\mathcal{E}$ exists across each resistor, and the currents $I_{1,2} = \mathcal{E}/R_{1,2}$ in the two resistors R_1 and R_2 may be different. Figure 17.12a shows that a total current I leaves the battery. Since there's no charge buildup in the circuit, the current splits at point P, some of it (I_1) going through R_1 and the rest (I_2) through R_2. Algebraically,

$$I = I_1 + I_2 \quad \text{(Kirchhoff's junction rule; SI unit: A)} \quad (17.8)$$

Equation 17.8 is an instance of **Kirchhoff's junction rule**:

> The net current entering a junction equals the net current leaving the junction.

Kirchhoff's junction rule follows directly from conservation of charge.

With the same potential difference $\mathcal{E}$ across each parallel resistor, the resistor currents are $I_1 = \mathcal{E}/R_1$ and $I_2 = \mathcal{E}/R_2$. For the equivalent circuit in Figure 17.12b, $I = \mathcal{E}/R_p$. Using these currents in the junction rule, Equation 17.8, gives

$$\frac{\mathcal{E}}{R_p} = \frac{\mathcal{E}}{R_1} + \frac{\mathcal{E}}{R_2}$$

Canceling the common factor $\mathcal{E}$,

$$\frac{1}{R_p} = \frac{1}{R_1} + \frac{1}{R_2} \quad \text{(Equivalent resistance of parallel resistors; SI unit; } 1/\Omega) \quad (17.9)$$

So here's the rule for the equivalent resistance of parallel resistors:

> For resistors in parallel, the reciprocal of the equivalent resistance is the sum of the reciprocals of the individual resistances.

As with the series rule, this one can be extended to n resistors in parallel:

$$\frac{1}{R_p} = \frac{1}{R_1} + \frac{1}{R_2} + \frac{1}{R_3} + \cdots + \frac{1}{R_n}$$

A display of multiple lights powered by a single source—such as the set of brake lights on your car or a string of holiday lights—should normally be placed in parallel. If they're in series, then when one light goes out, it interrupts the series, and no current flows—thus all the lights go out. But with lights in parallel, having one burn out doesn't affect the others. A "string" of lights may *appear* to be in series, but it's usually not!

Resistors and Capacitors

Chapter 16 introduced series and parallel capacitors; here we've just seen series and parallel resistors. How do they compare? Simple: The rules are opposite. Resistors in series and capacitors in parallel add algebraically. Resistors in parallel and capacitors in series add reciprocally. Combining resistors in series makes an equivalent resistance that's larger than any of the individual resistances; combining resistors in parallel makes an equivalent resistance that's smaller than any of the individual resistances.

EXAMPLE 17.6 **Parallel Resistors**

Three resistors $R_1 = 100\ \Omega$, $R_2 = 150\ \Omega$, and $R_3 = 300\ \Omega$ are connected in parallel. (a) Find the equivalent resistance. (b) When the parallel combination is connected across a 12.0-V battery, find the current in each resistor and the total current supplied by the battery

ORGANIZE AND PLAN Our sketch of the circuit is shown in Figure 17.13. The equivalent resistance follows from the rule for resistors in parallel:

$$\frac{1}{R_p} = \frac{1}{R_1} + \frac{1}{R_2} + \frac{1}{R_3}$$

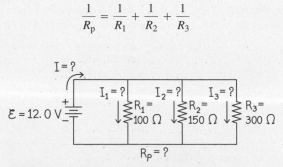

FIGURE 17.13 Our sketch for Example 17.6: circuit with three resistors in parallel.

Each of the parallel resistors has the same potential difference (12.0 V), which determines the current in each. The battery current is then the sum of the three resistor currents.

SOLVE The equivalent resistance follows from

$$\frac{1}{R_p} = \frac{1}{R_1} + \frac{1}{R_2} + \frac{1}{R_3} = \frac{1}{100\ \Omega} + \frac{1}{150\ \Omega} + \frac{1}{300\ \Omega} = \frac{1}{50\ \Omega}$$

Therefore, $R_p = 50\ \Omega$.
The three resistor currents are

$$I_1 = \frac{\mathcal{E}}{R_1} = \frac{12.0\ \text{V}}{100\ \Omega} = 0.120\ \text{A} = 120\ \text{mA}$$

$$I_2 = \frac{\mathcal{E}}{R_2} = \frac{12.0\ \text{V}}{150\ \Omega} = 0.080\ \text{A} = 80\ \text{mA}$$

and

$$I_3 = \frac{\mathcal{E}}{R_3} = \frac{12.0\ \text{V}}{300\ \Omega} = 0.040\ \text{A} = 40\ \text{mA}$$

cont'd.

The battery supplies a total current $I = I_1 + I_2 + I_3 = 120$ mA + 80 mA + 40 mA = 240 mA.

REFLECT There's a good "check" on that last step. The battery current should be the same as if a single resistor R_p is connected in place of the other three. Is it?

$$I = \frac{\mathcal{E}}{R_p} = \frac{12.0 \text{ V}}{50 \text{ }\Omega} = 0.240 \text{ A} = 240 \text{ mA}$$

Yes, the current I is the same calculated either way!

MAKING THE CONNECTION What's the battery current when the same battery is connected across R_1 alone? When it's across just R_1 and R_2 in parallel?

ANSWER With R_1 alone, $I = 120$ mA. With R_1 and R_2 in parallel, $I = 120$ mA + 80 mA = 200 mA. Putting more resistors in parallel always increases the total current.

PROBLEM-SOLVING STRATEGY 17.1 **Networks of Resistors**

ORGANIZE AND PLAN

- Draw a schematic diagram.
- Identify series and parallel resistor combinations. If necessary, redraw your diagram to make these clear. Some resistors may not be in either combination.

SOLVE

- Apply the rules for parallel and series combinations, one step at a time, until you've found the equivalent resistance for the entire circuit.
- To find currents or potential differences, apply Ohm's law as needed.

REFLECT

- Is your equivalent resistance reasonable, given the individual resistances? (Remember that adding resistors in parallel decreases resistance, while adding resistors in series increases resistance.)
- Do the currents or potential differences you've calculated make sense in terms of the battery emf and the total current supplied by the battery?

More Complex Resistor Networks

Knowing the rules for series and parallel resistors, you can treat more complex networks that contain both series and parallel combinations. The following examples show how you apply the rules sequentially, each time simplifying the circuit by replacing two or more resistors with a single equivalent resistance.

EXAMPLE 17.7 **A Resistor Network**

Consider the network of five resistors shown in Figure 17.14a, with $R_1 = 6 \text{ }\Omega$, $R_2 = 4 \text{ }\Omega$, $R_3 = 5 \text{ }\Omega$, $R_4 = 3 \text{ }\Omega$, and $R_5 = 2 \text{ }\Omega$. Find the equivalent resistance between the two ends of the network.

ORGANIZE AND PLAN At each step, all that's needed is the series rule or the parallel rule. Identify series and parallel combinations and proceed to simplify until there's just one resistance left. To keep things straight, it helps to redraw the circuit each time it's simplified.

Here we can simplify the group of three resistors R_3, R_4, and R_5 into a single resistance, which will then be in series with R_1 and R_2. Within the group of three, R_3 and R_4 are in series, and that series combination is in parallel with R_5.

SOLVE We replace R_3 and R_4 with a single resistance R_6 (Figure 17.14b), which by the series rule is $R_6 = R_3 + R_4 = 5 \text{ }\Omega + 3 \text{ }\Omega = 8 \text{ }\Omega$.

cont'd.

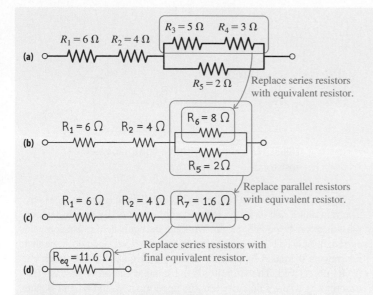

Replace series resistors with equivalent resistor.

Replace parallel resistors with equivalent resistor.

Replace series resistors with final equivalent resistor.

FIGURE 17.14 (a) Network of five resistors. (b) First intermediate step. (c) Second intermediate step. (d) Equivalent resistance of the network.

Next, combine the new R_6 with R_5 using the parallel rule, and call the result R_7:

$$\frac{1}{R_7} = \frac{1}{R_6} + \frac{1}{R_5} = \frac{1}{8\ \Omega} + \frac{1}{2\ \Omega} = \frac{1}{1.6\ \Omega}$$

so $R_7 = 1.6\ \Omega$ (Figure 17.14c). Now all that remains are R_1, R_2, and R_7, in series. Therefore, the equivalent resistance R_{eq} of the network is (Figure 17.14d)

$$R_{eq} = R_1 + R_2 + R_7 = 6\ \Omega + 4\ \Omega + 1.6\ \Omega = 11.6\ \Omega$$

REFLECT Reducing the network to a single equivalent resistance is possible whenever the circuit consists of series and parallel combinations.

MAKING THE CONNECTION If you connected the two ends of this network to a battery, which resistors would carry the most current? Which would carry the least?

ANSWER All the battery current goes through R_1 and R_2, so they carry the most current. Then the current divides between the two parallel branches. Of those two branches, the top one has more resistance, so the bottom one carries more current. Therefore, $I_1 = I_2 > I_5 > I_3 = I_4$.

EXAMPLE 17.8   **Resistor Network, Revisited**

For the circuit in the preceding example, find the current in each resistor when the network is connected across a 14.5-V power supply.

ORGANIZE AND PLAN The first step in finding the current in each resistor is to find the total current from the power supply (Figure 17.15). That's just $I_{total} = \mathcal{E}/R_{eq}$. When there are parallel branches, the potential difference V is the same across each branch. Find the potential difference V, and you know that a branch with resistance R carries a current $I = V/R$.

Known: $\mathcal{E} = 14.5\ V$ for the power supply.

SOLVE From the preceding example, you know that $R_{eq} = 11.6\ \Omega$. Therefore, the total current is

$$I_{total} = \mathcal{E}/R_{eq} = (14.5\ V)/(11.6\ \Omega) = 1.25\ A$$

The series resistors R_1 and R_2 carry this total current (Figure 17.15), so $I_1 = I_2 = 1.25\ A$. Then the potential differences across these two are

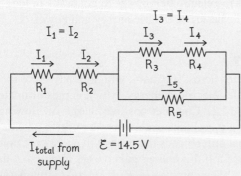

FIGURE 17.15 Our sketch for Example 17.8, showing currents in the circuit.

$$V_1 = I_1 R_1 = (1.25\ A)(6\ \Omega) = 7.5\ V$$

and

$$V_2 = I_2 R_2 = (1.25\ A)(4\ \Omega) = 5.0\ V.$$

These potential differences sum to 12.5 V. By Kirchhoff's loop rule, the potential difference across the combination of R_3, R_4, and R_5 is 14.5 V − 12.5 V = 2.0 V. This difference appears across the series combination of R_3 and R_4 (total 8 Ω), giving a current

$$I_3 = I_4 = V/R = (2.0\ V)/(8\ \Omega) = 0.25\ A$$

The same 2-V difference appears across R_5, since it's in parallel with the R_3, R_4 combination. Thus

$$I_5 = V/R = (2.0\ V)/(2\ \Omega) = 1.0\ A$$

REFLECT Our answers make sense, because the current in the two parallel branches adds up to the 1.25-A power supply current: 1.0 A + 0.25 A = 1.25 A.

MAKING THE CONNECTION For the parallel part of the network, discuss the statement "Current takes the path of least resistance."

ANSWER The larger current (1.0 A) flows through the branch with resistance 2 Ω, and the smaller current (0.25 A) flows through the 8-Ω branch. So the larger current flows through the smaller resistance. Quantitatively, the ratio of the currents is the reciprocal of the ratio of the resistances. This is because the potential difference is the same across parallel resistors. "Current takes the path of least resistance" does *not* mean that *all* the current flows through the smaller resistor; rather, current divides among the available paths according to their resistances.

CONCEPTUAL EXAMPLE 17.9 **Identical Lightbulbs**

Four identical lightbulbs are connected as in Figure 17.16. When this network is connected across a battery, currents heat the bulbs' filaments, making them glow. More current makes a bulb brighter. Here we'll ignore the change in resistance with temperature, taking the bulbs' resistances to be constant and equal. (a) With the battery connected as shown, rate the brightness of the four bulbs. (b) What happens to the other bulbs if B_1 is unscrewed from its socket? Repeat for each of the other bulbs, each time assuming the other three remain in place.

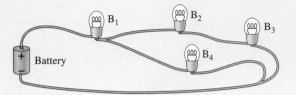

FIGURE 17.16 A battery and four lightbulbs.

SOLVE (a) It's helpful to redraw the circuit representing each lightbulb as a resistor (Figure 17.17). Bulb B_1 has the most current (and is brightest), because all the current from the battery flows through it. The current then reaches a junction point and splits between bulb B_4 and the two bulbs B_2 and B_3 in series. The series has a higher resistance, so more current flows through B_4 than through B_2 and B_3.

The results can be summarized in terms of the current (equivalently, brightness) in each bulb: $I_1 > I_4 > I_2 = I_3$.

(b) B_1 unscrewed: There's no path for current to reach other bulbs, so none lights.

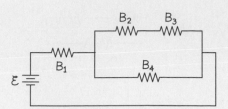

FIGURE 17.17 Circuit with the lightbulbs as resistors.

B_2 unscrewed: Current can't flow through B_2 and B_3, so B_3 goes out. The remaining network is a series combination of B_1 and B_4, which must therefore be equally bright. Work out the equivalent resistances and current in each bulb before and after B_2 is unscrewed, and you'll see that unscrewing B_2 makes B_1 dimmer than it was originally, B_4 brighter.

B_3 unscrewed: The situation is the same as the previous case (B_2 unscrewed). There's no current in the upper branch, and B_1 and B_4 are equally bright.

B_4 unscrewed: Now there's no current through the lower parallel branch (B_4 alone). What remains is a series combination of B_1, B_2, and B_3, which will be equally bright. Having three bulbs in series gives a higher equivalent resistance than with just two series bulbs, so the three are dimmer than the two lit bulbs in the previous cases.

REFLECT Analyzing connections of batteries and bulbs can help develop your feel for current and resistance. Try sketching some other combinations and analyzing them. Try building some of these circuits in lab and see if your intuition was correct.

GOT IT? Section 17.3 In the circuits shown, all the resistors are equal. Rank in decreasing order the equivalent resistances of the circuits.

(a)　　　(b)　　　(c)　　　(d)

17.4 Electric Energy and Power

Electric current entails a flow of energy. The rate at which energy is delivered—power—varies from milliwatts in battery-powered electronic devices to gigawatts in long-distance transmission lines. Here we'll cover the basics of electric energy and power.

Energy Supplied by a Battery

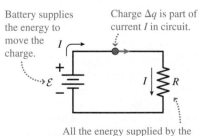

Battery supplies the energy to move the charge.　　Charge Δq is part of current I in circuit.

All the energy supplied by the battery is used up in the resistor.

FIGURE 17.18 Energy transfers in an electric circuit.

Figure 17.18 shows a resistor connected across a battery, with a steady current I flowing around the circuit. The battery supplies electrical energy, which comes ultimately from chemical reactions within it (Section 17.2). Consider positive charge Δq flowing around the circuit. Passing through the battery, this charge experiences a positive potential difference $\mathcal{E}$, the battery emf. Since potential difference is the change in energy per unit charge, our charge gains energy $\Delta U = (\Delta q)(\mathcal{E})$. If this energy gain occurs in time Δt, the power P—the *rate* at which the battery supplies energy—is $P = \mathcal{E}\Delta q/\Delta t$. But $\Delta q/\Delta t$ is the current I, so the battery power is

$$P = I\mathcal{E} \quad \text{(Power supplied by a battery; SI unit: W)} \quad (17.10)$$

Energy Dissipated in a Resistor

In Section 17.1 you saw how collisions in a resistive material dissipate electrons' energy, heating the material. That happens in the resistor of Figure 17.18. Furthermore, because there's no capacitor or other energy storage device, *all* the energy from the battery ends up dissipated in the resistor. So the power dissipated in the resistor is $P = I\mathcal{E}$ (Equation 17.10). But $\mathcal{E} = IR$ for the resistor, so the power dissipation is $P = I(IR)$ or

$$P = I^2R \quad \text{(Power dissipated in a resistor; SI unit: W)} \quad (17.11)$$

Equation 17.11 is a general result for any resistor carrying current I. Because $I = V/R$, equivalent expressions for the resistor power are

$$P = IV \quad \text{(Electric power; SI unit: W)} \quad (17.12)$$

and

$$P = \frac{V^2}{R} \quad \text{(Power dissipated in a resistor; SI unit: W)} \quad (17.13)$$

Equations 17.11 and 17.13 are specific to resistors. However, Equation 17.12 applies to *any* electrical component with current I through it and potential difference V across it. In a resistor, that power is dissipated as heat; in other devices the energy might be converted to other forms—for example, mechanical energy in a motor.

Converting electrical energy to heat isn't necessarily a bad thing. The heating element in an electric stove is a resistor in which you want a lot of power dissipation. The wire in the heating element must have higher resistance than the connecting wires, so most of the power dissipation I^2R occurs in the element, but that resistance must be low enough to pass a high current. The heating element transforms electrical energy into thermal energy, which is conducted through the bottom of your pan to cook the food.

EXAMPLE 17.10   **Lightbulb**

A lightbulb uses energy at the rate of 75 W with a 120-V potential difference across it. (a) Find the current in the bulb's tungsten filament and the filament's resistance. (b) Under these conditions, the filament temperature is about 2300 K. Use this fact along with your answer to part (a) to estimate the resistance of the filament at room temperature.

ORGANIZE AND PLAN This example involves relationships among the power, current, potential difference, and resistance. The resistance of tungsten, a metal, increases with increasing temperature. The white-hot filament should have a much higher resistance than at room temperature.

The power dissipation for a resistor, in this case the tungsten filament, is $P = IV$. Therefore, the current is $I = P/V$ and then the resistance is $R = V/I$. From Section 17.1, the temperature dependence of the resistance R (proportional to resistivity ρ) follows from Equation 17.5: $R = R_0[1 + \alpha(T - T_0)]$.

Known: $P = 75$ W; $V = 120$ V, $\alpha = 3.9 \times 10^{-3}\,°C^{-1}$ from Table 17.2.

SOLVE (a) First, find the filament current:

$$I = \frac{P}{V} = \frac{75\text{ W}}{120\text{ V}} = 0.625\text{ W/V} = 0.625\text{ A or }625\text{ mA}$$

Then compute the filament's resistance:

$$R = \frac{V}{I} = \frac{120\text{ V}}{0.625\text{ A}} = 192\ \Omega$$

(b) This resistance is at 2300 K; we want the resistance R_0 at $T_0 = 293$ K (room temperature). Solving the resistance temperature formula for R_0:

$$R_0 = \frac{R}{1 + \alpha(T - T_0)}$$

$$= \frac{192\ \Omega}{1 + (3.9 \times 10^{-3}\,°C^{-1})(2300°C - 293°C)} = 22\ \Omega$$

REFLECT As expected, the filament's resistance is much higher when it's hot. Tungsten is used in lightbulbs because it has a high melting point and can withstand repeated heating and cooling.

MAKING THE CONNECTION Would a 100-W bulb carry more or less current than that 75-W bulb?

ANSWER From the analysis in the example, $I = P/V$, so for the same V, the current is higher in the 100-W bulb. This assumes—as is the case for household lightbulbs—that they operate at the same potential difference.

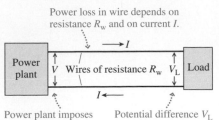

Power loss in wire depends on resistance R_w and on current I.

Power plant imposes potential difference V across transmission line.

Potential difference V_L across load is lower because of power lost in wires.

FIGURE 17.19 Power loss is I^2R, so a transmission line is most efficient with a low current and, correspondingly, a high potential difference V between its two wires.

Transmitting Electric Power

Electrical energy is generated from a variety of sources, including fossil fuels, nuclear fission, water, and wind. We'll discuss electric generators in Chapter 19 and nuclear energy in Chapter 25. Here, we're concerned with how electrical energy gets to you, the end user. Current I generated at a power plant flows through transmission lines (Figure 17.19), with potential difference V between them. We could transmit the same power $P = IV$ either with large I and small V, or vice versa. Which is preferable?

The transmission line wires have some resistance, R_w. By Equation 17.11, the power lost (P_{lost}) in transmission is $P_{lost} = I^2R_w$. Obviously, it's best to use transmission lines with the smallest possible resistance, to minimize power lost—but that means thick wires and high cost. More importantly, it's best to minimize the current, because the power loss scales as the *square* of the current. Since $P = IV$, we can have low current if we put a high potential difference between the two wires. This becomes clear if we write the current as $I = P/V$; then Equation 17.11 gives

$$P_{lost} = I^2R_w = \left(\frac{P}{V}\right)^2 R_w$$

Electric power lines typically have potential differences of up to hundreds of kilovolts, with the highest values for long-distance transmission. The potential difference is "stepped down" to a standard 120 V or 240 V for home and commercial use. That's easier to do with alternating current (AC), as you'll see in Chapter 19—one reason why AC is used in most electrical systems.

EXAMPLE 17.11 **Electric Bill**

Electric utilities measure energy in kilowatt-hours (kWh), where 1 kWh is the energy consumed if you use energy at the rate of 1 kW for 1 hour. If your monthly electric bill (30 days) is $100 and you pay 12.5¢/kWh, what's your home's average power consumption and average current, assuming a 240-V potential difference between the wires supplying your home?

ORGANIZE AND PLAN Power is energy per time. You can use the electric bill data to compute the total energy used in the month. Then divide by time to get the average power. Average current then follows from $P = IV$.

Your energy consumption for the month, in kWh, is your monthly bill divided by the cost per kWh. Divide by the number of hours in a month ($30 \times 24 = 720$) and you've got the power in kW. Then $I = P/V$.

Known: Cost = 12.5¢/kWh = $0.125/kWh; bill = $100 for 30 days.

SOLVE

$$\text{Total energy used} = \frac{\text{monthly energy cost}}{\text{cost per unit energy}}$$

$$= \frac{\$100}{\$0.125/\text{kWh}} = 800 \text{ kWh}$$

$$P_{average} = \frac{800 \text{ kWh}}{720 \text{ h}} = 1100 \text{ W} = 1.1 \text{ kW}$$

Then

$$I_{average} = \frac{P_{average}}{\Delta V} = \frac{1100 \text{ W}}{240 \text{ V}} = 4.6 \text{ A}$$

REFLECT This is a reasonable current for a whole house. Any of the large energy hogs (furnace, air conditioner, or electric dryer) can use more than this by itself. Of course, they don't run all the time, so 4.6 A is a reasonable average.

MAKING THE CONNECTION Compare the cost of electrical energy with the cost per unit energy of a 75-¢ candy bar that contains 200 kcal (200 food calories) of energy.

ANSWER Using conversion factors from Appendix C, 200 kcal = 0.232 kWh. The cost of this much electrical energy is (0.232 kWh)(12.5¢/kWh) = 2.9¢. Electrical energy is far cheaper than food energy! *Note:* We could have converted kWh to joules in this example, and divided by time in seconds. But there's no need; since 1 kWh is the energy associated with a power of 1 kW for 1 hour, 1 kWh/h is the same as 1 kW. By the way, you can show that 1 kWh = 3.6 MJ.

Fuses and Circuit Breakers

Current-carrying wires have resistance and so they heat up. To reduce the risk of fire, **fuses** and **circuit breakers** interrupt currents that exceed safe values. A fuse is just a thin wire that's designed to melt when the current exceeds the fuse's rated current. To restore the current, you have to replace the fuse. Circuit breakers offer the same protection but don't need replacement. Most breakers use electromagnets whose strength depends on the current. When the current exceeds the breaker's rating, the magnet pulls on a switch and interrupts the current; the switch then latches in the open state. You have to reset the breaker to restore the current. You'll see how electromagnets work in Chapter 18, but for now all that's important to know is that higher current makes a stronger magnet.

Fuses and circuit breakers can be manufactured to trip at any desired current. Your lab multimeter might have a fuse rated at 2 A. Household circuit breakers are typically rated at 15 A to 20 A. Circuit breakers on large motors may be rated in the hundreds of amperes.

EXAMPLE 17.12   **Tripping the Circuit Breaker**

Several students in the same dorm room want to dry their hair. Knowing about electricity, they've set their hair dryers to the "low," 1000-W settings. Assuming a standard 120-V line, how many hair dryers can they operate simultaneously without tripping the 20-A circuit breaker?

ORGANIZE AND PLAN Here you don't want the total current to exceed the breaker's 20-A rating. If you know the current through one hair dryer, you can multiply by the number of dryers to find the total current. The power in any device carrying current I with potential difference V is $P = IV$. Therefore, each hair dryer draws current $I = P/V$.

Known: $P = 1000$ W, $\Delta V = 120$ V.

SOLVE A single hair dryer using 1000 W at 120 V carries current

$$I = \frac{P}{\Delta V} = \frac{1000 \text{ W}}{120 \text{ V}} = 8.33 \text{ A}$$

With a maximum current of 20 A, only two dryers can be on without tripping the breaker.

REFLECT These students know their physics! If they had chosen the 1500-W "max power" rating, each hair dryer would draw more than 10 A, and only one could be used at a time. You can add currents here because devices plugged into the power line are all in parallel.

MAKING THE CONNECTION   How much current would two 1500-W hair dryers draw?

ANSWER Each one draws 12.5 A, so the total is 25 A.

GOT IT? Section 17.4   A 60-W lightbulb has a potential difference of 120 V across it. The current and resistance of the bulb are (a) 0.5 A, 240 Ω; (b) 1.0 A, 120 Ω; (c) 0.5 A, 480 Ω; (d) 1.0 A, 60 Ω.

17.5 *RC* Circuits

Chapter 16 introduced capacitors, and here you've met resistors. What happens when you combine them?

A simple ***RC* circuit** consists of a battery (or power supply), a resistor, and a capacitor in series (Figure 17.20a). A switch can be closed to complete the circuit or opened to interrupt the current. Assume the capacitor is initially uncharged. When the switch is closed, current begins to flow as shown in Figure 17.20b. This current flows through the resistor and starts to charge the capacitor. Let I and q represent the current through the resistor and charge on the capacitor, respectively. For this single-loop circuit, applying Kirchhoff's loop rule helps explain what happens. Trace your way clockwise around the loop, beginning with the battery. Then the potential differences are $+\mathcal{E}$ for the battery, $-IR$ for the resistor, and $-q/C$ for the capacitor. (Remember from Chapter 16 that $V = q/C$ for a capacitor.) So Kirchhoff's loop equation reads

$$\mathcal{E} - IR - q/C = 0 \qquad \text{(Kirchhoff's loop equation for an *RC* circuit; SI unit: V)} \qquad (17.14)$$

Consider all three components at the instant the switch is closed ($t = 0$). The battery's potential difference is fixed at $\mathcal{E}$. The capacitor is initially uncharged ($q = 0$). Then Kirchhoff's loop Equation 17.14 implies that the current in the circuit is $I = \mathcal{E}/R$. That's the same current that would flow if the capacitor were just a wire, at least initially. But as time passes, the capacitor begins to charge, and q/C increases. Then Kirchhoff's equation

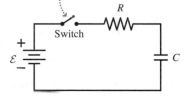

(a) *RC* circuit with open switch

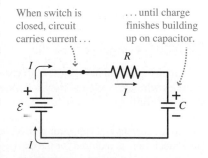

(b) After closing the switch

FIGURE 17.20 Analyzing an *RC* circuit used to charge a capacitor.

shows that the current I must decrease as the capacitor charge q increases. After a long time ($t \rightarrow \infty$), the charge on the capacitor approaches its maximum value. By Equation 17.14, this maximum charge is $q = \mathcal{E}C$, reached when $I = 0$, so those are the values of q and I as $t \rightarrow \infty$.

Between the extremes $t = 0$ and $t \rightarrow \infty$, the capacitor charge grows continually from zero toward its maximum value $q_{max} = \mathcal{E}C$, while the current decreases from its maximum $I_{max} = \mathcal{E}/R$ toward zero. Determining the exact time behavior follows from Kirchhoff's loop equation but requires calculus, so we'll omit the derivation. For the charging capacitor of Figure 17.20b, the capacitor charge $q(t)$ as a function of time is

$$q(t) = \mathcal{E}C(1 - e^{-t/RC}) = q_{max}(1 - e^{-t/RC}) \qquad \text{(Capacitor charge as a function of time; SI unit: C)} \qquad (17.15)$$

and the current $I(t)$ is

$$I(t) = \frac{\mathcal{E}}{R}(e^{-t/RC}) = I_{max}(e^{-t/RC}) \qquad \text{(Resistor current as a function of time; SI unit: A)} \qquad (17.16)$$

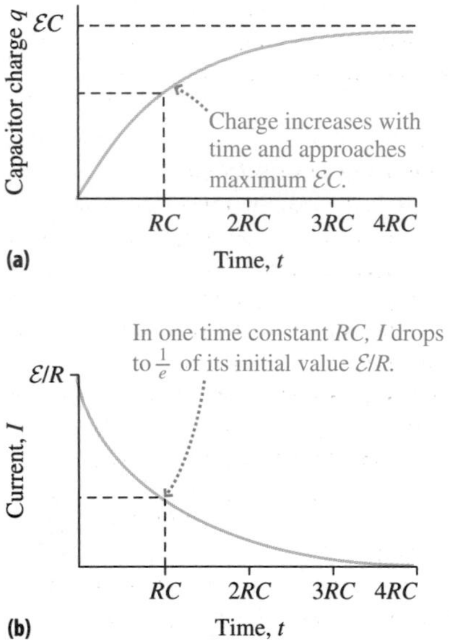

You should convince yourself that these formulas are consistent with the analysis above. A good way is to graph the functions $q(t)$ and $I(t)$, as shown in Figure 17.21. The exponential functions in Equations 17.15 and 17.16 mean the charge and current approach their limiting values asymptotically—a fact that wasn't obvious from looking at the circuit loop equation alone.

✓ **TIP**

To make your capacitor charge faster, decrease the series resistance R. To slow the charging process, increase R.

The *RC* Time Constant

The exponential function $e^{-t/RC}$ in Equations 17.15 and 17.16 takes a dimensionless argument, and therefore the quantity $-t/RC$ should be dimensionless. So RC, the product of resistance and capacitance, has units of time. You can check this: One ohm is 1 V/A, and 1 A = 1 C/s, so 1 Ω = 1 V·s/C. Capacitance is in farads, with 1 F = 1 C/V. Therefore, the product RC has units (V·s/C)(C/V) = s, the SI time unit.

The value RC is a specific time, characteristic of a given RC circuit. Does something special happen at this time? No: Figure 17.21 shows gradual variations with time—there's no sudden change when $t = RC$. Rather, the significance of RC lies in the exponential factor $e^{-t/RC}$ that appears in the equations for charge and current. When $t = RC$, $e^{-t/RC} = e^{-RC/RC} = e^{-1}$. The number e^{-1} (approximately 0.368) means that at that time, the current has fallen to about 36.8% of its initial (maximum) value, and the charge has risen to about $1 - 0.368 = 0.632$ or 63.2% of its maximum. In other words, the charging is nearly two-thirds complete. Because it provides this convenient scale, RC is called the **time constant** for the circuit and is given the special symbol τ (Greek tau). That is, $\tau = RC$.

Capacitor charge q

$\mathcal{E}C$ ----

Charge increases with time and approaches maximum $\mathcal{E}C$.

RC $2RC$ $3RC$ $4RC$

(a)   Time, t

Current, I

In one time constant RC, I drops to $\frac{1}{e}$ of its initial value $\mathcal{E}/R$.

$\mathcal{E}/R$

RC $2RC$ $3RC$ $4RC$

(b)   Time, t

FIGURE 17.21 (a) Charge versus time for the charging capacitor. (b) Current in the same circuit.

<table>
<tr><td>**EXAMPLE 17.13**</td><td>**Charging a Capacitor**</td></tr>
</table>

An RC circuit (as shown in Figure 17.20) consists of a 2.0-mF capacitor in series with a 6.4-kΩ resistor, and the pair connected across a 12-V battery. Initially the switch is open and the capacitor is uncharged. (a) After the switch is closed, what are the maximum values for the circuit current and capacitor charge, and when do they occur? (b) How much time is required for the capacitor to reach half its maximum charge?

ORGANIZE AND PLAN We've sketched the circuit with the appropriate numerical values in Figure 17.22. The maximum current occurs at $t = 0$

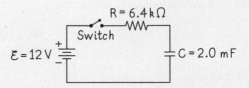

FIGURE 17.22 Our sketch of the circuit for Example 17.13.

cont'd.

and follows from the battery's emf and resistance of the resistor (see Figure 17.21). The maximum charge occurs after a long time and is a function of the battery's emf and the capacitance. The maximum current is $I_{max} = \mathcal{E}/R$, and the maximum charge is $q_{max} = \mathcal{E}C$. Equation 17.15 gives the charge at any time: $q(t) = \mathcal{E}C(1 - e^{-t/RC})$. This equation can be solved for the time t at which there's a specified charge q on the capacitor.

Known: $R = 6.4$ kΩ, $C = 2.0$ mF, $\mathcal{E} = 12$ V.

SOLVE (a) Maximum current:

$$I_{max} = \frac{12 \text{ V}}{6.4 \text{ k}\Omega} = 1.88 \text{ mA}$$

Maximum capacitor charge:

$$q_{max} = C\mathcal{E} = (2.0 \text{ mF})(12 \text{ V}) = 24 \text{ mC}$$

(b) When the charge is half the maximum, $q(t) = \mathcal{E}C/2$, so

$$\mathcal{E}C/2 = \mathcal{E}C(1 - e^{-t/RC}) \text{ or } \tfrac{1}{2} = 1 - e^{-t/RC}$$

Rearranging gives $e^{-t/RC} = \tfrac{1}{2}$.

To solve for time t, we "undo" the exponential by taking the natural logarithm (ln) of both sides:

$$-\frac{t}{RC} = \ln\left(\frac{1}{2}\right)$$

Solving for time t,

$$t = -RC \ln\left(\frac{1}{2}\right) = -(6.4 \text{ k}\Omega)(2.0 \text{ mF})(-0.693) = 8.9 \text{ s}$$

REFLECT The time required to reach half the maximum charge is just a little less than the time constant $\tau = RC$. You could track this charging capacitor by monitoring the potential difference across it, since q is proportional to V. Note in the last calculation how the k(10^3) in kΩ cancelled the m(10^{-3}) in mF to give the answer in s.

In many *RC* circuits, the action occurs much faster. For example, if $R = 100$ Ω and $C = 1.0$ μF (1.0×10^{-6} F), then $\tau = RC = (100\ \Omega)(1.0 \times 10^{-6}\text{ F}) = 100$ μs.

MAKING THE CONNECTION When the capacitor in this example has reached half its maximum charge, what are the potential differences across the capacitor and resistor?

ANSWER When $q = 12$ μC (half the 24-μC maximum charge), $V = q/C = 6.0$ V. Because the potential difference across the battery is 12 V, this leaves for the resistor 12 V $-$ 6.0 V $=$ 6.0 V. When the charge is half its maximum value, the potential difference across the capacitor is half the battery's emf and equal to the potential difference across the resistor.

Discharging a Capacitor

You can discharge a capacitor by connecting its two plates with a conductor. The excess electrons on one plate then flow to the other, leaving zero charge on each plate (Figure 17.23a). If the conductor has some resistance, that limits the current and makes the discharge gradual, much like the charging of the capacitor in an *RC* circuit. In fact, the discharge involves the same exponential factor as before:

$$q(t) = q_0 e^{-t/RC} \qquad \text{(Discharging capacitor; SI unit: C)} \qquad (17.17)$$

Here q_0 is the initial charge on the capacitor, and $t = 0$ is the time when we close the switch in Figure 17.23a. The decay of charge is exponential (Figure 17.23b), and the time constant $\tau = RC$ again sets the timescale. In this case τ is the time in which the charge falls to $1/e$ (just over one-third) of its initial value.

The exponential rise or decay of charge on a capacitor is one example where the exponential function arises in natural processes. Another is the exponential decay of radioactive materials (Chapter 25).

Applications

RC circuits are widely used in timing applications, where the charging of an *RC* circuit triggers a switch at a time that's selectable by varying R. Or they can favor one timescale over another, as in the tone controls on an audio system, which use *RC* circuits to boost bass or treble. A slow *RC* charging circuit stores energy gradually in a capacitor, without requiring high currents from a battery or other source. Then the capacitor can be discharged rapidly through a low resistance. A camera flash is a good example. So is a medical defibrillator. This device (see Section 16.4) uses an *RC* circuit to charge a capacitor that's then discharged across the patient's chest. The charging occurs fairly slowly; it may take half a minute. The slow buildup allows use of a modest battery. The discharge is much more rapid, because it's the sudden jolt of current that's needed to restart or regulate the heart. Another medical application is in the **pacemaker**, in which a capacitor discharge provides electrical stimulus to the heart with a regulated period.

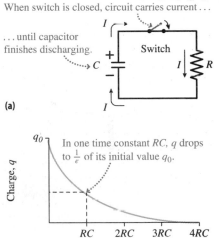

When switch is closed, circuit carries current . . .

. . . until capacitor finishes discharging.

(a)

In one time constant *RC*, q drops to $\frac{1}{e}$ of its initial value q_0.

(b)

FIGURE 17.23 (a) With the switch closed, the capacitor discharges. (b) Charge versus time for a discharging capacitor.

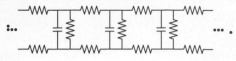

APPLICATION **Cable Model of the Axon**

Biologists and biophysicists studying the nervous system have developed an electrical model of the axons that carry information between nerve cells. In this model the axon consists of a long chain of repeating *RC* circuits (see diagram). The resistors represent electrical resistance of the fluid surrounding nerve cells. The axon membrane is normally maintained at a "rest potential" corresponding to charged capacitors; the transmission of a message occurs when the capacitors discharge to the "action potential."

GOT IT? Section 17.5 Rank in decreasing order the time constants of *RC* circuits with the following values of *R* and *C*: (a) 200 Ω, 1500 μF; (b) 1 kΩ, 1000 μF; (c) 100 Ω, 0.002 F; (d) 500 Ω, 5500 μF.

17.6 Semiconductors and Superconductors

Energy Bands and Semiconductors

Chapter 15 introduced conductors and insulators. In conductors, free electrons move easily in response to electric fields, making electric currents. An insulator has no free charges, so it can't carry a current. The distinct behavior of conductors and insulators follows from the quantum theory of atoms and solids. In atoms, electrons move about the massive, positively charged nucleus, bound by the attractive electric force. Quantum theory (Chapter 24) shows how electrons are organized in **shells** and **orbitals**. Electrons in the outermost shells are weakly bound, and the details of that binding determine a material's conducting properties.

In metals, the outermost electrons are so weakly bound that they constitute a "sea" of electrons that move freely throughout the metal. Electric fields readily accelerate these electrons, making metals excellent conductors. In contrast, electrons in insulators are bound tightly enough that electric fields can't pull them from their atoms, so these materials don't conduct. In between are **semiconductors**. Table 17.2 shows that their resistivities lie between those of conductors and insulators; they also have negative temperature coefficients—implying a different conduction mechanism than the one we described in Section 17.1 for metals.

According to quantum theory, energies of electrons in solids are restricted to specific ranges, called **energy bands**. The highest band occupied by electrons is the **valence band**; above this is the **conduction band** (Figure 17.24). In a metal, these bands overlap, so it takes only a tiny amount of energy to move an electron into a higher level within the continuous range of allowed energies. That's why electrons can gain energy from an electric field, and why the material is a conductor. In insulators and semiconductors, though, a **band gap** separates the valence and conduction bands. For an electron in the valence band to gain energy, it would have to "jump" the band gap into the unoccupied conduction band. That doesn't happen in insulators, but in semiconductors the band gap is small enough that thermal energy can promote some electrons to the conduction band. That gives these materials their limited conductivity. It also results in the negative temperature coefficient, as a temperature increase means more thermal energy, more electrons in the conduction band, and thus lower resistivity. Figure 17.24 compares the band structure of conductors, insulators, and semiconductors.

Adding even tiny quantities of impurities to semiconductors—a process called **doping**—lets engineers adjust band gaps and conduction properties of these versatile materials. The result is a myriad of electronic devices, including **diodes** that act as "one-way valves" for electric current and **transistors** in which current in one circuit can control a larger current in another. Transistors are at the heart of modern electronics, providing amplification in audio equipment and instrumentation, and switching between the "one" or "zero" states that constitute the basic language of computers. Even the simpler diodes have many uses, including **light-emitting diodes** (LEDs) that are rapidly replacing incandescent lights and **photovoltaic cells** that convert sunlight into electrical energy.

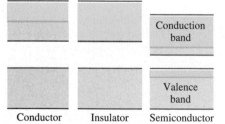

FIGURE 17.24 Band structure of conductors, insulators, and semiconductors.

Transistors, diodes, and other electronic components are manufactured on tiny silicon chips, interconnected to make complete circuits. Today these **integrated circuits** may hold several billion individual components—a number that has been growing exponentially, doubling every 18 to 24 months since the 1960s. That's why the computer you bought last month is already obsolete!

Superconductors

Superconductors are remarkable. They exhibit exactly zero resistance below a **transition temperature** T_c (Figure 17.25), which varies among superconducting materials (Table 17.3). In superconductors, a quantum-mechanical pairing among electrons allows them to move through the material without resistance.

Superconductors have fascinated scientists and engineers ever since Dutch physicist Heike Kamerlingh Onnes (1853–1926) discovered superconductivity in 1911. With zero resistance, superconductors dissipate no power and thus offer the potential for significant energy savings.

Superconductors have many applications, with more under development. One common application is high-strength electromagnets. The medical diagnostic technique known as magnetic resonance imaging (MRI) uses superconducting coils that surround the patient, and superconducting magnets guide charged particles in the largest particle accelerators used in high-energy physics research. Cell phone networks use superconducting filters to achieve sharp separation between nearby channels, and superconducting devices on the power grid help prevent blackouts. In urban underground power distribution systems, superconducting cables help increase capacity without taking up much space.

Superconductors also have the unusual property that they exclude magnetic fields. This effect results in the magnetic levitation shown in Figure 17.26. On a larger scale, magnetically levitated transportation systems (MAGLEV) may use superconducting electromagnets to lift vehicles just a few centimeters off a guideway, affording comfortable ground transportation at hundreds of kilometers per hour.

Their extreme sensitivity to magnetism makes superconductors excellent detectors of small magnetic fields. Devices called SQUIDs (**superconducting quantum interference devices**) detect fluctuations as small as 10^{-8} times Earth's magnetic field and find applications from scientific research to brain imaging.

Maintaining the low temperatures required for superconductivity is the main drawback of the known superconductors. The 1986 discovery of a new class of ceramic superconductors with T_c above 77 K (the boiling point of nitrogen, an inexpensive refrigerant) rekindled interest in superconductivity and led to the development of new superconducting devices. The ultimate dream of a room-temperature superconductor would enable many more applications.

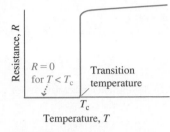

FIGURE 17.25 Resistance versus temperature for a typical superconductor.

TABLE 17.3 Some Selected Superconducting Transition Temperatures

Material	Type	Transition temperature T_c (K)
Mercury	Element	4.2
Lead	Element	7.2
Niobium	Element	9.3
Nb_3Ge	Intermetallic	23
$YBa_2Cu_2O_7$	Ceramic	93
TlBaCaCuO	Ceramic	110–125

Chapter 17 in Context

This is the last of three chapters based primarily on electric charge. Using the idea of electric potential (Chapter 16), you learned to analyze the flow of charge in *electric circuits*, and how that flow (*current*) depends on *electrical resistance*. You've also seen how to treat *series* and *parallel resistors*, as well as combinations of resistors and capacitors.

Looking Ahead In Chapters 15–17 you've seen the basics of electric charge and circuits. In Chapter 18 we'll show how electric current produces magnetism, and in Chapter 19 we'll take a look at alternating current (AC) circuits. Electric charge and current are of fundamental importance in physics, so you'll see them in other contexts, particularly in the modern physics of Chapters 23–26.

FIGURE 17.26 A magnet levitated above a superconductor.

CHAPTER 17 SUMMARY

Current and Resistance

(Section 17.1) Electric current flows from higher to lower electric potential. An ohmic material's resistance is constant, independent of potential difference, although it varies with temperature.

Electrical **resistivity** is a property of a material; resistance of a piece of material depends on size and shape as well as resistivity.

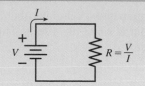

Electric current: $I = \dfrac{\Delta q}{\Delta t}$

Resistance: $R = \dfrac{V}{I}$

Temperature dependence of resistivity or resistance:
$\rho = \rho_0[1 + \alpha(T - T_0)]$

Batteries: Real and Ideal

(Section 17.2) Batteries have a fixed potential difference (emf) and supply current in closed circuits.

Kirchhoff's loop rule says that the sum of the potential differences around a closed loop is zero.

Combining Resistors

(Section 17.3) For resistors in series, the equivalent resistance is the sum of all the resistances.

Equivalent resistance of resistors in series: $R_s = R_1 + R_2$.

Kirchhoff's junction rule says that the net current entering a junction equals the net current leaving the junction. Resistors in parallel add reciprocally.

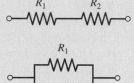

Reciprocal of equivalent resistance for resistors in parallel:
$\dfrac{1}{R_p} = \dfrac{1}{R_1} + \dfrac{1}{R_2}$

Electric Energy and Power

(Section 17.4) Resistors dissipate electrical energy.

The rate of energy dissipation (power) can be expressed in terms of any two of three quantities: potential difference, current, and resistance.

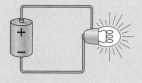

Electric power with resistor in a circuit: $P = IV = I^2R = \dfrac{V^2}{R}$

RC Circuits

(Section 17.5) In an **RC** circuit, the capacitor charges gradually, with **time constant** RC.

Charge and potential difference decay exponentially in a discharging capacitor, with time constant RC.

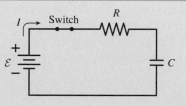

RC circuit charging: $q(t) = \mathcal{E}C(1 - e^{-t/RC})$

Discharging: $q(t) = q_0e^{-t/RC}$

NOTE: Problem difficulty is labeled as ■ straightforward to ■■■ challenging. Problems labeled BIO are of biological or medical interest.

Conceptual Questions

1. If resistance is due to collisions between conduction electrons and ions in a solid, why should resistance increase approximately linearly with temperature? *Hint:* Think about the oscillations of the ions in the solid.
2. Explain why the electron drift speed (Equation 17.2) is proportional to I/A.
3. Is thermal expansion (Chapter 12) an important factor in how a metal conductor's resistance changes as a function of temperature? Explain.
4. Which has the higher resistance, a 60-W lightbulb or a 100-W lightbulb? Each is designed to operate at 120 V.
5. How can $P = I^2R$ and $P = V^2/R$ both be true?
6. Express the ohm (Ω) in terms of kg, m, s, and C.

In Conceptual Questions 7–9, you have a battery and a box of identical lightbulbs. Let a single bulb attached to the battery serve as a reference (Figure CQ17.7).

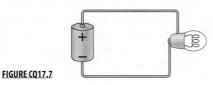

FIGURE CQ17.7

7. Three bulbs are connected in series across the battery. How do the bulbs' brightnesses compare with each other and with the reference bulb? What happens to the relative brightness if one bulb is unscrewed?
8. Three bulbs are connected in parallel across the battery. How do their brightnesses compare with each other and with the reference bulb? What happens to the relative brightness if one bulb is unscrewed?
9. Three bulbs are connected as shown (Figure CQ17.9), with a parallel connection of two bulbs in series with a third. How do their brightnesses compare with each other and with the reference bulb? What happens to the relative brightness if B_1 is removed while the others are left in place? Repeat for the case when B_2 alone is removed.

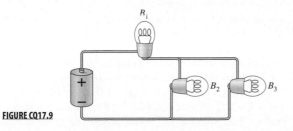

FIGURE CQ17.9

10. Show that the units on the right-hand side of $P = I^2R$ reduce to J/s or W.
11. Is the power consumed by a lightbulb greater immediately after it's turned on (when the filament is still cool) or later, after the filament has warmed?
12. Explain why the energy loss in power lines carrying a given power is lower if a higher potential is maintained between the wires.

13. Two cylindrical wires are made of the same material, but one has a larger diameter. Which wire is more likely to melt if they're each connected across the same battery?
14. If you double the length of the heating wire in your toaster, will you get more heating power, less, or the same amount? Explain.
15. Why is it safe for a bird to sit on a power line? Could you safely sit on the line next to the bird?
16. What's the potential difference between two ends of a superconducting wire carrying a current I? How does $I = V/R$ make sense here?
17. Are the electrical outlets in your home connected in series or parallel? How do you know?
18. Would a superconducting wire make a good lightbulb filament? Why or why not?

Multiple-Choice Problems

19. The resistance of a 100-m-long copper wire with diameter 2.0 mm is (a) 540 Ω; (b) 0.54 Ω; (c) 1.9 Ω; (d) 190 Ω.
20. The electron drift speed in a 0.75-mm-diameter copper wire carrying a current of 1.2 A is (a) 0.2 mm/s; (b) 2.0 mm/s; (c) 0.05 mm/s; (d) 5.0 mm/s.
21. What resistor would have a current of 1.5 mA when connected across a 1.5-V battery? (a) 10 Ω; (b) 15 Ω; (c) 1 kΩ; (d) 1.5 kΩ.
22. An 18.1-V battery is attached to a 25.4-Ω load, and the resulting current is 678 mA. The battery's internal resistance is (a) 0.013 Ω; (b) 0.77 Ω; (c) 1.3 Ω; (d) 1.7 Ω.
23. A 1.51-V battery with internal resistance 1.25 Ω delivers 0.125 A to its load. The load resistance is (a) 5.4 Ω; (b) 7.9 Ω; (c) 10.8 Ω; (d) 54 Ω.
24. The equivalent resistance of a 120-kΩ resistor and a 140-kΩ resistor in parallel is (a) 260 kΩ; (b) 20 kΩ; (c) 0.15 kΩ; (d) 65 kΩ.
25. For any two parallel resistors, (a) the current in each resistor is the same; (b) the power dissipated by each resistor is the same; (c) the potential difference across each resistor is the same; (d) none of the above is necessarily true.
26. A 320-Ω resistor and a 530-Ω resistor are in series. When this combination is connected across a 12-V battery, the current is (a) 14 mA; (b) 23 mA; (c) 38 mA; (d) 140 mA.
27. You have a 240-Ω resistor and wish to combine it with a second resistor to make an equivalent resistance of 200 Ω. You should (a) add a 40-Ω resistor is series; (b) add a 40-Ω resistor in parallel; (c) add a 440-Ω resistor in parallel; (d) add a 1200-Ω resistor in parallel.
28. A 9.0-V battery, 100-Ω resistor, and 240-Ω resistor are in series. The power delivered by the battery is (a) 0.24 W; (b) 4.2 W; (c) 0.026 W; (d) 0.09 W.
29. In an RC circuit with $R = 350$ kΩ and $C = 200 \, \mu$F, the time constant is (a) 1.75 s; (b) 70 s; (c) 700 s; (d) 7000 s.
30. The resistor you should put in series with a 250-pF capacitor to make the time constant 1 ms is (a) 250 kΩ; (b) 2.5 MΩ; (c) 4.0 MΩ; (d) 7.5 MΩ.
31. If you discharge a 60-μF capacitor through a 1.5-kΩ resistor, how much time does it take for 99.9% of the charge to leave the capacitor? (a) 0.62 s; (b) 0.09 s; (c) 1.75 s; (d) 150 s.

Problems

Section 17.1 Current and Resistance

32. ■ In a 0.50-mm-diameter copper wire, what current corresponds to an electron drift speed of 1.0 m/s? Do you think this is practical?

33. ■ A thin wire made of tin with diameter 2.5×10^{-5} m carries 0.15 mA of current. Find the electron drift speed.

34. ■ What diameter should you make a copper wire so its resistance is 0.010 Ω per meter of length?

35. ■ Find the resistance between the two ends of an aluminum wire 0.25 m long and 1.0 cm in diameter.

36. ■■ A platinum wire is originally at 20°C. Find the temperature required to raise its resistance by (a) 1% and (b) 10%.

37. ■ In a lightning strike lasting 45 ms, a total charge of 350 C is transferred to the ground. What's the average current?

38. ■■ An electron beam carries 1.0 mA. How many electrons pass a given point each second?

39. ■ Find the resistance of a 1.0-m length of silver wire with diameter (a) 5.0×10^{-5} m; (b) 5.0×10^{-4} m; (c) 5×10^{-3} m.

40. ■■ You have some copper wire that you want to connect from a power supply to a 150-Ω load. The wire is a standard 24-gauge (diameter 0.51 mm). What's the maximum length of wire you can use without its resistance exceeding 1% of the load resistance? *Note:* You're finding the total length of wire to and from the load.

41. ■■ A platinum wire with room-temperature resistance 58.5 Ω is used as a thermistor. If you want to measure the temperature to within 0.1°C in a range around room temperature, how precisely should the wire's resistance be measured?

42. ■■■ A nichrome toaster wire, initially at room temperature, carries 2.75 A immediately after it's turned on. After the wire heats, the current drops to 2.46 A. What's the temperature of the nichrome then?

43. ■■■ Two cylindrical metal wires, A and B, are made of the same material and have the same mass. Wire A is twice as long as wire B. What's the ratio R_A/R_B?

44. ■■ A 2.50-m-long copper wire has 1.50 V across its ends. (a) Find the electric field strength within the wire. (b) What's the acceleration of an electron in this field? (c) A typical drift speed for conduction electrons is 1 mm/s. How far would an electron travel before reaching this speed?

Section 17.2 Batteries: Real and Ideal

45. ■■ (a) For how much time must a 12.0-V battery be connected across a 1.50-kΩ load to move 1.0 C of charge through the load? (b) How much energy does the battery provide during that time?

46. ■■ A 12.0-V battery produces 230 mA when it's across a 50.0-Ω load. (a) What's the battery's internal resistance? (b) Find the potential difference across the battery's terminals.

47. ■■ When an 18.0-V battery is connected across a 130-Ω load, the potential difference between the battery terminals is 17.4 V. What's the internal resistance?

48. ■■ A 1.51-V battery has internal resistance 1.35 Ω. The battery powers a lightbulb whose filament (when lit) has resistance 8.55 Ω. Find (a) the current through the lightbulb and (b) the potential difference across the battery terminals with the bulb connected.

49. ■■ The nickel-metal hydride battery in your cell phone has a 1.2-V emf, and it can be recharged about 1000 times. Manufacturers often rate batteries in milliamphours (mA-h) they can supply between rechargings. (a) Show that the mA-h is a charge unit, and find the conversion factor from mA-h to C. (b) How much charge can a 1400-mA-h battery supply? (c) How much total energy can that battery supply between rechargings?

50. ■■ In a rechargeable battery, the charging process consists of reversing the normal current flow, forcing current into the positive battery terminal. Suppose it takes 3.2×10^5 C to recharge a 9-V battery. (a) How much work must be done to move that much charge against the battery's 9-V emf? (b) If the charging process takes 2.5 h, what's the average power required?

51. ■■ A battery produces 15.5 mA when it's connected to a 230-Ω load and 22.2 mA when it's connected to a 160-Ω load. Find the battery's emf and internal resistance.

Section 17.3 Combining Resistors

52. ■ A 34-Ω and a 45-Ω resistor are in series. (a) Find the equivalent resistance. (b) Find the current when this series combination is connected across a 7.5-V battery.

53. ■ Two resistors, 250 Ω and 370 Ω, are in parallel. (a) Find the equivalent resistance. (b) The parallel combination is connected across a 12-V battery. Find the current supplied by the battery and the current in each resistor.

54. ■■ You have three 100-Ω resistors. List all the possible equivalent resistance values you can make using one, two, or three resistors in combination.

55. ■■ Four resistors, 12 Ω, 15 Ω, 20 Ω, and 35 Ω, are in series. (a) Find the equivalent resistance. (b) Find the circuit current and the potential difference across each resistor when the series combination is across a 6.0-V battery. Verify that the sum of the four potential differences equals the battery emf.

56. ■■ The resistors of the preceding problem are now connected in parallel. (a) Find the equivalent resistance. (b) Find the current in each resistor when the parallel combination is across a 6.0-V battery. Verify that the sum of the resistor currents equals the battery current.

57. ■■ You're given a box of 10-Ω resistors. How would you make the following equivalent resistances: (a) 2 Ω; (b) 35 Ω; (c) 7 Ω; (d) 19 Ω?

Problems 58–61 refer to the accompanying diagram (Figure P17.58).

FIGURE P17.58

58. ■■ If $\mathcal{E} = 9.0$ V, $R_1 = 150$ Ω, $R_2 = 250$ Ω, and $R_3 = 1000$ Ω, find the current in each resistor.

59. ■■ Assume all the resistors are identical. If the battery's emf is 12 V and it supplies 200 mA, find each resistance.

60. ■■ Suppose $\mathcal{E} = 10.0$ V, $R_1 = 200$ Ω, and $R_2 = 300$ Ω. What value of R_3 would you select so the battery supplies 25.0 mA?

61. ■■ The three resistors each have resistance 50 kΩ. What should be the battery's emf if it's to supply 1.0 mA of current?

62. ■■ A 1.5-MΩ resistor and a 2.5-MΩ resistor are in series with a battery. The potential difference across the 1.5-MΩ resistor is measured at 6.0 V. What is the emf of the battery? (b) What is the circuit current?

63. ■■ A 20-kΩ resistor, a 30-kΩ resistor, and a 75-kΩ resistor are in parallel across a battery. The current in the 20-kΩ resistor is 0.125 mA. Find (a) the battery emf and (b) the current in each of the other resistors.

64. ■■ A series combination of four identical resistors is across a battery. The potential difference across each resistor is 2.2 V, and

the circuit current is 0.26 A. Find the value of each resistor and the battery emf.

Section 17.4 Electric Energy and Power

65. ■ A 150-Ω resistor is across a 24.0-V battery. Find the energy supplied by the battery in 1 min.

66. ■ What is the resistance (when lit) of a lightbulb rated at 120 V and 100 W?

67. ■ ■ Carbon resistors commonly come in ratings of 1/8 W, 1/4 W, 1/2 W, or 1 W, meaning that this is the maximum rate at which the resistor can dissipate energy at room temperature. If a 280-Ω resistor has a rating of 1/2 W, what's the maximum current it should carry?

68. ■ ■ (a) A battery supplies 100 W with a 20-Ω resistor across its terminals. Find (a) the battery emf and (b) the resistor current.

69. ■ ■ A hair dryer consumes 1750 W when connected to a 120-V source. (a) Find the current in the hair dryer. (b) How much energy is consumed if the dryer runs for 10 min? (c) Compute the monthly cost if the hair dryer runs for 10 min each day at a cost of 15¢ per kilowatt-hour.

70. ■ ■ A 0.10-mm-diameter copper wire is 15 m long. (a) Find its resistance. (b) What's the power dissipation in the wire if the potential difference across its ends is 12 V? (c) Repeat parts (a) and (b) if the wire is immersed in liquid nitrogen at 77 K.

71. ■ ■ A group of 200-W floodlights is wired in parallel to a 120-V electrical outlet to light an outdoor basketball court. What's the maximum number of lights that can be used without tripping the 20-A circuit breaker?

72. ■ ■ ■ A tungsten lightbulb filament is designed to dissipate 100 W at a 120-V potential difference. (a) What's the filament current, assuming negligible resistance in the rest of the circuit? (b) What is the filament's resistance when lit? (c) If the lit filament is at 1850 K, what's its resistance when off?

73. ■ ■ ■ An electric space heater draws 7.25 A at 120 V. (a) At what rate does it supply heat? (b) If the heater's energy goes into warming the air in an empty room (3.0 m by 3.0 m by 2.5 m), by how much does the air temperature rise in 1 min? (The specific heat capacity of air is about 1.0 kJ/(kg · K).)

Section 17.5 RC Circuits

74. ■ How large a capacitor should you combine with a 450-kΩ resistor to make an RC time constant of 25 ms?

75. ■ ■ A 300-pF capacitor, initially uncharged, is in series with a 175-MΩ resistor. The series combination is then connected across a 6.0-V battery. Find the charge on the capacitor after (a) 1.0 ms; (b) 10 ms; (c) 100 ms.

76. ■ ■ Find the number of time constants for the capacitor in an RC charging circuit to reach 99% of its maximum charge.

77. ■ ■ ■ How many time constants does it take for a charging capacitor to reach (a) half of its maximum charge? (b) half its maximum stored energy?

78. ■ ■ A 25-nF capacitor is in series with a 1.45-MΩ resistor. The capacitor is initially uncharged when the series combination is connected across a 24.0-V battery. (a) What's the maximum charge the capacitor will eventually attain? (b) How much time does it take for the capacitor to reach 90% of that maximum charge?

79. ■ ■ A series RC circuit consists of a 550-μF capacitor (initially uncharged) and a 12-kΩ resistor. The combination is connected across a 12-V battery; 3.0 s later, what's (a) the charge on the capacitor and (b) the current in the resistor?

80. ■ ■ A 150-μF capacitor charged initially to 0.28 mC is connected across a resistor, and 2.0 s later its charge has dropped to 0.21 mC. Find the resistance.

81. BIO ■ ■ **Defibrillator.** A defibrillator contains a 210-μF capacitor, charged by a 2.4-kV source through a 31-kΩ resistance. (a) What's the capacitor charge after 20 s? (b) What fraction of the maximum possible charge is your answer to part (a)? (c) How much charge would be added if the charging continued for another 20 s?

Section 17.6 Semiconductors and Superconductors

82. ■ ■ A section of copper transmission line carries a steady current of 250 A and has resistance 0.76 Ω. Suppose this line is replaced by a superconducting line. (a) Ignoring the energy needed to refrigerate the superconductor, how much energy will be saved in one year? (b) What's the cost of that energy, if the utility company's price is 11¢ per kilowatt-hour?

General Problems

83. ■ ■ A 2.5-kΩ resistor and 4.4-mF capacitor are in series, and the combination is connected across a 15-V power supply. Find the charge on the capacitor and current in the resistor after a time of (a) 1.0 s and (b) 10 s.

84. ■ ■ ■ You have four 100-Ω resistors, each capable of dissipating 1/2 W. How could you combine them to make a 100-Ω resistor capable of dissipating 2 W?

85. ■ ■ You are building a circuit and need a 3.44-kΩ resistor. You have three resistors: 1.50-kΩ, 3.30-kΩ, and 4.70-kΩ. How can you connect them to get the required resistance?

86. ■ ■ You're about to purchase a battery. Normally, batteries are rated in ampere-hours (A · h)— the total amount of charge they can deliver. The 6-V battery you see is rated at 50 watt-hours, meaning the total amount of energy it can deliver, but your application calls for a 5.0-A · h battery. Will this battery work?

87. ■ ■ A 50-Ω resistor is connected across a battery, and a 26-mA current flows. When the 50-Ω resistor is replaced with a 22-Ω resistor, a 43-mA current flows. What are the battery's emf and internal resistance?

88. ■ ■ Find the resistance needed in an RC circuit to bring a 20-μF capacitor from zero charge to 45% of its full charge in 140 ms.

89. ■ ■ A capacitor is charged until it holds 5.0 J of energy. It is then discharged by connecting it across a 10-kΩ resistor. In the first 8.6 ms after making the connection, the resistor dissipates 2.0 J. What is the capacitance?

90. BIO ■ ■ **Ion channel.** The potential difference across the ion channel described in Example 17.1 is 80 mV. (a) What's the resistance of the ion channel? (b) The ion channel can be modeled as a cylinder with a length of 6.0 nm and a diameter of 0.35 nm. What's the resistivity of the fluid filling the channel? (c) At what rate is energy dissipated while the ions are flowing?

91. ■ ■ ■ A camera's flash uses the energy stored in a 240-μF capacitor charged to 170 V. The current in the charging circuit is not to exceed 25 mA. (a) What's the resistance in the charging circuit? (b) How long do you have to wait between taking flash pictures? Consider that the capacitor is essentially at full charge in 5 time constants.

92. ■ ■ ■ Consider the circuit in Figure GP17.92. (a) Find the equivalent resistance of the entire network. (b) Find the current through each resistor and the potential difference across each resistor.

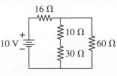

FIGURE GP17.92

93. ■ ■ ■ For the circuit in Figure GP17.93, find (a) the equivalent resistance of the network, (b) the current through each resistor, and (c) the power supplied by the battery.

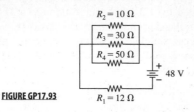

FIGURE GP17.93

94. ■ ■ ■ Repeat the preceding problem for the circuit in Figure GP17.94.

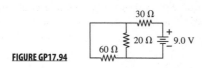

FIGURE GP17.94

95. BIO ■ ■ **Electric eel.** An electric eel uses a number of small cells called electrocytes, placed end to end in series, to produce a large potential difference. (a) If the eel produces a potential difference of 600 V, how many electrocytes, each with a potential difference of 125 mV, are required? (b) The eel can produce electrical energy at a rate of 600 W for a short time. What's the current required to do this?

96. ■ ■ ■ Figure GP17.96 shows the potential difference across a capacitor that's charging through a 4700-Ω resistor in the circuit of Figure 17.20. Use the graph to determine (a) the battery's emf; (b) the capacitance; (c) the total energy dissipated by the resistor during the 10 ms interval shown.

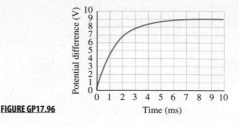

FIGURE GP17.96

Answers to Chapter Questions

Answer to Chapter-Opening Question
The apparatus measures the body's electrical resistance by passing a small current between electrodes attached to the skin. Fat has a higher resistance than lean tissue, so the result gives an indication of the body's fat content.

Answers to GOT IT? Questions
Section 17.1   (b) > (a) = (d) > (c)
Section 17.2   (d) 4 Ω
Section 17.3   (a) > (c) > (d) > (b)
Section 17.4   (a) 0.5 A, 240 Ω
Section 17.5   (d) > (b) > (a) > (c)

■ Why do auroras occur predominantly at high northern and southern latitudes?

This chapter introduces magnetism. We'll start with familiar examples—bar magnets and magnetic compasses—and discuss Earth's magnetism. Then you'll see how magnetic fields affect moving electric charges. Next, we'll consider magnetic forces on the charges in a current-carrying wire and common applications like electric motors that use these forces. Finally, we'll discuss magnetic materials.

18.1 Magnets, Poles, and Dipoles

You're familiar with everyday applications of magnets. Magnets hold notes to your refrigerator and magnetic compasses show direction. But magnetism is far more than that! It's a fundamental interaction with an intimate connection to electricity. We begin with familiar aspects of magnetism, and then we'll move on to develop a deeper understanding.

Magnetic Poles

Find a bar magnet with its **north pole** and **south pole**. Bring a second magnet near and you'll discover that *like poles repel, and unlike poles attract*—just as do like and unlike

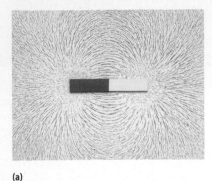

(a)

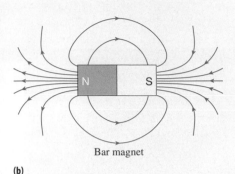

Bar magnet

(b)

FIGURE 18.1 (a) Iron filings align with the magnetic field, tracing out the field of a bar magnet. (b) Magnetic field lines around a bar magnet.

Repeatedly cut a magnet in half . . .

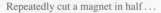
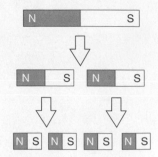

. . . and each half still has two magnetic poles.

FIGURE 18.2 Cutting a bar magnet in half results in smaller magnets, not magnetic monopoles.

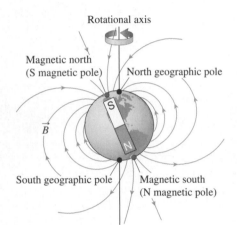

FIGURE 18.3 Earth's magnetic field is essentially that of a dipole. Magnetic north is actually the south pole of a magnet, so that the north pole of a compass needle will be attracted toward it and thus point approximately north.

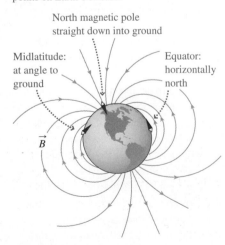

FIGURE 18.4 Magnetic inclination is due to the shape of the dipole magnetic field along Earth's surface.

electric charges. Experiment more, and you'll see that the closer the magnets, the stronger the force.

Figure 18.1a shows the distinctive pattern resulting when iron filings are sprinkled on a sheet of paper over a bar magnet. Each individual filing becomes a temporary magnet that aligns with the **magnetic field**; later we'll see just why this occurs. The orientation of the filings reveals the direction of the magnetic field. We'll use the symbol $\vec{B}$ for magnetic field. Figure 18.1b shows the field around a bar magnet, using field lines similar to those we introduced for the electric field in Section 15.4. Remember that (1) the field strength follows from the density of field lines and (2) the direction of the field is tangent to the field line at any point.

Magnetic Dipoles and Monopoles

Compare the bar-magnet field of Figure 18.1 with the electric field of an electric dipole (Figure 15.18). They look similar because both are *dipole* fields. The electric dipole consists of separated positive and negative point charges. The bar magnet acts as a **magnetic dipole**, with the north and south poles at the ends analogous to the positive and negative point charges, respectively. Thus magnetic field lines emerge from north poles and go in to south poles.

But there's a big difference between electric and magnetic dipoles. Cut an electric dipole in half, and you've got two separate point charges. Cut a bar magnet in half and, as Figure 18.2 shows, you get two smaller bar magnets, each with a north and a south pole. Keep cutting, but you'll never find separate north and south poles. Such isolated poles are called **magnetic monopoles**. Figure 18.2 shows a basic fact of magnetism: *There are no magnetic monopoles.* We'll explore this point further in Section 18.5.

Magnetic Earth

The ancient Greeks observed that lodestone, an iron ore now known as magnetite (Fe_3O_4), attracted small pieces of iron. By the twelfth century, European and Chinese sailors were using magnetic compasses for navigation. In 1600, the English physician William Gilbert (1544–1603) reported numerous experiments on magnets. Gilbert discovered that cutting magnets in half didn't produce separated poles. He also used a spherical magnetic stone to show how Earth behaves like a magnetic dipole (Figure 18.3).

We now know that Earth and many other astronomical bodies produce magnetic fields. As shown in Figure 18.3, Earth's dipole field aligns approximately, but not exactly, with the planet's rotation axis. The difference between the directions of *magnetic north* and *true north* is called **declination**. It varies with location, and you'd better take it into account if you're using a magnetic compass to find your way. Figure 18.4 shows that the magnetic field at Earth's surface has both vertical and horizontal components, with the vertical component generally increasing toward the poles. The angle the magnetic field makes with the horizontal is called **inclination**.

Furthermore, Earth's magnetic poles wander about, moving substantially from year to year. The north magnetic pole is currently in the Arctic Ocean north of central Canada and has been moving northwest at an average rate of about 10 km/year for the last century. In Chapter 19 we'll explore the origin of Earth's magnetism.

✓ **TIP**

Your compass needle points toward magnetic north, not geographic north. Declination is the east/west deviation of the compass needle. Inclination is the vertical deviation.

18.2 Magnetic Force on a Moving Charge

So far we've described the interaction between magnets. But the fundamental magnetic interaction is with *moving electric charge*. You'll see later how this relates to the magnets with which you're familiar. Here we'll explore the magnetic force on a moving charge.

The Magnetic Force

Figure 18.5a shows a positive electric charge q moving with velocity $\vec{v}$ through a magnetic field $\vec{B}$. This moving charge experiences a magnetic force $\vec{F}$, as shown. Note that the magnetic force is *not* in the direction of the magnetic field; rather, *the magnetic force is perpendicular to both the velocity $\vec{v}$ and the magnetic field $\vec{B}$*. The magnitude of the force depends on the charge magnitude $|q|$, the speed v, and the field magnitude B, and on the angle between the vectors $\vec{v}$ and $\vec{B}$. We'll summarize the force law as follows:

1. The magnitude of the force is

$$F = |q|vB \sin \theta \quad \text{(Magnetic force law; SI unit: N)} \tag{18.1}$$

   where θ is the angle between the velocity and the magnetic field.
2. The direction of the force is perpendicular to both the velocity and magnetic field, as given by the right-hand rule (Figure 18.5b; this is the same rule we used in Section 8.6 to compute torque). If the charge is negative, then the force is opposite the right-hand rule direction.

The SI unit of magnetic field is the **tesla** (T). If you use SI units in Equation 18.1—coulombs for charge, meters per second for speed, and tesla for magnetic field—then the force comes out in newtons. Solving Equation 18.1 for B gives $B = F/|q|v \sin \theta$. This shows that $1 \text{ T} = 1 \text{ N}/(\text{C} \cdot \text{m/s})$, or $1 \text{ T} = 1 \text{ N}/(\text{A} \cdot \text{m})$. A 1-T magnetic field is rather large. A strong laboratory electromagnet might generate a field as large as 5 T. By comparison, the field at Earth's surface is around 50 μT.

The factor $\sin \theta$ in Equation 18.1 shows that the magnetic force is greatest when the charge velocity and the magnetic field are perpendicular, because then $\sin \theta = 1$. With velocity and field in the same direction, $\sin \theta = \sin 0° = 0$, so there's no force. Similarly, when the velocity and magnetic field are opposite, $\sin \theta = \sin 180° = 0$, and again there's no force.

✓ **TIP**

Before applying the right-hand rule, check to see if the angle between the velocity and magnetic field is zero or 180°. If it is, then the force is zero, and you're done.

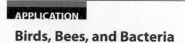
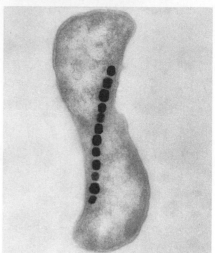

Many birds and bees use Earth's magnetic field to help navigate. In their heads are small particles of magnetite (Fe_3O_4), which respond to the field. Although birds and bees also use visual clues to navigate, magnetism helps them find their way in the dark or on foggy days. Some bacteria, like the one shown here, contain magnetite (the dark specks in the photo), which guides them to swim into desirable low-oxygen environments. Scientists aren't sure exactly how this works, but it may involve sensing magnetic inclination and navigating vertically into oxygen-free mud.

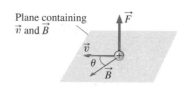

(a) Force on a charge moving at an angle θ to a magnetic field $\vec{B}$

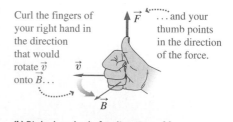

(b) Right-hand rule for direction of force

FIGURE 18.5 Determining the direction of the force on a moving charge.

CONCEPTUAL EXAMPLE 18.1 **Motion of Charges in Earth's Field**

You're at the equator, in a place where Earth's magnetic field is horizontal and points due north. A positively charged particle is moving horizontally. Find the direction of the magnetic force on the charge when it's moving (a) north, (b) south, (c) east, (d) west.

SOLVE Figure 18.6 shows the situations.

(a) North is the field direction, so the angle between $\vec{v}$ and $\vec{B}$ is zero, and there's no force.

(b) Moving south, the angle is now 180°, and again there's no force.

(c) Moving east gives a 90° angle between $\vec{v}$ and $\vec{B}$, so the force isn't zero. Applying the right-hand rule shows that the force is vertically upward.

(d) Again the angle between force and field is 90°, and now the right-hand rule shows its direction is vertically downward.

REFLECT What if the charge were negative? In parts (a) and (b), the force is zero regardless of the charge. In general, the force on a negative charge is opposite that on a positive charge, so the answers to parts (c) and (d) reverse.

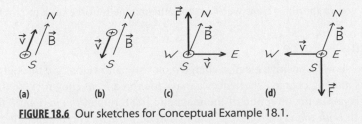

(a)   (b)   (c)   (d)

FIGURE 18.6 Our sketches for Conceptual Example 18.1.

You've seen that there's a three-dimensional relation among velocity, magnetic field, and force. It can be difficult to visualize vectors in three dimensions when you're forced to draw them on two-dimensional paper. Tactic 18.1 shows one way to represent the third dimension. We'll often use this technique to illustrate vectors in three dimensions.

TACTIC 18.1 **Drawing Vectors in Three Dimensions**

Here's a way to represent vectors perpendicular to your paper (Figure 18.7):

1. Use a cross (×) for a vector going straight into the paper.
2. Use a dot (·) for a vector coming straight out of the paper.

These make sense if you think of a vector as an arrow—the kind you shoot from a bow. If the arrow is going directly away from you, you see its tail feathers, forming the cross. If it's coming toward you, then you see the sharp point as a dot.

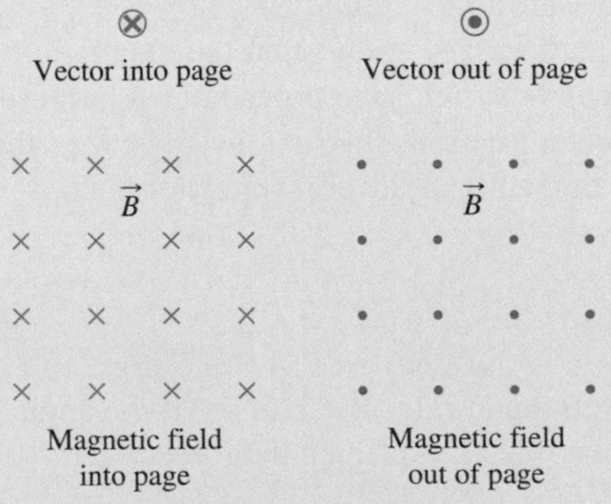

$\otimes$ $\odot$

Vector into page   Vector out of page

× × × ×   · · · ·
$\vec{B}$ $\vec{B}$
× × × ×   · · · ·

× × × ×   · · · ·

× × × ×   · · · ·

Magnetic field   Magnetic field
into page   out of page

FIGURE 18.7 Representing vector fields that are perpendicular to the page.

The same conventions apply to magnetic field lines as well as individual vectors. Figure 18.7 shows two different uniform magnetic fields, one going into the page and one out. How can you tell they're uniform?

EXAMPLE 18.2   **Saved by Earth's Magnetism!**

Earth's magnetic field deflects high-energy particles from the Sun, protecting terrestrial life from harmful radiation. Consider a solar electron moving straight toward Earth's equator at 450 km/s. It encounters the planet's northward-pointing magnetic field about 10 Earth radii above the planet, where the field is essentially uniform with strength 30 nT. Find the direction and magnitude of the magnetic force on the electron.

ORGANIZE AND PLAN The direction of the force on a charge moving in a magnetic field follows from the right-hand rule, and the magnitude comes from the force law, Equation 18.1: $F = |q|vB \sin \theta$. The particle is an electron, so its charge is $q = -e = -1.60 \times 10^{-19}$ C. The angle θ between the velocity and magnetic field is 90°.

Known: $v = 450$ km/s; $B = 30$ nT; $q = -e = -1.60 \times 10^{-19}$ C.

SOLVE Figure 18.8 shows the situation, as viewed from above the solar system. In this orientation, the northward-directed magnetic field is out of the page, represented by dots. The force is perpendicular to the electron's velocity and the magnetic field. With the electron moving toward Earth and the field pointing northward, the force must be

in the east-west direction. Start with your fingers to the right ($\vec{v}$). Then curl them out of the page ($\vec{B}$), and your thumb points eastward. Since the electron is *negative*, the force is westward, as shown. The magnitude of the force is

$$F = |q|vB \sin \theta = (1.60 \times 10^{-19}\,\text{C})(4.5 \times 10^{5}\,\text{m/s})(30\,\text{nT})$$
$$= 2.2 \times 10^{-21}\,\text{N}$$

REFLECT Even in this nano-scale magnetic field, the magnetic force is still many orders of magnitude larger than the gravitational force on the electron. In general, magnetic forces on subatomic particles, like electric forces, are typically so large that gravity normally can be neglected.

Some of those solar electrons become trapped in Earth's magnetic field, which channels them downward near the poles. The electrons collide with nitrogen and oxygen molecules in the upper atmosphere, exciting the molecules to higher energy states. The excited molecules then emit energy as visible light. Red and green come mostly from oxygen, blue and violet from nitrogen. The result (see the photo on the first page of the chapter) is the brilliant display called aurora borealis in the north and aurora australis in the south. Aurora activity peaks at the maximum of the 11-year solar activity cycle.

MAKING THE CONNECTION Find the magnitude of the electron's acceleration in this example. Compare with the acceleration due to gravity.

ANSWER The electron's mass is 9.1×10^{-31} kg. With $F = ma$, the acceleration is $a = F/m = 2.4 \times 10^{9}$ m/s². Since the electron is 10 Earth radii out, the gravitational acceleration is $g/10^2$, or about 0.1 m/s². So the magnetic acceleration is more than 10^{10} times larger! But there's a difference: Gravity can change the electron's speed, while the magnetic force, being perpendicular to $\vec{v}$, can change only its direction—as you'll see next.

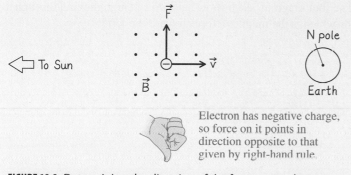

Electron has negative charge, so force on it points in direction opposite to that given by right-hand rule.

FIGURE 18.8 Determining the direction of the force on an electron.

Circular Trajectories

In Example 18.2, the electron felt a westward force. What happens next? Its velocity acquires a westward component, and, because $\vec{v}$ has changed direction, so does the direction of the magnetic force (Figure 18.9a, next page). So what's the electron's trajectory?

As long as the electron remains in this uniform magnetic field, its trajectory will be a *circle*. In fact, any charged particle moving perpendicular to a uniform magnetic field undergoes uniform circular motion—that is, circular motion with constant speed. The magnetic force on the charged particle is consistent with what you know about uniform circular motion, for two reasons:

1. The force on a particle in uniform circular motion is toward the center of the circle (Chapter 4), which is perpendicular to the particle's velocity.
2. The work done by any force is proportional to $\cos \theta$, where θ is the angle between force and displacement. Because the magnetic force is always perpendicular to the motion ($\theta = 90°$), there's never any work done, and hence no change in the particle's kinetic energy or speed.

Thus you can see (Figure 18.9a, next page) that the electron moves in a circle, counterclockwise as viewed with the magnetic field pointing out of the page. In the same field, a

As magnetic force changes the electron's direction of motion, direction of force also changes. By the right-hand rule, . . .

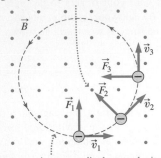

. . . force remains perpendicular to velocity, so the electron moves in a counterclockwise circle.

(a) Motion of negative charge in magnetic field

Magnetic force on a positive charge results in clockwise circular motion.

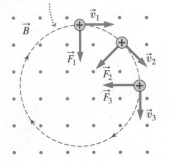

(b) Motion of positive charge in magnetic field

FIGURE 18.9 The motion of electric charges in a uniform magnetic field.

positive particle experiences a force in the opposite direction and therefore describes a clockwise circle (Figure 18.9b).

The radius R of the circle for any charged particle is related to its charge q, mass m, speed v, and magnetic field B. You saw in Chapter 4 that $F = mv^2/R$ for the force holding a particle in circular motion. Here that force is magnetic, and Equation 18.1 gives its magnitude as $F = |q|vB \sin\theta = |q|vB$. (Here we're assuming motion in a plane perpendicular to the magnetic field, so $\sin\theta = 1$.) Therefore, $|q|vB = mv^2/R$, which simplifies to give

$$R = \frac{mv}{|q|B} \quad \text{(Radius of charged particle's circular path; SI unit: m)} \quad (18.2)$$

In the next section, we'll describe some applications where charged particles undergo circular motion in magnetic fields.

Reviewing New Concepts

- A charged particle moving through a magnetic field experiences a magnetic force, perpendicular to the magnetic field and the particle's velocity.
- A charged particle moving perpendicular to a uniform magnetic field has a circular trajectory.

GOT IT? **Section 18.2** The four diagrams each show a positive charge moving through a uniform magnetic field. For which is the magnetic force toward the left?

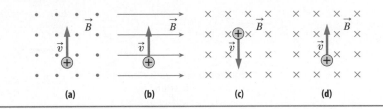

(a) (b) (c) (d)

18.3 Applications of Magnetic Forces

Here we'll consider some applications of magnetic forces on moving charges. Most involve the circular motion described by Equation 18.2.

EXAMPLE 18.3 **Bubble Chamber**

High-energy physicists use liquid-hydrogen-filled **bubble chambers** to study charged particles produced in particle accelerators and nuclear reactions. When a fast-moving charged particle passes through, it vaporizes liquid along its path, leaving a visible trail. The chamber is in a magnetic field, and thus the particle trajectories are curved, as shown in Figure 18.10. In a bubble chamber experiment with a uniform magnetic field of 1.45 T, a positive pion moving at 9.49×10^5 m/s perpendicular to the field follows a path with curvature radius 1.02 mm. (a) From these data, what is the ratio of the pion's charge to mass? (b) If the charge on the pion is $+e$, what's its mass?

ORGANIZE AND PLAN Equation 18.2 involves the mass and charge of a particle in circular motion in a magnetic field. Rearranging allows you to solve for the ratio $|q|/m$. With $|q|/m$ known, the mass m of the pion can be computed using its charge, $q = +e$.

Known: $v = 9.49 \times 10^5$ m/s; $B = 1.45$ T; $R = 1.02$ mm.

FIGURE 18.10 Charged particle tracks in a bubble chamber.

cont'd.

SOLVE (a) Solving Equation 18.2 for the charge-to-mass ratio gives

$$\frac{|q|}{m} = \frac{v}{RB} = \frac{9.49 \times 10^5 \text{ m/s}}{(1.02 \times 10^{-3} \text{ m})(1.45 \text{ T})} = 6.42 \times 10^8 \text{ C/kg}$$

(b) The pion's charge is $q = +e = 1.60 \times 10^{-19}$ C. Therefore, its mass is

$$m = \frac{|q|}{|q|/m} = \frac{1.60 \times 10^{-19} \text{ C}}{6.42 \times 10^8 \text{ C/kg}} = 2.49 \times 10^{-28} \text{ kg}$$

REFLECT In part (a), do the units work out to C/kg? Yes: $1 \text{ T} = 1 \text{ N}/(\text{A} \cdot \text{m})$, and $1 \text{ N} = 1 \text{ kg} \cdot \text{m/s}^2$. The numerical result 6.42×10^8 C/kg is not an unreasonable charge-to-mass ratio for a subatomic particle, as you'll see in subsequent examples. A table of subatomic particle masses will confirm the result in part (b), to within rounding errors.

This example suggests that magnetic fields are useful for measuring particles' charge to-mass ratios. You'll see this approach used again in the *mass spectrometer*.

MAKING THE CONNECTION How does the trajectory of an electron, moving with the same speed in the same magnetic field, differ from that of the pion?

ANSWER The electron's charge is $q = -e$, and its mass is 9.11×10^{-31} kg. With a smaller mass, the electron moves in a circle of smaller radius. Also, the electron's negative charge means its path curves in the opposite way.

CONCEPTUAL EXAMPLE 18.4 **Velocity Selector**

A charged particle moving horizontally with speed v passes through a uniform electric field E directed vertically downward. What magnitude and direction of magnetic field in the same region would result in the charged particle passing undeflected? Assume the particle is traveling fast enough to neglect gravity.

SOLVE For a positive charge q, the force due to the electric field (Chapter 15) is $F = qE$, downward. In order for the particle to pass undeflected, the net force on it must be zero. Therefore, the required magnetic force is upward, and the magnetic field is perpendicular to the electric field (Figure 18.11), perpendicular to the electric field. From Equation 18.1, the magnetic force has a magnitude $F = qvB$, which must equal that of the oppositely directed electric force. Thus $qE = qvB$, or

$$B = \frac{E}{v}$$

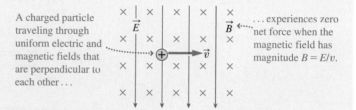

FIGURE 18.11 Electric and magnetic fields in a velocity selector.

REFLECT What happens with a negative charge? Now the electric force is upward and the magnetic force downward. Both forces have the same magnitudes as before, so again the particle is not deflected. This device is called a **velocity selector**, because for any choice of E and B, there's only one velocity, $v = E/B$, for a charged particle to pass without deflection. This velocity doesn't depend on the charge's magnitude or sign.

Mass Spectrometer

The **mass spectrometer** is an instrument for separating and identifying atoms and molecules by mass. It relies on magnetic forces on charged particles, as illustrated in Figure 18.12 (next page).

First, the atoms or molecules are ionized, then accelerated through a potential difference. They pass through a velocity selector with fields E and B_1, resulting in a beam of ions with speed $v = E/B_1$. Next they enter a uniform magnetic field B_2, where they describe circular arcs with radii given by Equation 18.2: $R = mv/|q|B_2$. Because $v = E/B_1$, this becomes

$$R = \frac{mE}{|q|B_1B_2}$$

The ions strike a detector that indicates their position and thus path radius R. Their charge-to-mass ratio is then

$$\frac{|q|}{m} = \frac{E}{B_1B_2R} \quad \text{(Charge-to-mass ratio in a mass spectrometer; SI unit: C/kg)} \quad (18.3)$$

Given the ions' charges, their mass follows. Particles with the same charge will be separated according to their masses, with larger masses describing larger circles.

FIGURE 18.12 (a) Operation of a mass spectrometer. (b) Output from a mass spectrometer. The graph shows the numbers of ions as a function of mass.

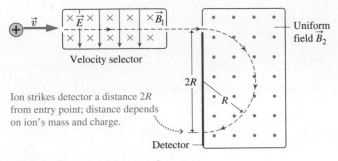

(a) How a mass spectrometer works

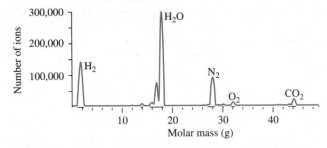

(b) Typical output of mass spectrometer. Curve shows number of ions as a function of mass.

Francis Aston invented the mass spectrometer in 1919. Immediately it became a useful tool for sorting and identifying different isotopes of the same element, a crucial step in understanding nuclear properties (Chapter 25). Chemists routinely use mass spectrometers to identify unknown atoms and compounds.

EXAMPLE 18.5   **Isotope Separation**

A mass spectrometer is being used to separate the helium isotopes ^{3}He and ^{4}He, which have masses of 5.01×10^{-27} kg and 6.65×10^{-27} kg, respectively. Both are singly ionized with charge $+e$, and they first pass through a velocity selector with $E = 2800$ N/C and $B_1 = 6.10$ mT. In the mass spectrometer, $B_2 = 1.20$ T. (a) What is the ions' speed? (b) How far apart do the ^{3}He and ^{4}He beams strike the detector?

ORGANIZE AND PLAN From the analysis of the velocity selector, the ion speeds are $v = E/B_1$. Then the curvature radius of each beam follows from solving Equation 18.3 for R.

As shown in Figure 18.12a, each ion beam strikes the detector at a distance $2R$ from the entry point. Thus, the difference in positions on the detector is the difference between the values of $2R$ for the two ions.

Known: $m\left(^3\text{He}\right) = 5.01 \times 10^{-27}$ kg, $m\left(^4\text{He}\right) = 6.65 \times 10^{-27}$ kg, $B_1 = 6.10$ mT $= 6.10 \times 10^{-3}$ T, $B_2 = 1.20$ T, $E = 2800$ N/C.

SOLVE (a) Using the numerical values given for the electric and magnetic field in the velocity selector,

$$v = \frac{E}{B_1} = \frac{2800 \text{ N/C}}{6.10 \times 10^{-3} \text{ T}} = 4.59 \times 10^5 \text{ m/s}$$

(b) Solving Equation 18.3 for R,

$$R = \frac{mE}{|q|B_1B_2}$$

Therefore, the radius for the ^{3}He isotope is

$$R_3 = \frac{mE}{|q|B_1B_2} = \frac{(5.01 \times 10^{-27} \text{ kg})(2800 \text{ N/C})}{(1.60 \times 10^{-19} \text{ C})(6.10 \times 10^{-3} \text{ T})(1.20 \text{ T})}$$

$$= 0.0120 \text{ m} = 1.20 \text{ cm}$$

A similar calculation gives $R_4 = 1.59$ cm for ^{4}He. Therefore, the two beams are separated on the detector by a distance

$$\Delta x = 2R_4 - 2R_3 = 0.78 \text{ cm}$$

REFLECT In part (a), the units reduce to m/s. This follows because $1 \text{ T} = 1 \text{ N}/(\text{A} \cdot \text{m})$, and $1 \text{ N} = 1 \text{ kg} \cdot \text{m/s}^2$. The resulting ion speed is reasonable for subatomic particles. In part (b), the difference in position means these isotopes are easily distinguished.

MAKING THE CONNECTION You're using a mass spectrometer to distinguish two isotopes that differ in mass by only 1%. By what percentage do the two radii differ?

ANSWER The radius is proportional to the mass, so these radii differ by only 1%—making them harder to distinguish.

Charge-to-Mass Ratio for Electrons

A device similar to the mass spectrometer can be used to measure the charge-to-mass ratio of the electron—the subatomic particle with the largest such ratio. Figure 18.13 shows the schematic for this experiment. Electrons "boil off" a heated wire and are accelerated through a potential difference ΔV to give them kinetic energy $K = \frac{1}{2}mv^2 = e\Delta V$. Rearranging gives $v^2 = 2e\Delta V/m$. When electrons with this speed pass through a magnetic field B, then by Equation 18.2

$$v^2 = \frac{e^2 B^2 R^2}{m^2}$$

where we've used $|q| = e$ for electrons. Equating our two expressions for v^2 leads to

$$\frac{e}{m} = \frac{2\Delta V}{B^2 R^2}$$

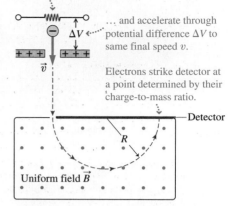

Electrons boil off hot filament …

… and accelerate through potential difference ΔV to same final speed v.

Electrons strike detector at a point determined by their charge-to-mass ratio.

Detector

Uniform field $\vec{B}$

FIGURE 18.13 Schematic diagram of an apparatus used to measure electrons' charge-to-mass ratio.

EXAMPLE 18.6 **Charge-to-Mass Ratio for Electrons**

An experiment using the apparatus described above produces the following results. Electrons accelerated through a potential difference of 43 V pass through a 2.2-mT magnetic field. The electron beam curves with a radius measured at 1.0 cm. What's the experimental charge-to-mass ratio for electrons?

ORGANIZE AND PLAN From the analysis in the text, the charge-to-mass ratio is

$$\frac{e}{m} = \frac{2\Delta V}{B^2 R^2}$$

The three quantities on the right side of this equation are measured in the experiment.

Known: $\Delta V = 43$ V, $B = 2.2$ mT, $R = 1.0$ cm.

SOLVE Using the measured quantities,

$$\frac{e}{m} = \frac{2\Delta V}{B^2 R^2} = \frac{2(43 \text{ V})}{(2.2 \times 10^{-3} \text{ T})^2 (0.010 \text{ m})^2} = 1.8 \times 10^{11} \text{ C/kg}$$

REFLECT You should verify that the SI units reduce to C/kg, using $1 \text{ T} = 1 \text{ N}/(\text{A} \cdot \text{m})$, $1 \text{ V} = 1 \text{ J/C}$, and $1 \text{ N} = 1 \text{ kg} \cdot \text{m/s}^2$. Our result agrees with the accepted value, to two significant figures.

Historically, measurement of e/m for electrons was a crucial step in understanding not only electrons but also atoms. The first good measurement of e/m, made by the English physicist J.J. Thomson in 1897, showed that the electron's charge-to-mass ratio was more than 1000 times larger than that of the hydrogen ion. This indicated that electrons comprise a tiny fraction of an atom's mass. In Chapter 25 you'll see how this fact helped lead to our current understanding of the atom as a massive, positively charged nucleus surrounded by lighter electrons.

✓ **TIP**

The electron is about 1800 times less massive than the proton.

Hall Effect

The **Hall effect** is another instance of moving charges deflected by a magnetic field. Figure 18.14a (next page) shows a current I flowing through a rectangular metallic bar of width a and thickness b. The moving charges in most conductors are electrons, which experience a magnetic force that drives them to one side of the bar, leaving an equal positive charge on the other side. This charge separation creates an electric field, soon resulting in equilibrium with a balance between the oppositely directed electric and magnetic forces (Figure 18.14b, next page). For electrons, the magnitudes of the electric and magnetic forces are eE and evB, respectively, where v is the electron drift speed introduced in Chapter 17. In equilibrium $eE = evB$, or $v = E/B$.

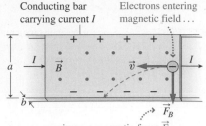

...experience a magnetic force $\vec{F}_B$ toward bottom of bar. Negative charge builds up on bottom of bar and positive

(a) charge on top.

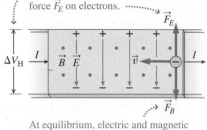

At equilibrium, electric and magnetic

(b) forces exactly cancel each other.

FIGURE 18.14 (a) Origin of the Hall effect. (b) The resulting charge buildup generates the Hall potential difference ΔV_H.

Charged particles gain kinetic energy each time they cross the potential difference ΔV between the dees, so they follow a spiral path within the magnetic field.

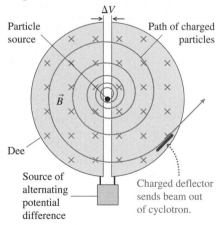

FIGURE 18.15 Operation of a cyclotron.

Associated with the electric field is the **Hall potential difference**, ΔV_H (Figure 18.14b). Electric field and potential difference are related by $E = \Delta V_H / a$, so the drift speed v can be rewritten

$$v = \frac{\Delta V_H}{Ba}$$

Equation 17.2 shows that $v = I/neA$, where A is the conductor's cross-sectional area and n the electron number density. Equating our two expressions for speed and solving for ΔV_H gives $\Delta V_H = IBa/neA$. With $A = ab$ for our conducting bar, ΔV_H becomes

$$\Delta V_H = \frac{IB}{neb} \quad \text{(Hall potential difference; SI unit: V)} \quad (18.4)$$

Equation 18.4 shows that the Hall potential difference is directly proportional to the magnetic field strength, which makes the Hall effect a means of measuring magnetic fields. However, metallic conductors have a high charge carrier density ($n = 8.47 \times 10^{28} \, \text{m}^{-3}$ for copper), so the Hall potential difference is small. Semiconductors, with lower charge carrier densities n, therefore make more practical Hall effect measuring devices. Another application of the Hall effect is determining the charge carrier density and sign for different materials. For example, the Hall effect confirms Chapter 17's description of different semiconductors as having either positive charge carriers (holes) or negative charge carriers (electrons).

Cyclotrons and Synchrotrons

An important use of magnetic forces is to "steer" charged particles in particle accelerators, forcing them to follow circular paths. Accelerators are used to explore the fundamental properties of matter and in practical applications such as producing radioisotopes for medical diagnostics.

The **cyclotron** (Figure 18.15) is a particle accelerator developed in the 1930s by the American physicist Ernest O. Lawrence (1901–1958). A charged particle q is injected from the source S near the center of the cyclotron. It starts with low speed and describes a circular motion in the cyclotron's magnetic field. Recall from Equation 18.2 that the curvature radius is proportional to speed, so initially the circle is small. Within the cyclotron are two hollow D-shaped metal structures called "dees." There's a potential difference ΔV between them, and each time a particle crosses the gap between the dees, it gains kinetic energy $\Delta K = |q|\Delta V$ (as you know from Chapter 16). The potential difference reverses periodically, so each time a particle crosses the gap, the direction of the potential difference is the right way to *increase* the particle's energy.

Here's the clever part. Solving Equation 18.2 for v gives $v = |q|BR/m$. We can also write v as the distance—the circumference $2\pi R$—divided by the period T of the circular motion:

$$v = \frac{|q|BR}{m} = \frac{2\pi R}{T}$$

Solving for T gives

$$T = \frac{2\pi m}{|q|B}$$

The **cyclotron frequency** f is the reciprocal of this period, so

$$f_c = \frac{|q|B}{2\pi m} \quad \text{(Cyclotron frequency; SI unit: Hz)} \quad (18.5)$$

Note that f_c is independent of the radius R. So as long as the device reverses the dees' potential difference with frequency f, charged particles will always get an energy boost when they cross the dee gap. As the particles gain energy they move in ever larger circles, but the cyclotron frequency remains unchanged.

APPLICATION   **Medical Cyclotrons**

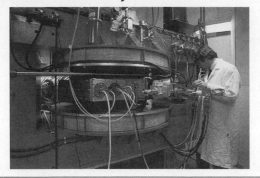

The medical imaging technique known as PET scanning (for *positron emission tomography*) often uses radioactive isotopes with such short lifetimes that they have to be produced just prior to a PET scan. Bombarding the appropriate materials with high-energy particles from a cyclotron can induce nuclear reactions that produce the desired isotopes. Larger hospitals often have in-house cyclotrons for just that purpose.

EXAMPLE 18.7   **A Medical Cyclotron**

A cyclotron used to produce radioisotopes for PET scanning is 54.0 cm in diameter. It accelerates alpha particles (helium nuclei, charge $+2e$, mass 6.64×10^{-27} kg) in a uniform magnetic field of 1.2 T. (a) Find the cyclotron frequency. (b) Find the maximum speed and kinetic energy of the alpha particles.

ORGANIZE AND PLAN Equation 18.5 gives the cyclotron frequency. The particles reach maximum speed at the outer edge of the cyclotron, where they're moving in circles of radius $R = 27.0$ cm (half the cyclotron's diameter). Their speed can then be found from the kinematics of circular motion.

Known: $R = 27.0$ cm $= 0.270$ m, $q = 2e = 3.2 \times 10^{-19}$ C, $m = 6.64 \times 10^{-27}$ kg, $B = 1.2$ T.

SOLVE (a) Using the values given, the cyclotron frequency is

$$f_c = \frac{|q|B}{2\pi m} = \frac{(3.2 \times 10^{-19}\ \text{C})(1.2\ \text{T})}{2\pi(6.64 \times 10^{-27}\ \text{kg})} = 9.20\ \text{MHz}$$

(b) Speed is circumference divided by period, or $v = 2\pi R/T$, or, since $T = 1/f_c$,

$$v = 2\pi R f_c = 2\pi(0.270\ \text{m})(9.20 \times 10^6\ \text{Hz})$$
$$= 1.56 \times 10^7\ \text{m/s}$$

Then kinetic energy is

$$K = \frac{1}{2}mv^2 = \frac{1}{2}(6.64 \times 10^{-27}\ \text{kg})(1.56 \times 10^7\ \text{m/s})^2$$
$$= 8.08 \times 10^{-13}\ \text{J}$$

Particle energies are often given in electron-volts; here $K = 5.05$ MeV.

REFLECT The frequency in MHz is reasonable for an electronic oscillator. (More on this in Chapter 20.) The final speed is about 5% of the speed of light—low enough that we can get away with disregarding relativity.

MAKING THE CONNECTION How would your answers change if this cyclotron were used to accelerate protons?

ANSWER Protons have half the charge but one-fourth the mass of alpha particles. Therefore, the cyclotron frequency and speed would both double.

Relativity applies when particle speeds are appreciable compared with the speed of light, so neither our Newtonian analysis nor the cyclotron itself will work for very high particle energies. The alternative is the **synchrotron**, in which charged particles move in a circular ring of fixed radius. As particle energy increases, the magnetic field is adjusted to keep the radius constant. The world's largest particle accelerator, the Large Hadron Collider at the CERN facility on the Swiss/French border, is a synchrotron that accelerates protons to 7 TeV.

The cyclotron frequency isn't just for cyclotrons. One everyday application is your microwave oven. Here, electrons undergo circular motion in the magnetic field of a so-called *magnetron*. Their motion generates the microwaves that jostle water molecules to cook your food. The field strength is chosen to give a frequency of 2.45 GHz, which is particularly effective in exciting water molecules. As with the cyclotron, a uniform magnetic field works because the required frequency is independent of the electrons' energy.

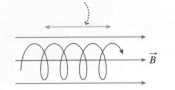

Motion parallel to the field is not affected by the magnetic force.

(a) Charged particle following helical path in uniform magnetic field

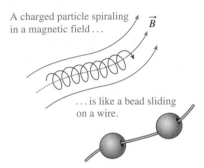

A charged particle spiraling in a magnetic field . . .

. . . is like a bead sliding on a wire.

(b) Moving charged particles "frozen" to a magnetic field

FIGURE 18.16 A charged particle with velocity components perpendicular and parallel to the magnetic field spirals through the field.

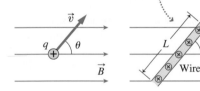

You can think of a current as a stream of positive charges having the same speed and direction.

Magnetic force on a moving charge:

$F = qvB \sin\theta$

(a)

Magnetic force on a current-carrying wire:

$F = ILB \sin\theta$

(b)

FIGURE 18.17 (a) A single charge moving through a magnetic field. (b) A current-carrying wire represented as a row of moving charges.

Particle Trajectories in Three Dimensions

So far we've considered particles moving perpendicular to a uniform magnetic field. If that isn't the case, we can resolve the motion into components perpendicular and parallel to the magnetic field. The perpendicular component gives circular motion, just as before. And, because there's no magnetic force on a particle moving parallel to a magnetic field, the parallel component is unaffected by the field. The result is spiral motion (called a **helix**) about the magnetic field (Figure 18.16a). This remains approximately true even if the magnetic field isn't uniform, with the result that charged particles are effectively "frozen" to the magnetic field, able to move easily along the field but held in tight circles perpendicular to it (Figure 18.16b). This spiraling, "frozen-to-the-field" motion is important in technological devices and in astrophysics. For example, the auroras discussed in Section 18.2 occur at high latitudes because charged particles spiral along Earth's magnetic field lines, entering the atmosphere where the field lines do—mostly near the poles.

GOT IT? Section 18.3 In a synchrotron electrons move counterclockwise as viewed from above. What direction of magnetic field is required to keep the electrons in the ring? (a) Vertically up; (b) vertically down.

18.4 Magnetic Forces on Conducting Wires

You've seen how electric charges experience magnetic forces as they move through magnetic fields. Electric currents consist of many moving charges, so there can be substantial magnetic forces on current-carrying wires in magnetic fields.

The single positive charge q in Figure 18.17a experiences a force $F = qvB \sin\theta$, with direction given by the right-hand rule—here into the page. Figure 18.17b shows a conducting wire in the same magnetic field, with a steady current I resulting from a stream of charges. Suppose a length L of the wire carries a total moving charge Q. This charge covers the distance L in time t, with $v = L/t$ being the speed of the moving charges. Thus, the magnetic force is

$$F_{\text{on }Q} = QvB \sin\theta = \frac{QLB \sin\theta}{t}$$

The current I is the charge flow per time, Q/t. Then the magnetic force becomes

$$F = ILB \sin\theta \qquad \text{(Magnetic force on a conducting wire; SI unit: N)} \qquad (18.6)$$

Equation 18.6 gives the magnetic force on a length L of straight wire carrying current I through a magnetic field B. The angle θ is the angle the wire makes with the magnetic field, and the direction of the force follows from the right-hand rule. As for single charges, the force is perpendicular to both the wire and the magnetic field. In Figure 18.17b, the force is into the page.

✓**TIP**

The right-hand rule for current-carrying wires uses the flow of positive current, not electron flow.

CONCEPTUAL EXAMPLE 18.8   **Magnetic Force on a Straight Wire**

Each of the four diagrams (Figure 18.18) depicts a straight wire carrying current in the direction shown, within a magnetic field. In each case determine whether there's a magnetic force on the wire, and if so, give its direction.

SOLVE Equation 18.6 gives the magnetic force on a straight wire: $F = ILB \sin\theta$. The force is zero if $\theta = 0$ or $180°$. Otherwise, there's a nonzero force, with direction given by the right-hand rule.

(a) The current is in the same direction as the magnetic field. Therefore, $\theta = 0$, and the magnetic force on this wire is zero.

(b) The angle between wire and magnetic field is between $90°$ and $180°$, so there is a magnetic force. The force is perpendicular to both wire and field, which makes it perpendicular to the page. By the right-hand rule, the force is out of the page.

(c) The angle between the wire and the magnetic field is $90°$. Applying the right-hand rule gives a force direction straight up, perpendicular to the wire.

(d) Again, the angle between the wire and the magnetic field is $90°$, but here the right-hand rule gives a force that's down and to the right, again perpendicular to the wire.

REFLECT It's important to check the angle between the wire and magnetic field first. If the angle is 0 or $180°$, then the magnetic force is zero. Otherwise, you can apply the right-hand rule to find the direction of the force.

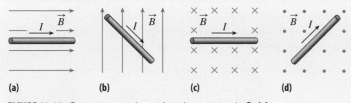

FIGURE 18.18 Current-carrying wires in magnetic fields.

EXAMPLE 18.9   **Defying Gravity?**

Can the magnetic force on a current-carrying wire be large enough to suspend the wire against gravity? To find out, (a) sketch a horizontal wire that's carrying current I from right to left. In what direction should a magnetic field point in order to produce the maximum upward force on this wire? (b) Consider a 14-gauge copper wire, with cross-sectional area 2.0 mm² and density 8920 kg/m³. A 3.5-cm length of this wire runs horizontally, perpendicular to a 2.4-T magnetic field. How large a current must the wire carry in order to be suspended against gravity?

ORGANIZE AND PLAN The maximum magnetic force occurs with the wire perpendicular to the magnetic field. The field direction for a vertical force follows from the right-hand rule, as shown in our sketch (Figure 18.19).

The mass of the wire is given by mass = density × volume, or $m = \rho V = \rho AL$. To support the wire, the upward magnetic force ILB must balance the downward gravitational force mg, or $ILB = mg$.

Known: $L = 3.5$ cm, $\rho = 8920$ kg/m³, $A = 2.0$ mm² $= 2.0 \times 10^{-6}$ m², $B = 2.4$ T.

SOLVE (a) With current from right to left, the magnetic field has to be perpendicular—either into or out of the page—to produce the maximum vertical force. By the right-hand rule, the correct choice is out of the page.

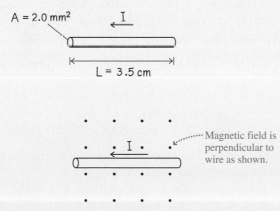

FIGURE 18.19 Determining the magnetic field needed to suspend the wire.

cont'd.

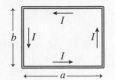

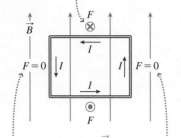

(a) Current-carrying wire loop

Sides at right angles to $\vec{B}$ experience forces of magnitude $F = IaB$ into and out of page . . .

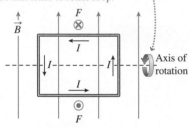

. . . while sides parallel to $\vec{B}$ experience zero force.

(b) Forces on loop when placed in a magnetic field oriented parallel to two of its sides

Net force on loop is zero, but forces exert a net torque that tends to rotate loop.

(c) Net torque on loop

FIGURE 18.20 A current-carrying wire in a magnetic field experiences torque.

After rotating 90°, the loop is perpendicular to the magnetic field.

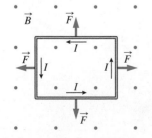

Now the forces on the four sides point as shown, and the net torque on the loop is zero.

FIGURE 18.21 After a 90° rotation, the loop is perpendicular to the magnetic field. The forces on the four sides are as shown, and the net force and net torque are both zero.

(b) The wire's mass is $m = \rho V = \rho AL = (8920 \text{ kg/m}^3)(2.0 \times 10^{-6} \text{ m}^2)(0.035 \text{ m}) = 6.24 \times 10^{-4}$ kg. Equating the magnetic and gravitational forces, $ILB = mg$, and solving for current I gives

$$I = \frac{mg}{LB} = \frac{(6.24 \times 10^{-4} \text{ kg})(9.80 \text{ m/s}^2)}{(0.035 \text{ m})(2.4 \text{ T})} = 0.073 \text{ A} = 73 \text{ mA}$$

REFLECT From Chapter 17, you know that 73 mA is a modest current for a copper wire of this size, so the suspension is easily accomplished. This is an example of how the magnetic force can equal or exceed the force of gravity—another sign that gravity is generally weak compared with electric and magnetic forces.

MAKING THE CONNECTION How would the required current change if other conditions were kept the same but the wire diameter doubled?

ANSWER Doubling the diameter increases the wire's cross-sectional area (and mass) by a factor of 4, so four times as much current is required.

Torque on a Current Loop

We'll now consider magnetic forces on a loop of current, a configuration found in many applications. Figure 18.20a shows a rectangular wire loop carrying current I. What happens to this loop in a uniform magnetic field (Figure 18.20b)? Two of the sides are parallel to the field, so there's no force on them. Figure 18.20b and Equation 18.6 show that each of the other sides experiences a force of magnitude $F = IaB$. Those forces are in opposite directions, so the net force on the loop is zero. However, the two forces don't act along the same line, so there's a net torque, shown in Figure 18.20c. (Recall the discussion of torque in Chapter 8.) The top and bottom of the loop are each a distance $b/2$ from the rotation axis, so the net torque on the loop is

$$\tau = \frac{b}{2}IaB + \frac{b}{2}IaB = IabB = IAB$$

where $A = ab$ is the loop area.

The torque tends to rotate the loop until it's oriented as shown in Figure 18.21. Apply the right-hand rule and you can see that the force on each segment of wire is now pointing toward the center of the loop. In this orientation the net force and net torque are both zero.

To see how the torque varies as the loop rotates, consider the three-dimensional view in Figure 18.22. There's an angle θ between the magnetic field and the perpendicular to the loop plane. Based on the rules for calculating torque (Chapter 8), you can show that the net torque on the loop is now $\tau = IAB \sin \theta$. Although we've considered a rectangular loop, this equation works for any planar loop with area A. Making the loop a compact coil with multiple turns increases the torque; for an N-turn coil, we have

$$\tau = NIAB \sin \theta \quad \text{(Torque on a current loop; SI unit: N·m)} \quad (18.7)$$

Magnetic Moments

The magnetic moment $\vec{\mu}$ of a current loop is a vector defined as follows:

- The magnitude of $\vec{\mu}$ of an N-turn loop with area A carrying current I is $\mu = NIA$. Thus the SI unit for magnetic moment is A·m^2.
- The direction of $\vec{\mu}$ is perpendicular to the loop plane, as given by the following right-hand rule: Curl your fingers in the direction of current flow around the loop. Then your thumb points in the direction of the magnetic moment $\vec{\mu}$ (Figure 18.23).

In terms of the magnetic moment, the torque on a current loop (Equation 18.7) becomes $\tau = \mu B \sin \theta$, with θ the angle between the magnetic moment $\vec{\mu}$ and magnetic field $\vec{B}$.

Consider again the current loops in Figure 18.20b and Figure 18.21. In Figure 18.20b, the magnetic moment $\vec{\mu}$ points straight out of the page and makes a 90° angle with the magnetic field $\vec{B}$. Thus $\theta = 90°$ and $\sin \theta = 1$, its maximum value. In Figure 18.21, $\vec{\mu}$ and $\vec{B}$ are in the same direction, so $\tau = \mu B \sin \theta = \mu B \sin (0°) = 0$. These examples illustrate how you use the magnetic moment vector to analyze the torque on current loops, without having to consider individual forces on the loop.

Applications of Torque on Current Loops

The torque on a current loop is what makes **electric motors** run, as seen in Figure 18.24. Electric motors are everywhere, from washing machines to hybrid cars and subway trains to the hard drives in your computer and iPod. A motor's **armature** is a wire coil wound on an axle and mounted between the poles of a magnet. Passing current through the coil results in a torque that rotates the armature. Figure 18.24 shows another key element in the motor, the **commutator**. This rotating electrical contact reverses the direction of the current after every 180° turn of the armature, keeping it rotating continually in the same direction. Without the commutator, the coil would stop with its magnetic moment aligned with the field.

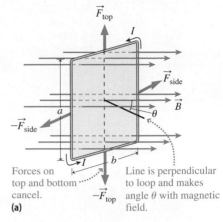

Forces on top and bottom cancel.

Line is perpendicular to loop and makes angle θ with magnetic field.

(a)

EXAMPLE 18.10   **Washing Machine Motor**

A washing machine motor has a 100-turn square coil 8.0 cm on a side, in a 0.65-T magnetic field. During the "spin" cycle, the coil carries 5.5 A. Find the maximum torque on the armature.

ORGANIZE AND PLAN By Equation 18.7, the torque on a current loop is $\tau = NIAB \sin \theta$. Here A is the 64-cm^2 area of the square coil (Figure 18.25). The maximum torque occurs when $\theta = 90°$, so $\sin \theta = 1$ and $\tau = NIAB$.

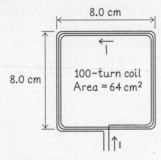

8.0 cm

8.0 cm

100-turn coil
Area = 64 cm^2

I

I

FIGURE 18.25 Armature coil in a washing machine motor.

Known: $A = 64$ cm$^2 = 0.0064$ m^2, $I = 5.5$ A, $N = 100$, $B = 0.65$ T.

SOLVE The maximum torque is

$$\tau = NIAB = (100)(5.5\text{ A})(0.0064\text{ m}^2)(0.65\text{ T}) = 2.3\text{ N}\cdot\text{m}$$

REFLECT Check that the units work out. With $1\text{ T} = 1\text{ N}/(\text{A}\cdot\text{m})$, the units for torque are $\text{A}\cdot\text{m}^2\cdot\text{T} = \text{A}\cdot\text{m}^2\cdot\text{N}/(\text{A}\cdot\text{m}) = \text{N}\cdot\text{m}$, the SI units for torque.

MAKING THE CONNECTION   To get more torque, why can't you just make the armature coil as large as you want?

ANSWER The entire rotating coil has to fit between the magnet poles—see Figure 18.24. If you put the poles too far apart, that reduces the magnetic field. In designing a motor, you need to maximize both the current and magnetic field strength.

Electric motors are a large-scale application of magnetic torque on a current loop. There are much smaller current loops—even as small as an atom! Figure 18.26 (next page) shows how an electron orbiting a nucleus constitutes a current loop, resulting in a magnetic moment. Individual electrons also have "spin" that, with the electron's charge, results in an intrinsic magnetic moment. Atoms in magnetic fields experience torques because of these orbital and spin magnetic moments. Most of the magnetic behavior you see in materials is due to electron spin.

Forces on sides also cancel to give zero net force, but produce a net torque $\tau = IaB \sin \theta$.

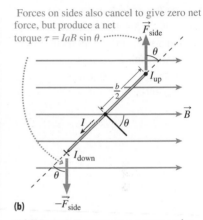

(b)

FIGURE 18.22 Torque on a rectangular current loop.

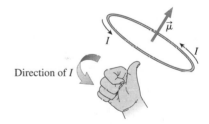

Direction of I

FIGURE 18.23 Magnetic moment of a current loop.

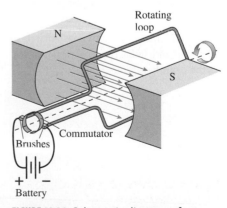

Rotating loop

N

S

Commutator

Brushes

Battery

FIGURE 18.24 Schematic diagram of an electric motor.

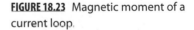

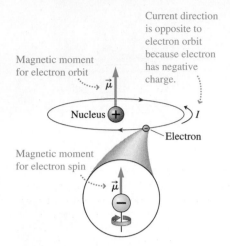

FIGURE 18.26 The atom has a magnetic moment due to the electron's orbital motion and a separate magnetic moment due to electron spin.

Atomic nuclei, too, have magnetic moments and experience magnetic torques. An important application is **nuclear magnetic resonance (NMR)**, used in studying the structures of materials, and in the medical technique called **magnetic resonance imaging (MRI)**. In NMR/MRI, magnetic torques cause nuclei to precess at a frequency that depends on both the applied magnetic field and the fields associated with orbiting electrons. The field is varied slowly, causing nuclei in different environments to absorb energy at different field strengths—a condition of "resonance." These resonance frequencies depend strongly on the material and its environment, which makes possible the amazingly clear MRI images you see. For example, the resonance is quite different in bone versus muscle tissue, and in normal cells versus cancer cells.

GOT IT? Section 18.4   What magnetic field direction will make the coil rotate as shown? (a) Directly out of page; (b) directly into page; (c) right; (d) left; (e) up; (f) down.

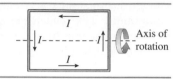

18.5 The Origin of Magnetism

The relationship between electricity and magnetism is at the foundations of physics, hence the term **electromagnetism**. So far, you've seen that magnetic fields affect moving electric charges. In fact, the relation between moving electric charge and magnetic fields goes both ways: moving electric charge is also the source of magnetic field. It was the Danish physicist Hans Christian Ørsted (1777–1851) who first recognized this fact, when he noticed that an electric current caused a nearby compass needle to deflect. Although there's a general rule for computing the magnetic field arising from moving charge, we'll consider just the special cases where the moving charges are associated with currents in straight wires and circular loops.

A current-carrying coil designed specifically to produce a magnetic field is an **electromagnet**. Today, most magnets used in science, medicine, and industry are electromagnets, because they produce magnetic fields that are larger and more easily controlled than fields of permanent magnets.

Magnetic Field of a Straight Wire

Figure 18.27 shows a long, straight wire carrying a steady current I. Outside the wire, at a distance d from its center, the magnetic field has magnitude

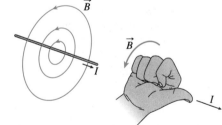

FIGURE 18.27 Magnetic field lines encircle a straight wire (left). The direction is given by the right-hand rule (right).

$$B = \frac{\mu_0 I}{2\pi d} \qquad \text{(Magnetic field of a long straight wire; SI unit: T)} \qquad (18.8)$$

where μ_0 is a constant called the **permeability of free space**, with value $4\pi \times 10^{-7}\,\text{T}\cdot\text{m/A}$. With current in amperes and distance in meters, the field is in tesla. The magnetic field lines are circular, as shown in Figure 18.27, with the direction given by another right-hand rule:

Point your thumb along the direction of the current, and curl your fingers. Your fingers show the direction of the magnetic field.

The magnetic field of a single wire is fairly small, even with substantial current. Just 1 cm from the center of a wire carrying 10 A, the field is

$$B = \frac{\mu_0 I}{2\pi d} = \frac{(4\pi \times 10^{-7}\,\text{T}\cdot\text{m/A})(10\,\text{A})}{2\pi(0.01\,\text{m})} = 2 \times 10^{-4}\,\text{T}$$

This is a fairly small magnetic field, but it's a bit larger than Earth's field, so it would disrupt a compass reading.

CONCEPTUAL EXAMPLE 18.11   **Magnetic Forces on a Pair of Wires**

Two long parallel wires carry currents I_1 and I_2 in the same direction. Explain why the wires exert forces on one another and determine the directions of those forces.

SOLVE Consider the magnetic field produced by the current I_1. Using the right-hand rule of Figure 18.27, you can see that the magnetic field of this wire points up ($\vec{B}_1$) in the vicinity of the second wire (Figure 18.28). From the right-hand rule of Section 18.4, you know that this magnetic field acting on the current I_2 produces a force $\vec{F}_{12}$ toward wire 1. The force on wire 1 follows by similar reasoning, using the magnetic field of I_2: it's $\vec{F}_{21}$ as shown. So the two wires attract each other—and Newton's third law is satisfied.

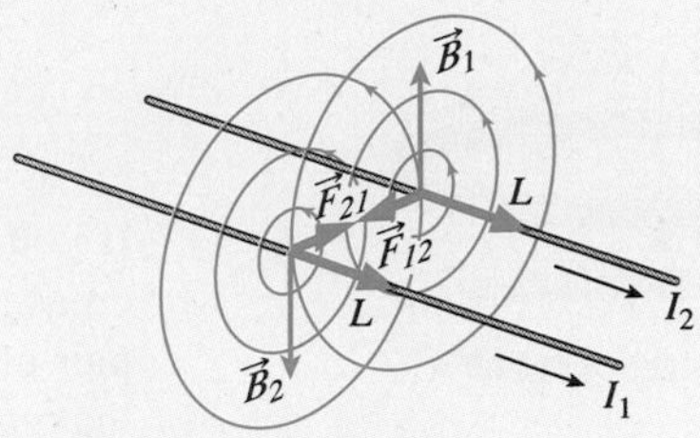

REFLECT If the wires carried opposite currents, they would repel. You can convince yourself of this by retracing the solution above.

FIGURE 18.28 Parallel wires carrying current in the same direction attract one another.

EXAMPLE 18.12   **Attractive Forces on Parallel Wires**

The parallel wires of the preceding example are separated by a center-to-center distance of 1.0 cm, and each carries a 20-A current. Find the force per unit length on each wire.

ORGANIZE AND PLAN Consider the force that wire 1 in Figure 18.28 exerts on wire 2. Wire 1 carries current I_1, so by Equation 18.8 the magnetic field a distance d from its center is $B = \mu_0 I_1 / 2\pi d$. Equation 18.5 shows that the force exerted on a length L of the second wire, carrying current I_2, is $F = I_2 LB \sin \theta = I_2 LB$, where we've used the fact that the angle between the magnetic field and the wire is 90°. Putting together these results,

$$F = I_2 LB = \frac{\mu_0 I_1 I_2 L}{2\pi d}$$

The force per unit length is then just F/L.

Known: $I_1 = I_2 = 20$ A, $d = 1.0$ cm.

SOLVE Solving for force per unit length F/L and inserting numerical values,

$$\frac{F}{L} = \frac{\mu_0 I_1 I_2}{2\pi d} = \frac{(4\pi \times 10^{-7}\,\text{T} \cdot \text{m/A})(20\,\text{A})^2}{2\pi(0.010\,\text{m})} = 8.0 \times 10^{-3}\,\text{N/m}$$

Thus the force on a 1.0-m length of wire is 8.0 mN, directed toward the other wire.

REFLECT This is indeed a small force, as Making the Connection below shows.

MAKING THE CONNECTION A 12-gauge copper wire (diameter 2.05 mm) can safely handle the 20-A current of this example. Compare the magnetic force on each meter of wire with the wire's weight. The density of copper is 8920 kg/m³.

ANSWER Mass = density × volume, and the volume of a cylindrical wire of radius r and length L is $\pi r^2 L$. This gives a weight $mg =$ 29 mN, about three and one half times the magnetic force.

The force between parallel wires serves as the basis for the definition of the SI unit ampere (A). If two long parallel wires 1 m apart carrying equal currents experience a force of 2×10^{-7} N, then the current is defined to be 1 ampere. This definition of the ampere, along with the second (s), then defines the coulomb: 1 C = 1 A · s.

Magnetic Field of a Circular Coil

Figure 18.29 shows a circular wire loop of radius r. If the loop carries a current I, then the magnetic field at the loop's center has magnitude $B = \mu_0 I / 2r$. For an N-turn coil, the field is N times larger:

$$B = \frac{\mu_0 N I}{2r} \quad \text{(Magnetic field at the center of a circular coil; SI unit: T)} \quad (18.9)$$

The direction of the field at the coil's center is perpendicular to the plane of the coil and is given by another right-hand rule, shown in Figure 18.29.

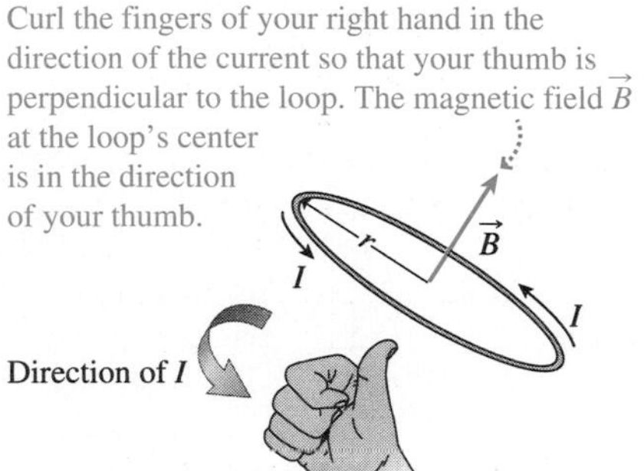

Curl the fingers of your right hand in the direction of the current so that your thumb is perpendicular to the loop. The magnetic field $\vec{B}$ at the loop's center is in the direction of your thumb.

Direction of I

FIGURE 18.29 Finding the direction of the magnetic field at the center of the loop.

Between the coils, the magnetic fields combine to produce a large field.

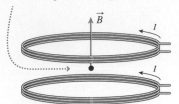

FIGURE 18.30 Parallel coils produce a strong magnetic field in the region between them.

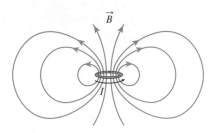

FIGURE 18.31 A more complete view of the magnetic field around a circular coil.

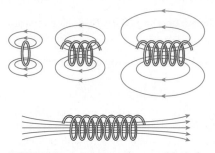

FIGURE 18.32 As more coils are added, the interior magnetic field becomes more uniform and the exterior field weakens. The result is a nearly uniform magnetic field near the solenoid's center.

Each box can hold a pair of electrons, one with spin up and one with spin down.

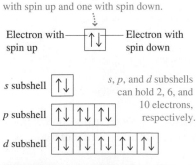

FIGURE 18.33 Illustration of how the s, p, and d electron subshells fill.

A coil with many turns can produce a large magnetic field. For example, the field at the center of a 1.0-cm-radius single-turn loop carrying 10 A is

$$B = \frac{\mu_0 I}{2r} = \frac{(4\pi \times 10^{-7}\,\text{T}\cdot\text{m/A})(10\,\text{A})}{2(0.01\,\text{m})} = 6 \times 10^{-4}\,\text{T}$$

If you wind 1000 turns into a coil of the same compact size, then the magnetic field increases by a factor of 1000 to a substantial 0.6 T. In a common electromagnet design, a pair of coils is mounted along the same axis, with a gap between them (Figure 18.30). In that gap their fields superpose to produce an even larger field.

So far we've only considered the field at the center of a coil. Figure 18.31 shows the field surrounding the coil. You might recognize this as a dipole field, similar to the magnetic field of a bar magnet (Figure 18.1). That's not a coincidence: The macroscopic magnetic field of the bar magnet comes from the combined dipole fields of individual atoms, which can be thought of as small current loops (as you saw in Figure 18.26; more in Section 18.6). Absent magnetic monopoles, *every* magnetic field has its source in some kind of electric current, even if at the atomic level. Ultimately those currents always form closed loops, whether in wires or due to spinning electrons, and so the most fundamental magnetic field is the dipole field. Ampère first made this macro/micro connection in 1822, stating "... (a) magnet should be considered an assemblage of electric currents, which flow in planes perpendicular to its axis ...." It's remarkable that Ampère understood magnetism on this level nearly a century before the nuclear atom was established, and more than a century before the discovery of electron spin.

Magnetic Field Inside a Solenoid

Extending our compact circular coil into a long, tightly wound cylindrical coil makes a *solenoid* (Figure 18.32). The magnetic field deep in a solenoid's interior is essentially uniform, points along the axis, and has magnitude

$$B = \mu_0 n I \qquad \text{(Magnetic field inside a solenoid; SI unit: T)} \qquad (18.10)$$

where n is the number of turns per unit length. Curiously, the magnetic field doesn't depend on the solenoid's radius, so you can make the inside radius large—as you do with the solenoid that encircles your body in an MRI scanner.

The nonuniform field near the ends of a solenoid results in magnetic materials like iron being pulled into the solenoid. Therefore, solenoids are often used in applications requiring straight-line motion. The valves that admit water to your washing machine are solenoid-actuated.

GOT IT? **Section 18.5** A single circular loop of wire has radius 5.0 cm. You wish to produce a magnetic field at the coil's center equal in magnitude to Earth's field, 5×10^{-5} T. The required current is (a) 1.3 A; (b) 2.0 A; (c) 4.0 A; (d) 12.6 A.

18.6 Magnetic Materials

The atomic magnetic moments depicted in Figure 18.26 are the basis of magnetism in materials like iron and the magnets you're familiar with. Here we'll discuss the several types of magnetism observed in materials.

We've noted that electron spin is primarily responsible for the magnetic behavior of materials. Here's a preview of Chapter 24's discussion of atoms; you may also have seen these ideas in chemistry. Each atom has a positive nucleus, surrounded by electrons in different **shells**, numbered $n = 1, 2, 3, \ldots$. The shells are broken into **subshells**, with one subshell for the $n = 1$ level, two for $n = 2$, and so on. The subshells are named $1s$; $2s$ and $2p$; $3s$, $3p$, and $3d$; $4s$, $4p$, $4d$, and $4f$; and so on.

An s subshell holds at most two electrons, a p subshell six, a d subshell 10, and so on. Each electron takes one of two possible spin orientations, called "up" and "down." Figure 18.33 illustrates a common scheme for visualizing the electron spins in each subshell. Figure 18.34 shows two examples, helium and iron. With two electrons, helium has a filled

1s subshell. The spins have to be opposite, so the electrons' magnetic moments cancel. It's like putting two small bar magnets together pointing in opposite directions. Iron, on the other hand, has 26 electrons, with filled subshells up to the 3d level, where six electrons fill the subshell as shown. Note that the majority of the 3d electrons have the same spin. The result is that iron has a net magnetic moment, which makes it a good candidate for magnetic behavior on a macroscopic scale.

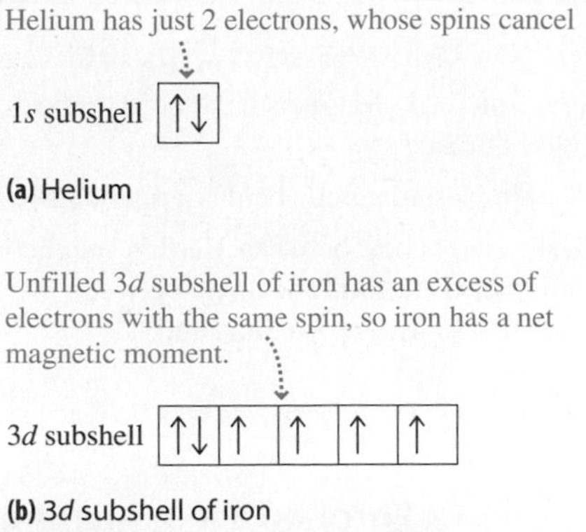

Ferromagnetism

In iron, atomic magnetic moments tend to align with thousands of their neighbors to form **magnetic domains**. Domains vary in size but are typically about 0.1 mm across—large enough to see with a microscope. Note that the magnetic moments of the neighboring domains don't generally align in the same direction. In iron, however, an applied magnetic field can flip domains into a common alignment so the iron acquires a net magnetic moment. The domains remain aligned even when the applied field is removed—and there's your permanent magnet!

Iron is an example of a **ferromagnetic** material—one that can have a substantial net magnetic moment. Ferromagnetism is rare; there are only five ferromagnetic elements (Fe, Ni, Co, Gd, and Dy). More compounds are ferromagnetic, including some that don't contain ferromagnetic elements. In the most strongly ferromagnetic compounds, the magnetic field at the material surface can exceed 1 tesla.

Ferromagnetism is rare because it's a highly ordered state, with a majority of magnetic moments pointing in one direction. Random thermal motions threaten this highly ordered or low-entropy state. (Recall from Chapter 14 that entropy is a measure of disorder.) Not surprisingly, therefore, ferromagnetism is temperature dependent. Ferromagnetism ceases abruptly at the so-called **Curie temperature**—an example of a phase transition analogous to the melting of a solid. Table 18.1 lists some Curie temperatures.

Paramagnetism and Diamagnetism

Most materials with unfilled d and f subshells are **paramagnetic**. Paramagnetic materials have a net magnetic moment only in the presence of an applied magnetic field. That net moment aligns with the field, and its magnitude increases as the applied field strengthens. Eventually the material reaches saturation, where the magnetic dipoles are as aligned as possible. Like ferromagnetism, paramagnetism is temperature dependent. Increasing the temperature reduces the net magnetic moment, as the dipole alignment becomes more random. Paramagnetism is much weaker and less obvious than ferromagnetism. Among the most strongly paramagnetic materials is liquid oxygen, which is sufficiently paramagnetic that it can be suspended between the poles of a magnet.

Materials such as helium with filled electronic subshells have little response to an applied magnetic field. They're called **diamagnetic**. Diamagnetic materials actually develop a small magnetic dipole moment in the opposite direction to an applied magnetic field. Diamagnetism is best explained using quantum mechanics, and we won't pursue it further here.

Chapter 18 in Context

This chapter introduced *magnetism*, starting with familiar bar magnets and compasses. You learned about *magnetic forces* on moving electric charges and saw numerous applications. By considering the moving charges comprising electric currents, you learned about the magnetic forces on current-carrying conductors. You saw how these forces lead to *torques on current loops*, with applications from *electric motors* to *MRI imaging*. You then saw how moving electric charge—specifically in the form of electric currents—is the ultimate source of magnetism. Finally, you learned about the magnetic properties of materials, including *ferromagnetism*, *paramagnetism*, and *diamagnetism*.

Looking Ahead Chapter 19 will explore yet another relationship between electricity and magnetism, whereby changing magnetic fields become a source of electric field. This so-called electromagnetic induction is a fundamental process in nature and is at the heart of the technology we use to generate electricity.

Helium has just 2 electrons, whose spins cancel

1s subshell $\boxed{\uparrow\downarrow}$

(a) Helium

Unfilled 3d subshell of iron has an excess of electrons with the same spin, so iron has a net magnetic moment.

3d subshell $\boxed{\uparrow\downarrow}\,\boxed{\uparrow}\;\boxed{\uparrow}\;\boxed{\uparrow}\;\boxed{\uparrow}$

(b) 3d subshell of iron

FIGURE 18.34 Examples of how the positions of electrons in subshells affect magnetic behavior.

TABLE 18.1 Curie Temperatures for Selected Ferromagnetic Materials

Material	Curie temperature (K)
Fe	1043
Co	1388
Ni	627
Gd	293
Dy	85
$CrBr_3$	37
Au_2MnAl	200
Cu_2MnAl	630
Cu_2MnIn	500
EuO	77
EuS	16.5
MnAs	318
MnBi	670
$GdCl_3$	2.2

Magnets, Poles, and Dipoles

(Section 18.1) Magnets have north and south poles; like poles repel, unlike poles attract.

The simplest **magnetic field** is a **dipole field**; there are no **magnetic monopoles**.

The discrepancy between Earth's magnetic north and geographic north is called **declination**. The angle between the magnetic field vector and Earth's surface is called the **inclination**.

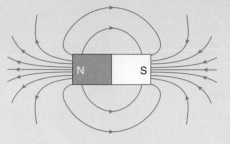

Magnetic Force on a Moving Charge

(Section 18.2) A charged particle in a magnetic field experiences a force perpendicular to the magnetic field and to the particle's velocity.

A charged particle moving perpendicular to a uniform magnetic field follows a circular trajectory.

The direction of the charged particle's motion around the circle depends on the orientation of the magnetic field and the sign of the charge.

Magnetic force: $F = |q|vB \sin \theta$

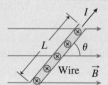

Plane containing $\vec{v}$ and $\vec{B}$

Radius of circle of charged particle in circular motion:

$$R = \frac{mv}{|q|B}$$

Applications of Magnetic Forces

(Section 18.3) In a **velocity selector**, for any choice of E and B, there is only one velocity of charged particle that passes through without deflection.

The **mass spectrometer** allows for measurement of a particle's **charge-to-mass ratio** and is good for separating isotopes.

The **Hall effect** can be used to measure magnetic fields or charge carrier density.

Cyclotrons and **synchrotrons** are used to accelerate subatomic particles to high energies.

Charge-to-mass ratio in a mass spectrometer: $\dfrac{|q|}{m} = \dfrac{E}{B_1 B_2 R}$

Hall potential difference: $\Delta V_H = \dfrac{IB}{neb}$

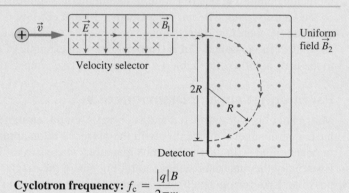

Velocity selector

Uniform field $\vec{B_2}$

Detector

Cyclotron frequency: $f_c = \dfrac{|q|B}{2\pi m}$

Magnetic Forces on Conducting Wires

(Section 18.4) A current-carrying wire in a magnetic field experiences a force perpendicular to the direction of current and to the magnetic field.

A current loop in a magnetic field experiences torque and has a **magnetic moment**, with a direction given by the right-hand rule.

One application using atomic magnetic moments is **nuclear magnetic resonance (NMR)**, also used in the medical diagnostic technique called **magnetic resonance imaging (MRI)**.

Magnetic force on a conducting wire: $F = ILB \sin \theta$

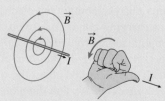

Wire

Torque on a current loop: $\tau = NIAB \sin \theta = \mu B \sin \theta$

The Origin of Magnetism

(Section 18.5) Moving charge is the source of a magnetic field, so an electric current produces a magnetic field. The field of a straight wire encircles the wire. A current loop produces a dipole field.

A long cylindrical coil is a **solenoid**, which produces a nearly uniform magnetic field in its interior.

Magnetic field of a long straight wire: $B = \dfrac{\mu_0 I}{2\pi d}$

Magnetic field at the center of a circular coil: $B = \dfrac{\mu_0 NI}{2r}$

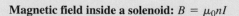

Magnetic field inside a solenoid: $B = \mu_0 nI$

Magnetic Materials

(Section 18.6) Atoms have magnetic properties due to their electrons' orbital motion and spins.

In **ferromagnetic** materials like iron, many atomic moments align to form **magnetic domains**; such materials can have strong magnetic moments.

Paramagnetic materials have a net magnetic moment only in the presence of an applied magnetic field.

Diamagnetic materials respond only weakly to an applied magnetic field, with their magnetic moment opposite the field direction.

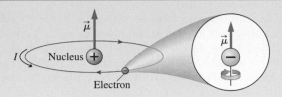

NOTE: Problem difficulty is labeled as ▪straightforward to ▪▪▪challenging. Problems labeled BIO are of biological or medical interest.

Conceptual Questions

1. Explain how and why you expect the magnetic declination to change as you travel eastward from California to New York.
2. Explain how and why you expect magnetic inclination to change as you travel southward from Canada to Mexico.
3. Suppose that a compass needle in the magnetic field of a strong electromagnet points northeast. What's the field direction at that point?
4. The electric field is defined to be in the direction of the electric force on a positively charged particle. Why isn't the magnetic field defined to be in the direction of the magnetic force on a positively charged particle?
5. A negatively charged particle is moving horizontally in a region with a magnetic field directed vertically upward. Describe the particle's motion.
6. Discuss the trajectory of a charged particle in a magnetic field when the particle's velocity is *not* perpendicular to the field. Explain why the motion is a *helix*, such that the particle's velocity component in the direction of the magnetic field is constant, and the motion in the perpendicular direction is circular.
7. A charged particle is injected into a uniform magnetic field. If the angle between the particle's velocity and the magnetic field is 45°, describe the particle's trajectory.
8. Does the magnetic force ever do work on a moving charged particle?
9. If magnetic fields do no work, is it also true that they don't accelerate charged particles?
10. An uncharged particle will always pass through a velocity selector with no deflection. Describe how you could use the velocity selector to determine whether you have a beam of uncharged particles.
11. Why are synchrotrons rather than cyclotrons used to accelerate charged particles to extremely high energies?
12. Explain why all charged particles with a certain speed pass through a velocity selector undeflected, regardless of their charge and mass.
13. You're looking down on a positively charged particle moving in a clockwise circle. In what direction is the resulting magnetic moment? What if it's a negatively charged particle moving in the same direction?

14. Explain why a charged particle moving initially perpendicular to a uniform electric field follows a parabolic trajectory, while a charged particle moving perpendicular to a magnetic field follows a circular trajectory.
15. A wire carries electric current horizontally from west to east. In what direction should a magnetic field point so that the force on the wire is vertically downward?
16. You're at a location in which the magnetic declination is 0° and the inclination is 60°. In what direction is the force on a horizontal wire carrying current (a) from south to north and (b) from west to east?
17. A current loop in the x-y plane carries a clockwise current. For each of the following magnetic fields, tell whether the loop will rotate, and if so, in what direction. (a) Magnetic field in $+x$-direction; (b) magnetic field in $+y$-direction; (c) magnetic field in $+z$-direction.
18. Describe the force of interaction between two nearby long straight wires that are perpendicular and carry the same current.
19. What makes an element a good candidate for each of the following: ferromagnetism, paramagnetism, and diamagnetism?
20. Considering the electronic configuration of the five ferromagnetic elements, justify why they are ferromagnetic. Why are some other elements in the same columns of the periodic table not ferromagnetic? Do you expect these other elements to be paramagnetic or diamagnetic?
21. You have a fixed length L of wire and wish to make a single loop that will experience the maximum torque in a magnetic field. Is it better to make the coil circular or square, or doesn't it matter?

Multiple-Choice Problems

22. A 0.75-T magnetic field points west. Find the magnitude and direction of the force on a $+2.5$-μC charge moving southward at 5.5 m/s. (a) 1.0×10^{-5} N, straight up; (b) 1.0×10^{-5} N, straight down; (c) 1.0×10^{-2} N, straight up; (d) 1.0×10^{-2} N, straight down.
23. A negative particle traveling northward is deflected toward the east by a magnetic force. The direction of the magnetic field is (a) straight up; (b) straight down; (c) south; (d) west.

24. A charged particle is moving eastward. This particle experiences zero force in a magnetic field directed (a) vertically up; (b) vertically down; (c) north; (d) west.

25. A -50-μC charge moves vertically upward at 75 m/s. Which magnetic field will produce a northward magnetic force of 15 mN? (a) 2 T, east; (b) 2 T, west; (c) 4 T, east; (d) 4 T, west.

26. A proton moves at 2.0 Mm/s in a bubble chamber, perpendicular to a magnetic field. The radius of its trajectory is 6.0 cm. The magnetic field strength is (a) 0.25 T; (b) 0.35 T; (c) 0.50 T; (d) 0.60 T.

27. A cyclotron with radius 0.35 m and a uniform 0.14-T magnetic field can accelerate alpha particles to a kinetic energy of (a) 750 eV; (b) 14 keV; (c) 42 keV; (d) 116 keV.

28. A 50-cm-long straight wire carrying 2.5 A makes a 30° angle with a 0.30-T magnetic field. The strength of the magnetic force on the wire is (a) zero; (b) 0.19 N; (c) 0.28 N; (d) 0.38 N.

29. A straight conducting wire carries 0.25 A in the $+y$-direction. Find the magnetic field that results in a force per unit length on the wire of 0.20 N/m in the $+z$-direction. (a) 1.25 T, $-x$-direction; (b) 1.25 T, $+x$-direction; (c) 0.80 T, $-x$-direction; (d) 0.80 T, $+x$-direction.

30. A circular wire loop with diameter 24 cm carries 2.5 A. Find the torque on the loop in a 1.5-T magnetic field directed perpendicular to the loop. (a) 0; (b) 0.08 N · m; (c) 0.12 N · m; (d) 0.17 N · m.

31. A 20-cm-diameter circular wire loop carries 2.3 A. Find the torque on the loop in a 1.4-T magnetic field directed in the plane of the loop. (a) 0; (b) 0.08 N · m; (c) 0.12 N · m; (d) 0.17 N · m.

32. A long straight wire carries 4.5 A. At a distance of 1.0 cm from the wire's center, the magnetic field strength is (a) 4.5×10^{-5} T; (b) 7.7×10^{-5} T; (c) 9.0×10^{-5} T; (d) 2.8×10^{-4} T.

33. A circular coil has 50 turns of wire and carries 2.1 A. If the coil's radius is 3.5 cm, the magnetic field at the coil's center is (a) 3.8×10^{-5} T; (b) 7.6×10^{-5} T; (c) 9.4×10^{-4} T; (d) 1.9×10^{-3} T.

Problems

Section 18.2 Magnetic Force on a Moving Charge

34. ■ A 0.45-T magnetic field points due east. (a) Find the magnitude and direction of the force on a $+0.10$-μC charge moving southward at 3.5 m/s. (b) Repeat for a -0.10-μC charge with the same velocity.

35. ■ A 0.950-T magnetic field points due north. (a) Find the magnitude and direction of the force on a $+1.40$-μC charge moving westward at 1.50 m/s. (b) Repeat for a -1.40-μC charge with the same velocity.

36. ■ A uniform magnetic field points vertically upward. In what direction should a positive charge move so that it experiences a magnetic force toward the (a) north and (b) east?

37. ■■ A 1.50-T magnetic field points northeast. Find the magnitude and direction of the force on a $+0.50$-C charge that is moving 10 m/s toward the (a) north; (b) east; (c) south; (d) west; (e) southwest.

38. ■■ A -200-μC charge moves vertically downward at 12.3 m/s. Find the magnetic field (magnitude and direction) required to produce a northward magnetic force of 1.40 mN.

39. ■■ Near Peoria, IL the magnetic declination is zero and the inclination is 70°. (a) Find (a) the magnitude and direction of the force on a proton moving horizontally northward at 2.0×10^5 m/s and (b) the proton's acceleration.

40. ■■ A proton in the upper atmosphere moves at 9.0×10^5 m/s, perpendicular to Earth's magnetic field, which at that point has magnitude 5.1×10^{-5} T. (a) Find the curvature radius of the proton's trajectory. (b) Draw the resulting curve with the correct spatial orientation.

Section 18.3 Applications of Magnetic Forces

41. ■■ An electron is moving in the $+y$-direction at 2.0×10^5 m/s through a uniform magnetic field of 3.40 T in the $-x$-direction. (a) What are the magnitude and direction of the force on the electron? (b) Find the electric field that would give zero net force on the electron.

42. ■■ A velocity selector has its electric field directed vertically downward, with magnitude 150 N/C. (a) Find the magnitude of the horizontal magnetic field required to select charged particles moving at 300 m/s. (b) This velocity selector is being used on protons. Compute explicitly the magnitudes of the electric and magnetic forces and show that they're equal.

43. ■■ A proton moving at 4.50 km/s in the $+y$-direction enters a region containing a uniform magnetic field of 450 mT in the $+z$-direction. Find the electric field that must be applied to give zero net force on the proton.

44. ■■ The positron is a particle with the same mass as an electron but charge $+e$. A positron in a bubble chamber moves perpendicular to a 4.0-T magnetic field. If the positron's kinetic energy is 25 eV, what's the radius of its trajectory in the bubble chamber? Sketch the trajectory, assuming the magnetic field is directed vertically upward.

45. ■ A muon is a subatomic particle with a charge $-e$. A muon moves at 1.5×10^6 m/s through a bubble chamber, its velocity perpendicular to a 1.2-T magnetic field. If the radius of the muon's path is 1.46 mm, what is its mass? Compare your answer with the masses of the electron and proton.

46. ■■ A mass spectrometer is used to separate the uranium isotopes ^{235}U and ^{238}U. (a) What is the ratio of the radii of their circular paths? (b) In this spectrometer, singly ionized uranium ions pass first through a 140-m/s velocity selector, and then into a 2.75-T magnetic field. Find the diameters of the arcs traced out by the two isotopes.

47. ■ A cyclotron of 15.0-cm diameter operates at a frequency of 400 kHz. (a) To what speed can this cyclotron accelerate alpha particles? (b) What's the cyclotron's magnetic field?

48. ■ A singly ionized oxygen atom moves at 9200 m/s in a circular path perpendicular to a 0.75-T magnetic field. Find the radius and period of its circular motion.

49. ■■ A proton takes 95.0 μs to complete one orbit around a 14.5-m-diameter cyclotron. What's the cyclotron's magnetic field?

50. ■■ In a mass spectrometer, singly ionized iron atoms are accelerated through a 120-V potential difference and then injected into a region with a perpendicular magnetic field of 0.42 T. Describe the ions' subsequent motion.

51. ■■ You're designing a bubble chamber with a 1.25-T magnetic field. How large should it be to observe the complete circular path of these particles, each having 1.0 keV kinetic energy: (a) electrons; (b) protons; (c) alpha particles?

52. ■■ You wish to make a mass spectrometer with a chamber 50 cm on a side. How large a magnetic field is needed to keep singly ionized helium atoms (^{4}He) with kinetic energy 250 eV within the chamber?

53. ■■ What's the magnetic field strength in a microwave oven's magnetron? *Hint:* See the discussion of microwave ovens in Section 18.3.

54. ■■ In 1897, the English physicist J.J. Thomson used crossed electric and magnetic fields to measure the electron's charge-to-mass

ratio. (a) Calculate this ratio, in units of C/kg. (b) Find the charge-to-mass ratio of singly ionized hydrogen, which has the largest such ratio of any ionized atom. (c) Comparing your answers to parts (a) and (b), discuss the implications for the mass of the electron.

55. ■■ The conduction electron density in germanium (a semiconductor) is 2.01×10^{24} m^{-3}. The Hall effect is observed in germanium using a 0.150-mm-thick bar in a magnetic field of 1.25 T. Find the current required in the bar in order to produce a Hall potential difference of 1.0 mV.

56. ■■ A Hall effect experiment uses a silver bar 3.50 μm thick. When the bar carries a current of 1.42 A, a perpendicular magnetic field of 0.155 T results in a Hall potential difference of 6.70 μV. (a) Use these data to determine the density of conduction electrons in silver. (b) How many conduction electrons are there per atom of silver? [*Note:* The density of silver is 10,490 kg/m^3.]

57. ■■ A Hall effect probe used to measure magnetic fields is made with a copper bar of thickness 2.00 μm. If the current in the bar is 500 mA, find the perpendicular magnetic field present when the Hall potential difference is (a) 1.0 V; (b) 1.0 mV.

Section 18.4 Magnetic Forces on Conductin g Wires

58. ■ A 7.70-cm length of straight wire runs horizontally along a north-south line. The wire carries a 3.45-A current flowing northward through a uniform 1.25-T magnetic field directed vertically upward. Find the magnitude and direction of the magnetic force on the wire.

59. ■■ A copper wire has diameter 0.150 mm and length 10.0 cm. It's in a horizontal plane and carries a current of 2.15 A. Find the magnitude and direction of the magnetic field needed to suspend the wire against gravity.

60. ■■ A square wire loop 20 cm on a side lies in the x-y plane, its sides parallel to the x- and y-axes. The loop has 15 turns and carries a current of 300 mA, clockwise around the loop. Find the net force on the loop when there is a uniform magnetic field of strength 0.50 T (a) in the $+z$-direction; (b) in the $+x$-direction; (c) along a diagonal of the square, from lower left to upper right.

61. ■■■ Find the net torque on the loop of the preceding problem for each case described.

62. ■■ A straight conducting wire carries a 7.5-A current in the $+y$-direction. Find the magnetic field (magnitude and direction) that results in a force per unit length on the wire of 0.14 N/m in the $-z$-direction.

63. ■■■ A closed wire loop is in the shape of a right triangle, with sides 10 cm, 10 cm, and $10\sqrt{2}$ cm. The loop lies in the x-y plane, with the one corner at the origin and the shorter sides along the x- and y-axes. The loop carries 250 mA in the clockwise direction. A uniform magnetic field of 0.75 T points in the $+z$-direction. Find the force on each of the three sides of the loop, and add them to get the net force.

64. ■■■ Repeat the preceding problem if the magnetic field is in the $+y$-direction.

65. ■■ You're at a place where Earth's magnetic field has magnitude 5.0×10^{-5} T with declination 0° and inclination 70°. Find the force on a 1.0-m length of wire carrying 1.5 A horizontally (a) from south to north and (b) from west to east.

66. ■■ A straight wire carries 2.25 A through a uniform magnetic field of 0.725 T. What angle should the wire make with the magnetic field so that the wire experiences a force per unit length of 1.40 N/m?

67. ■ A 150-turn circular wire loop with a 25-cm diameter carries 1.5 A. Find the coil's magnetic dipole moment.

68. ■■ A 12-cm-diameter circular wire loop carries a current of 2.3 A. Find the torque on the loop in a 1.4-T magnetic field directed (a) perpendicular to the loop and (b) in the plane of the loop.

69. ■■ You have a 1.50-m length of conducting wire, and you wish to make a circular coil with magnetic dipole moment 9.70×10^{-3} A·m^2 when the loop current is 650 mA. Find the radius of the coil and the number of turns.

70. ■■ An electric motor has a 900-turn coil on its armature. The coil is circular with diameter 12 cm and carries 12.5 A. Find the maximum torque on the coil when it rotates in a 1.50-T magnetic field.

71. ■■ An electric motor has a 1500-turn, 15.0-cm-diameter circular coil on its armature. Find the magnetic field needed to produce a maximum torque of 25.0 N·m when the coil current is 12.0 A.

Section 18.5 The Origin of Magnetism

72. ■ A long straight wire carries 2.5 A along the $+x$-axis. Find the magnitude and direction of the magnetic field at (a) a point 15 cm away on the $+y$-axis and (b) a point 15 cm away on the $-y$-axis.

73. ■■ Two closely spaced parallel wires carry currents of 1.25 A and 1.98 A in opposite directions. Find the magnetic field a distance of 5.0 cm from the pair of wires.

74. ■■ Two long parallel wires each carry 2.5 A in the same direction, with their centers 1.5 cm apart. (a) Find the magnetic field halfway between the wires. (b) Find the magnetic field at a point in the same plane as the wires, 1.5 cm from one wire and 3.0 cm from the other. (c) Find the force of interaction between the wires, and tell whether it's attractive or repulsive.

75. ■■ Repeat the preceding problem if the currents flow in opposite directions.

76. ■■■ A long straight wire carries 5.2 A along the x-axis, in the $+x$-direction. A second wire carries 5.2 A along the y-axis, in the $+y$-direction. Find the magnetic field (magnitude and direction) (a) at the point (0.10 m, 0.10 m) and (b) at the point (−0.10 m, −0.10 m).

77. ■■ Two parallel wires with centers separated by 8.6 mm each carry 12 A in the same direction. Find the magnitude and direction of the force per unit length on each wire.

78. ■■ Repeat the preceding problem if the currents are in opposite directions.

79. ■■■ Four long parallel wires pass through the corners of a square 2.0 cm on a side. The wires each carry 2.5 A in the same direction. (a) What is the magnetic field at the center of the square? (b) What is the force per unit length on each wire?

80. ■■■ A long straight wire is held fixed in a horizontal position. A second parallel wire is 2.4 mm below the first but is free to fall under its own weight. The second wire is copper (density 8920 kg/m^3), with diameter 1.0 mm. What equal current in both wires will suspend the lower wire against gravity?

81. ■ Two 8.0-cm-diameter circular wire loops lie directly atop one another. Find the magnetic field at the center of the common circle if each loop carries a current of 7.5 A (a) in the same direction and (b) in the opposite direction.

82. ■ A 25-cm-long solenoid consists of 5000 turns of wire wound uniformly. What current is needed to produce an 8.0-mT magnetic field inside the solenoid?

83. ■■■ In Figure P18.83, the current is 2.0 A. Find the net force on (a) the rectangular loop and (b) the straight wire. (c) Repeat parts

(a) and (b) if the direction of the current in the rectangular loop is reversed.

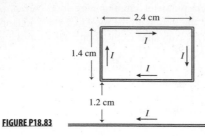

FIGURE P18.83

84. ■■ For the situation in Figure P18.84, find the net magnetic field at the center of the circular loop.

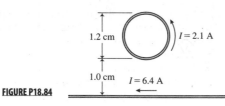

FIGURE P18.84

85. ■■ A 20-cm-long solenoid has 4000 turns of wire. Inside this solenoid is a second one, also 20 cm long but with 2500 turns of wire. (a) If a 2.5-A current flows in each solenoid in the same direction, what is the magnetic field inside the inner solenoid? (b) What is the magnetic field inside the inner solenoid if the two carry 2.5-A currents in opposite directions? (c) In each case, what's the magnetic field in the space between the two solenoids?

General Problems

86. ■■ A small particle carrying a $+50$-μC charge moves at 35 m/s through a uniform 0.75-T magnetic field. The velocity and magnetic field are both in the xy-plane, with the velocity making a $25°$ angle and the magnetic field a $262°$ angle, both measured counterclockwise from the $+x$-axis. Find the magnitude and direction of the magnetic force on the particle.

87. ■■ A particle carrying a -1.4-μC charge moves at 185 m/s through a uniform 45-mT magnetic field pointing in the $+y$-direction. The resulting force on the particle is 9.3 μN in the $+z$-direction. In what direction is the particle moving?

88. **BIO** ■■ **Medical cyclotron.** A medical cyclotron with a 2.0-T magnetic field is intended to accelerate deuterium nuclei (deuterons), consisting of one proton and one neutron. (a) At what frequency should the potential difference between the dees be alternated? (b) If the vacuum chamber has diameter 0.90 m, what is the maximum kinetic energy of the deuterons? (c) If the potential difference is 1500 V, how many orbits do the deuterons complete before achieving the maximum energy?

89. ■■ A wire of negligible resistance is bent to a rectangle, and a battery and resistor are connected as shown in Figure GP18.89. The right-hand side of the circuit extends into a region containing a uniform 0.15-T magnetic field pointing into the page. Find the magnitude and direction of the net force on the circuit.

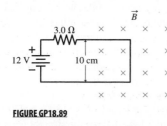

FIGURE GP18.89

90. ■■■ A rectangular copper bar measures 1.0 mm in the direction of a uniform 2.4-T magnetic field. When the bar carries a 6.8-A current at right angles to the field, the Hall potential difference across it is 1.2 μV. Find the number density of free electrons in copper.

91. ■■ A single-turn wire loop of radius 3.6 cm carries a 12-A current. It experiences a torque of 0.015 N·m when the normal to the loop plane makes a $25°$ angle with a uniform magnetic field. Find the magnetic field strength.

92. ■■ A simple electric motor consists of a 100-turn coil 3.0 cm in diameter, mounted between the poles of a magnet that produces a 0.12-T magnetic field. When a 5.0-A current flows in the coil, what are (a) its magnetic dipole moment and (b) the maximum torque developed by the motor?

93. ■■■ A power line carries a 500-A current toward magnetic north and is suspended 10 m above the ground. The horizontal component of Earth's magnetic field at the power line's latitude is 24 μT. If a magnetic compass is placed on the ground directly below the power line, in what direction will it point?

94. ■■ You have 10 m of 0.50-mm-diameter copper wire and a power supply capable of passing 15 A through the wire. What magnetic field strengths would you obtain (a) inside a 2.0-cm-diameter solenoid with the wire spaced as closely as possible and (b) at the center of a single circular loop made from the wire?

95. ■■ The largest lightning strikes have peak currents around 250 kA, flowing in essentially cylindrical channels of ionized air. How far from such a flash would the resulting magnetic field be equal to Earth's magnetic field strength, about 50 μT?

96. ■■ You're taking a class in space weather, which involves the far upper atmosphere and magnetic regions surrounding Earth. You're preparing a term paper on the van Allen belts, regions where high-energy particles are trapped within Earth's magnetic field. Your textbook says that the magnetic field strength at the belts is 10 μT. To impress your professor, you calculate the radii of the spiral paths of 0.1-MeV, 1.0-MeV, and 10-MeV protons in the van Allen belts. What values do you get?

Answers to Chapter Questions

Answer to Chapter-Opening Question

Auroras result when charged particles from the Sun are trapped in Earth's magnetic field and then enter the atmosphere. Auroras occur because Earth's magnetic field lines emerge predominantly at one pole and return to Earth at the other pole.

Answers to GOT IT? Questions

Section 18.2   (d)
Section 18.3   (a) Vertically up
Section 18.4   (f) Down
Section 18.5   (c) 4.0 A

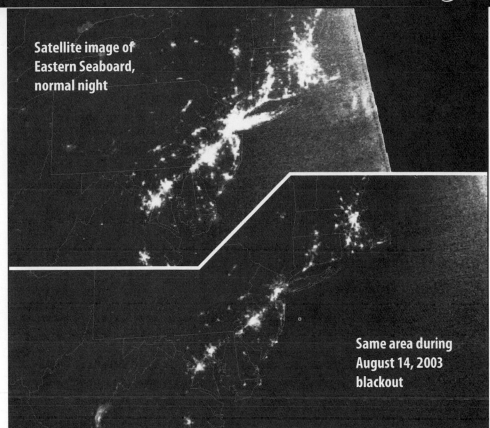

Satellite image of Eastern Seaboard, normal night

Same area during August 14, 2003 blackout

■ What's cos ϕ, and what does it have to do with the 2003 blackout that left 50 million people without power?

To Learn

By the end of this chapter you should be able to

■ Describe electromagnetic induction qualitatively.

■ Describe induction quantitatively using Faraday's law.

■ Explain how Lenz's law relates to conservation of energy.

■ Apply Faraday's law to motional emf.

■ Explain electric generators and transformers.

■ Understand inductance and inductors.

■ Describe the behavior of resistors, capacitors, and inductors in AC circuits.

■ Describe the resonant behavior of *RLC* circuits.

This chapter introduces the process of electromagnetic induction, which involves a fundamental relation between electricity and magnetism. You'll see practical uses, including electrical generators, transformers, and inductors. The latter join resistors and capacitors as basic circuit components. We'll finish this chapter with an introduction to alternating current (AC) circuits, showing the distinctive behavior of all three circuit components first separately and then in combination.

19.1 Induction and Faraday's Law

In Chapter 18 you saw how electric and magnetic phenomena are intimately related. Electric current gives rise to magnetic fields, and magnetic fields in turn affect moving electric charges. Given these close connections, it's logical to ask another question: If electric current produces a magnetic field, can a magnetic field produce electric current? The answer is yes, and this phenomenon is **electromagnetic induction**. In this section we'll explore how induction works, and then we'll examine some consequences and applications of induction.

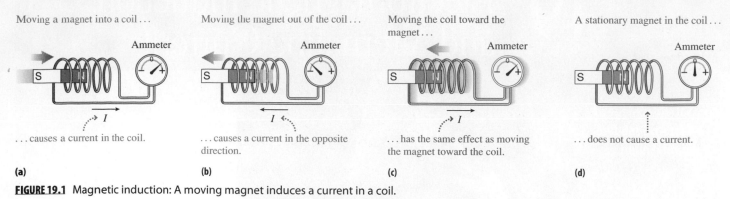

Moving a magnet into a coil . . .

Moving the magnet out of the coil . . .

Moving the coil toward the magnet . . .

A stationary magnet in the coil . . .

. . . causes a current in the coil.

. . . causes a current in the opposite direction.

. . . has the same effect as moving the magnet toward the coil.

. . . does not cause a current.

(a)

(b)

(c)

(d)

FIGURE 19.1 Magnetic induction: A moving magnet induces a current in a coil.

Electromagnetic induction was observed in 1831 by Michael Faraday (1791–1867) in England and Joseph Henry (1797–1878) in the United States. Both were aware of Ørsted's earlier discovery that magnetism results from electric current (Section 18.5), and they wanted to see if a magnetic field could generate electric current. Faraday's studies were more extensive, and he published first, so his name is most often associated with induction. Henry went on to study some practical uses of induction and is considered among the most significant figures in nineteenth-century American physics. Henry was the first Secretary of the Smithsonian Institution and was a founding member of the National Academy of Sciences.

Demonstrating Electromagnetic Induction

To observe induction, all you need is a magnet, a coil of wire, and an ammeter. You'll find that moving the magnet through the coil causes an **induced current** to flow in the coil (Figure 19.1). Only the relative motion of magnet and coil matters, and when the motion is reversed, the induced current reverses direction. With no relative motion of magnet and coil, there's no induced current.

Relative motion—or, more generally, *changing* magnetism—is essential to electromagnetic induction. To understand why, and to describe induction quantitatively, we'll first introduce the concept of magnetic flux.

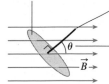

Surface with area A

Line perpendicular to surface

θ

θ = angle between $\vec{B}$ and line perpendicular to surface.

$\vec{B}$

(a)

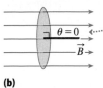

Flux Φ is greatest when surface is perpendicular to magnetic field. Then

$\theta = 0$

$\vec{B}$

$\theta = 0$ and $\cos\theta = 1$, so $\Phi = BA$.

(b)

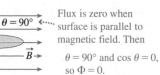

Flux is zero when surface is parallel to magnetic field. Then

$\theta = 90°$

$\vec{B}$

$\theta = 90°$ and $\cos\theta = 0$, so $\Phi = 0$.

(c)

FIGURE 19.2 Magnetic flux.

Magnetic Flux

Figure 19.2 illustrates **magnetic flux**. Flux is about vector fields passing through surfaces; the idea works for electric fields, gravitational fields, and the velocity fields of fluid flows, as well as for magnetic fields. The flux of a uniform magnetic field of magnitude B through a flat surface of area A is

$$\Phi = BA\cos\theta \quad \text{(Magnetic flux in a uniform field; SI unit: Wb)} \quad (19.1)$$

where θ is the angle between the magnetic field and the normal to the area (Figure 19.2a). "Flux" comes from the Latin "to flow." You can visualize magnetic flux as the number of magnetic field lines "flowing" though the surface. That's why the factor $\cos\theta$ appears in the expression for flux. For a given magnetic field, the flux is maximum with the field perpendicular to the surface (Figure 19.2b). In that case the angle between the field and normal to the surface is 0. Then $\cos\theta = \cos(0) = 1$, and the flux is $\Phi = BA$. When the field is parallel to the surface, $\cos\theta = \cos(90°) = 0$, and the flux is zero (Figure 19.2c). This makes sense, because then no field lines pass through the surface.

For example, if a 4.0-T magnetic field passes perpendicularly through a square surface 0.10 m on a side, the magnetic flux is

$$\Phi = BA\cos\theta = (4.0\text{ T})(0.10\text{ m})^2\cos(0) = 0.040\text{ T}\cdot\text{m}^2$$

The SI unit of flux is the weber (Wb), with $1\,\text{Wb} = 1\,\text{T} \cdot \text{m}^2$, so our answer can be expressed as 0.040 Wb. If the same field makes a 45° angle with the normal, then the flux is reduced to

$$\Phi = BA \cos\theta = (4.0\,\text{T})(0.10\,\text{m})^2 \cos(45°) = 0.028\,\text{T} \cdot \text{m}^2 = 0.028\,\text{Wb}$$

✓ **TIP**

Magnetic flux depends on the magnetic field strength, the surface area, and the angle between the field and the surface normal.

Faraday's Law

Given magnetic flux, we now present **Faraday's law**—the quantitative description of electromagnetic induction. Picture again the magnet and coil of Figure 19.1. The simplest expression of Faraday's law relates the emf $\mathcal{E}$ induced in the coil to the changing magnetic flux through the coil:

$$\mathcal{E} = -N\frac{\Delta\Phi}{\Delta t} \qquad \text{(Faraday's law of induction; SI unit: V)} \qquad (19.2)$$

Here Φ is the flux through each turn of the coil, and N is the number of turns. Remember that emf is similar to potential difference and is measured in volts.

The minus sign in Faraday's law is crucial. It shows that the induced emf *opposes the change in magnetic flux*. This fact is important enough to have its own name: **Lenz's law**. Figure 19.3 illustrates how Lenz's law works, and Tactic 19.1 helps you apply it.

Faraday's law says there's an induced emf whenever magnetic flux changes. Recall that flux is the product of three factors: $\Phi = BA \cos\theta$. Thus, it changes when any of those factors changes:

- a change (increase or decrease) in the magnetic field strength B
- a change in the area A
- a change in the angle θ

You'll see examples of each of these throughout this chapter. Note that Faraday's law is fundamentally about emf; if there's a complete circuit present, then induced current flows. It's given by $I = \mathcal{E}/R$ (Chapter 18), with R the circuit resistance. In finding the magnitude of the current, you can neglect the minus sign in Faraday's law, and then use Tactic 19.1 to get the current direction.

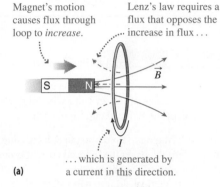

Magnet's motion causes flux through loop to *increase*.   Lenz's law requires a flux that opposes the increase in flux...

(a)   ...which is generated by a current in this direction.

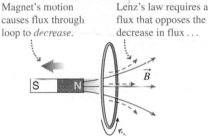

Magnet's motion causes flux through loop to *decrease*.   Lenz's law requires a flux that opposes the decrease in flux...

(b)   ...which is generated by a current in this direction.

FIGURE 19.3 Finding the direction of the induced current.

TACTIC 19.1 Applying Lenz's Law

Lenz's law states that any current induced by a changing magnetic flux opposes that change in flux. Recall from Chapter 18 that current flowing in a planar coil produces a magnetic field perpendicular to the coil plane, as given by the right-hand rule (Figure 19.4). The magnetic flux from this induced current is what opposes the original change in magnetic flux.

To apply Lenz's law, first determine whether the magnetic flux is increasing or decreasing. Then:

1. If the magnetic flux is increasing, the induced current generates a magnetic flux in the direction opposite the applied flux, to counteract the increase. Think of the induced current as trying to fight against the flux increase (Figure 19.3a).
2. If the magnetic flux is decreasing, the induced current generates a magnetic flux in the same direction as the applied flux, to counteract the decrease. Think of the induced current as trying to restore the flux that's being lost (Figure 19.3b).

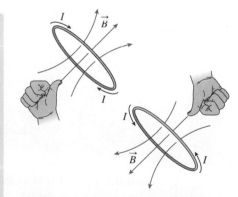

FIGURE 19.4 The right-hand rule used to find direction of magnetic field due to a current-carrying coil.

Card Swipe

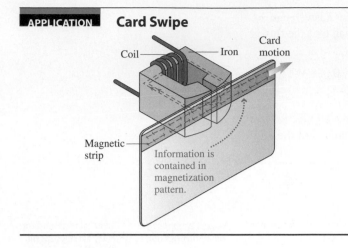

On the back of your credit card is a strip of magnetic material whose magnetization pattern encodes the card number and other information. The diagram shows that when you swipe the card, you move the magnetic strip past a coil wound on an iron core. The changing magnetization pattern results in an induced current in the coil, which electronic circuitry decodes to reveal the information on the strip.

CONCEPTUAL EXAMPLE 19.1 **Lenz's Law**

Consider each of the cases pictured in Figure 19.5, with a magnetic field perpendicular to a circular wire loop. Use Lenz's law to find the direction of the induced current in each.

SOLVE (a) Magnetic field into page, increasing. To oppose the increase in flux into the page, the induced current is counterclockwise, producing a flux out of the page.

(b) Magnetic field into page, decreasing. To oppose this decrease, the induced current is clockwise, producing a flux into the page and thus countering the flux decrease.

(c) Magnetic field out of page, increasing. To oppose the increase in flux out of the page, the induced current is clockwise, producing a flux into the page.

(d) Magnetic field out of page, decreasing. To oppose the decrease in flux out of the page, the induced current is counterclockwise,

producing a flux out of the page and thus countering the flux decrease.

REFLECT The induced current can be in either direction, depending on whether the flux increases or decreases.

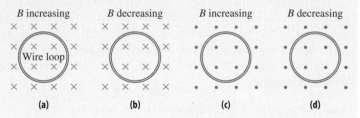

FIGURE 19.5 Four situations involving Lenz's law.

EXAMPLE 19.2 **Induced Current**

A 15-turn coil is in a uniform magnetic field of 1.25 T, perpendicular to the coil's plane. The coil has resistance 7.40 Ω and area 0.0120 m². If the magnetic field steadily increases to 2.85 T over 3.50 s, what's the induced current in the coil?

ORGANIZE AND PLAN Our sketch is shown in Figure 19.6. The induced emf $\mathcal{E}$ is given by Faraday's law (Equation 19.2): $\mathcal{E} = -N\Delta\Phi/\Delta t$.

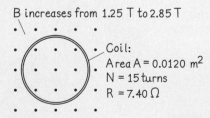

FIGURE 19.6 Our sketch for Example 19.2.

Here we're only interested in the magnitude of the current, which is $I = |\mathcal{E}|/R$.

Known: $R = 7.40\ \Omega$; $\Delta t = 3.50$ s; $A = 0.0120$ m²; $B_0 = 1.25$ T; final $B = 2.85$ T.

SOLVE Using Faraday's law, the current becomes

$$I = \frac{|\mathcal{E}|}{R} = \frac{N\Delta\Phi}{R\Delta t}$$

Magnetic flux is $\Phi = BA\cos\theta = BA$, because here $\theta = 0$ so $\cos\theta = 1$. The coil's area isn't changing, so $\Delta\Phi = \Delta(BA) = A\Delta B$. Therefore,

$$I = \frac{NA\Delta B}{R\Delta t} = \frac{(15)(0.0120\ \text{m}^2)(2.85\ \text{T} - 1.25\ \text{T})}{(7.40\ \Omega)(3.50\ \text{s})}$$

$$= 0.0111\ \text{A} = 11.1\ \text{mA}$$

cont'd.

REFLECT The units must come out in amperes (A), because we used SI units consistently. Just to be sure, note that the units initially read $m^2 \cdot T/\Omega \cdot s$. Now remember from Chapter 18 that $1\,T = 1\,N/(A \cdot m)$. With this substitution, our units become

$$\frac{m^2 \cdot (N/(A \cdot m))}{\Omega \cdot s} = \frac{N \cdot m}{A \cdot \Omega \cdot s}$$

But $1\,A \cdot \Omega = 1\,V$ by Ohm's law (Chapter 17), and $1\,N \cdot m = 1\,J$ (Chapter 5), reducing our units to $J/V \cdot s$. Finally, Chapter 16 shows that $1\,V = 1\,J/C$, so the units become $J/(J/C) \cdot s = C/s = A$, as expected.

MAKING THE CONNECTION What would happen if the magnetic field now *decreased* from 2.85 T to 1.25 T, over another 3.50-s interval?

ANSWER The induced current would have the same magnitude but would flow in the opposite direction, according to Lenz's law.

GOT IT? Section 19.1 Rank in decreasing order the magnetic flux through the four loops.

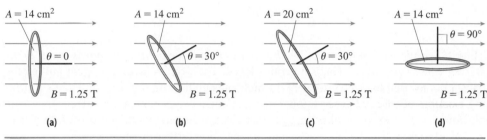

(a) $A = 14\ cm^2$, $\theta = 0$, $B = 1.25\ T$
(b) $A = 14\ cm^2$, $\theta = 30°$, $B = 1.25\ T$
(c) $A = 20\ cm^2$, $\theta = 30°$, $B = 1.25\ T$
(d) $A = 14\ cm^2$, $\theta = 90°$, $B = 1.25\ T$

19.2 Motional emf

Motional emf is the emf induced in a conductor that moves through a magnetic field. You've already seen one example: the coil moving near the magnet in Figures 19.1c and 19.1d. Here we'll explore other examples and some practical applications.

Induced emf, Current, and Energy Conservation

Figure 19.7 shows two parallel conducting rails a distance L apart, electrically connected by a resistance R. The rails support a conducting bar that's free to slide; here it's moving to the right with speed v. A uniform magnetic field $\vec{B}$ points into the page, perpendicular to the plane of the rails.

As the bar moves to the right, the area of the enclosed rectangle increases, and so therefore does the magnetic flux through the circuit. This induces an emf $\mathcal{E}$, which drives an induced current $I = \mathcal{E}/R$. Lenz's law requires the induced current to flow counterclockwise around the loop, to produce a magnetic flux that opposes the flux increase due to the increasing area. The induced current depends on the rate of change of magnetic flux, which in turn depends on the bar's speed v. When the bar is a distance x from the left end of the rails, the flux is $\Phi = BA = BLx$. The change in flux when the bar slides a distance Δx is thus $\Delta\Phi = BL\Delta x$. If that takes time Δt, then Faraday's law gives

$$\mathcal{E} = -\frac{\Delta\Phi}{\Delta t} = -\frac{BL\Delta x}{\Delta t} = -BLv$$

where $v = \Delta x/\Delta t$. Therefore, the induced current is

$$\boxed{I = \frac{|\mathcal{E}|}{R} = \frac{BLv}{R}} \qquad \text{(Induced current due to motional emf; SI unit: A)} \qquad (19.3)$$

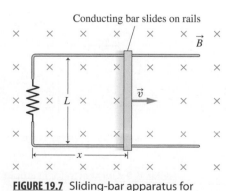

The posts in this metal detector contain parallel coils, one of which carries a known time-varying current. If a piece of metal passes between the coils, eddy currents induced in the metal change the flux through the second coil. The resulting change in the induced current in the second coil sets off the detector.

FIGURE 19.7 Sliding-bar apparatus for analyzing motional emf.

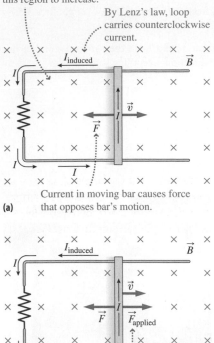

Motion of bar causes flux through this region to increase.

By Lenz's law, loop carries counterclockwise current.

$I_{induced}$

$\vec{B}$

$\vec{v}$

$\vec{F}$

I

(a) Current in moving bar causes force that opposes bar's motion.

$I_{induced}$

$\vec{B}$

$\vec{v}$

$\vec{F}$ $\vec{F}_{applied}$

(b) To keep bar moving with constant speed, we must apply a force to the right.

FIGURE 19.8 Analyzing the forces on the bar.

Remember that the current flows counterclockwise, by Lenz's law. Thus, the current in the moving bar is upward, as shown in Figure 19.8a. Recall from Section 18.4 that a current-carrying wire in a magnetic field experiences a force. By the right-hand rule, that force is to the left, and Equation 18.6 gives its magnitude: $F = ILB$, with current I given by Equation 19.3. This leftward force would slow the bar, unless it's balanced by a rightward force of equal magnitude (Figure 19.8b).

It's instructive to compare the *mechanical power* expended by the agent pulling the bar with the *electrical power* associated with the induced current. From Chapter 5 (Equation 5.23), the power used when a force F acts on an object with speed v is $P = Fv$. Here, with $F = ILB$,

$$P = Fv = ILBv = \frac{B^2 L^2 v^2}{R} \quad \text{(Mechanical power)}$$

where we've used Equation 19.3 for the current I. Meanwhile, you know from Chapter 17 (Equation 17.11) that electric power can be written $P = I^2 R$. Again using Equation 19.3 for I, the electric power is

$$P = I^2 R = \frac{B^2 L^2 v^2}{R} \quad \text{(Electric power)}$$

Thus, *the mechanical power supplied by the external force pulling the bar is equal to the electric power generated.* That's no coincidence: The work done by the agent pulling the bar gets converted to electrical energy, which ends up heating the resistor. Here's a striking example of energy conservation, one that connects mechanical and electromagnetic phenomena. This example also suggests that electromagnetic induction is useful for generating electric power. We'll explore power generation further in Section 19.3.

EXAMPLE 19.3 **Induced Current, Force, and Power**

Suppose the rails in Figure 19.7 have spacing $L = 10 \text{ cm}$ and $R = 1.2 \ \Omega$, with the apparatus in a uniform 2.2-T magnetic field. (a) Find the induced current when the bar moves at 2.0 cm/s and the force needed to keep it moving with this speed. (b) What mechanical power is required to maintain this speed, and at what rate is electrical energy generated?

ORGANIZE AND PLAN Our sketch of the situation is shown in Figure 19.9. The induced current follows from Equation 19.3: $I = \mathcal{E}/R = BLv/R$. With the current known, the required force is $F = ILB$. The mechanical power and electrical power are the same, as discussed in the text; both are given by $P = B^2 L^2 v^2 / R$.

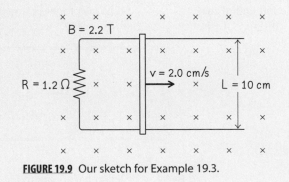

$\times$ $\times$ $\times$ $\times$ $\times$ $\times$ $\times$

B = 2.2 T

$\times$ $\times$ $\times$ $\times$ $\times$ $\times$ $\times$

v = 2.0 cm/s

R = 1.2 Ω $\times$ $\times$ $\longrightarrow$ $\times$ L = 10 cm

$\times$ $\times$ $\times$ $\times$ $\times$ $\times$ $\times$

$\times$ $\times$ $\times$ $\times$ $\times$ $\times$ $\times$

FIGURE 19.9 Our sketch for Example 19.3.

Known: $R = 1.2 \ \Omega$; $L = 10 \text{ cm}$; $v = 2.0 \text{ cm/s}$; $B = 2.2 \text{ T}$.

SOLVE (a) The induced current is

$$I = \frac{BLv}{R} = \frac{(2.2 \text{ T})(0.10 \text{ m})(0.020 \text{ m/s})}{1.2 \ \Omega} = 0.0037 \text{ A} = 3.7 \text{ mA}$$

Then the force is

$$F = ILB = (0.0037 \text{ A})(0.10 \text{ m})(2.2 \text{ T}) = 8.1 \times 10^{-4} \text{ N}$$

(b) The mechanical power and electrical power are both

$$P = \frac{B^2 L^2 v^2}{R} = \frac{(2.2 \text{ T})^2 (0.10 \text{ m})^2 (0.020 \text{ m/s})^2}{1.2 \ \Omega} = 1.6 \times 10^{-5} \text{ W}$$

REFLECT That's only 16 microwatts! You'd need a much larger apparatus to generate significant electrical power.

MAKING THE CONNECTION Describe in words how the motion would change if you stopped pulling the bar.

ANSWER The induced current creates a leftward force on the sliding rail, so it slows. This in turn reduces the induced current. The bar still slows, but at an ever-slower rate, and only gradually does it approach a full stop.

Applications of Motional emf

A solid conducting material moving into a magnetic field experiences a changing magnetic flux, inducing so-called **eddy currents** in the material (Figure 19.10). Electrical resistance results in heating, with the thermal energy coming ultimately from the motion of the material. So, just like the bar in Making the Connection in Example 19.3, the moving material slows unless an external force keeps it going.

Magnetic brakes utilize this principle, stopping systems ranging from roller coasters to circular saws to subway trains. Normally eddy currents convert the energy of motion into random thermal energy (loosely, "heat"), but it's also possible to capture the electrical energy and store it. That's what happens when braking a hybrid car; the energy of motion is converted to electrical energy that's stored in a battery. Some subway systems use the same approach, capturing the energy as a train slows when approaching a station, storing it briefly in a capacitor, and then using that energy to accelerate the train as it leaves the station.

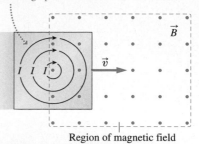

As metal plate enters magnetic field, increasing flux through plate induces clockwise currents.

Region of magnetic field

FIGURE 19.10 As the solid metal plate enters the magnetic field, eddy currents arise that sap its energy of motion. By Lenz's law, the currents flow clockwise.

CONCEPTUAL EXAMPLE 19.4 **The Falling Magnet**

In a classic demonstration, a bar magnet is dropped through a nonmagnetic metal tube. Explain why the magnet falls slowly through the tube. Does the magnet's orientation matter?

SOLVE This is essentially the same induction process as in Figure 19.1, but with the magnet moving vertically. As the magnet falls, the magnetic flux in the tube changes, inducing a current in the conducting tube (Figure 19.11). By Lenz's law, the induced current sets up an opposing magnetic field that produces an upward force on the falling magnet, retarding its motion. It doesn't matter which pole is on the bottom; either way, the induced current opposes the flux change and provides an upward force.

REFLECT This is an impressive demonstration. It can take 10 seconds for a strong magnet to fall through a 2-m-long tube. Falling freely through the same drop would take less than a second!

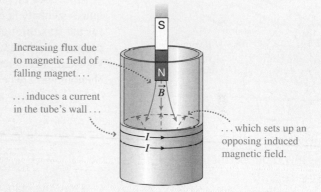

Increasing flux due to magnetic field of falling magnet . . .

. . . induces a current in the tube's wall . . .

. . . which sets up an opposing induced magnetic field.

FIGURE 19.11 Induced current in the tube due to the falling magnet.

The apparatus of Figure 19.7 converts mechanical energy into electrical energy. Reverse it, as in Figure 19.12, and you convert electrical energy into mechanical energy. This is an example of a **linear electric motor**, a more advanced version of which drives some experimental high-speed trains. You can also use the device as a **rail gun**, with the force on the sliding bar accelerating it to the right. Such devices have been proposed for launching back toward Earth materials mined on the Moon. However, the acceleration occurs for a limited time: once the bar is moving fast enough, the induced current just cancels the current from the battery, and from then on the bar's speed is constant.

Reviewing New Concepts

- Changing magnetic flux induces an emf, which may drive an induced current (Faraday's law).
- The direction of the induced current serves to oppose the change in magnetic flux (Lenz's law).
- Motional emf is generated in a conductor moving through a magnetic field.
- The induction process can convert mechanical energy into electrical energy.

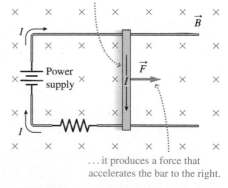

When clockwise current from power supply flows through sliding bar . . .

. . . it produces a force that accelerates the bar to the right.

FIGURE 19.12 Principle of the linear motor or rail gun.

GOT IT? Section 19.2 Suppose that the sliding rail in Figure 19.7 is moving to the left. Then the induced current in the loop is (a) clockwise; (b) counterclockwise; (c) zero.

19.3 Generators and Transformers

Induction plays a central role in electric power systems. Here we'll see how induction generates essentially all our electric power, as well as converting potential differences and thus enabling long-distance power transmission.

Generators

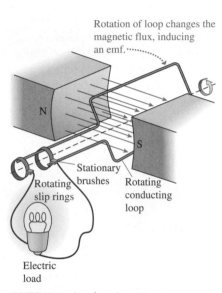

Rotation of loop changes the magnetic flux, inducing an emf.

Stationary brushes

Rotating slip rings

Rotating conducting loop

Electric load

FIGURE 19.13 An electric generator.

Our bar-on-rails apparatus converts mechanical energy into electrical energy. We presented that example first, because it's straightforward conceptually and shows directly that energy is conserved. That device wouldn't make a practical electric generator, however, because you'd need to keep sliding the bar back and forth. Figure 19.13 shows a better design for a generator. Here, mechanical energy turns a wire coil in a magnetic field. Mechanical energy sources include falling water in hydroelectric power plants; steam turbines in fossil-fueled, biomass, and nuclear plants; and, increasingly, wind.

How does the generator work? You know from Faraday's law that induced current requires a changing magnetic flux. With a uniform field, Equation 19.1 gives $\Phi = BA \cos \theta$ for the flux through each turn of the coil. The field B and area A are constant, but as the coil rotates, the angle θ varies. If the coil turns with constant angular velocity ω, then you know from Chapter 8 that $\theta = \omega t$. Then by Faraday's law the emf is

$$\mathcal{E} = -N\frac{\Delta \Phi}{\Delta t} = -N\frac{\Delta (BA \cos \theta)}{\Delta t} = -NBA\frac{\Delta (\cos (\omega t))}{\Delta t}$$

for a coil with N turns. Calculus shows that this reduces to an instantaneous value

$$\mathcal{E} = NBA\omega \sin(\omega t) \quad \text{(emf induced in rotating coil; SI unit: V)} \quad (19.4)$$

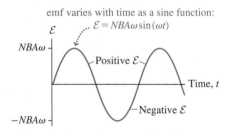

emf varies with time as a sine function:

$\mathcal{E} = NBA\omega \sin(\omega t)$

$\mathcal{E}$

$NBA\omega$

Positive $\mathcal{E}$

Time, t

Negative $\mathcal{E}$

$-NBA\omega$

FIGURE 19.14 emf from a rotating-coil generator.

Thus, the induced emf varies sinusoidally with time as shown in Figure 19.14, changing sign periodically. This results in **alternating current** (AC), which changes direction periodically. The electrical energy is transmitted through power lines to homes and businesses. Sections 19.5 and 19.6 consider alternating-current circuits.

✓**TIP**

Alternating current generally follows a sinusoidal (sine or cosine) function of time.

The generator in Figure 19.13 looks a lot like the electric motor in Figure 18.25. In fact, the same device can serve both purposes: Put in electrical energy, and you get out mechanical energy—that's a motor. Put in mechanical energy, and you get out electrical energy—that's a generator. In some applications—hybrid cars, for example—the same physical device may work sometimes as a generator and sometimes as a motor.

Your car's *alternator* is a small AC generator used to recharge the car's battery. A belt from a pulley on the engine shaft drives the alternator's rotating coil. Because the battery needs direct current (DC), a *rectifier* is used to convert the AC into DC, which then flows through the battery to charge it. Energy from the battery starts your car, powers lights and electronics, and energizes the spark that ignites gasoline in the engine. If it weren't recharged, these energy demands would soon drain the battery.

EXAMPLE 19.5 **Generating AC Current**

The AC available from standard wall outlets in North America has peak emf about 170 V and frequency 60 Hz. (You may know that most devices are designed to run at 120 V; that's the average over one cycle, when the peak is 170 V. You'll see why in Section 19.5.) If you want to produce this emf in a coil with area 0.024 m² rotating in a uniform magnetic field of 0.15 T, how many turns of wire should the coil have?

ORGANIZE AND PLAN Equation 19.4 gives the emf of the rotating coil: $\mathcal{E} = NBA\omega \sin(\omega t)$. The peak emf (here 170 V) occurs when the sine function has its maximum value, 1. Recall from Chapter 8 that angular velocity ω and frequency f are related by $\omega = 2\pi f$. Therefore, the maximum emf is $\mathcal{E}_{max} = 2\pi NBAf$.

Known: $\mathcal{E}_{max} = 170$ V; $A = 0.024$ m²; $f = 60$ Hz; $B = 0.15$ T.

SOLVE Solving for the number of turns N gives

$$N = \frac{\mathcal{E}_{max}}{2\pi BAf} = \frac{170 \text{ V}}{2\pi (0.15 \text{ T})(0.024 \text{ m}^2)(60 \text{ Hz})} = 125$$

REFLECT Here all the SI units cancel, leaving a dimensionless quantity for the number of turns.

MAKING THE CONNECTION If the magnetic field or area were increased, how would that affect the number of turns required?

ANSWER Both factors B and A are in the denominator of the expression for N. Therefore, fewer turns would be required.

Transformers

Another common device that uses electromagnetic induction is the **transformer**, illustrated in Figure 19.15. A **primary coil** and **secondary coil** are wrapped around an iron core; they have, respectively, N_p and N_s turns. When a time-varying electric current passes through the primary coil, the induced emf in that coil is $\mathcal{E}_p = -N_p \Delta\Phi/\Delta t$, where Φ is the magnetic flux in each turn of the primary coil. The iron core concentrates the field in the iron, thus ensuring that the same flux Φ passes through each turn of the secondary. Then the emf induced in the secondary is $\mathcal{E}_s = -N_s \Delta\Phi/\Delta t$. Both emfs contain the factor $-\Delta\Phi/\Delta t$, so the emf ratio $\mathcal{E}_s/\mathcal{E}_p$ is equal to the turns ratio N_s/N_p. Therefore,

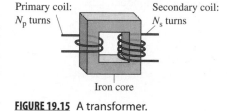

Primary coil: N_p turns Secondary coil: N_s turns

Iron core

FIGURE 19.15 A transformer.

$$\mathcal{E}_s = \mathcal{E}_p \frac{N_s}{N_p} \quad \text{(Transformer emf; SI unit: V)} \qquad (19.5)$$

Equation 19.5 shows that you can "step up" or "step down" an alternating emf by choosing the coil ratio N_s/N_p. Many electronic devices have built-in transformers, because the 120-V emf at the outlet is too high. For example, if you need 12-V power, then you can use a step-down transformer with $N_s/N_p = 1/10$.

You've probably seen transformers atop power poles. Electric power is generally transmitted at high potentials and then stepped down to 120 V for home and office use. The following example explains why.

CONCEPTUAL EXAMPLE 19.6 **Why Use High Potentials?**

Electric transmission lines normally have potentials of kilovolts and higher. What's the advantage of high potentials for power transmission?

SOLVE The answer follows from Chapter 17's relations among power, current, potential, and resistance. Suppose your city gets power P from a distant power plant. If the current in the wires carrying the power is I and there's a potential difference V between them, then $P = IV$ and $I = P/V$. Within the transmission line, which has some resistance R, energy is dissipated at a rate $P_{lost} = I^2 R$. Using the fact that $I = P/V$,

$$P_{lost} = I^2 R = \left(\frac{P}{V}\right)^2 R = \frac{P^2 R}{V^2}$$

Thus the rate of energy loss is proportional to $1/V^2$, so losses drop with higher potentials. Increasing V by a factor of 10 cuts the transmission loss by a factor of 100. This analysis also gives the expected result that power loss drops with decreasing transmission line resistance.

REFLECT Typically, about 8% of electrical energy is lost in transmission, depending on transmission line lengths and other factors. Long-distance transmission lines operate at 345 kV and up, while lines within cities and towns are typically between 3 kV and 115 kV.

GOT IT? Section 19.3 The primary coil in a transformer has 100 turns and an emf of 120 V. If you want a 60-V emf in the secondary, then the number of turns in the secondary coil should be (a) 25; (b) 50; (c) 100; (d) 200; (e) 400.

19.4 Inductance

In a transformer, changing current in one circuit induces current in another. The effect is reciprocal: If circuit A affects circuit B, then B affects A—hence the term **mutual inductance**. However, it's not necessary to have two circuits. As Figure 19.16 illustrates, a circuit can induce a current in itself—that's **self-inductance**. Here we'll consider **inductors**, devices designed specifically to exploit self-inductance. They're usually multiturn coils of wire.

We'll first describe qualitatively how inductors work. Suppose an external source supplies current to the coil of Figure 19.16. If the current increases, so does the magnetic flux through the coil. Then there's an induced emf that, by Lenz's law, opposes this change. The effect is to slow the rate of increase in current. In contrast, if the external current decreases, then so does the flux. In order to oppose this change, the induced emf now "helps" the current to keep flowing. In each case, the results are similar: Induction limits the rate at which the current changes.

Current in wire creates magnetic flux through coil.

If current is *changing*, flux through coil will change . . .

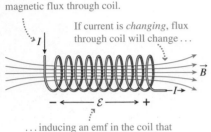

. . . inducing an emf in the coil that *opposes* the change in current.

FIGURE 19.16 Self-inductance.

Self-Inductance

Self-inductance L is defined as the ratio of flux to current. For an N-turn coil:

$$L = \frac{N\Phi}{I} \quad \text{(Definition of inductance; SI unit: H)} \quad (19.6)$$

where Φ is the flux through each turn and I is the current in the coil. For a coil with fixed shape, flux is proportional to current; doubling the current doubles the flux. Therefore, self-inductance L is a constant. The SI unit of inductance is the henry (H), named for Joseph Henry. Its definition follows from the definition of inductance as the ratio of flux to current:

$$1\,\text{H} = 1\frac{\text{Wb}}{\text{A}} = 1\frac{\text{T}\cdot\text{m}^2}{\text{A}}$$

We can apply Faraday's law to find the emf induced in an inductor. From Equation 19.6, the change in magnetic flux through a coil is related to the change in current: $N\Delta\Phi = L\Delta I$. Dividing through by the time Δt gives $N(\Delta\Phi/\Delta t) = L(\Delta I/\Delta t)$. But Faraday's law (Equation 19.2) shows that $N(\Delta\Phi/\Delta t) = -\mathcal{E}$, with $\mathcal{E}$ the induced emf. Therefore,

$$\mathcal{E} = -N\frac{\Delta\Phi}{\Delta t} = -L\frac{\Delta I}{\Delta t} \quad \text{(Inductor emf; SI unit: V)} \quad (19.7)$$

You now have two ways to think of inductance: as a ratio of magnetic flux to current (Equation 19.6) or as a constant that relates induced emf to changing current (Equation 19.7).

Inductance of a Solenoid

The solenoid (Section 18.5) is a common inductor design. A solenoid's inductance follows from its physical properties: number of turns per unit length n, cross-sectional area A, and length d. Recall from Equation 18.10 that the magnetic field inside a solenoid carrying current I is $B = \mu_0 nI$. Then from the definition of inductance,

$$L = \frac{N\Phi}{I} = \frac{NBA}{I} = \frac{N(\mu_0 nI)A}{I} = \mu_0 nNA$$

But $n = N/d$, so $N = nd$, so

$$L = \mu_0 n^2 A d \quad \text{(Inductance of a solenoid; SI unit: H)} \quad (19.8)$$

The inductance of a solenoid therefore depends only on how tightly it's wound and the geometrical factors A and d; in particular, inductance doesn't depend on the inductor current. Notice that the inductance of a solenoid is proportional to n^2, so doubling the density of turns raises the inductance by a factor of 4.

EXAMPLE 19.7 **An MRI Solenoid**

A solenoid used in magnetic resonance imaging (MRI) is 2.4 m long and 94 cm in diameter, with 1200 turns of superconducting wire. Find (a) its inductance and (b) the magnitude of the induced emf in the solenoid during the 30 s it takes to "ramp up" the current from zero to its operating value of 2.3 kA.

ORGANIZE AND PLAN Our sketch of the coil is shown in Figure 19.17. Equation 19.8 gives the inductance of a solenoid: $L = \mu_0 n^2 A d$. Then Faraday's law (Equation 19.7) gives the emf: $\mathcal{E} = -L(\Delta I/\Delta t)$.

Known: $d = 2.4$ m; $N = 1200$; $r = 47$ cm.

SOLVE (a) The number of turns per unit length is

$$n = \frac{N}{d} = \frac{1200 \text{ turns}}{2.4 \text{ m}} = 500 \text{ turns/m}$$

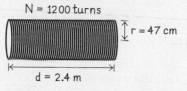

N = 1200 turns

r = 47 cm

d = 2.4 m

FIGURE 19.17 Our sketch for Example 19.7.

Then the inductance is

$$L = \mu_0 n^2 A d$$
$$= (4\pi \times 10^{-7} \, \text{T} \cdot \text{m/A})(500 \, \text{m}^{-1})^2 (\pi (0.47 \, \text{m})^2)(2.4 \, \text{m})$$
$$= 0.5232 \text{ H}$$

(b) The magnitude of the induced emf is

$$|\mathcal{E}| = \left| -L \frac{\Delta I}{\Delta t} \right| = (0.5232 \text{ H}) \frac{2.3 \times 10^3 \text{ A}}{30 \text{ s}} = 40.1 \text{ V}$$

REFLECT Here the current change occurs at a rate you could achieve by turning a knob on a power supply. But closing or opening a switch, or using AC current, gives a much faster change—and a correspondingly higher emf.

MAKING THE CONNECTION People have been electrocuted by opening switches in circuits with large inductors. Why?

ANSWER Opening a switch quickly drops the current to zero, meaning $|\Delta I/\Delta t|$ is large. With a large L, Equation 19.7 can then give a dangerously high emf. This can happen even if batteries or power supplies in the circuit have modest emfs.

Energy Storage in Inductors

What happens when you close the switch in Figure 19.18, with a battery and resistor in series with an inductor? The current can't start flowing right away, because an abrupt change in current implies an infinite $\Delta I/\Delta t$, and thus an infinite inductor emf—impossible! Instead, the current increases gradually, with the inductor emf opposing the increase. The battery does work against the inductor emf to build up that current. Calculus shows that the work required to establish current I in an inductance L is $\frac{1}{2}LI^2$. What happens to this work? Unlike the work done in forcing current through a resistor—which is dissipated as heat—the work done building current in an inductor is stored as **magnetic energy** in the magnetic field associated with the current. So the energy U stored in the inductor is

$$U = \tfrac{1}{2}LI^2 \quad \text{(Energy stored in an inductor; SI unit: J)} \quad (19.9)$$

For example, at its full current the MRI solenoid of Example 19.7 contains stored energy $U = \frac{1}{2}LI^2 = \frac{1}{2}(0.5232)(2.3 \times 10^3 \text{ A})^2 = 1.384$ MJ.

Recall that for a solenoid the magnetic field is $B = \mu_0 n I$, and the inductance is $L = \mu_0 n^2 A d$. Using these relationships, the stored energy is

$$U = \frac{1}{2}LI^2 = \frac{1}{2}(\mu_0 n^2 A d)\left(\frac{B}{\mu_0 n}\right)^2 = \frac{B^2 A d}{2\mu_0}$$

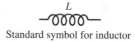

L

Standard symbol for inductor

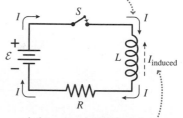

The current that begins to flow when the switch is closed …

… induces a current in the inductor that opposes the change in the current.

FIGURE 19.18 Starting the current in an LR circuit.

The **energy density** (symbol u_B) is the stored energy per unit volume. The volume of the solenoid is Ad, so $u_B = U/(Ad)$, or

$$u_B = \frac{B^2}{2\mu_0} \quad \text{(Energy density in a magnetic field; SI unit: J/m}^3\text{)} \quad (19.10)$$

Even though we derived this equation for a solenoid, it's true for all magnetic fields. If you worked Problem 80 in Chapter 16, you found a similar expression for the energy density in an electric field: $u_E = \frac{1}{2}\epsilon_0 E^2$. In each case, the energy stored is proportional to the field strength *squared*. This will be important when you learn about electromagnetic waves in Chapter 20.

Our results show that the behavior of inductors is analogous to that of capacitors. You learned in Chapter 16 that capacitors store energy, with the energy proportional to the capacitor charge squared. Here you've seen that inductors store energy, with the energy proportional to the inductor current squared. These are instances of more general energy storage in electric and magnetic fields, as summarized in Table 19.1.

TABLE 19.1 Energy Storage in Electric and Magnetic Fields

Electric fields	Magnetic fields
Energy stored in capacitor: $U = \dfrac{Q^2}{2C}$	Energy stored in inductor: $U = \dfrac{1}{2}LI^2$
Energy density in an electric field: $u_E = \dfrac{1}{2}\epsilon_0 E^2$	Energy density in a magnetic field: $u_B = \dfrac{B^2}{2\mu_0}$

EXAMPLE 19.8 **Inside the MRI Solenoid**

You're having an MRI scan inside the solenoid of Example 19.7. But you're not alone: There's also magnetic energy in there. Find the energy density by (a) dividing the stored energy U by the solenoid's volume and (b) using Equation 19.10.

ORGANIZE AND PLAN We already found the stored energy: $U = 1.384$ MJ. The solenoid's volume is area times length, and its interior magnetic field is $B = \mu_0 nI$. That value of magnetic field is used in Equation 19.10 to find u_B.

Known: $d = 2.4$ m; $n = 500$ m^{-1}; $r = 47$ cm; $I = 2.3$ kA; $U = 1.384$ MJ.

SOLVE Dividing energy U by the solenoid's volume,

$$u_B = \frac{U}{Ad} = \frac{U}{\pi r^2 d} = \frac{1.384 \times 10^6 \text{ J}}{\pi (0.47 \text{ m})^2 (2.4 \text{ m})} = 831 \text{ kJ/m}^3$$

The magnetic field in the solenoid is

$$B = \mu_0 nI = (4\pi \times 10^{-7} \text{ T}\cdot\text{m/A})(500 \text{ m}^{-1})(2.3 \times 10^3 \text{ A})$$
$$= 1.445 \text{ T}$$

Using this field in Equation 19.10,

$$u_B = \frac{B^2}{2\mu_0} = \frac{(1.445 \text{ T})^2}{2(4\pi \times 10^{-7} \text{ T}\cdot\text{m/A})} = 831 \text{ kJ/m}^3$$

REFLECT The results agree. Although that energy density seems large, it's far less than, say, gasoline's 34-GJ/m^3 energy density.

MAKING THE CONNECTION How large an electric field is required for the same energy density?

ANSWER Using the electric energy density $u_E = \frac{1}{2}\epsilon_0 E^2$ gives $E = 4.3 \times 10^8$ N/C. That's more than 100 times the electric field that produces sparks in air!

RL Circuits

Figure 19.18 shows an ***RL* circuit**, with a battery, resistor, and inductor all in series. There's also a switch, which we'll close to get things going. As you've seen, the induced emf in an inductor acts to oppose changes in the inductor current. In Figure 19.18 all three components are in series, so the same current flows through each, and the inductor acts to oppose changes in that current. The result is that the current only gradually rises to the value $\mathcal{E}/R$ you'd expect for a battery with emf $\mathcal{E}$ across a resistor R. Calculus shows that the current as a function of time is

$$I = \frac{\mathcal{E}}{R}(1 - e^{-Rt/L})$$

where $t = 0$ when the switch closes. Figure 19.19 graphs this time-dependent current.

The slow rise of the current in Figure 19.19 results from the inductor "fighting" the change in current. This effect is similar to the charging capacitor in the *RC* circuits of Chapter 17. There we found a characteristic *time constant RC* required for significant

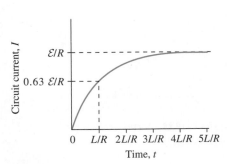

FIGURE 19.19 Current versus time in the *LR* circuit.

changes in the charge on a capacitor. For the inductor, we have an analogous time constant $\tau = L/R$ for significant changes in current. When $t = \tau = L/R$, the exponential factor $e^{-Rt/L} = e^{-1} \approx 0.37$, so the current has reached about 63% of its final value (because $1 - 0.37 = 0.63$). Increasing the inductance L lengthens the time constant τ, consistent with the inductance delaying the rise of the current.

✓ **TIP**

The RL time constant is $\tau = L/R$, analogous to the RC time constant $\tau = RC$.

LC Circuits

The ***LC* circuit**, shown in Figure 19.20, consists of a capacitor and inductor. There's no battery to drive current, but we'll assume the capacitor initially carries charge Q_0 (Figure 19.20a). At this point the capacitor "wants" to discharge, pushing current through the inductor. But as you've seen, the inductor opposes the change in current. Therefore, the capacitor can't discharge immediately. Again, the current rises gradually (Figure 19.20b).

Once the capacitor is completely discharged, the inductor current is at a maximum (Figure 19.20c). This makes sense from energy conservation: The fully charged capacitor

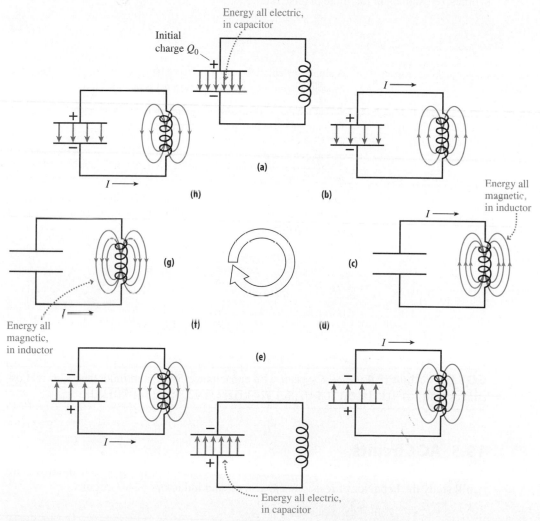

FIGURE 19.20 Flow of charge in an *LC* circuit.

contained stored energy. There's no resistance in the circuit to dissipate energy, so once the capacitor is discharged, that energy must be in the inductor. As current continues though the inductor, charge again builds on the capacitor. As the capacitor charges, its stored energy increases, so the stored energy $\frac{1}{2}LI^2$ and therefore the inductor current I must drop. Eventually the capacitor is fully charged, with polarity opposite the original configuration (Figure 19.20e). The process repeats, continuing indefinitely as long as energy isn't lost.

The result is an **electromagnetic oscillation**, analogous to the mechanical oscillations of simple harmonic motion in Chapter 7. In fact, the capacitor charge as a function of time is analogous to position in SHM:

$$Q = Q_0 \cos(\omega t) \quad \text{(Charge oscillation in } LC \text{ circuit; SI unit: C)} \quad (19.11)$$

where the angular frequency ω is $\omega = 1/\sqrt{LC}$, in analogy with the mechanical oscillator's $\omega = \sqrt{k/m}$. The "inertia" in the electric circuit is supplied by the inductor, and the "force" comes from the capacitor. This analogy will be useful later in the chapter, when we add a resistor to the LC circuit.

Older radios with analog tuning dials use LC circuits to select the desired station. Normally there's a fixed inductance and a variable capacitance, which you change by tuning the dial. The value of L and the range of C are chosen to match the desired band, for example, 540 to 1600 kHz for AM, or 88 to 108 MHz for FM. Modern radios use digitally synthesized signals to facilitate precise tuning.

EXAMPLE 19.9 **Tuning the Radio**

A radio has a 1.20-μH inductor and a variable capacitor. What capacitance range is needed to cover the FM band?

ORGANIZE AND PLAN The angular frequency of the LC oscillator is $\omega = 1/\sqrt{LC}$. Frequency f and angular frequency ω are related by $\omega = 2\pi f$. As stated above, the FM band covers 88 to 108 MHz.

Known: $L = 1.20\,\mu\text{H} = 1.20 \times 10^{-6}\,\text{H}$.

SOLVE With $\omega = 2\pi f = 1/\sqrt{LC}$, we solve for C:

$$C = \frac{1}{4\pi^2 f^2 L}$$

At the low-frequency end of the FM band,

$$C = \frac{1}{4\pi^2 f^2 L} = \frac{1}{4\pi^2 (88 \times 10^6\,\text{Hz})^2 (1.20 \times 10^{-6}\,\text{H})}$$

$$= 2.7 \times 10^{-12}\,\text{F} = 2.7\,\text{pF}$$

A similar calculation with $f = 108$ MHz gives $C = 1.8$ pF. Thus, the capacitance must be variable between 1.8 pF and 2.7 pF.

REFLECT The range of FM frequencies is fairly narrow, so a modest variation in capacitance covers the whole band.

MAKING THE CONNECTION Would a larger or smaller capacitance be needed for AM radio?

ANSWER AM frequencies are lower than FM. As this example shows, capacitance is proportional to $1/f^2$, so a larger capacitance is needed if we keep the same inductance.

GOT IT? Section 19.4 An LC circuit with capacitance 50 pF and inductance 1.0 mH oscillates at (a) 4470 kHz; (b) 50 kHz; (c) 124 kHz; (d) 920 kHz; (e) 710 kHz.

19.5 AC Circuits

An **AC circuit** is one in which potential differences and currents vary sinusoidally. Here you'll study the behavior of resistors, capacitors, and inductors in AC circuits.

Resistors in AC Circuits

Figure 19.21a shows an AC emf connected across a resistor R. As in Section 19.3, we'll describe the emf by $\mathcal{E} = \mathcal{E}_{max}\sin(\omega t)$, where $\mathcal{E}_{max}$ is the emf's peak value. As usual, ω is the angular frequency $\omega = 2\pi f$, corresponding to frequency f in Hz (cycles per second). The potential difference V_R across the resistor is equal to the emf, and the current through the resistor follows from Ohm's law, $V_R = IR$:

$$I = \frac{\mathcal{E}}{R} = \frac{\mathcal{E}_{max}\sin(\omega t)}{R} = I_{max}\sin(\omega t)$$

where $I_{max} = \mathcal{E}_{max}/R$ is the peak current. This equation shows that the current and potential difference are **in phase**, meaning they peak together (Figure 19.21b).

In Chapter 17 we found the power dissipated in a resistor: $P = IV = I^2R = V^2/R$. In an AC circuit, current and potential difference usually vary at a high frequency, so it's more important to know the *average power* over many cycles. Using $P = I^2R$, the average power is $\overline{P} = \overline{I^2R} = I_{max}^2 R\sin^2(\omega t)$, where I_{max} is the peak current. The function $\sin^2$ swings symmetrically from 0 to 1, and its average over one or more complete cycles is $\frac{1}{2}$. Therefore, $\overline{P} = \frac{1}{2}I_{max}^2 R$. It's convenient to define the **root-mean-square current** I_{rms} (similar to root-mean-square speed in Chapter 12): $I_{rms} = \sqrt{\overline{I^2}} = \sqrt{I_{max}^2/2} = I_{max}/\sqrt{2}$. Thus

$$I_{rms} = \frac{1}{\sqrt{2}}I_{max} \qquad \text{(rms current; SI unit: A)} \qquad (19.12)$$

With this definition, average power becomes $\overline{P} = I_{rms}^2 R$, like the familiar I^2R. Similarly, the emf has root-mean-square value $\mathcal{E} = \mathcal{E}_{max}/\sqrt{2}$, showing that the source supplies energy at the average rate $\overline{P} = I_{rms}\mathcal{E}_{rms}$. The current and emf listed on AC devices are rms values. Most home appliances are rated at 120 V rms, but some use 240 V rms.

AC power is often supplied as three distinct sinusoidal emfs, each separated in time by one-third of a cycle ($120°$). This *three-phase power* helps smooth the highs and lows of current, making large motors run more smoothly and reducing mechanical stress. Three-phase power is common in industrial and institutional settings, while household power is single phase.

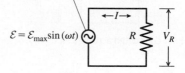

Standard symbol for AC power supply

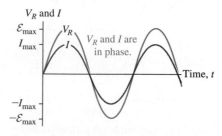

(a) Resistor connected across AC source

V_R and I — V_R and I are in phase. — Time, t

(b) Current in circuit and potential difference across resistor as functions of time

FIGURE 19.21 (a) Circuit with an AC source and a single resistor. (b) Current and potential difference are in phase in the resistor.

EXAMPLE 19.10 **Current and Power Use**

A wall outlet supplies AC at emf $\mathcal{E}_{rms} = 117$ V. Find (a) the peak emf and (b) the rms and peak current in an 1800-W hair dryer.

ORGANIZE AND PLAN The rms and peak emf are related by $\mathcal{E}_{rms} = \mathcal{E}_{max}/\sqrt{2}$. That 1800 W is the hair dryer's average power consumption, given by $\overline{P} = I_{rms}\mathcal{E}_{rms}$. Finally, rms and peak currents are related by $I_{rms} = I_{max}/\sqrt{2}$.

Known: $\overline{P} = 1800$ W.

SOLVE With $\mathcal{E}_{rms} = \mathcal{E}_{max}/\sqrt{2}$, the peak emf is $\mathcal{E}_{max} = \sqrt{2}\mathcal{E}_{rms} = \sqrt{2}(117\text{ V}) = 165$ V. With average power of 1800 W, the rms and peak current are then

$$I_{rms} = \frac{\overline{P}}{\mathcal{E}_{rms}} = \frac{1800\text{ W}}{117\text{ V}} = 15.4\text{ A}$$

and $I_{max} = \sqrt{2}I_{rms} = 21.8$ A.

REFLECT Many home circuit breakers are rated at 20 A. Will this hair dryer blow a 20-A breaker? No, because that 20-A rating is for rms current, not peak.

MAKING THE CONNECTION You're traveling in Europe, and you connect this hair dryer to a European 240-V rms outlet. How much power does the dryer consume?

ANSWER The rms current is proportional to the emf, so the current doubles. With current and emf both doubled, the average power $\overline{P} = I_{rms}\mathcal{E}_{rms}$ quadruples to 7200 W. What's the result? Your hair dryer gets fried—and could start a fire. Fortunately, European outlets have a different prong configuration to prevent just such problems. You can buy a transformer that plugs into the 240-V outlet and reduces the emf to a safe 120 V.

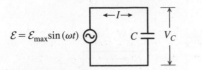

(a) Capacitor connected across AC source

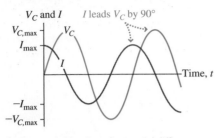

(b) Current in circuit and potential difference across capacitor as functions of time

FIGURE 19.22 Current leads potential difference by 90° in a capacitor.

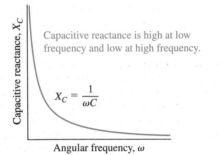

FIGURE 19.23 Capacitive reactance is a function of frequency.

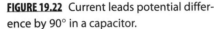

(a) Low-pass filter   (b) High-pass filter

FIGURE 19.24 Two kinds of filters.

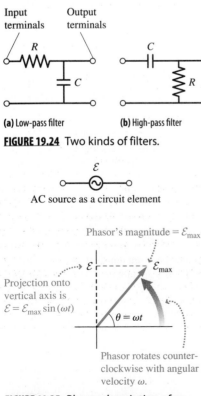

AC source as a circuit element

FIGURE 19.25 Phasor description of an AC emf.

AC Circuits with Capacitors

Figure 19.22a shows a capacitor connected across an AC emf. As with the resistor of Figure 19.21, the potential difference V_C across the capacitor equals the source emf. But now the current and potential difference are *not* in phase. Figure 19.22b plots I and V_C over one cycle. Recall that the capacitor's potential difference V_C is related to the capacitor charge Q by $Q = CV_C$. In this case it's current that moves charge on and off the capacitor plates. When the charge—and therefore V_C—are changing most rapidly, the current must be at its peak. And when the charge peaks, the current is zero. This explains why potential difference and current are out of phase by one-fourth of a cycle, or 90°.

The exact relation between the capacitor's potential difference and current flowing to the capacitor follows from calculus: With potential difference $V_C = V_{C,max} \sin(\omega t)$, the current is $I = \omega C V_{C,max} \sin(\omega t + 90°)$. This equation, along with the graphs in Figure 19.22b, reflect an important fact about capacitors:

> The current to a capacitor leads the capacitor's potential difference by 90°.

Capacitive Reactance

Because the maximum value of the sine function is 1, the equation for the current $I = \omega C V_{C,max} \sin(\omega t + 90°)$ shows that the peak current I_{max} is $I_{max} = \omega C V_{C,max}$. We can make this relationship look like the familiar relationship $I = V/R$ by defining **capacitive reactance** X_C as

$$X_C = \frac{1}{\omega C} \quad \text{(Capacitive reactance; SI unit: } \Omega) \quad (19.13)$$

Then the peak current becomes $I_{max} = V_{C,max}/X_C$.

The units of reactance are ohms, just like resistance. But reactance isn't quite the same as resistance. Using $I_{max} = V_{C,max}/X_C$, reactance relates the *peak* values of potential difference and current. But it doesn't tell the whole story, because it doesn't describe the 90° phase difference. Therefore, $I_{max} = V_{C,max}/X_C$ does *not* relate instantaneous values.

Unlike resistance, reactance depends on frequency. Figure 19.23 plots capacitive reactance as a function of the angular frequency ω. The reactance is low at high frequencies and high at low frequencies. Reactance approaches infinity as the frequency approaches zero. This makes sense, because zero frequency means a constant potential difference, thus no change in the capacitor's charge, and therefore no current flowing to or from the capacitor.

Filters are an important application of frequency-dependent reactance. These circuits preferentially pass either lower or higher frequencies. In the low-pass filter of Figure 19.24a, little current flows to the capacitor at low frequencies because of its high reactance. So there's not much potential difference across the resistor, making the potential difference at the output nearly the same as at the input. At higher frequencies, current flows to the capacitor, making a greater drop across the resistor and a lower potential difference at the output terminals. The converse applies to the high-pass filter (Figure 19.24b), where the capacitor effectively blocks low frequencies. Filters are widely used to eliminate electrical "noise"; for example, a high-pass filter can remove the pervasive 60-Hz noise from AC power wiring in any circuit where you're more interested in higher frequencies. Filters with variable resistances enable the bass and treble controls on audio equipment. Finally, a combination of low-pass and high-pass filters lets you select a narrow range of frequencies.

Phasor Analysis

Phasors provide a graphical approach to AC circuits. A phasor is a vector representing potential difference or current. Its magnitude corresponds to the peak value, and the phasor rotates counterclockwise with the angular frequency ω of the AC signal. The instantaneous value of the potential difference or current is then the projection of the phasor onto the vertical axis, as shown in Figure 19.25. This analysis should remind you of Chapter 7's comparison between simple harmonic motion and uniform circular motion, in which the projection of uniform circular motion onto an axis gave simple harmonic motion.

Figure 19.26a shows an AC emf in series with a resistor and capacitor. Since all components are in series, the current is the same throughout the circuit. Potential difference and current are in phase in the resistor, so the corresponding phasors are aligned (Figure 19.26b). Current leads potential difference by 90° in the capacitor, so the phasor representing the capacitor's potential difference is 90° behind the current phasor. This 90° angle remains fixed as the phasors rotate.

By Kirchoff's loop rule, the emf of the AC source equals the sum of the potential differences across the resistor and capacitor. Because of the phase difference, you can't just add the peak potential differences. Instead, as Figure 19.26c shows, you add them as vectors. Thus, as Figure 19.26d shows, the relation between the peak potential differences is $\mathcal{E}_{max}^2 = V_{R,max}^2 + V_{C,max}^2$. Now, $V_{R,max} = I_{max}R$ and $V_C = I_{max}X_C$ for resistor and capacitor, respectively, so we can write $\mathcal{E}_{max}^2 = I_{max}^2(R^2 + X_C^2)$. Thus

$$I_{max} = \frac{\mathcal{E}_{max}}{\sqrt{R^2 + X_C^2}}$$

Multiplying this current by R and X_C then gives the individual potential differences across the two components. For the capacitor, the result is

$$V_{C,max} = I_{max}X_C = \frac{\mathcal{E}_{max}X_C}{\sqrt{R^2 + X_C^2}}$$

At low frequencies, X_C is huge, and you can convince yourself that this makes $V_{C,max}$ very nearly equal to $\mathcal{E}_{max}$. But at high frequencies X_C is small, giving a very low potential difference across the capacitor. That's a quantitative analysis of the low-pass filter discussed earlier.

Our phasor diagram (Figure 19.26d) shows something else: The current in our series circuit leads the source emf, but not by 90°. As you can see from the right triangle, the phase difference is $\tan^{-1}(V_{C,max}/V_{R,max})$. Because the capacitor's reactance and therefore V_C change with frequency, so does this phase relation.

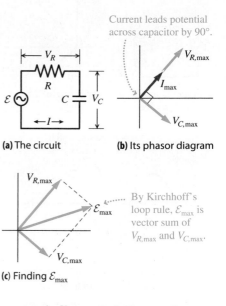

Current leads potential across capacitor by 90°.

(a) The circuit   (b) Its phasor diagram

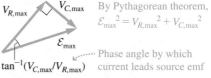

By Kirchhoff's loop rule, $\mathcal{E}_{max}$ is vector sum of $V_{R,max}$ and $V_{C,max}$.

(c) Finding $\mathcal{E}_{max}$

By Pythagorean theorem, $\mathcal{E}_{max}^2 = V_{R,max}^2 + V_{C,max}^2$

Phase angle by which current leads source emf

$\tan^{-1}(V_{C,max}/V_{R,max})$

(d) Magnitude of $\mathcal{E}_{max}$ and phase angle

FIGURE 19.26 (a) Series *RC* circuit with AC power supply. (b)–(d) Phasor analysis of the series *RC* circuit.

EXAMPLE 19.11 Tweeter!

Your stereo speakers contain a large "woofer" to reproduce low-frequency sound and a "tweeter" for high frequencies. A series capacitor helps "steer" high-frequency power to the tweeter. Treat the tweeter as having an 8.0-Ω resistance and take $C = 25\,\mu F$. If your amplifier develops a peak potential difference of 50 V at its output, what's the peak current in the speaker when the frequency f is (a) 5.0 kHz and (b) 60 Hz?

ORGANIZE AND PLAN The circuit looks like Figure 19.26a, with the tweeter being R and the amplifier being the source emf. Frequency f and angular frequency ω are related by $\omega = 2\pi f$. Our phasor analysis gave the peak current: $I_{max} = \mathcal{E}_{max}/\sqrt{R^2 + X_C^2}$, where $X_C = 1/\omega C$.

Known: $\mathcal{E}_{max} = 50\,V$, $R = 8.0\,\Omega$, $C = 25\,\mu F$, $f = 5.0\,kHz$ and 60 Hz.

SOLVE (a) The capacitive reactance depends on frequency; at 5.0 kHz:

$$X_C = \frac{1}{\omega C} = \frac{1}{2\pi f C} = \frac{1}{2\pi(5.0 \times 10^3\,Hz)(25 \times 10^{-6}\,F)} = 1.27\,\Omega$$

Then the peak current at 5.0 kHz is

$$I_{max} = \frac{\mathcal{E}_{max}}{\sqrt{R^2 + X_C^2}} = \frac{50\,V}{\sqrt{(8.0\,\Omega)^2 + (1.27\,\Omega)^2}} = 6.2\,A$$

(b) Repeating the method of part (a),

$$X_C = \frac{1}{\omega C} = \frac{1}{2\pi f C} = \frac{1}{2\pi(60\,Hz)(25 \times 10^{-6}\,F)} = 106\,\Omega$$

$$I_{max} = \frac{\mathcal{E}_{max}}{\sqrt{R^2 + X_C^2}} = \frac{50\,V}{\sqrt{(8.0\,\Omega)^2 + (106\,\Omega)^2}} = 0.47\,A$$

REFLECT The capacitor is doing its job, providing a higher tweeter current at high frequencies. In the next section you'll see how an inductor steers low-frequency power to the woofer.

MAKING THE CONNECTION How does increasing the capacitance affect the peak current?

ANSWER Increasing the capacitance decreases the reactance X_C, resulting in a larger current. In this case the effect is more pronounced at low frequencies, because the capacitive reactance at 5 kHz is already much smaller than the resistance R.

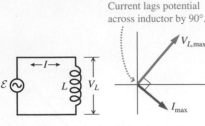

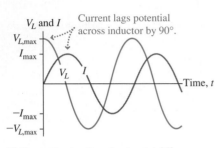

(a) Inductor connected across AC source   **(b)** Its phasor diagram

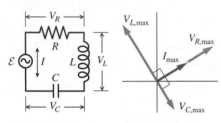

(c) Current in circuit and potential difference across resistor as functions of time

FIGURE 19.27 Potential difference leads current in an *RL* circuit.

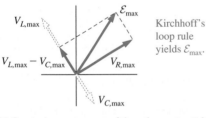

(a) Series *RLC* circuit   **(b)** Phasor diagram

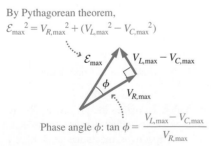

(c) $\mathcal{E}_{max}$ is the vector sum of the other potentials

By Pythagorean theorem,

$\mathcal{E}_{max}{}^2 = V_{R,max}{}^2 + (V_{L,max}{}^2 - V_{C,max}{}^2)$

Phase angle ϕ: $\tan \phi = \dfrac{V_{L,max} - V_{C,max}}{V_{R,max}}$

(d) Magnitude of $\mathcal{E}_{max}$ and phase angle ϕ

FIGURE 19.28 Analysis of a series *RLC* circuit. Just as in any series circuit, the order of the circuit elements is interchangeable.

AC Circuits with Inductors

Figure 19.27a shows an inductor connected across an AC emf. You know from Section 19.4 that the induced emf in an inductor opposes the change in magnetic flux. Thus the induced emf is greatest when the current is changing rapidly, and it's zero when the current isn't changing. Thus, the current lags 90° behind the potential difference, as shown in the phasor diagram (Figure 19.27b). Figure 19.27c shows the resulting current and induced emf as functions of time; mathematically, they're given by

$$V_L = V_{L,max} \cos (\omega t)$$

$$I = \frac{V_{L,max}}{\omega L} \cos (\omega t - 90°)$$

Graphs and equations both show that

> The current in an inductor lags the inductor's potential difference by 90°.

In analogy with capacitors, we define **inductive reactance** X_L as

$$X_L = \omega L \qquad \text{(Inductive reactance; SI unit: } \Omega\text{)} \qquad (19.14)$$

For example, if a standard 60-Hz AC power supply is connected across a 0.10-H inductor, the inductive reactance is $X_L = \omega L = 2\pi f L = 2\pi (60 \text{ Hz})(0.10 \text{ H}) = 38 \ \Omega$. Given inductive reactance, the peak current and potential difference are related by

$$I_{max} = \frac{V_{L,max}}{X_L}$$

Note that inductive reactance increases with increasing frequency. This makes sense, given Faraday's law. A higher frequency implies a more rapid change in magnetic flux through the inductor, thus inducing a larger emf for a given peak current.

GOT IT? Section 19.5 Rank in decreasing order the reactance values of the following capacitors operating at these frequencies: (a) $C = 25 \ \mu F$ at 60 Hz; (b) $C = 50 \ \mu F$ at 120 Hz; (c) $C = 30 \ \mu F$ at 90 Hz.

19.6 *RLC* Circuits and Resonance

Section 19.5 introduced basic AC circuits, including phasor analysis. We'll now apply what you've learned to the more complex *RLC* series circuit (Figure 19.28a). We'll then discuss the important practical applications of this circuit.

As with any series circuit, the current in our *RLC* circuit is the same everywhere at a given instant. You also know that for a series circuit the source emf equals the sum of the potential differences across the rest of the circuit. But as with the single-capacitor circuit of Section 19.5, the current and potential difference are out of phase. Therefore, we need phasor analysis to find the peak current.

Phasor Analysis of the *RLC* Circuit

Figure 19.28b shows current and potential difference phasors for the series *RLC* circuit. It's like the *RC* phasor diagram in Section 19.5, but with the additional V_L phasor for the inductor. Phasor addition (Figure 19.28c) gives $\mathcal{E}_{max}^2 = V_{R,max}^2 + (V_{L,max} - V_{C,max})^2$. We've assumed arbitrarily that $V_{L,max} > V_{C,max}$. That's not necessarily so, but it doesn't matter in our analysis, because $V_{L,max} - V_{C,max}$ is squared. The potential difference and current in the resistor, capacitor, and inductor are related by $V_{R,max} = I_{max}R$,

$V_{C,\text{max}} = I_{\text{max}}X_C$, and $V_{L,\text{max}} = I_{\text{max}}X_L$. Thus $\mathcal{E}_{\text{max}}^2 = I_{\text{max}}^2 R^2 + I_{\text{max}}^2 (X_L - X_C)^2$, so the peak current is

$$I_{\text{max}} = \frac{\mathcal{E}_{\text{max}}}{\sqrt{R^2 + (X_L - X_C)^2}} \quad \begin{array}{l}\text{(Peak current in a series } RLC \\ \text{circuit; SI unit: A)}\end{array} \quad (19.15)$$

Impedance and Phase

To make this result look like the familiar $I = V/R$, we define the **impedance**, Z, of the *RLC* series combination as $Z = \sqrt{R^2 + (X_L - X_C)^2}$. Like resistance and reactance, impedance is measured in ohms. Given impedance, the peak current follows from the Ohm's-law-like relation $I_{\text{max}} = \mathcal{E}_{\text{max}}/Z$. Impedance is similar to resistance, in that higher impedance implies lower current.

Figure 19.28d shows the **phase angle** ϕ between source emf and current. Trigonometry gives $\tan \phi = (V_{L,\text{max}} - V_{C,\text{max}})/V_{R,\text{max}}$. In terms of resistance and reactance, $\tan \phi = (I_{\text{max}}X_L - I_{\text{max}}X_C)/I_{\text{max}}R$, or

$$\tan \phi = \frac{X_L - X_C}{R} \quad \text{(Phase angle, series } RLC) \quad (19.16)$$

Equation 19.16 shows that the phase angle is positive when $V_{L,\text{max}} > V_{C,\text{max}}$ (as in Figure 19.28d) and negative when $V_{L,\text{max}} < V_{C,\text{max}}$. A positive phase angle means that the current lags the source emf, and a negative phase angle means that the current leads.

CONCEPTUAL EXAMPLE 19.12   **Capacitive and Inductive Circuits**

Use the results above to describe the phase angle and the peak current in the following circuits: (a) an AC emf connected across a single capacitor C and (b) an AC emf connected across a single inductor L.

SOLVE (a) For a capacitor, $X_C = 1/\omega C$. There's no resistor or inductor, so $R = X_L = 0$. By Equation 19.16, $\tan \phi = (X_L - X_C)/R = -\infty$, so $\phi = -90°$. This agrees with our analysis in Section 19.5, where we found that the current leads potential difference by 90° in a capacitor. The maximum current is $I_{\text{max}} = \mathcal{E}_{\text{max}}/Z$. With just a capacitor, $Z = X_C$, so $I_{\text{max}} = \mathcal{E}_{\text{max}}/X_C$, which is also the result for the single-capacitor circuit in Section 19.5.

(b) For the inductor alone, $X_L = \omega L$, and $R = X_C = 0$, so $\tan \phi = (X_L - X_C)/R = \infty$, so $\phi = 90°$. Now the current lags the potential difference, which was our result for an inductor in Section 19.5. With $Z = X_L$ for a single inductor, the peak current is $I_{\text{max}} = \mathcal{E}_{\text{max}}/Z = \mathcal{E}_{\text{max}}/X_L$, again consistent with Section 19.5.

REFLECT This example suggests that the impedance and phase angle are general quantities that apply to any combination of resistors, inductors, and capacitors. For a circuit with just one of these components, the simple results of Section 19.5 apply; more complex circuits, like the series *RLC* circuit, have phase and peak current determined by the combined impedance.

Power in AC Circuits

The reactance of capacitors and inductors is like resistance, but not exactly, because of the phase difference. With potential difference and current 90° out of phase, the power $P = IV$ in these components averages to zero over one cycle (Figure 19.29a). So, unlike resistors (Figure 19.29b), they don't dissipate energy; instead, they alternately store and release it. More generally, the power dissipation in a circuit depends on the phase angle. Our *RLC* phasor triangle shows that $\cos \phi = V_{R,\text{max}}/\mathcal{E}_{\text{max}} = I_{\text{max}}R/I_{\text{max}}Z = R/Z$. The only power dissipation is in the resistor, which, as you learned in Section 19.5, is $\overline{P} = I_{\text{rms}}^2 R$. Given $\cos \phi = R/Z$, we can solve for R to write the power as

$$\overline{P} = I_{\text{rms}}^2 Z \cos \phi \quad \text{(Average power in an AC circuit; SI unit: W)} \quad (19.17)$$

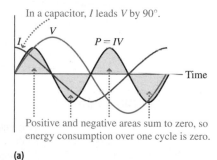

In a capacitor, I leads V by 90°.

Positive and negative areas sum to zero, so energy consumption over one cycle is zero.

(a)

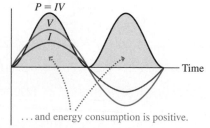

In a resistor, I and V are in phase …

…and energy consumption is positive.

(b)

FIGURE 19.29 (a) Average energy consumption in a capacitor (or inductor) is zero, but (b) a resistor always dissipates energy.

The factor $\cos \phi$ is the **power factor**. Absent the resistor, you saw in Conceptual Example 19.12 that $\phi = \pm 90°$. Since $\cos(90°) = \cos(-90°) = 0$, this confirms that neither the capacitor nor the inductor dissipates power. On the other hand, the average power is maximized when the power factor $\cos \phi = 1$, which corresponds to $\phi = 0$. This occurs when $R = Z$, which, with $Z = \sqrt{R^2 + (X_L - X_C)^2}$, means that capacitive and inductive reactances must be equal: $X_C = X_L$.

AC circuits are most efficient with power factors close to 1. Equation 19.17 shows that a lower power factor means higher currents are needed to deliver a given power, meaning more losses in transmission lines. That's played a role in major power failures, where current and potential difference got too far out of phase, thus lowering the power factor. Adjustable capacitances in the power grid can mitigate this effect.

Resonance

Something else happens when $X_C = X_L$: The denominator in Equation 19.15 takes its minimum value, resulting in the largest possible peak current. With $X_C = 1/\omega C$ and $X_L = \omega L$, this condition is $1/\omega C = \omega L$, or

$$\omega_0 = \sqrt{\frac{1}{LC}} \qquad \text{(Resonant frequency; SI unit: Hz)} \qquad (19.18)$$

Equation 19.18 gives the circuit's **resonant frequency**, designated ω_0. If you adjust the frequency of the source in Figure 19.28a, you'll see the current rise to a maximum at resonance and then decline (Figure 19.30). The resonant frequency is independent of the circuit resistance, but lower resistance leads to higher peak current, which steepens the resonance curve.

Resonance in the RLC circuit should remind you of resonance in the simple harmonic oscillator of Chapter 7. In Section 19.4, we made an analogy between the LC circuit and the harmonic oscillator. The RLC series circuit is analogous to the damped, driven oscillator of Section 7.5. Adding resistance R is like adding a frictional force to the mechanical oscillator, because both dissipate energy. To keep the oscillations going, the mechanical oscillator needs energy supplied by an external force. Similarly, the RLC circuit needs energy from the source of AC emf. Both the harmonic oscillator and the driven RLC circuit have maximum response at the same frequency as their undamped natural oscillations, $\omega = \sqrt{k/m}$ for the mechanical oscillator and $\omega = 1/\sqrt{LC}$ for the electrical oscillator.

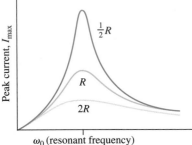

FIGURE 19.30 Resonance curves for RLC circuits with three different values of resistance.

Powering a Motor

A motor operates at 120 V and 60 Hz. It has negligible capacitance, has a power factor of 0.90, and uses 1250 W of power. (a) What's the rms current? (b) What's the motor's resistance?

ORGANIZE AND PLAN Equation 19.17 relates average power and rms current: $\overline{P} = I_{rms}^2 Z \cos \phi$. The impedance Z is also $Z = \mathcal{E}_{max}/I_{max} = \mathcal{E}_{rms}/I_{rms}$, so $\overline{P} = I_{rms}\mathcal{E}_{rms}\cos\phi$. To find the motor's resistance, recall that $\overline{P} = I_{rms}^2 R$.

Known: $\mathcal{E}_{rms} = 120$ V, $f = 60$ Hz, $\overline{P} = 1250$ W.

SOLVE (a) Using $\overline{P} = I_{rms}\mathcal{E}_{rms}\cos\phi$, the rms current is

$$I_{rms} = \frac{\overline{P}}{\mathcal{E}_{rms}\cos\phi} = \frac{1250 \text{ W}}{(120 \text{ V})(0.90)} = 11.6 \text{ A}$$

(b) With $\overline{P} = I_{rms}^2 R$, the motor's resistance is

$$R = \frac{\overline{P}}{I_{rms}^2} = \frac{1250 \text{ W}}{(11.6 \text{ A})^2} = 9.3 \text{ }\Omega$$

REFLECT That 11.6 A is more current than a 1.25-kW motor would draw with a power factor of 1. The wires supplying the motor carry this additional current, and that results in greater power loss.

MAKING THE CONNECTION What's the inductance in this circuit?

ANSWER We're told there's no capacitance, so $\cos \phi < 1$ means there must be inductance. Note that $Z = R/\cos\phi = 10.3$ Ω. Then with $Z = \sqrt{R^2 + (X_L - X_C)^2}$ and $X_L = \omega L$, you can solve for L to find $L = 0.14$ mH. That's enough to reduce the power factor in this circuit to 0.90.

GOT IT? Section 19.6   To double the resonant frequency of an *RLC* circuit, you should change the inductance by a factor of (a) 0.25; (b) 0.50; (c) 2; (d) 4.

Chapter 19 in Context

This chapter's big idea is *electromagnetic induction*. It's quantified in *Faraday's law*, which shows how changing *magnetic flux* induces an emf that, in turn, can drive a current. One example is *motional emf*, resulting from conductors moving in magnetic fields; it's the basis of *electric generators*. Changing magnetic flux also enables *transformers* that step AC potential differences up or down. Another practical application is the *inductor*, a device whose inductive properties complement the properties of capacitors in electric circuits. You've seen how resistors, inductors, and capacitors—alone and in combination—behave in *AC circuits*. In particular, inductors and capacitors introduce frequency-dependent behavior.

Looking Ahead The AC circuits you studied here lead naturally to electromagnetic waves, the subject of Chapter 20. Understanding electromagnetic waves, including visible light, the foundation for special relativity, also covered in Chapter 20. Your study of light continues with geometrical and physical optics in Chapters 21 and 22. Light plays a significant role in atomic and quantum phenomena, discussed in Chapters 23–26.

CHAPTER 19 SUMMARY

Induction and Faraday's Law

(Section 19.1) In **electromagnetic induction**, a **changing magnetic flux** induces an emf, as described by **Faraday's law**. The emf gives rise to a current if a complete circuit is present. **Lenz's law** states that the induced current opposes the change in magnetic flux.

Surface with area A

Line perpendicular to surface

θ = angle between $\vec{B}$ and line perpendicular to surface.

Magnetic flux: $\Phi = BA \cos \theta$

Faraday's law of induction: $\mathcal{E} = -N\dfrac{\Delta \Phi}{\Delta t}$ for an N-turn coil

Motional emf

(Section 19.2) **Motional emf** is the emf induced when there is relative motion between a conductor and a magnetic field. If the conductor is a solid piece, induced **eddy currents** develop and dissipate energy. This effect is the basis of magnetic braking and other applications.

Current induced by motional emf in closed loop: $I = \dfrac{|\mathcal{E}|}{R} = \dfrac{BLv}{R}$

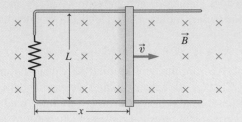

Generators and Transformers

(Section 19.3) An electric generator converts mechanical energy to electrical energy. The **rotating-coil generator** consists of conducting coils rotating in a magnetic field; it produces alternating current at frequency equal to its rotation rate. **Transformers** use electromagnetic induction to change emf in AC circuits.

emf induced in a rotating N-turn coil: $\mathcal{E} = NBA\omega \sin(\omega t)$

Transformer emf: $\mathcal{E}_s = \mathcal{E}_p \dfrac{N_s}{N_p}$

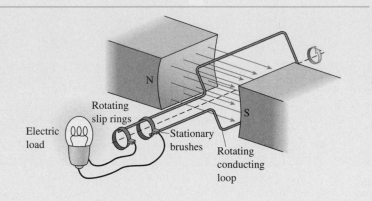

Rotating slip rings

Electric load

Stationary brushes

Rotating conducting loop

Inductance

(Section 19.4) Through **inductance**, a changing current in one circuit induces current in another circuit. An **inductor** is a coil designed for its inductive properties. The reciprocal effect between two circuits is known as **mutual inductance**. A single coil of wire can induce a current in itself through **self-inductance**. Inductors store energy in the magnetic field. When an inductor is placed in a circuit with a resistor (an **RL circuit**), inductance delays the growth of current in the circuit. When an inductor is placed in series with a capacitor (an **LC circuit**), the charge oscillates.

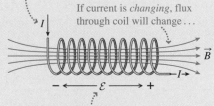

Current in wire creates magnetic flux through coil.

If current is *changing*, flux through coil will change...

...inducing an emf in the coil that *opposes* the change in current.

Definition of inductance: $L = \dfrac{N\Phi}{I}$

Energy stored in an inductor: $U = \frac{1}{2}LI^2$

Energy density in a magnetic field: $u_B = \dfrac{B^2}{2\mu_0}$

Charge oscillation in LC circuit: $Q = Q_0 \cos(\omega t)$

AC Circuits

(**Section 19.5**) The AC current to a capacitor leads the capacitor's potential difference by 90°. The AC current to an inductor lags the inductor's potential difference by 90°. Capacitors and inductors have **reactance**, which acts somewhat like resistance but also introduces a phase difference between current and potential difference. Phasors allow graphical analysis of AC circuits.

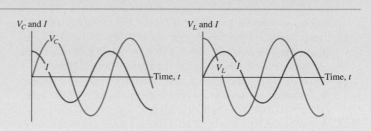

rms current: $I_{rms} = I_{max}/\sqrt{2}$

Capacitive reactance: $X_C = 1/\omega C$

Inductive reactance: $X_L = \omega L$

RLC Circuits and Resonance

(**Section 19.6**) An *RLC* series circuit has an **impedance** Z, analogous to the resistance of a DC circuit. The **phase angle** ϕ tells how much the current is ahead of or behind the emf. **Resonance** occurs when $X_C = X_L$ and is analogous to resonance in a mechanical oscillator.

Peak current in *RLC* series circuit: $I_{max} = \dfrac{\mathcal{E}_{max}}{\sqrt{R^2 + (X_L - X_C)^2}}$

Resonant frequency: $\omega_0 = \sqrt{\dfrac{1}{LC}}$

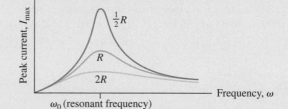

NOTE: Problem difficulty is labeled as ■ straightforward to ■ ■ ■ challenging. Problems labeled BIO are of biological or medical interest.

Conceptual Questions

1. Show that with SI units on the right side of Faraday's law (Equation 19.2), the emf comes out in volts.
2. Explain why magnetic flux decreases as the angle θ increases from 0 to 90°.
3. Is the magnetic flux through a horizontal loop larger at the equator or the North Pole?
4. Explain why the magnetic brake shown in Figure 19.10 works equally well when the metal plate is moving into or out of the magnetic field but doesn't work when the plate is entirely within the uniform magnetic field.
5. In the dropping-magnet demonstration (Conceptual Example 19.4), the magnet has very little kinetic energy when it exits the tube. What happened to most of the gravitational potential energy it had at the top?
6. Suppose in Figure 19.7 that the magnetic field is out of the page instead of in. What are (a) the direction of the induced current, (b) the induced emf, and (c) the direction of the force on the sliding bar?
7. Suppose the bar in Figure 19.7 moves leftward instead of rightward. What are (a) the direction of the induced current, (b) the induced emf, and (c) the direction of the force on the sliding bar?
8. In an *RL* circuit, is there more energy in the inductor just after the switch is closed or much later?
9. How is an *LC* circuit like a simple harmonic oscillator?
10. How is an *RLC* circuit like a damped harmonic oscillator?
11. Suppose you know the resistance, inductive reactance, and capacitive reactance for a series *RLC* circuit. Is the total impedance for the circuit equal to the sum of these three quantities? Why or why not?

12. For which of the following AC circuits does impedance depend on frequency? A circuit with (a) only a resistor; (b) a resistor and inductor; (c) a resistor and capacitor.
13. In a series *RLC* circuit, is energy dissipated in the inductor or capacitor? Explain.
14. A series *RLC* circuit has $X_C > X_L$. To bring the circuit into resonance, should you increase or decrease the capacitance?
15. An *RLC* circuit has power factor 0.90. From this information, can you tell whether the current leads or lags the emf?

Multiple-Choice Problems

16. A 2.3-T magnetic field lies perpendicular to a 35-cm-diameter circular wire loop. The magnetic flux through the loop is (a) 0.07 T · m²; (b) 0.14 T · m²; (c) 0.22 T · m²; (d) 0.28 T · m².
17. A 1.9-T magnetic field makes a 60° angle with a square loop 50 cm on a side. The magnetic flux through the loop is (a) 0.24 T · m²; (b) 0.41 T · m²; (c) 0.48 T · m²; (d) 0.75 T · m².
18. A 130-turn circular coil has diameter 5.2 cm. A magnetic field perpendicular to the coil is changing at 0.75 T/s. The induced emf in the coil is (a) 0.07 V; (b) 0.14 V; (c) 0.21 V; (d) 0.28 V.
19. A 50-turn circular coil has diameter 6.2 cm and resistance 0.75 Ω. A magnetic field perpendicular to the coil is changing at 0.50 T/s. The induced current in the coil is (a) 10 mA; (b) 30 mA; (c) 100 mA; (d) 300 mA.
20. For the apparatus shown in Figure 19.7, the magnetic field strength is 4.5 T and the rail separation is $L = 25$ cm. What's the induced emf when the bar moves at 35 cm/s? (a) 0.39 V; (b) 0.45 V; (c) 0.53 V; (d) 0.65 V.

21. A generator has a coil with area 0.12 m², rotating at 60 Hz in a 0.47-T magnetic field. If the generator's peak emf is 340 V, the number of turns in the coil is (a) 16; (b) 32; (c) 50; (d) 100.

22. The primary coil in a transformer has 100 turns and a 120-V emf. If you want a secondary emf of 240 V, the number of turns in the secondary coil should be (a) 25; (b) 50; (c) 200; (d) 400.

23. You have a 320-mH inductor. Over what time must you increase the inductor current by 2.0 A in order to induce a 5.0-V emf? (a) 0.13 s; (b) 0.26 s; (c) 0.39 s; (d) 0.52 s.

24. A radio receiver has a 25-μH inductance. The capacitance required to tune in a 1470-kHz radio station is (a) 0.47 nF; (b) 1.88 nF; (c) 3.76 nF; (d) 18.8 nF.

25. An AC power source produces a 170-V peak emf at 60 Hz. What's the average power dissipated in a 480-Ω resistor connected across this source? (a) 75 W; (b) 60 W; (c) 42 W; (d) 30 W.

26. A variable capacitor is connected across a variable-frequency AC source. The combination that gives the highest current is: (a) 23 μF, 60 Hz; (b) 33 μF, 55 Hz; (c) 30 μF, 70 Hz; (d) 25 μF, 50 Hz.

Problems

Section 19.1 Induction and Faraday's Law

27. ■ A 1.5-T magnetic field lies perpendicular to a 25-cm-diameter circular wire loop. What's the magnetic flux through the loop?

28. ■ A 2.9-T magnetic field makes a 40° angle with a square loop 25 cm on a side. What's the magnetic flux through the loop?

29. ■■ At one location, Earth's magnetic field has a magnitude of 5.4×10^{-5} T with an inclination of 72° to the horizontal. Find the magnetic flux through a horizontal rectangular roof measuring 35 m by 20 m.

30. ■■ A circular wire loop of wire has radius 2.0 cm and resistance 0.050 Ω. A magnetic field is perpendicular to the loop. At what rate must the field change in order to induce a current of 1.0 mA in the loop?

31. ■■ A 150-turn circular coil has diameter 5.25 cm and resistance 1.30 Ω. A magnetic field perpendicular to the coil is changing at 1.15 T/s. Find the induced current in the coil.

32. ■■ A circular loop of wire with area 0.015 m² lies in the x-y plane. Initially there's a magnetic field of 4.0 T in the $-z$-direction. The field remains constant for 10 s, then decreases gradually to zero in 10 s, and then remains zero for 10 s. Find the magnitude and direction of the induced emf in the loop for each of the three 10-s intervals.

33. ■■ A square wire loop 15 cm on a side lies in the x-y plane. Its resistance is 0.55 Ω. A magnetic field points in the $+z$-direction and increases at 15 mT/s. (a) What's the direction of the induced current in the loop? (b) Find the magnitudes of the induced emf and induced current in the loop.

34. ■■■ A 75-turn circular coil has diameter 4.0 cm and resistance 2.6 Ω. This coil is placed inside a solenoid, with coil and solenoid axes aligned. The solenoid has 5000 turns of wire and is 24 cm long. If the solenoid current increases steadily from 0 to 10 A in 2.5 s, find the induced emf and induced current in the coil.

35. ■■ A circular wire coil with resistance 1.4 Ω and area 5.0×10^{-3} m² lies perpendicular to a magnetic field that's increasing at 2.0 T/s. If the induced current is 250 mA, how many turns are in the coil?

Section 19.2 Motional emf

36. ■ In Figure 19.7, take $B = 3.5$ T and $L = 12$ cm. Find the induced emf when the bar moves at 0.25 m/s.

37. ■ In Figure 19.7, take $B = 1.8$ T, $R = 2.3$ Ω, and $L = 12$ cm. Suppose the bar moves at 0.80 m/s to the *left*. (a) Find the magnitude and direction of the induced current. (b) At what rate is electric power generated?

38. ■■ Suppose the rectangular wire loop with a resistance of 3.5 Ω enters the magnetic field as shown in Figure P19.38, moving to the right at 25 cm/s. (a) What are the magnitude and direction of the induced current? (b) Later the loop is fully in the field, still moving at 25 cm/s. What are the magnitude and direction of the induced current? (c) Eventually the loop begins to emerge from the field, still at 25 cm/s. What are the magnitude and direction of the induced current?

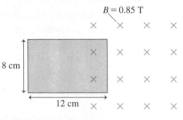

$B = 0.85$ T

8 cm

12 cm

FIGURE P19.38

39. ■■ The apparatus in Figure 19.7 is rotated 90°, so the bar falls vertically while maintaining electrical contact with the rails. The rail spacing is $L = 30$ cm, and the field strength is $B = 1.2$ T. If the bar drops from rest, what's the magnitude of the induced emf after it's been falling for (a) 0.5 s and (b) 1.0 s? (You may assume that the magnetic force on the falling bar is much less than the bar's weight.)

40. ■■ In the rail gun of Figure 19.12, the field strength is 3.0 T and the rail spacing is 50 cm. A discharging capacitor supplies a current of 100 kA. (a) What's the acceleration of the 0.25-kg bar? (b) What's the bar speed at the end of the 2.75-m track?

41. ■■■ A horizontal metal bar of length L falls vertically through a horizontal magnetic field of strength B. (a) Show that if the bar falls with speed v, the induced emf between its ends is $\mathcal{E} = BLv$. (b) Suppose $B = 0.25$ T and $L = 0.50$ m. What's the induced emf 1.0 s after the bar is dropped from rest?

42. ■■■ An airplane flies due north in a region where Earth's magnetic field has magnitude 5.3×10^{-5} T, declination 0°, and inclination 75°. What's the induced emf between the wingtips, 35 m apart, when the plane flies at 220 m/s? *Hint:* See the preceding problem.

Section 19.3 Generators and Transformers

43. ■ (a) A 20-turn generator coil with area 0.016 m² rotates at 50 Hz in a 0.75-T magnetic field. Find the peak induced emf. (b) Graph the induced emf as a function of time from $t = 0$ to $t = 40$ ms.

44. ■ A 10-turn generator coil with area 0.150 m² rotates in a 1.25-T magnetic field. At what frequency should the coil rotate to produce a peak emf of 150 V?

45. ■■ A generator has a 150-turn square coil, 5 cm on a side. Find the peak emf in this coil when it's rotating at 100 Hz (a) in Earth's field, $B = 5 \times 10^{-5}$ T, and (b) in a strong 6.4-T field.

46. ■■ Figure P19.46 shows the induced emf in a 5-turn square coil, 8 cm on a side, rotating in a uniform magnetic field. Find (a) the rotation frequency, (b) the angular velocity, and (c) the magnetic field strength.

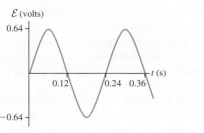

FIGURE P19.46

47. ■ (a) A 120-V emf is across a transformer's 200-turn primary coil. How many turns should the secondary have in order to produce a 30-V emf? (b) A 240-V emf is across a transformer's 140-turn primary coil. What's the emf in the 250-turn secondary coil?

48. ■■ A home receives 1.5 kW of electric power from a transformer on a nearby pole. The line from the substation to the transformer is at 10 kV and has 0.13 Ω resistance. What fraction of the power is lost in this transmission line?

49. ■■ A typical coal-fired power plant produces 1000 MW of electrical energy. Seven percent of the power is lost in a 500-kV transmission line. (a) What's the line's resistance? (*Note:* This is actually the equivalent resistance of all the parallel lines in the network, so your answer should be quite small.) (b) What would be the power loss if the line were at 100 kV?

Section 19.4 Inductance

50. ■ A 240-turn coil carries 350 mA. If the magnetic flux through the coil due to this current is 0.75 Wb, what's the coil's inductance?

51. ■ You have a 120-mH inductor. Over what time must you increase the current through the coil by 3.0 A to induce a 6.0-V emf?

52. ■■ A solenoid has diameter 2.5 cm, length 30 cm, and inductance 1.34 mH. (a) How many turns of wire does it have? (b) What's the energy stored in the solenoid when it carries a 6.0-A current?

53. ■■ A solenoid has the following dimensions: diameter = 5.0 cm, n = 45 turns/cm, length = 23 cm. (a) What's its inductance? (b) Find the current when the solenoid stores 0.50 J of energy.

54. ■ Find the inductance required to store 1.0 J of energy with current of 10 A through the inductor.

55. ■■ A 48-V battery is in series with a switch, a 0.50-H inductor, and a 10-Ω resistor. (a) What's the maximum current in this circuit? (b) What's the current at t = 0.10 s after the switch is closed? (c) When does the current reach half the maximum value? (d) What's the current at time $t = 2\tau$?

56. ■■ A 12.0-V battery is in series with a switch, a 0.25-H inductor, and a 20-Ω resistor. (a) What's the time constant τ? (b) Find the potential differences across the inductor and resistor at $t = \tau$ and $t = 2\tau$.

57. ■■ A 10.0-mH inductor is in series with a capacitor that can be varied from 200 nF to 400 nF. What's the range of possible oscillation frequencies?

58. ■ You have a variable capacitor ranging from 5.9 pF to 8.8 pF. What inductor should be used with this capacitor in an FM radio receiver covering the entire FM band, 88 MHz to 108 MHz?

59. ■■ A charged, 250-μF capacitor is connected across a 450-mH inductor. (a) What's the oscillation period? (b) How should you change the capacitance in order to double the period?

60. ■■■ A 4700-μF capacitor is charged to 9.0 V and then connected across a 1.50-H inductor. (a) What's the energy in this circuit? (b) What's the maximum current? (c) At what time after the circuit is connected are there equal energies in both components?

61. ■■ A 500-μF capacitor is connected across a 1.25-H inductor. At a certain time, the charge on the capacitor is zero and the current is 0.342 A. (a) How much later will the capacitor charge reach its peak? (b) What's the total energy in the circuit? (c) What is the peak charge on the capacitor?

62. ■■■ (a) Consider the circuit shown in Figure P19.62. Initially there's no current in the circuit. Find the current in each resistor immediately after the switch is closed and a long time later. (b) After the switch has been closed for a long time, it's opened. Find the current in each resistor immediately after it opens.

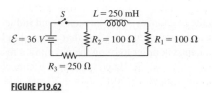

FIGURE P19.62

Section 19.5 AC Circuits

63. ■ An AC power supply operates with peak emf 340 V and frequency 120 Hz. A 3.50-kΩ resistor is connected across the supply. Find (a) the rms current and (b) the average power dissipated in the resistor. (c) Graph the potential difference across and current through the resistor, both as functions of time.

64. ■■ A capacitor is connected to a 200-V (peak), 50-Hz power supply. If the maximum current is 400 mA, what is the capacitance?

65. ■■ You wish to design an AC power supply that can drive 2.0 A rms current through a 140-Ω load. (a) What should be the rms value of the power supply's emf? (b) At what rate does this power supply deliver energy?

66. ■■ When a 750-Ω resistor is connected across an AC power supply, the resistor dissipates energy at a rate of 12.5 W. (a) Find the rms current and maximum current through the resistor. (b) What is the rms value of the power supply's emf?

67. ■■■ An AC power supply delivers 120 V rms. The time interval between peak emf in one direction and the opposite direction is 10.2 ms. Find (a) the peak emf and (b) the frequency. (c) Write an expression for the power supply's emf as a function of time, including all appropriate numerical values.

68. ■■ A resistor connected across an AC power supply has a current given by $I = (1.20 \text{ A}) \cos(300t)$ when connected to a power supply with emf 100 V rms. Find (a) the rms current, (b) the resistance, and (c) the average power delivered to the resistor.

69. ■ (a) At what frequency does a 100-μF capacitor have 50-Ω reactance? What would be its reactance if the frequency were tripled? (b) At what frequency does a 100-mH inductor have 50-Ω reactance? What would be the reactance of this inductor if the frequency were tripled?

70. ■■ A 240-V (rms), 50-Hz power supply is connected to a 4700-μF capacitor. Find (a) the capacitor's reactance and (b) the peak current. (c) What's the peak current if the power supply's frequency is changed to 60 Hz?

Section 19.6 *RLC* Circuits and Resonance

71. ■■ A solenoid has resistance 13.5 Ω and inductance 410 mH. When it's connected across a 50-Hz AC power supply, what's the circuit's impedance?

72. ■■ An *RC* series circuit has R = 75.0 Ω and C = 580 nF. The series combination is connected across a 200-V (rms), 60-Hz

power supply. Find (a) the phase angle, (b) the impedance of the *RC* combination, and (c) the peak current.

73. ■ ■ An *RC* series circuit has $R = 500\ \Omega$ and $C = 200\ \mu F$. The series combination is connected across a 120-V (rms) power supply. Find the peak current when the frequency is (a) 60 Hz and (b) 120 Hz.

74. ■ ■ An *RL* series circuit has $R = 75.0\ \Omega$ and $L = 1.90$ mH. The series combination is connected across a 150-V (rms), 60-Hz power supply. Find (a) the phase angle, (b) the impedance of the *RL* combination, and (c) the peak current.

75. ■ ■ ■ An *RLC* series circuit has $R = 1.35$ kΩ, $L = 225$ mH, and $C = 2.50\ \mu F$ and is connected across a 60-Hz power supply. Find (a) the reactances of the capacitor and inductor and (b) the impedance of the circuit. (c) To what value should you change the capacitance in order to bring the circuit into resonance?

76. ■ ■ ■ A 120-V (rms), 60-Hz power supply is connected to an *RLC* series circuit with $R = 150\ \Omega$, $L = 1.20$ mH, and $C = 33.5\ \mu F$. Find (a) the reactance of the capacitor and inductor, (b) the circuit impedance, and (c) the peak current.

77. ■ ■ For the circuit in the preceding problem, find (a) the phase angle, (b) the power factor, and (c) the average power consumption.

78. ■ ■ An *RLC* series circuit is in resonance at 60 Hz. If you triple both the inductance and capacitance, what's the new resonance frequency?

79. ■ ■ An *RLC* series circuit has $R = 920\ \Omega$, $L = 15.0$ mH, and $C = 250\ \mu F$. (a) What's its resonance frequency? (b) If the circuit is connected to a 100-V rms power source at the resonance frequency, what's the average power consumption?

80. ■ ■ ■ An *RLC* series circuit has $R = 1.10$ kΩ, $L = 190$ mH, and $C = 4.50\ \mu F$. (a) What frequencies will give a power factor of (a) 0.5; (b) 0.75; (c) 1.0?

General Problems

81. ■ ■ A car alternator produces about 14 V peak output to charge the car's 12-V battery. If an alternator coil is 15 cm in diameter and is spinning at 1200 revolutions per minute in a 0.15-T magnetic field, how many turns must it have to produce a 14-V peak output?

82. ■ ■ A square loop 3.0 m on a side is perpendicular to a uniform 2.0-T magnetic field as shown in Figure GP19.82. A 6.0-V lightbulb is in series with the loop. The magnetic field is reduced steadily to zero during a time interval Δt. (a) Find Δt such that the bulb will shine at full brightness during this time. (b) In which direction does the loop current flow?

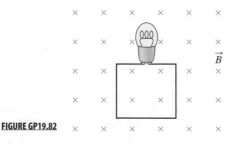

FIGURE GP19.82

83. ■ ■ Opening switches in highly inductive circuits can be dangerous, even with low-voltage power sources. To see why, consider a 6-V battery supplying 3.0 A to an electromagnet with inductance 2.5 H. A switch in this circuit opens and the current drops to zero in a mere 6.8 ms. What's the emf induced in the electromagnet during this time?

84. BIO ■ ■ **MRI scanner.** An MRI scanner uses a large solenoid with inductance 0.53 H, carrying 2.4 kA of current. (a) What's the energy stored in the solenoid? (b) The solenoid coils are normally superconducting (zero resistance), but if superconductivity fails they revert to a resistance of 0.31 mΩ. (b) Given that the current can't change instantaneously, find the power dissipation in the solenoid immediately after a sudden loss of superconductivity.

85. ■ ■ ■ A 3.2-cm-diameter circular coil with 30 turns of wire has resistance 1.3 Ω. This coil is placed inside a solenoid, with coil and solenoid axes aligned. The solenoid has 3000 turns and is 25 cm long. (a) At what rate must the current in the solenoid be increased to induce a 1.0-mA current in the coil? (b) Is the induced current in the same direction or opposite the solenoid current?

86. ■ ■ An electric toothbrush rests on a stand containing a 50-turn coil connected across the 120-V rms AC power line. Inside the brush handle is another coil whose induced current charges a 6-V battery. How many turns should this coil contain if its peak emf is to be 7.1 V?

87. ■ ■ ■ Earth's magnetic field has an approximate value of 50 μT. (a) Estimate the total magnetic energy contained in the first 100 km above Earth's surface. (b) If we could tap this energy, how long would it supply humankind's power demand of 15 TW?

Answers to Chapter Questions

Answer to Chapter-Opening Question

The power factor $\cos \phi$ describes the phase difference between current and potential difference in AC circuits. A heavy air conditioning load during a hot August week in 2003 lowered the power factor, resulting in higher-than-normal currents to deliver a given power. High ambient temperatures and the high current caused power lines to overheat, sag, and short-circuit to trees. Disruptions spread, taking more than 500 generators offline, including 22 nuclear power plants.

Answers to GOT IT? Questions

Section 19.1   (c) > (a) > (b) > (d)
Section 19.2   (a) Clockwise
Section 19.3   (b) 50
Section 19.4   (e) 710 kHz
Section 19.5   (a) > (c) >(b)
Section 19.6   (a) 0.25

20 Electromagnetic Waves and Special Relativity

■ Nearly all our information about the universe beyond Earth comes via one physical phenomenon. What is that?

This chapter applies your knowledge of electromagnetism to electromagnetic waves. Visible light and other electromagnetic radiation cover a wide range of wavelength and frequencies, but they're all similar and all travel at the speed of light. Deep questions involving light helped lead Einstein to his theory of special relativity. Relativity shows that measures of time and space depend on your frame of reference and also reveals new relationships between energy, momentum, and mass.

20.1 Electromagnetic Waves

Chapters 18 and 19 revealed deep connections between electricity and magnetism. Electric currents produce magnetic fields, and changing magnetic flux induces emf and current. Here you'll see another connection between electricity and magnetism in the oscillating electric and magnetic fields that constitute **electromagnetic waves**.

Producing Electromagnetic Waves

In Figure 20.1 (next page) an AC power supply drives electric charge back and forth along a metal rod. The rod is an **antenna**, which serves as a source of electromagnetic waves. At any moment the charge in the rod is dipole-like, with positive charge on one half and negative on the other. This charge distribution generates an electric field in the neighboring space, as shown. The field changes as the charge in the rod oscillates.

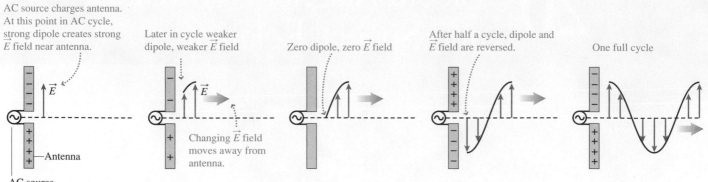

AC source charges antenna. At this point in AC cycle, strong dipole creates strong $\vec{E}$ field near antenna.

Later in cycle weaker dipole, weaker $\vec{E}$ field

Zero dipole, zero $\vec{E}$ field

After half a cycle, dipole and $\vec{E}$ field are reversed.

One full cycle

Antenna

AC source

Changing $\vec{E}$ field moves away from antenna.

FIGURE 20.1 An electromagnetic wave originating from alternating current in an antenna.

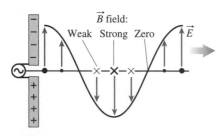

Oscillating electric field induces oscillating magnetic field at right angles to itself.

$\vec{B}$ field:
Weak Strong Zero $\vec{E}$

FIGURE 20.2 The wave's magnetic field is perpendicular to the electric field. The sinusoidal curve represents the electric part of the electromagnetic wave. The smaller dots and crosses represent weaker magnetic fields, and larger dots and crosses stronger magnetic fields.

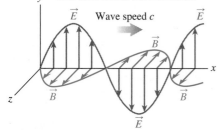

A 3-D view showing $\vec{E}$ and $\vec{B}$ fields in phase, perpendicular to each other and to wave direction.

Wave speed c

FIGURE 20.3 Three-dimensional representation of an electromagnetic wave. The arrows are field vectors, and their lengths represent the variation of the fields in space. They do *not* represent the height of a mechanical wave, such as a water wave.

In the 1860s, Scottish physicist James Clerk Maxwell (1831–1879) had a profound insight that led to our understanding of electromagnetic waves. Maxwell knew Faraday's law, which says that a changing magnetic field can induce electric current (Section 19.1). You know that electric current ultimately results from an electric field within a conductor (Section 17.1). Thus, Maxwell reasoned that Faraday's law implies that a changing magnetic flux produces an electric field. Based on observed symmetries between electric and magnetic phenomena, Maxwell argued that a changing electric flux should produce a magnetic field, as shown in Figure 20.2. That magnetic field is changing, so it produces more electric field. These changing fields, continually regenerating each other, travel through space as an electromagnetic wave. In fact, any time electric charge accelerates—as in our antenna—then electromagnetic waves are produced.

Properties of Electromagnetic Waves

Maxwell not only predicted the existence of electromagnetic waves, but also determined their properties. As shown in Figures 20.2 and 20.3, the wave's electric magnetic fields are perpendicular to each other. They're also perpendicular to the direction of the wave's propagation—making electromagnetic waves **transverse waves** (recall from Section 11.1 the distinction between transverse and longitudinal waves). Figure 20.3 shows clearly the spatial relationship among the electric field, the magnetic field, and the propagation direction. The figure also shows that the wave's electric and magnetic fields are in phase. That is, they peak at the same point. Similarly, where one field is zero, so is the other. The fields remain locked in phase as the wave travels.

A key prediction of Maxwell's theory is that all electromagnetic waves travel with the same speed c in a vacuum, where

$$c = \sqrt{\frac{1}{\mu_0\epsilon_0}} \approx 3.00 \times 10^8 \, \text{m/s} \quad \text{(Speed of light in vacuum; SI unit: m/s)} \quad (20.1)$$

Using the exact values of ϵ_0 and μ_0 in Equation 20.1 gives $c = 299{,}792{,}458 \, \text{m/s}$. There's no uncertainty here: In SI, c is *defined* to have this value, and, as you saw in Chapter 1, the meter is then defined in terms of c.

By Maxwell's time it was clear that light was a wave, and measurements of its speed were within about 1% of Maxwell's theoretical prediction. The happy coincidence of Maxwell's electromagnetic wave speed and measured values for the speed of light hinted strongly that light must be an electromagnetic wave. In one of the greatest syntheses in all of physics, Maxwell had brought the entire science of optics under the umbrella of electromagnetism! We'll explore optics and light further in Chapters 21–22.

Equation 20.1 gives the speed of electromagnetic waves *in vacuum*. In other media, such as water or glass, electromagnetic waves travel slower. In Chapter 21 we'll discuss how the speed of light in different media is related to refraction, the bending of light rays passing from one medium into another.

✓ **TIP**

The speed of light in water is about three-quarters of its speed c in vacuum, and the speed of light in common glass is about two-thirds of c.

Speed, Wavelength, and Frequency

To understand better the properties of electromagnetic waves, recall the relationship between speed v, wavelength λ, and frequency f of a periodic wave (Equation 11.1): $v = \lambda f$. Electromagnetic waves have speed $v = c$ in vacuum, so for electromagnetic waves

$$c = \lambda f \quad \text{(EM wave speed, wavelength, frequency; SI unit: m/s)} \quad (20.2)$$

The speed of light c in Equation 20.1 doesn't depend on wavelength or frequency. Thus you can see from Equation 20.2 that electromagnetic waves with higher wavelengths have lower frequencies, and vice versa. For example, light in the middle of the visible spectrum has wavelength $\lambda = 550\,\text{nm} = 5.50 \times 10^{-7}\,\text{m}$, so its frequency is

$$f = \frac{c}{\lambda} = \frac{3.00 \times 10^8\,\text{m/s}}{5.50 \times 10^{-7}\,\text{m}} = 5.45 \times 10^{14}\,\text{s}^{-1} = 5.45 \times 10^{14}\,\text{Hz}$$

Maxwell's theory places no restrictions on possible wavelengths and frequencies of electromagnetic waves. In Section 20.2 you'll see the electromagnetic spectrum, covering a wide range of wavelengths and corresponding frequencies.

Experimental Evidence for Electromagnetic Waves

In 1886 the German physicist Heinrich Hertz (1857–1894) performed a definitive experiment confirming Maxwell's ideas. He built a transmitter and receiver, demonstrating the propagation of electromagnetic waves through space. The basic component in Hertz's transmitter was the LC oscillator we described in Section 19.4, operating at a frequency on the order of 100 MHz. The LC circuit contained a spark gap, over which charges jumped back and forth, analogous to the charge oscillation in Figure 20.1's antenna. Several meters away Hertz placed a receiver loop with a gap, where the passing wave induced oscillations, evidenced by sparks at the gap. Hertz also showed that these radio waves exhibited the same reflection, refraction, and other properties shown by much higher frequency visible light. He even determined, using interference techniques, that the wave speed was consistent with the speed of light.

Although the term "radio" did not exist in Hertz's time, the frequency and wavelength of Hertz's waves put them squarely in what we now call the radio portion of the electromagnetic spectrum. Radio waves are easy to produce using electronic oscillators because practical values of inductance and capacitance result in radio-frequency oscillations. Hertz's discovery had monumental practical applications. In 1895, Guglielmo Marconi demonstrated radio as a communication tool, and by 1901 he transmitted radio waves across the Atlantic. The 20th century saw the development of a vast array of wireless communication using radio frequencies. These include AM and FM radio, television, radar, cell phones, wireless computer networks, and radio frequency identification tags. All these technologies we now take for granted followed from Hertz's curiosity about the laws of physics!

EXAMPLE 20.1   **Electromagnetic Wave Frequencies**

What is the wavelength of Hertz's 100-MHz radio waves? Compare the radio frequency and wavelength with those of 430-nm-wavelength violet light.

ORGANIZE AND PLAN Equation 20.2 relates speed, wavelength, and frequency: $c = \lambda f$.

Known: $c = 3.00 \times 10^8\,\text{m/s}$; $f = 100\,\text{MHz}$ (radio wave), and $\lambda = 430\,\text{nm}$ (violet light).

SOLVE Solving Equation 20.2 for the radio wavelength,

$$\lambda = \frac{c}{f} = \frac{3.00 \times 10^8\,\text{m/s}}{100 \times 10^6\,\text{Hz}} = 3.00\,\text{m}$$

cont'd.

For the 430-nm light, the frequency is

$$f = \frac{c}{\lambda} = \frac{3.00 \times 10^8 \text{ m/s}}{430 \times 10^{-9} \text{ m}} = 6.98 \times 10^{14} \text{ Hz}$$

REFLECT The light's frequency is some 7 million times higher, and its wavelength 7 million times shorter, than for the 100-MHz radio wave (which happens to lie in the middle of the FM radio band). In Section 20.2 you'll see that the complete electromagnetic spectrum covers an even wider range of frequencies.

MAKING THE CONNECTION Compare the frequency computed for visible light with the frequency of audible sound (Chapter 11).

ANSWER Normal audible frequencies range from 20 Hz to 20 kHz—much lower than the frequency of light. Sound is also an entirely different type of wave. It's a mechanical, longitudinal wave that propagates through air at speeds far below that of light. Light is a transverse electromagnetic wave that can travel through vacuum.

Energy and Momentum in Electromagnetic Waves

Waves transport energy, and electromagnetic waves are no exception; here it's the energy stored in the wave's electric and magnetic fields. In Section 19.4 you saw that the energy densities in electric and magnetic fields are $u_E = \frac{1}{2}\epsilon_0 E^2$ and $u_B = B^2/2\mu_0$. Maxwell showed that these energy densities are equal in an electromagnetic wave in vacuum, so $\frac{1}{2}\epsilon_0 E^2 = u_B = B^2/2\mu_0$. Rearranging gives $E/B = \sqrt{1/\mu_0\epsilon_0}$. But Equation 20.1 shows that the expression on the right is just the speed of light c. Therefore,

$$c = \frac{E}{B} \qquad \text{(Electric and magnetic fields in EM wave; SI unit: m/s)} \qquad (20.3)$$

for electromagnetic waves in vacuum.

In addition to energy, electromagnetic waves carry momentum. This isn't surprising; in classical mechanics, a particle of mass m moving with speed v has kinetic energy $K = \frac{1}{2}mv^2$ and momentum of magnitude $mv = \sqrt{2mK}$. Electromagnetic waves don't have particle mass, so their energy-momentum relation is different:

$$p = \frac{U}{c} \qquad \text{(EM wave momentum; SI unit: kg} \cdot \text{m/s)} \qquad (20.4)$$

for the momentum p of a wave carrying energy U.

✓ TIP

We'll use the variable U for energy here, rather than E, so you won't confuse energy with electric field.

Light's momentum is generally small and therefore difficult to observe. The first good measurements were made in 1901 by E.F. Nichols and G.F. Hull, and these agreed with Equation 20.4. Light striking an object delivers momentum, and therefore by Newton's second law ($F = \Delta p/\Delta t$), it exerts a force. Figure 20.4a shows light absorbed by a black target, so its full momentum is transferred to the target, giving the target a momentum change $\Delta p = U/c$. For the reflective target of Figure 20.4b, the light's momentum is reversed; by momentum conservation, the momentum transfer to the target is therefore doubled: $\Delta p = 2U/c$. You can think of light as exerting *pressure* on the target, because pressure is force per unit area. Therefore, it's commonly said that the electromagnetic waves exert **radiation pressure** on the target.

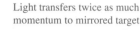

Black target absorbs light

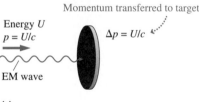

Momentum transferred to target

Energy U
$p = U/c$

$\Delta p = U/c$

EM wave

(a)

Mirrored target reflects light

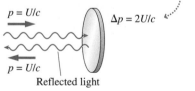

Light transfers twice as much momentum to mirrored target

$p = U/c$

$\Delta p = 2U/c$

$p = U/c$

Reflected light

(b)

FIGURE 20.4 A dark target absorbs all the wave's momentum. A reflective target reverses the wave's momentum, absorbing twice as much momentum in the process.

EXAMPLE 20.2   **Solar Sailing**

One propulsion scheme proposed for interplanetary spacecraft is to use radiation pressure of sunlight to accelerate the craft. Find the force on a reflective solar sail with area $A = 1.0\ \text{km}^2$ located at Earth's orbit, where the average intensity of sunlight is $1.4\ \text{kW/m}^2$.

ORGANIZE AND PLAN   Our sketch of the situation is shown in Figure 20.5. Reflected light with energy U causes a momentum change $\Delta p = 2U/c$, as shown in the text. In terms of momentum, Newton's second law is $F = \Delta p/\Delta t$. Here we know the light intensity—the power P per square meter. Multiplying by area gives the total power, or energy U per unit time. From that we can find $F = \dfrac{\Delta p}{\Delta t} = \dfrac{2}{c}\dfrac{\Delta U}{\Delta t}$.

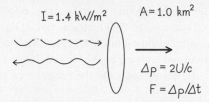

$I = 1.4\ \text{kW/m}^2$ $A = 1.0\ \text{km}^2$

$\Delta p = 2U/c$

$F = \Delta p/\Delta t$

FIGURE 20.5 Our sketch for Example 20.2.

Known: $I = 1.4\ \text{kW/m}^2$; $A = 3.00 \times 10^8\ \text{m/s}$; $c = 1.0\ \text{km}^2 = 1.0 \times 10^6\ \text{m}^2$

SOLVE Given the intensity $I = 1.4\ \text{kW/m}^2$, the rate at which the sunlight delivers energy is $\Delta U/\Delta t = IA$. Then our expression for force becomes

$$F = \frac{\Delta p}{\Delta t} = \frac{2}{c}\frac{\Delta U}{\Delta t} = \frac{2IA}{c} = \frac{(2)(1.4 \times 10^3\ \text{W/m}^2)(1.0 \times 10^6\ \text{m}^2)}{3.00 \times 10^8\ \text{m/s}}$$

$$= 9.3\ \text{N}$$

REFLECT This is a rather small force, enough to give a tiny 10-kg spacecraft a modest acceleration of about $1\ \text{m/s}^2$—much smaller than with conventional rockets despite the square kilometer of sail area. Nevertheless, space engineers are actively working to develop "sailing" spacecraft. In earthbound labs, the intense beams of high-powered lasers are capable of levitating small particles.

MAKING THE CONNECTION   What would happen to the spacecraft's acceleration if the sail were black rather than reflective?

ANSWER From the analysis in the text, the force and therefore the acceleration would be halved.

GOT IT? Section 20.1   Rank in order, from lowest to highest, the frequencies of electromagnetic waves in vacuum with the following wavelengths: (a) 550 nm; (b) $1.05\ \mu\text{m}$; (c) 434 nm; (d) 780 nm.

20.2 The Electromagnetic Spectrum

In Section 20.1 we noted that Maxwell's theory placed no restrictions on the wavelength or frequency of electromagnetic waves. The range of known wavelengths covers many orders of magnitude, from radio waves with continent-sized wavelengths to gamma rays with wavelengths shorter than $10^{-25}\ \text{m}$—10 orders of magnitude smaller than a proton! Figure 20.6 shows this vast **electromagnetic spectrum**.

There aren't distinct boundaries between the different categories of electromagnetic waves in Figure 20.6. All electromagnetic waves are essentially similar, but their different wavelengths mean they interact very differently with matter. The categories are a matter of convention; the behavior of electromagnetic waves generally changes only gradually with wavelength. So a 1.1-mm "microwave" and a 0.9-mm "infrared wave" are no more different

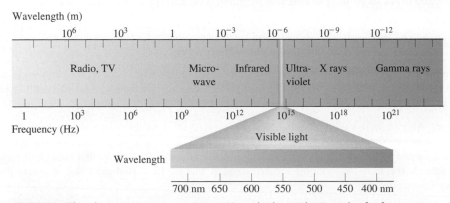

FIGURE 20.6 The electromagnetic spectrum. Note the logarithmic scales for frequency and wavelength.

than a 1.1-mm wave and a 1.3-mm wave. Even the 400-nm to 700-nm range for visible light isn't firm, because individual people perceive slightly different wavelength ranges.

Radio Waves

The "radio" part of the spectrum includes not only AM and FM radio but also television, cell phone, and other forms of wireless communication. In order to avoid disruptive interference (such as a cell phone signal interfering with TV reception), governments regulate the wavelengths allocated for different uses. Commercial broadcasters and telecommunications companies are licensed to operate only within specific narrow wavelength bands.

Two principal ways information is carried by radio waves are **amplitude modulation** (AM) and **frequency modulation** (FM). AM begins with a "carrier signal," a pure sine wave with the wavelength and frequency specified for that broadcaster (e.g., a 1470-kHz radio station). The sound to be broadcast is then added to the carrier, varying its amplitude. Frequency modulation encodes information as variations in the frequency of the carrier signal. This gives FM the advantage of being nearly static-free, because it isn't sensitive to variations in the signal amplitude caused by electrical noise, lightning, and similar sources. Increasingly, AM, FM, and other modulation schemes are used to encode information digitally. In 2009, all television broadcasting in the United States became digital, and many radio stations "piggyback" digital signals onto their regular carriers.

✓ TIP

Radio waves are electromagnetic waves, not sound waves. Electromagnetic waves carry information from transmitter to receiver, and electronic circuits in the receiver "decode" the signal to produce audible sound in a speaker.

Microwaves

Microwaves generally have shorter wavelengths than radio waves, but there's some overlap between radio and longer-wavelength microwaves, and between infrared and shorter-wavelength microwaves. Microwaves are used widely in communications and in radar.

You're probably most familiar with the microwaves that cook food in your microwave oven. Water molecules are strong electric dipoles (Chapters 15 and 16), so they respond readily to electric fields. Fields oscillating at frequencies in the gigahertz range flip water molecules back and forth; as the molecules jostle against each other, they convert electric field energy into heat. Most other molecules aren't strong dipoles, so they don't respond nearly as much; that's why you can heat water in a paper cup in a microwave oven. The oscillating electric field in microwave ovens comes from microwaves with frequency 2.45 GHz, corresponding to a wavelength around 12 cm. Reflections off the oven walls produce standing waves, with nodes separated by $\lambda/2 = 6$ cm. That's why there will be uncooked areas in your food unless it's rotated as it cooks.

CONCEPTUAL EXAMPLE 20.3   Aluminum in the Microwave

Microwave oven instructions warn against putting aluminum-foil-covered food containers in the oven. Why do they include this warning, even though a microwave oven's walls are usually made of aluminum?

SOLVE Aluminum is a good reflector of microwaves. That's why the oven's aluminum walls set up a standing-wave pattern. If you cover your food with aluminum, the microwaves will be reflected and won't reach the food. In addition, sharp edges of the foil or other metal will accumulate charge, as in the lightning rod discussed in Section 15.4. That can result in sparks that may damage the oven.

REFLECT You might wonder why microwaves don't escape through the glass door. If they did, they could certainly harm you—effectively cooking the water inside your body! In Chapter 22, you'll see how the metal mesh in the door minimizes microwave leakage.

Infrared (IR)

Infrared radiation has shorter wavelengths than radio and microwaves, ranging from the red end of the visible spectrum at 700 nm up to about 1 mm. Infrared results when atoms and molecules undergo transitions from higher to lower energy states (discussed in Chapter 24).

At typical temperatures on Earth (around 300 K) jostling of molecules due to random thermal motion results predominantly in infrared emission. The emission rate rises rapidly with temperature, making IR imaging useful in detecting heat loss from buildings as well as tumors in the body, which tend to be warmer than surrounding tissues. Astronomers use infrared telescopes to explore cool stars, interstellar gas clouds, and extrasolar planets. With IR, they can also peer through clouds of galactic dust that are opaque to visible light. Closer to home, Earth itself emits infrared radiation. Atmospheric **greenhouse gases**—notably carbon dioxide and water vapor—absorb some of this infrared, resulting in a higher planetary surface temperature. That temperature continues to climb as we humans add carbon dioxide to the atmosphere by burning fossil fuels.

Visible Light

Visible light covers the range from about 400 nm to 700 nm (Figure 20.6; Table 20.1). Light's behavior is the subject of **optics**, covered in Chapters 21 and 22. Also, much of what we know about atoms and molecules comes from studying their emissions of visible light, along with emissions in other parts of the electromagnetic spectrum.

Although the human eye can see only from 400 nm to 700 nm, this represents about half the Sun's energy emission (most of the rest is IR). So we humans have evolved to make good use of the available light. Some animals have adapted to other wavelength ranges; for example, many snakes that hunt primarily at night can see well into the infrared range.

TABLE 20.1 Approximate Wavelength Ranges for Colors of Visible Light

Color	Wavelength range (nm)
Violet	400–440
Blue	440–480
Green	480–560
Yellow	560–590
Orange	590–630
Red	630–700

CONCEPTUAL EXAMPLE 20.4   **Range of Sight and Hearing**

Compare the range of wavelengths (or frequencies) humans can see with the range of sound wavelengths (or frequencies) humans can hear.

SOLVE It doesn't matter whether you compare frequency or wavelength. Because $v = \lambda f$, a factor of 2 increase in frequency is equivalent to a factor of 2 decrease in wavelength.

Hearing: Remember from Chapter 11 that a normal range of hearing is considered to be from 20 Hz to 20 kHz. That's a factor of 1000 in frequency. In terms of octaves (where one octave represents a doubling of the frequency), a factor of 1000 is almost 10 octaves (because $2^{10} = 1024$).

Vision: Visible light varies in wavelength from 400 nm to 700 nm. That's less than a factor of 2, or less than one octave.

REFLECT This comparison seems to imply that the ear is a much more broadband instrument than the eye. On the other hand, visual images contain a lot more information than sounds—just think about the difference in storage capacity of a DVD versus a CD.

Ultraviolet (UV)

Ultraviolet radiation has wavelengths shorter than visible light, from 400 nm down to approximately 1 nm. Ultraviolet waves come primarily from atomic transitions.

In Chapter 23, you'll see how electromagnetic waves deliver energy in "bundles," called **photons**, with higher frequencies corresponding to higher-energy photons. That means the photons of ultraviolet radiation have more energy than those of visible light. Sunlight includes a significant amount of UV, so that's why you need to avoid excessive exposure to sunlight. UV protection in sunglasses prevents damage to your eyes. When UV radiation strikes your skin, it causes melanocyte cells to produce more pigment, darkening the skin. It also damages skin cell DNA, causing premature aging and increased cancer risk.

Atmospheric gases—principally ozone (O_3)—shield us from much of the Sun's UV radiation. In recent decades, emissions of chlorofluorocarbons ("CFCs," mainly in refrigeration and air conditioning) have destroyed some of the protective ozone. An international treaty banning CFCs is now in effect, and the ozone layer will gradually recover.

Multiwavelength Astronomy

Our understanding of the universe expanded greatly throughout the 20th century, as astronomers learned to detect emissions throughout the electromagnetic spectrum. Ordinary stars emit mostly IR, visible light, and UV, but more exotic objects like black holes and neutron stars emit x rays and even gamma rays. These short-wavelength EM waves don't penetrate Earth's atmosphere, so some modern astronomy is done from space. The photo shown here is from the orbiting Chandra X-Ray Observatory, with the bright spots revealing strong x-ray emissions from black holes and neutron stars in the galaxy M101. Astrophysical systems also emit radio waves, which we detect with giant radio telescopes.

X rays and Gamma Rays

X rays and gamma rays have shorter wavelengths even than UV. X rays generally result when high-energy electrons interact with matter. If an electron slows or otherwise accelerates suddenly, it can emit x rays. A high-energy electron can also knock one of the innermost electrons out of an atom; when another atomic electron drops into the vacant state, energy is emitted as x rays. We'll explain both these mechanisms in Chapter 23. Gamma rays have even shorter wavelength than x rays, so it's not surprising that they come from energetic transitions in even smaller systems—namely, atomic nuclei (Chapter 25).

X rays penetrate soft tissue and are therefore important in medical diagnosis. They've been used for over a century to detect bone fractures and other structural maladies. The CT (computerized tomography) scan is a newer diagnostic technique in which x rays are scanned across layers of tissue. The resulting signals are processed by computer to reveal much more detail than a single x-ray image. Medical practitioners weigh x rays' diagnostic value against damage they may cause. X-ray exposure disrupts cells and increases cancer risk. On the other hand, high-energy x-ray and gamma-ray beams treat cancer by targeting cancer cells while limiting the damage to surrounding tissue.

Reviewing New Concepts

- Electromagnetic waves are transverse waves with electric and magnetic fields perpendicular to each other and to the propagation direction.
- The speed of electromagnetic waves in vacuum is $c \approx 3.00 \times 10^8$ m/s, regardless of wavelength and frequency.
- Speed c, wavelength λ, and frequency f of electromagnetic waves in vacuum are related by $c = \lambda f$.
- Electromagnetic radiation with energy U carries momentum $p = U/c$.
- The electromagnetic spectrum (Figure 20.6) covers a wide range of wavelengths and frequencies.

GOT IT? Section 20.2 The frequency of visible light is on the order of magnitude of (a) 10^{10} Hz; (b) 10^{12} Hz; (c) 10^{14} Hz; (d) 10^{16} Hz.

20.3 The Fundamental Speed c

By the end of the 19th century, physicists had measured the speed of light to within 1% of its actual value. Maxwell's electromagnetic theory showed convincingly that light is an electromagnetic wave, and that theory correctly predicted the measured speed. As you'll see in Chapter 22, the wave nature of light was used to understand other optical phenomena, including diffraction, interference, and polarization.

But in 1900 one puzzle remained. All known waves seemed to require a medium in which to propagate. For example, sound waves propagate in gases, liquids, and solids, but not in vacuum. That's because they're disturbances of the gas, liquid, or solid medium. Similarly, water waves are disturbances of water; it makes no sense to think of a water wave without the water. But what, physicists wondered, is the medium for electromagnetic waves? Although nothing was known about this supposed medium, it was given a name: luminiferous (meaning light-carrying) ether, or simply **ether**.

Attempts to Detect the Ether

Ether must be an unusual substance: It has to fill all space, because light reaches us from distant stars, and it should also fill substances like glass that transmit light. Other waves tend to travel faster through denser media, so ether must be very dense to carry electromagnetic waves at speed c. On the other hand, planets moving in their orbits must pass through the ether with negligible resistance. That's a strange combination of properties.

Around 1880, the American physicist Albert Michelson proposed measuring Earth's speed through the ether. Michelson knew that Earth's orbital speed is about 30 km/s, or

$10^{-4} c$. He reasoned that one should measure different speeds for light propagating in different directions relative to Earth's orbital velocity. Figure 20.7 illustrates this concept using a mechanical analogy, with Earth representing a moving boat. Based on this analogy, light waves sent with and against the direction of Earth's orbital motion should have different speeds, $c + v$ and $c - v$, where v is Earth's orbital speed. Although $v = 10^{-4} c$, Michelson built an apparatus that used wave interference to detect that small speed difference.

Michelson first performed his experiment in 1881 and then repeated it using a much more precise apparatus in 1887. This was the famous the **Michelson-Morley experiment**, so named because of Michelson's collaboration with Edward Morley. The results were always the same: There was no detectable motion of Earth relative to the ether. Physicists were puzzled by these results for two decades. The ether seemed impossible to detect, yet it was difficult to conceive of a wave without a medium to carry it.

Physicists proposed various solutions. One idea was that the motion through the ether somehow changed the size of the apparatus. This idea was impossible to check, because any device used to measure size would be similarly affected. Another idea was that Earth dragged the neighboring ether with it, masking its motion. This idea, however, was contradicted by an astronomical effect known as stellar aberration. At the start of the 20th century, the idea of ether and the results of the Michelson-Morley experiment were simply irreconcilable.

Einstein's Relativity

In 1905, Albert Einstein was a young patent clerk working in Bern, Switzerland. He had spent years thinking about the behavior of light. He finalized his ideas and published them in 1905. He started with two simple postulates:

Einstein's Postulates

1. Principle of relativity: The laws of physics are the same in all inertial reference frames.
2. Constant speed of light: The speed of light c is the same in all inertial reference frames, regardless of the relative velocity of the source and receiver of the light.

Remember from Section 4.2 that an *inertial reference frame* is one in which Newton's first law of motion (the law of inertia) holds. An inertial frame can be considered at rest or moving with a constant velocity. A non-inertial frame is an accelerated one, such as a car rounding a curve. Our discussion here, like Einstein's 1905 paper, is restricted to inertial frames. Einstein's theory concerning inertial frames is the theory of **special relativity**. Later Einstein developed **general relativity**, applicable also to accelerated reference frames.

Einstein's postulates might seem so simple as to be pointless. Yet they have profound, sometimes startling, consequences. To begin with, Postulate 2 accounts for the Michelson-Morley results. No matter what their direction, Michelson's light beams had speed c—not $c + v, c - v$, or any other speed. Therefore, travel time for a light beam doesn't depend on its direction, relative to the direction that Earth or anything else is moving. Why? Because there is no ether: Electromagnetic waves, unlike mechanical waves, require no medium.

Although Einstein introduced relativity with two postulates, you can think of the second postulate as subsumed by the first. Einstein believed that Maxwell's theory of electromagnetism was correct—as we still do in the 21st century! Einstein's first postulate—the principle of relativity—then implies that Maxwell's theory must be valid in all inertial reference frames. Maxwell's theory predicts electromagnetic waves propagating in vacuum at speed c—a result that must therefore be true in all inertial frames. And that's Einstein's second postulate.

So far, this all sounds simple and reasonable. But the notion that light travels with the same speed regardless of the motions of its source and observer seems to violate common sense. Figure 20.8a shows an analogy: A baseball pitcher stands in the back of a pickup truck going 10 m/s relative to the ground. The pitcher throws the ball in the direction of the truck's motion, at 30 m/s relative to the truck. Then a person on the ground should

Boat moving toward waves

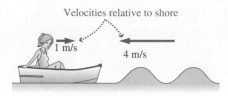

Velocities relative to shore

1 m/s   4 m/s

Person on boat sees waves approaching at 4 m/s + 1 m/s = 5 m/s.

Boat moving away from waves

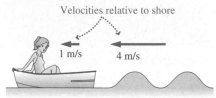

Velocities relative to shore

1 m/s   4 m/s

Person on boat sees waves approaching at 4 m/s − 1 m/s = 3 m/s.

FIGURE 20.7 A wave's apparent speed should depend on the observer's motion.

The ball's speed in each direction relative to the ground depends on the truck's velocity.

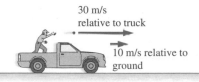

30 m/s
relative to truck

10 m/s relative to ground

Ball's speed: 30 m/s + 10 m/s = 40 m/s

30 m/s

10 m/s

Ball's speed: 30 m/s − 10 m/s = 20 m/s

(a)

The speed of light relative to an observer on Earth does not depend on the spaceship's velocity.

$v = 0.2c$ $v = 0.2c$

c c

Light's speed in either direction is c.

(b)

FIGURE 20.8 (a) How we expect velocities to add. (b) The observer always receives light at speed c, regardless of the motion of the light source.

measure the ball's speed as 30 m/s + 10 m/s = 40 m/s. If the pitch is opposite the truck's motion, then the ball's speed relative to the ground should be 30 m/s − 10 m/s = 20 m/s.

Now consider a similar experiment with light; this one is a "thought experiment" because we can't really do it as described. A spaceship traveling at 0.2c toward Earth (the spaceship on the left in Figure 20.8b) sends a light beam earthward. Your experience with the baseball suggests that an Earth-based observer should measure the speed of the light signal fired from the rocket as c + 0.2c = 1.2c. If the spaceship's motion were in the opposite direction (the spaceship on the right in Figure 20.8b), then it should be c − 0.2c = 0.8c. But that's not what happens. What does happen—and what's consistent with relativity—is that the Earth-based observer measures the light's speed as c. This has been verified by numerous experiments in more than a century since Einstein proposed the relativity principle. Although we don't have rockets that can go at 0.2c, we do have elementary particles that move much faster, and they verify relativity with exquisite precision.

The speed of light c is the same in all inertial frames. In the language of special relativity, that makes c an **invariant** quantity. Invariants are the true absolutes of relativity; all observers agree on their values, even if different observers are in motion relative to each other. Surprisingly, as you'll see next, measures of space and time are *not* invariants.

20.4 Relativity of Time and Space

An **event** is a physical happening that occurs at a specific position and time—for example, your birth. Events are key to understanding space and time in relativity. Comparing positions and times of the same events in different inertial frames shows how measurements of space and time differ from one inertial frame to another.

Simultaneous Measurements of Events

We'll begin with a thought experiment, suggested by Einstein and involving two inertial reference frames (Figure 20.9). In one frame, Sam is sitting on a bench next to a railroad track. Sharon is seated in the middle of a train as it passes Sam. The train's velocity is constant, so Sharon's frame, like Sam's, is inertial.

Suppose lightning bolts strike the front and the rear of the train car as Sharon passes Sam (Figure 20.10). The timing is such that the two light flashes reach Sam simultaneously. Because the light flashes traveled the same distance at the same speed c in Sam's inertial frame, he concludes that the two lightning strikes occurred simultaneously. But for Sharon the situation is different. As the second frame in Figure 20.10 shows, the flash from the front of the car reaches her first. In her inertial frame, she was also an equal distance from the two lightning strikes, and relativity asserts that the speed of both light flashes is also c in Sharon's inertial frame. So Sharon concludes that the lightning strike to the front of the car occurred first, because the flash from that strike reaches her first.

This simple thought experiment shows that:

> Distinct events that are simultaneous in one inertial frame need not be simultaneous in another inertial frame.

This seems to violate common sense, yet it follows simply from Einstein's principle of relativity. In relativity, measures of time and position aren't absolute, but depend on one's inertial frame. You might try to argue that Sharon's perception in Figure 20.10b is an illusion resulting from her motion—but to do so would imply that there's something special about Sam's reference frame, in that what he perceives is correct while Sharon's perception is an illusion. That would be a gross violation of the relativity principle, which states that all inertial reference frames are equal before the laws of physics. Any time you find

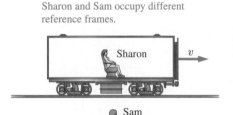

Sharon and Sam occupy different reference frames.

FIGURE 20.9 Sharon approaching Sam in a high-speed train. This diagram is shown from Sam's reference frame.

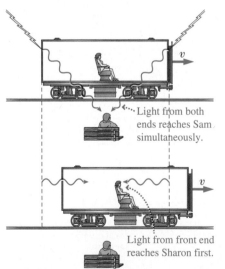

FIGURE 20.10 Why events simultaneous in Sam's reference frame aren't simultaneous in Sharon's. Both diagrams are shown from Sam's reference frame.

yourself doubting relativity, it's probably because you're clinging to the notion that one reference frame is actually the "right" one. It isn't, because relativity denies that favored status to any reference frame.

Time Dilation

If events simultaneous in one inertial frame aren't simultaneous in another, then observers in different frames won't agree about the time between events. To explore this, we'll use the same two inertial frames, but with light signals sent vertically instead of horizontally. Sharon has a "light clock," which keeps time using a light flash as it travels from a source on the floor up to a mirror and back again to the floor (Figure 20.11a). Let Δt_0 be the time for this round trip as measured by Sharon. In her inertial frame the light goes straight up and down, a total distance $2d$. The speed of light is c, so the time is simply

$$\Delta t_0 = \frac{2d}{c}$$

Sam also views the light signal's departure and return. Because the train is moving relative to him with speed v, Sam sees the light signal follow the diagonal path in Figure 20.11b. From the geometry of the right triangle in Figure 20.11c, the Pythagorean theorem gives

$$\left(\frac{c\Delta t}{2}\right)^2 = \left(\frac{v\Delta t}{2}\right)^2 + d^2$$

Solving for Δt,

$$\Delta t = \frac{2d}{c\sqrt{1 - v^2/c^2}}$$

Comparing Sharon's measure of the time interval (Δt_0) with Sam's (Δt),

$$\Delta t = \frac{\Delta t_0}{\sqrt{1 - v^2/c^2}}$$

The factor $1/\sqrt{1 - v^2/c^2}$ appears frequently in relativity and is given the symbol γ (lowercase Greek letter gamma). Thus,

$$\gamma = \frac{1}{\sqrt{1 - v^2/c^2}} \qquad \text{(Relativistic factor } \gamma \text{; SI unit: dimensionless)} \qquad (20.5)$$

and

$$\Delta t = \gamma \Delta t_0 \qquad \text{(Time dilation; SI unit: s)} \qquad (20.6)$$

For material objects, v is always less than c, giving $\gamma > 1$ for $v \neq 0$. Therefore, $\Delta t > \Delta t_0$, so more time elapses between two events for Sam than for Sharon. This effect is **time dilation**.

Our light clock example may seem rather contrived. You might well wonder whether time dilation really happens and whether you could measure it with another kind of clock. Consider the relativity principle: *All the laws of physics are the same in all inertial frames.* That means any other clock must keep the same time as the light clock. The effect is real, and it's been observed repeatedly with many different clocks.

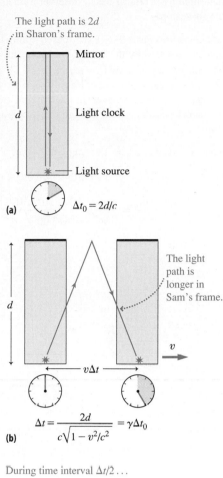

(a) $\Delta t_0 = 2d/c$

(b) $\Delta t = \dfrac{2d}{c\sqrt{1-v^2/c^2}} = \gamma \Delta t_0$

The light path is $2d$ in Sharon's frame.

Mirror

Light clock

Light source

The light path is longer in Sam's frame.

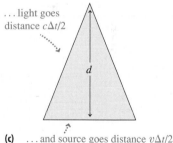

During time interval $\Delta t/2$ ...

...light goes distance $c\Delta t/2$

(c)   ...and source goes distance $v\Delta t/2$

FIGURE 20.11 A light clock, consisting of a light source and mirror, shown (a) in Sharon's reference frame (at rest with respect to the clock) and (b, c) in Sam's frame (with respect to which the clock moves at speed v).

✓ **TIP**

Time dilation works no matter what kind of clock you're using: pendulum, electronic, or even human metabolism. That's because it's about time itself, not the instrument used to measure time.

EXAMPLE 20.5 **Earth Time, Spaceship Time**

A 25th-century spaceship travels at $0.8c$. (a) How much time does it take this spaceship to cover the 150 million km from Earth to Sun, as measured by (a) Earth-based observer Sam and (b) Sharon, riding on the spaceship?

ORGANIZE AND PLAN For Earth-based Sam, the time is simply the distance divided by the speed. The spaceship-based clock runs slow by a factor of $\gamma = 1/\sqrt{1 - v^2/c^2}$.

Note: Because it rotates, Earth's surface is accelerated, so it isn't an inertial frame. However, in most applications, including this one, that acceleration is small enough that Earth *approximates* an inertial frame.

Known: Distance $d_0 = 150$ Mkm $= 1.50 \times 10^{11}$ m; $v = 0.8c$.

SOLVE (a) The time for the Earth-based clock is

$$\Delta t = \frac{d_0}{v} = \frac{d_0}{0.8c} = \frac{1.50 \times 10^{11} \text{ m}}{0.8\,(3.00 \times 10^8 \text{ m/s})} = 625 \text{ s}$$

or 10.4 min.

(b) Sharon's spaceship time Δt_0 differs by the factor $1/\gamma$ (Equation 20.6). Therefore,

$$\Delta t_0 = \frac{\Delta t}{\gamma} = \sqrt{1 - v^2/c^2}\,\Delta t = \sqrt{1 - 0.8^2}\,(625 \text{ s}) = 375 \text{ s}$$

or just 6.3 min. The trip is over 4 min shorter, according to the spaceship's clock.

REFLECT Although spaceships don't yet travel this fast, time dilation has been confirmed using precise clocks in satellites and even airplanes. The Global Positioning System would be off by several kilometers if time dilation and other relativistic effects weren't accounted for.

MAKING THE CONNECTION How fast must the spaceship travel for the time on its clock to be only half the time measured on Earth?

ANSWER This requires $\gamma = 2$, which implies $v = \sqrt{3}c/2 \approx 0.87c$.

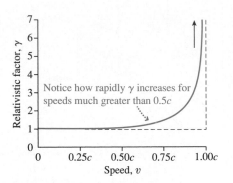

Notice how rapidly γ increases for speeds much greater than $0.5c$

FIGURE 20.12 The relativistic factor γ increases rapidly as relative speed v approaches c.

The factor γ is all-important in determining the size of relativistic effects. Figure 20.12 plots γ as a function of the relative speed v, and Table 20.2 lists numerical values at some benchmark speeds. Note that γ is extremely close to 1 at everyday speeds; that's why you don't normally notice relativistic effects. It's also why many consequences of relativity seem counterintuitive—like different measures of the time between two events. If we grew up moving around relative to our surroundings at speeds approaching c, relativity would be obvious! At speeds approaching c, γ grows large, and relativistic effects become significant. Although we don't have spaceships that go anywhere near $0.8c$, we can easily accelerate subatomic particles to more than 99% of light speed. In these applications, the effects of special relativity are obvious and Einstein's theory is routinely confirmed.

✓**TIP**

For many practical purposes, relative speeds up to $0.1c$ are considered "nonrelativistic," because the γ isn't much above 1; relative speeds $v > 0.1c$ are considered "relativistic."

TABLE 20.2 Relativistic Factor γ as a Function of Relative Speed v

Speed (m/s)	Speed (v/c)	$\gamma = \dfrac{1}{\sqrt{1 - v^2/c^2}}$
300 (jet aircraft)	1.0×10^{-6}	$1 + 5 \times 10^{-13}$
30,000 (Earth in orbit around Sun)	1.0×10^{-4}	$1 + 5 \times 10^{-9}$
3.0×10^6	0.01	1.00005
3.0×10^7	0.10	1.005
1.5×10^8	0.50	1.15
2.7×10^8	0.90	2.3
—	0.99	7.1
—	0.999	22.4

"Running Slow"

You may have heard time dilation described with the phrase: "Moving clocks run slow." But that's misleading: Since all inertial frames are equivalent, no inertial observer can claim to be "at rest" while another is "moving." Only relative motion matters. Using Figure 20.11, we concluded that Sharon's clock runs slow when viewed from Sam's inertial reference frame. But Sharon's is also an inertial frame, and Sam moves at v relative to her. So Sharon would see Sam's clock running slow by the same factor γ.

In Example 20.5 Sam remains on Earth while Sharon travels from Earth to Sun. Do you claim that Sam is "really" at rest since he's on Earth? Then you're denying the relativity principle, which—again—asserts that all inertial reference frames are equally valid for doing physics. Each observer does see the other's clocks "running slow," and yet there's no contradiction. How can that be?

Figure 20.13 shows the Earth-Sun trip from Sam's Earth-based perspective. Both observers measure the time between two events. Event 1 is the spaceship passing Earth, and Event 2 is the spaceship reaching the Sun. In Sam's reference frame those events occur at different places, so his reference frame needs two clocks, one at Earth and one at the Sun. But in Sharon's frame the events happen at the *same* place—which is obvious to Sharon since she sits in her spaceship and watches first Earth go by (Event 1) and then the Sun go by (Event 2). And for Sharon the time is shorter. This description provides a clearer way to talk about time dilation:

> Time dilation: The time between two events is shortest in a reference frame where the events occur at the same place.

That shortest time is called the **proper time** between the events—but that doesn't mean it's the "right" time. There is no "right" time between events, because all inertial frames are equally good for doing physics, and yet different frames measure different times between events.

It's the asymmetry between Sam's two clocks at different locations versus Sharon's one clock present at both events that means their experiments aren't identical. If we established two separated clocks in Sharon's reference frame and Sam watched them move past him, then he would measure a shorter time than Sharon for the two events of Sharon's first clock passing him, then the second. And remember something else: Events that are simultaneous in one reference frame aren't in another. That means the two clocks in Figure 20.13 are synchronized as far as Sam is concerned—both read $t = 0$ when Sam and Sharon coincide at Earth—but they aren't synchronized for Sharon. As a result, she can see them "running slow" and still agree that the elapsed time between the two events is longer for Sam!

Length Contraction

The Earth-Sun trip of Example 20.5 took 625 s in Earth's reference frame but only 375 s in the spaceship's frame. Yet observers in both frames agree that the relative speed is $0.8c$. Speed is distance per time, so how can they measure different times? This implies that the distances are also different. In the Earth frame the Earth-Sun distance was $d_0 = 150$ Mkm (Figure 20.14a). With a relative speed of $0.8c$, *speed = distance/time* gave $\Delta t = 625$ s in Earth's frame. In the spaceship's frame, we found the time

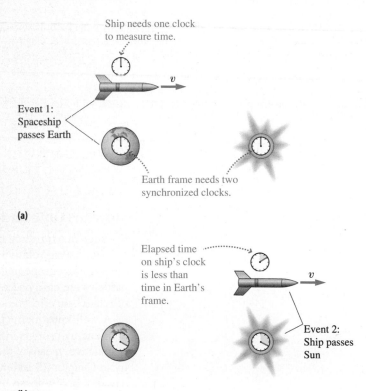

Ship needs one clock to measure time.

Event 1: Spaceship passes Earth

Earth frame needs two synchronized clocks.

(a)

Elapsed time on ship's clock is less than time in Earth's frame.

Event 2: Ship passes Sun

(b)

FIGURE 20.13 The time between two events is shortest when measured with a clock that's present at both events. Thus, in this example, the clock on the spaceship measures the shortest time.

Earth as reference frame

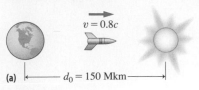

(a) $\vdash\!\!-\!\!-\!\!-\!\!-\ d_0 = 150 \text{ Mkm} \!\!-\!\!-\!\!-\!\!-\!\dashv$

Rocket as reference frame

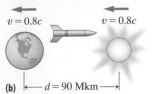

(b) $\vdash\!\!-\!\!-\!\ d = 90 \text{ Mkm} \!\!-\!\!-\!\dashv$

FIGURE 20.14 Earth and spaceship frames measure different values for the Earth-Sun distance.

$\Delta t_0 = 375$ s. Given the relative speed, we write *distance = speed × time*, giving $d = 90$ Mkm. Symbolically, the spaceship covers the distance d_0 in time Δt as measured in Earth's frame, with $d_0 = v\Delta t$. But the spaceship's time is $\Delta t_0 = \Delta t/\gamma$, so for the spaceship the Earth-Sun distance is $d = v\Delta t_0 = v\Delta t/\gamma = d_0/\gamma$ (Figure 20.14b). More generally, the distance between two objects or the length of a single object is smaller when measured by an observer for whom the object(s) is (are) in motion:

$$d = \frac{d_0}{\gamma} \qquad \text{(Length contraction; SI unit: m)} \qquad (20.7)$$

This is **length contraction**. The contraction depends on γ, which in turn depends on the relative speed v. Greater speeds increase γ, which makes the contraction more severe.

Length contraction is often summarized: "Moving objects appear shorter." But again that flaunts the relativity principle, since no inertial frame can claim to be "at rest" or "moving." Only relative motion matters. Stated more clearly, an object is longest in a reference frame relative to which it's at rest. The length in that frame is its **proper length**. The same length or distance measured in another inertial frame is shorter by a factor $\gamma = 1/\sqrt{1 - v^2/c^2}$, where v is the relative speed of the two frames. For the spaceship traveling from Earth to Sun, the proper length is the Earth-Sun distance as measured by an observer on Earth—the distance we've called d_0. That distance is shorter when measured by the ship-based observer.

EXAMPLE 20.6   A Shrinking Spaceship

The spaceship in Example 20.5 has proper length 50 m. While it's flying toward the Sun at $0.8c$, what's its length as measured by an Earth-based observer?

ORGANIZE AND PLAN The spaceship is moving relative to the Earth-based observer, who therefore measures it to be shorter than its proper length by the factor $1/\gamma$: $d = d_0/\gamma$.

Known: $d_0 = 50$ m; $v = 0.8c$.

SOLVE The spaceship's contracted length as measured by an Earth-based observer is

$$d = \frac{d_0}{\gamma} = \sqrt{1 - v^2/c^2}\,d_0 = \sqrt{1 - 0.8^2}\,(50.0 \text{ m}) = 30 \text{ m}$$

REFLECT The contraction factor here is the same as the one by which the spaceship-based observer measures the Earth-Sun distance to be contracted. Considered together, Examples 20.5 and 20.6 show what happens to measures of both space and time. Because distances and times are shortened by the same factor, the relation speed = distance/time is preserved.

MAKING THE CONNECTION Suppose you'd like to see the solar system shrunk to 1/10 of its proper length. What speed does this require?

ANSWER This requires $\gamma = 10$, which implies $v \approx 0.995c$.

More on Time and Space

Rocket ships traveling close to the speed of light provide interesting thought experiments for illustrating relativistic effects, but real rockets don't go that fast. However, there's plenty of experimental data confirming time dilation and length contraction. Subatomic particles are easily accelerated to speeds near c, and they offer most dramatic illustrations of relativistic effects.

A well-known effect involves subatomic particles called **muons** that are created when high-energy protons from the Sun collide with particles in Earth's atmosphere. Muons are radioactive, meaning they decay into other subatomic particles. (We'll discuss radioactivity in Chapter 25 and muons, with other elementary particles, in Chapter 26.) Muons decay quickly, with an average lifetime of about 2.2×10^{-6} s ($2.2\ \mu$s). A typical muon created in the upper atmosphere has a speed of about $0.99c$, so it should travel a distance $(0.99c)(2.2\ \mu\text{s}) = 650$ m before decaying. That's not far enough to reach Earth's surface. However, detectors at the surface record substantial numbers of muons. Time dilation and length contraction explain this discrepancy.

Relative to an inertial frame attached to Earth, the muon is moving at $0.99c$, and its time is running slow by a factor of $1/\gamma = 1/7.1$ (Table 20.2). That stretches its lifetime as measured in Earth's reference frame, allowing it to travel 7.1 times farther, or about 4.6 km on average. That's enough for many muons to reach Earth's surface—just what's observed. Now consider the situation from a muon's inertial frame. There's nothing special about that frame, so the muon decays with average lifetime 2.2 μs. But now the distance to Earth's surface is shortened by the factor 7.1 (Figure 20.15). The shorter distance allows many muons to reach the surface.

Note in this example that Earth measures the proper length (the muon's distance traveled), and the muon clock measures the proper time. The two inertial frames don't agree on distance and time measurements, and yet they must agree on the reality of how many muons reach Earth's surface. Time dilation in one inertial frame coordinates with length contraction in the other, so that both observers ultimately agree on the outcome.

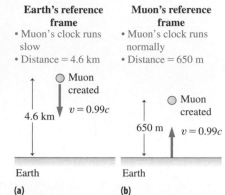

FIGURE 20.15 The muon's journey viewed (a) in Earth's reference frame and (b) in the muon's frame.

CONCEPTUAL EXAMPLE 20.7   Twin Paradox

Alice and Bob are twins, living in the 30th century. Alice becomes an astronaut and flies at a high, constant speed to a nearby star system. She finds nothing interesting and promptly returns to Earth. Compare the twins' ages when she returns.

SOLVE Because "moving clocks run slow," your first guess might be that Alice is younger. But from Alice's perspective, Bob moves relative to her and therefore he should be younger. Yet when the twins reunite, they're standing side by side and they have to agree on who's older. So which argument is correct? Or are they in fact the same age? This situation, with the incompatible solutions we've presented, is called the **twin paradox**.

There's an asymmetry in the situation that negates Alice's argument about Bob's being the "moving" twin: She doesn't stay in a single inertial reference frame. In order to stop, turn around, and return, Alice undergoes acceleration, and that acceleration separates two distinct inertial frames, one on her outbound trip and the other on her return. Bob, in contrast, remains in a single inertial reference frame the whole time, and he rightly concludes that time in Alice's reference frames is "running slow" relative to his frame. In his inertial frame, Bob can apply concepts and equations of special relativity. Alice, in her not-always-inertial frame, can't.

REFLECT This effect has been verified using atomic clocks accurate to a nanosecond. One remains on Earth while the other is flown in an airplane. When the clocks are reunited, the "traveling" clock shows several hundred nanoseconds less elapsed time. In special relativity, motion itself is meaningless; no one has the right to say "I'm at rest and you're moving." But *changes* in motion do matter in special relativity, and it's the *change* in her motion that makes the traveling twin's situation different.

EXAMPLE 20.8   Twin Paradox: Quantitative

Suppose Alice flies to the Alpha Centauri system, a distance of 4.0×10^{16} m. She travels at $0.8c$ both out and back, with negligible turnaround time. Find the round-trip time as measured by both Alice and Bob.

ORGANIZE AND PLAN For Bob the time computation is straightforward. It's Alice's total travel distance traveled divided by her speed. Alice's time then follows from the phenomenon of time dilation, described in Equation 20.6.

Known: $d_0 = 4.0 \times 10^{16}$ m; $v = 0.8c$.

SOLVE The round-trip distance, as measured by Earth-based Bob, is $2d_0$. Therefore, Bob measures the round-trip time as

$$t = \frac{2d_0}{v} = \frac{2(4.0 \times 10^{16} \text{ m})}{(0.8)(3.0 \times 10^8 \text{ m/s})} = 3.3 \times 10^8 \text{ s} \approx 10.5 \text{ years}$$

The elapsed time for Alice is shorter by the factor $1/\gamma$:

$$\Delta t_0 = \frac{\Delta t}{\gamma} = \sqrt{1 - v^2/c^2}\,\Delta t$$

Here $v/c = 0.8$, so $v^2/c^2 = 0.8^2$ and

$$\Delta t_0 = \sqrt{1 - v^2/c^2}\,\Delta t = \sqrt{1 - 0.8^2}\,(3.3 \times 10^8 \text{ s})$$
$$= 2.0 \times 10^8 \text{ s} \approx 6.3 \text{ years}$$

REFLECT The age difference is significant at this high relative speed. We worked this example from Bob's perspective; from Alice's perspective, her clock is running at a normal rate, but the distance is shortened by $1/\gamma$ each way.

MAKING THE CONNECTION What's the one-way distance for the trip in Alice's frame of reference? Use this distance to compute her elapsed round-trip time.

ANSWER The distance is shortened by a factor of $1/\gamma$ to 2.4×10^{16} m. The round trip is then 4.8×10^{16} m. Her time elapsed is that distance divided by her speed, 2.4×10^8 m/s, giving 2.0×10^8 s. This agrees with the previous result.

20.5 Relativistic Velocity and the Doppler Effect

You've seen that the speed of light as measured by an inertial observer doesn't depend on the motion of the light source—in contrast to what happens with material objects. What does change is the light's measured wavelength and frequency. Here we'll explore both these matters further.

Velocity Addition

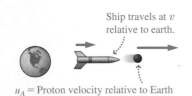

u_A = Proton velocity relative to Earth
u_B = Proton velocity relative to ship

FIGURE 20.16 Relativistic velocity addition.

The velocity of a light source doesn't "add" to that of the light, thus keeping the speed of light the same for all observers. But it isn't just light for which velocity addition is counterintuitive. Even material objects' velocities don't add as you'd expect—an effect that becomes substantial at high relative speeds. Figure 20.16 shows a spaceship moving away from Earth at speed v and firing a proton beam in the forward direction. Because the ship moves relative to Earth, observers on Earth and ship measure different speeds for the protons. Let u_A be the protons' velocity in Earth's reference frame, u_B their velocity in the spaceship's frame.

> ✓ **TIP**
>
> To avoid confusing different velocities, we'll reserve v for the relative velocity of two inertial frames and let a subscripted u be the velocity of an object in either frame.

Classical physics and common sense suggest that $u_A = u_B + v$. That could lead to speeds greater than light—something that never happens. Furthermore, you've seen that measures of time differ in different inertial frames. That means we can't simply add velocities measured in different frames. We won't derive it, but the correct relativistic velocity relationship is:

$$u_A = \frac{u_B + v}{1 + u_B v/c^2} \quad \text{(Relativistic velocity addition; SI unit: m/s)} \quad (20.8)$$

Suppose the spaceship in Figure 20.16 is moving at $v = 0.8c$, and the proton beam is fired at $u_B = 0.9c$ relative to the ship. Then the protons' velocity relative to Earth is

$$u_A = \frac{u_B + v}{1 + u_B v/c^2} = \frac{0.9c + 0.8c}{1 + (0.8c)(0.9c)/c^2} = \frac{1.7c}{1 + 0.72c^2/c^2} = \frac{1.7}{1.72}c \approx 0.988c$$

You can make the velocities u_B and v as close to c as you'd like, but there's no way to get the third velocity u_A to exceed c. We stress that this is an experimental fact:

Material objects never travel at c or greater relative to any inertial frame.

$v = 0.8c$

u_B = Proton velocity relative to spaceship = −0.9c
u_A = ?

FIGURE 20.17 What's the proton velocity relative to Earth?

Equation 20.8 adds *velocities*, not speeds, and thus accounts for direction. Suppose our spaceship now fires its proton beam toward Earth at $0.9c$ as measured by the spaceship (Figure 20.17). Take the direction of the ship's velocity to be positive, $v = 0.8c$. The protons are going the opposite way, so $u_B = -0.9c$. Now Equation 20.8 gives the protons' velocity in Earth's reference frame as:

$$u_A = \frac{u_B + v}{1 + u_B v/c^2} = \frac{-0.9c + 0.8c}{1 + (0.8c)(-0.9c)/c^2} = \frac{-0.1c}{1 - 0.72} = \frac{-0.1}{0.28}c \approx -0.36c$$

The *negative* sign shows that observers on Earth see the protons *approaching* at 0.36c.

✓**TIP**

Velocity involves a speed and direction. For motion along the x-axis, velocity is positive in the +x-direction and negative in the −x-direction.

EXAMPLE 20.9   **Approaching Spaceships**

Two spaceships approach Earth from opposite directions, at speeds of 0.5c and 0.8c. What's the speed of each ship as seen from the other?

ORGANIZE AND PLAN The velocity addition law involves two inertial frames in relative motion. In Figure 20.18 we show the situation in Earth's frame. We want the velocity u_A of the left-hand ship relative to the right-hand one, so the other frame is that of the right-hand ship. In that frame, Earth moves rightward with velocity $v = 0.8c$. We know that the left-hand ship's velocity in Earth's frame is $u_B = 0.5c$, so the velocity addition law gives the velocity u_A we seek.

Known: $v = 0.8c$; $u_B = 0.5c$.

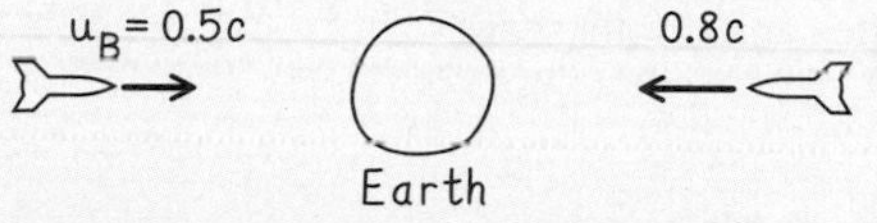

FIGURE 20.18 The situation described in Example 20.9 as viewed from Earth's frame of reference, with two spaceships approaching Earth from opposite directions.

SOLVE Equation 20.8 gives

$$u_A = \frac{u_B + v}{1 + u_B v/c^2} = \frac{0.5c + 0.8c}{1 + (0.8c)(0.5c)/c^2}$$
$$= \frac{1.3c}{1 + 0.40} = \frac{1.3}{1.4}c \approx 0.93c$$

So the left-hand ship approaches the right-hand ship at 0.93c. By symmetry, the right-hand ship approaches the left-hand one at −0.93c.

REFLECT As you'd expect, the approaching ships have a relative velocity larger than either ship's velocity relative to Earth. However, the velocity addition law keeps the relative velocity under c.

MAKING THE CONNECTION Suppose the left-hand ship fires a light beam toward Earth. What's the light's speed as measured on Earth?

ANSWER The relativistic velocity equation (20.8) still applies, now with $v = 0.5c$ for the relative velocity of ship and Earth, and $u_B = c$ for the light. Then the equation yields $u_A = c$, regardless of v. Of course: Light has the same speed c in all inertial reference frames!

What about that baseball in Figure 20.8, thrown at 30 m/s relative to a truck moving at 10 m/s? It, too, obeys Equation 20.8. But here both u_B and v are so small compared with c that the denominator is essentially 1, making the correct relativistic result indistinguishable from the common sense $u_A = u_B + v$.

Relativistic Doppler Effect

In the Making the Connection in Example 20.9, the left-hand spaceship fired a light beam with speed c. Despite the relative motion of ship and Earth, the light reached Earth at c. As relativity requires, its speed is the same—c—in both reference frames. But its wavelength differs. This is the **Doppler effect**, discussed for sound in Chapter 11. Figure 20.19b shows the origin of the Doppler effect for light, with the source approaching the receiver. The spaceship travels some distance during the time between emissions of successive wave fronts. Therefore, the wave fronts are closer than if the light source were stationary, and the receiver measures a reduced wavelength. For a receding source, in contrast, the received wavelength increases (Figure 20.19c). We won't derive it, but the Doppler effect for light is

$$\lambda = \sqrt{\frac{1 \mp v/c}{1 \pm v/c}}\lambda_0 \qquad \text{(Doppler effect for light; SI unit: m)} \qquad (20.9)$$

When source is stationary...

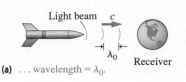

(a) ...wavelength $= \lambda_0$.

When source moves toward receiver...

(b) ...wavefronts are closer together, so $\lambda < \lambda_0$.

When source moves away from receiver...

(c) ...wavefronts are farther apart, so $\lambda > \lambda_0$.

FIGURE 20.19 (a) Light is sent from a stationary source with wavelength λ_0. (b) Doppler effect for an approaching source. (c) Doppler effect for a receding source.

Here the upper signs apply when the source approaches the receiver, the lower signs when it recedes. Because $f\lambda = c$ for light, there's always a corresponding frequency shift.

Suppose that ship approaching Earth at $v = 0.5c$ emits green laser light with wavelength $\lambda_0 = 532$ nm. Then the light received at Earth has wavelength

$$\lambda = \sqrt{\frac{1 - v/c}{1 + v/c}}\,\lambda_0 = \sqrt{\frac{1 - 0.5}{1 + 0.5}}\,(532 \text{ nm}) = 307 \text{ nm}$$

Here the shift has taken the waves out of the visible part of the spectrum—in this case to shorter-wavelength ultraviolet.

The Doppler formula for light is different from that for sound—a reflection of the fact that light has no medium and that only relative motion matters. But for speeds v much less than the wave speed (c or the sound speed), formulas for light and sound become essentially undistinguishable.

Applications of the Doppler Effect

The Doppler effect is crucial for astrophysics. In the early 20th century, American astronomer Edwin Hubble noticed that spectral lines from distant galaxies were **redshifted**—that is, shifted toward longer wavelengths. From these observations, Hubble deduced that the universe is expanding. We'll consider Hubble's work and the expanding universe in Chapter 26.

Meteorologists use radar signals reflected from water droplets to form images of clouds and precipitation. The newest systems also measure wind speeds within weather systems, detecting dangerous wind shear around airports or the swirling motions that can lead to tornadoes.

Radar guns contain a microwave transmitter and receiver. An approaching or receding object reflects the waves, and the measured shift to higher or lower frequencies yields the object's speed. Police use radar guns to catch speeders, and they're used in athletics to measure ball speeds, such as baseball pitches and tennis serves.

CONCEPTUAL EXAMPLE 20.10 **The Doppler Effect: It's Relative**

For the situation in Figure 20.19b, suppose a transmitter on Earth sends light with wavelength λ_0 toward the spaceship. What's the light's wavelength as received at the ship?

SOLVE From ship's viewpoint in Figure 20.19, Earth is approaching at speed v. Therefore, the Doppler shift formula gives

$$\lambda = \sqrt{\frac{1 - v/c}{1 + v/c}}\,\lambda_0$$

That's the same result we'd get for the ship's signals received at Earth, as in Figure 20.19b. Thus if ship and Earth carry identical transmitters, each receives signals from the other at the same reduced wavelength.

REFLECT This example shows again the truly "relative" nature of relativity. There's no privileged inertial frame attached to Earth or anywhere else. Each inertial frame has the same view of physics.

GOT IT? Section 20.5 A spaceship approaches a space station at $0.6c$. Another ship approaches the station from the opposite direction, also at $0.6c$. Each ship measures the other's speed as (a) less than $0.6c$; (b) $0.6c$; (c) more than $0.6c$ but less than c; (d) $1.2c$.

20.6 Relativistic Momentum and Energy

You've just seen how relativity affects space and time. Here we treat two other important physical quantities: momentum and energy. Relativity revises our classical notions of both, and experiments in high-energy particle accelerators validate relativistic concepts of energy and momentum.

Relativistic Momentum

In Chapter 6 you learned that momentum is a vector quantity with magnitude $p = mv$. Because no material object can go at c or faster, this expression implies an upper limit mc for momentum. But we can keep applying a force to an object—as we do with subatomic particles in particle accelerators—and, as Newton showed, force results in momentum change. So the expression $p = mv$ can't be consistent with relativity and the idea that force changes momentum. Instead, momentum must increase indefinitely as an object's speed approaches c, as suggested in Figure 20.20. In fact, relativistic momentum is given by

$$p = \frac{mv}{\sqrt{1 - v^2/c^2}} = \gamma mv \quad \text{(Relativistic momentum; SI unit: kg·m/s)} \quad (20.10)$$

Equation 20.10 follows from the relativity principle and has been confirmed by measurements on high-energy particles. The relativistic factor $\gamma = 1/\sqrt{1 - v^2/c^2}$ is always greater than 1 for a moving particle, and γ grows large when the particle's speed approaches c (Figure 20.12). That's why momentum continues to grow indefinitely as a particle's speed approaches c.

We've been a little sloppy with wording here, talking about "the particle's speed" and momentum increasing "as speed approaches c." Since only relative motion matters, we always mean the speed measured relative to some inertial reference frame. Observers in different inertial frames will measure different values for a particle's speed and therefore its momentum—a situation that's also true in classical physics.

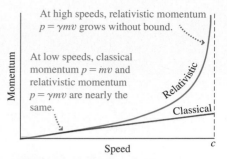

At high speeds, relativistic momentum $p = \gamma mv$ grows without bound.

At low speeds, classical momentum $p = mv$ and relativistic momentum $p = \gamma mv$ are nearly the same.

FIGURE 20.20 Relativistic and classical momentum, as functions of speed v.

✓ TIP

The relativistic factor $\gamma \approx 1$ for speeds much less than c. In this nonrelativistic limit, $p = \gamma mv \rightarrow mv$, in agreement with the classical formula.

Relativistic Kinetic Energy

The same reasoning we applied to momentum shows that classical kinetic energy $K = \frac{1}{2}mv^2$ must be revised for relativistic speeds. Again, relativity theory provides the correct expression:

$$K = (\gamma - 1)mc^2 \quad \text{(Relativistic kinetic energy: SI unit: J)} \quad (20.11)$$

Since γ increases indefinitely as a particle's speed approaches c, there's no limit to kinetic energy—as Figure 20.21 shows. You can show in Problem 94 that Equation 20.11 reduces to the nonrelativistic limit $K = \frac{1}{2}mv^2$ for speeds much less than c. Since γ becomes 1 when $v = 0$, the relativistic kinetic energy is clearly zero for a particle at rest. This is consistent with kinetic energy being "energy of motion."

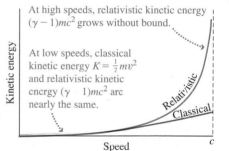

At high speeds, relativistic kinetic energy $(\gamma - 1)mc^2$ grows without bound.

At low speeds, classical kinetic energy $K = \frac{1}{2}mv^2$ and relativistic kinetic energy $(\gamma - 1)mc^2$ are nearly the same.

FIGURE 20.21 Relativistic and classical kinetic energies, as functions of speed v.

✓ TIP

The formula for kinetic energy (Equation 20.11) reduces to the classical result $K = \frac{1}{2}mv^2$ in the nonrelativistic limit $v \ll c$.

Rest Energy and Total Energy

Another way to write the relativistic kinetic energy is $K = \gamma mc^2 - mc^2$. The second term here, mc^2, is independent of speed. Einstein recognized the importance of this quantity and called it **rest energy** (E_0):

$$E_0 = mc^2 \quad \text{(Rest energy: SI unit: J)} \quad (20.12)$$

Anything with mass m has rest energy mc^2, whether or not it's in motion. Mass m and the speed of light c are both invariant quantities, so rest energy is also invariant.

Through Equation 20.12, Einstein recognized an intimate relation between mass and energy. If a system loses energy, its mass decreases—and vice versa. For example, energy-releasing reactions—whether chemical or nuclear—result in a decrease Δm in mass of the reacting particles. The energy equivalent, Δmc^2, shows up as the energy released. This mass loss is negligibly small in chemical reactions because of the relatively low energies involved, amounts to somewhat under 1% of the reacting masses in typical nuclear reactions, and can be dramatic in interactions involving subatomic particles. For example, an electron annihilates with its antimatter counterpart, a positron, to produce a pair of gamma-ray photons whose combined energy equals the entire rest energy of the original particles.

It's useful to define **total relativistic energy** E as the sum of rest energy and kinetic energy: $E = K + E_0$. An expression for E follows from Equations 20.11 and 20.12: $E = K + E_0 = (\gamma - 1)mc^2 + mc^2 = \gamma mc^2 - mc^2 + mc^2$, or

$$E = \gamma mc^2 \quad \text{(Total relativistic energy; SI unit: J)} \quad (20.13)$$

Energy conservation holds in relativity, but only if we account for rest energy as well as kinetic energy. In everyday interactions we can neglect rest energy because it's essentially the same before and after an interaction. But, as electron-positron annihilation suggests, rest energy may change significantly in high-energy interactions. We'll discuss such reactions further in Chapters 25 and 26.

EXAMPLE 20.11 **Mass Needed to Run a City**

A medium-sized city uses about 3×10^{14} J of energy per day. How much mass, in the form of electrons and positrons undergoing annihilation, would be required to supply that energy?

ORGANIZE AND PLAN Equation 20.12 gives the energy content of mass m: $E_0 = mc^2$. Here we know the energy we want, so we can solve for the mass.

Known: $E_0 = 3 \times 10^{14}$ J.

SOLVE Rearranging the equation gives

$$m = \frac{E_0}{c^2} = \frac{3 \times 10^{14}\,\text{J}}{(3 \times 10^8\,\text{m/s})^2} = 0.003\,\text{kg}$$

REFLECT That's just 3 g, about the mass of three raisins! Practical energy conversion schemes are much less efficient at extracting rest energy. A city powered by burning coal or fissioning uranium would reduce the mass of the reactants (coal and oxygen, or uranium) by this same 3 g, but the coal plant would burn many tons of coal in the process, and the nuclear plant would fission about 3 kg of uranium.

MAKING THE CONNECTION What mass of matter-antimatter fuel would be needed to supply the world's yearly energy demand, about 4×10^{20} J?

ANSWER Using the same method as in Example 20.10, the required mass is about 4400 kg.

EXAMPLE 20.12 **A High-Energy Electron**

(a) Find the electron's rest energy. (b) An electron is accelerated from rest through a potential difference of 800 kV. Find its kinetic energy, total energy, speed, and momentum. Give energies in electron volts as well as joules.

ORGANIZE AND PLAN Rest energy is $E_0 = mc^2$. The electron's mass can be found in Appendix D ($m = 9.11 \times 10^{-31}$ kg). From Chapter 16, the kinetic energy gained when a charge q falls through potential difference ΔV is $K = |q\Delta V| = e\Delta V$. Total energy is the sum of kinetic and rest energy: $E = K + E_0$. Then Equation 20.13 relates total energy and speed $E = \gamma mc^2$, where $\gamma = 1/\sqrt{1 - v^2/c^2}$. Knowing both γ and the speed v gives the momentum via Equation 20.10: $p = \gamma mv$.

Known: $m = 9.11 \times 10^{-31}$ kg; $\Delta V = 800$ kV; $e = 1.60 \times 10^{-19}$ C.

SOLVE The electron's rest energy is

$$E_0 = mc^2 = (9.11 \times 10^{-31}\,\text{kg})(3.00 \times 10^8\,\text{m/s})^2$$
$$= 8.20 \times 10^{-14}\,\text{J}.$$

With 1 eV equal to 1.6×10^{-19} J, that's 511 keV. The electron volt is *defined* as the energy gained by an elementary charge in a potential difference of 1 volt, so it's easiest to calculate the kinetic energy in electron volts:

$$K = e\Delta V = (e)(800\,\text{kV}) = 800\,\text{keV or } 1.28 \times 10^{-13}\,\text{J}$$

The total energy is

$$E = K + E_0 = 800\,\text{keV} + 511\,\text{keV}$$
$$= 1.31\,\text{MeV or } 2.10 \times 10^{-13}\,\text{J}$$

cont'd.

Finding the electron's speed requires the relativistic factor γ. The total energy is $E = \gamma m c^2$, so

$$\gamma = \frac{E}{mc^2} = \frac{1.31\ \text{MeV}}{0.511\ \text{MeV}} = 2.56$$

With $\gamma = 1/\sqrt{1 - v^2/c^2}$, we solve for v/c:

$$\frac{1}{\gamma^2} = 1 - \frac{v^2}{c^2}$$

so

$$\frac{v}{c} = \sqrt{1 - \frac{1}{\gamma^2}} = \sqrt{1 - 0.153} = 0.920$$

Therefore, $v = 0.920c = 0.920\,(3.00 \times 10^8\ \text{m/s}) = 2.76 \times 10^8\ \text{m/s}$

Finally, the electron's momentum is

$$p = \gamma m v = (2.56)(9.11 \times 10^{-31}\ \text{kg})(2.76 \times 10^8\ \text{m/s})$$
$$= 6.44 \times 10^{-22}\ \text{kg} \cdot \text{m/s}$$

REFLECT Note that the electron's kinetic energy is larger than its rest energy—a sure sign that this electron is relativistic, with speed close to c. Today's particle accelerators bring subatomic particles to even more relativistic energies and speeds, with the Large Hadron Collider, on the French-Swiss border, achieving 7 TeV in colliding beams of protons and antiprotons. You can explore the LHC quantitatively in Problem 84.

MAKING THE CONNECTION Would a proton with the same kinetic energy be moving faster or slower than the electron?

ANSWER The proton is nearly 2000 times more massive. Its kinetic energy would be much less than its rest energy, and it would be moving much more slowly.

The Energy-Momentum Relation

Classical physics provides a simple relation between an object's kinetic energy and momentum (Chapter 6): $K = p^2/2m$. In relativity, the corresponding relationship is different, and follows from the relativistic expressions for momentum and kinetic energy. That relationship is

$$E^2 = p^2 c^2 + E_0{}^2 = p^2 c^2 + m^2 c^4 \qquad \begin{array}{l}\text{(Energy-momentum relationship;}\\ \text{SI unit: J}^2)\end{array} \qquad (20.14)$$

As an object's speed increases, both its momentum p and total energy E increase. The rest energy $E_0 = mc^2$, however, is invariant. So if you measure the energy and momentum of a particle, then Equation 20.14 yields its rest energy and mass:

$$E_0 = mc^2 = \sqrt{E^2 - p^2 c^2}$$

In Problem 93 you can verify that this expression gives the rest energy we found in Example 20.11.

CONCEPTUAL EXAMPLE 20.13   Massless Particles

Discuss the implications of the energy-momentum relationship for particles with zero mass. Relate the results to electromagnetic waves (Section 20.1).

SOLVE With $m = 0$, Equation 20.14 becomes $E^2 = p^2 c^2 + m^2 c^4 = p^2 c^2$ or $E = pc$. Thus, the energy of a massless particle is directly proportional to its momentum.

From Section 20.1 (Equation 20.4), the energy U of an electromagnetic wave with momentum p is $U = pc$. These results are consistent with the idea that electromagnetic radiation consists of massless particles, called *photons*. We'll discuss photons in Chapter 23.

REFLECT How can a particle be massless and still exist? It must be moving at the speed of light. Massless particles, of which the photon is the only example that's actually been observed, *must* travel at exactly c. They can't be at rest with respect to any inertial observer.

GOT IT? Section 20.6   A proton and electron have the same momentum. Then the total energy E is (a) larger for the proton; (b) larger for the electron; (c) the same for the proton and electron.

Chapter 20 in Context

This chapter explored *electromagnetic waves* and *special relativity*. EM waves consist of oscillating electric and magnetic fields. They have a wide range of wavelengths and frequencies, but all travel at the same speed *c* in vacuum. Questions about the speed of light led Einstein to articulate the *relativity principle*: that the laws of physics are the same in all inertial reference frames. That simple idea has surprising consequences for the nature of space and time. You've seen relativistic *time dilation*, *length contraction*, and *velocity addition*, and how wavelength and frequency are altered by the *Doppler effect*. Finally, you learned about relativistic momentum and energy.

Looking Ahead Visible light is an especially important electromagnetic wave. In Chapters 21 and 22, we'll explore *optics*, the study of light's behavior and its uses. You'll learn about reflection and refraction of light and how mirrors and lenses form images. The properties of light and special relativity will be important in the study of atomic, nuclear, and quantum phenomena in Chapters 23–26. There you'll also learn more about how atoms and nuclei produce electromagnetic waves.

CHAPTER 20 SUMMARY

Electromagnetic Waves

(Section 20.1) An **electromagnetic wave** is a transverse wave, with oscillating electric and magnetic fields perpendicular to each other and to the direction of the wave's propagation. Electromagnetic waves include light, and the speed of all EM waves in vacuum is the **speed of light**, $c = 299{,}792{,}458$ m/s. Electromagnetic waves carry energy and momentum and therefore exert **radiation pressure** on objects.

Speed of light in vacuum: $c = \sqrt{\dfrac{1}{\mu_0 \varepsilon_0}} \approx 3.00 \times 10^8$ m/s

Speed, wavelength, frequency: $c = \lambda f$

EM wave momentum p and energy U: $p = \dfrac{U}{c}$

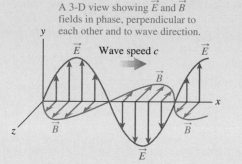

A 3-D view showing $\vec{E}$ and $\vec{B}$ fields in phase, perpendicular to each other and to wave direction.

The Electromagnetic Spectrum

(Section 20.2) The **electromagnetic spectrum** classifies EM waves according to frequency or wavelength. Radio waves have the longest waves, gamma rays the shortest. Visible light wavelengths range from 400 to 700 nm. Frequency is the inverse of wavelength—the shortest waves have the highest frequencies.

Wave type	Frequency (Hz)
Radio	$10^6 - 10^9$
Microwaves	$10^9 - 10^{12}$
Infrared	$10^{12} - 4.3 \times 10^{14}$
Visible light	$4.3 \times 10^{14} - 7.5 \times 10^{14}$
UV light	$7.5 \times 10^{14} - 10^{17}$
X rays	$10^{17} - 10^{20}$
Gamma	$> 10^{20}$

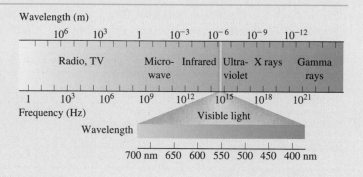

The Fundamental Speed c

(Section 20.3) The principle of relativity states that the laws of physics are the same in all inertial frames. That includes the fact that electromagnetic waves propagating in vacuum have speed c relative to all inertial frames. Einstein's theory of **special relativity** is based on this principle. Einstein's later theory of **general relativity** applies to accelerated reference frames.

Relativity of Time and Space

(Section 20.4) Distinct events that are simultaneous in one inertial frame aren't necessarily simultaneous in another. **Time dilation** and **length contraction** describe how measures of time between events and lengths of objects differ among different inertial reference frames.

Relativistic factor γ: $\gamma = \dfrac{1}{\sqrt{1 - v^2/c^2}}$

Time dilation: $\Delta t = \gamma \Delta t_0$

Length contraction: $d = \dfrac{d_0}{\gamma}$

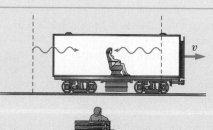

475

Relative Velocity and the Doppler Effect

(Section 20.5) Velocities transform from one reference frame to another using the relativistic velocity addition law. Although light's velocity never varies, its wavelength and frequency are Doppler shifted when source and receiver are in relative motion.

Relativistic velocity addition: $u_A = \dfrac{u_B + v}{1 + u_B v/c^2}$

Doppler effect: $\lambda = \sqrt{\dfrac{1 \mp v/c}{1 \pm v/c}} \lambda_0$

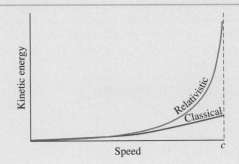

(a) Light beam c Receiver

(b) v $\lambda < \lambda_0$

(c) v c $\lambda > \lambda_0$

Relativistic Momentum and Energy

(Section 20.6) **Relativistic momentum** and **relativistic kinetic energy** grow without bound as an object's speed approaches c. An object also has **rest energy**, which is directly proportional to its mass. Rest and kinetic energies sum to the **total relativistic energy**.

Relativistic momentum: $p = \gamma m v$

Relativistic kinetic energy: $K = (\gamma - 1)mc^2$

Rest energy: $E_0 = mc^2$

Total relativistic energy: $E = \gamma mc^2$

Energy-momentum relationship: $E^2 = p^2 c^2 + E_0^2$

NOTE: Problem difficulty is labeled as ■ straightforward to ■■■ challenging. Problems labeled BIO are of biological or medical interest.

Conceptual Questions

1. When light enters water, its frequency remains the same but its speed drops to about three-quarters of c. What happens to its wavelength?
2. Discuss similarities and differences between light waves and sound waves.
3. A torsion pendulum has a mirrored target on one side and a black target on the other. If the same light source strikes both the black and mirrored sides which way does the torsion pendulum rotate?
4. You're in empty space, far from any sources of gravitational or other forces. If you turn on a flashlight and shine it in a fixed direction, what happens to you?
5. Why would spacecraft propelled by "solar sails" need large sail areas?
6. An electromagnetic wave is traveling from east to west. If its electric field oscillates in the north-south direction, in what direction does the magnetic field oscillate?
7. The speed of light in water is about $0.75c$. If an electron is traveling at $0.90c$ in water, does this violate relativity's prohibition against speeds greater than c?
8. The "light clock" discussed in Section 20.4 seems rather contrived. Would other clocks keep time at the same rate?

9. How would our lives be different if the speed of light were only 3×10^6 m/s? 30,000 m/s? 300 m/s? What if it were infinite?
10. The star Sirius is 8.7 light years from Earth. Yet an astronaut can reach Sirius in less than 8.7 years (as measured by the astronaut) without traveling faster than light. How is this possible?
11. An electron and proton have the same total energy E. Which has the larger kinetic energy? The larger momentum?
12. The most distant galaxies recede from us with speeds close to c. How does this affect the (a) speed and (b) wavelength of light we receive from those galaxies?

Multiple-Choice Problems

13. An electromagnetic wave with wavelength 570 nm in vacuum has frequency (a) 5.3×10^{11} Hz; (b) 5.3×10^{12} Hz; (c) 5.3×10^{13} Hz; (d) 5.3×10^{14} Hz.
14. An electromagnetic wave whose peak electric field is 150 N/C has a peak magnetic field (a) 1.0×10^{-7} T; (b) 3.0×10^{-7} T; (c) 5.0×10^{-7} T; (d) 7.5×10^{-7} T.
15. A 1360-nm electromagnetic wave is in what part of the electromagnetic spectrum? (a) Microwave; (b) ultraviolet; (c) visible; (d) infrared.

16. A 432-nm laser beam passes through a frequency-doubling device. What's its new wavelength? (a) 216 nm; (b) 432 nm; (c) 864 nm; (d) 1728 nm.

17. A rod with a proper length 1.0 m lies along the x-axis and is moving in the $+x$-direction at $0.8c$ relative to you. You measure its length to be (a) 0.4 m; (b) 0.6 m; (c) 0.8 m; (d) 0.9 m.

18. A spaceship flies at $0.8c$ from Earth to Moon, a distance of 400,000 km. A clock on board the spaceship measures the trip time to be (a) 1.0 s; (b) 1.3 s; (c) 1.7 s; (d) 2.8 s.

19. A spaceship makes the 300-million-km round-trip journey to the Sun in just 20 min, as measured by clocks on Earth. How much time elapses on the spaceship's clock? (a) 20 min; (b) 17 min; (c) 14 min; (d) 11 min.

20. A spaceship approaching Earth at $0.5c$ fires a particle beam toward Earth. The particles move at $0.6c$ relative to the spaceship. At what speed does Earth receive the particles? (a) $0.75c$; (b) $0.8c$; (c) $0.85c$; (d) $0.90c$.

21. A spaceship approaching Earth at $0.5c$ fires a light beam toward Earth. At what speed does Earth receive the light beam? (a) $0.75c$; (b) $0.85c$; (c) $0.95c$; (d) c.

22. Two spacecraft approach Earth from opposite directions, each at $0.5c$ relative to Earth. Each spacecraft's speed as measured by the other is (a) $0.5c$; (b) $0.8c$; (c) $0.9c$; (d) c.

23. A spacecraft heading to the outer solar system at $0.20c$ sends radio signals back to Earth. The spacecraft transmits at a wavelength of 2.75 cm. What's the wavelength received on Earth? (a) 3.37 cm; (b) 3.75 cm; (c) 4.13 cm; (d) 5.50 cm.

24. The speed of a particle with kinetic energy equal to its rest energy is (a) $0.50c$; (b) $0.67c$; (c) $0.75c$; (d) $0.87c$.

Problems

Section 20.1 Electromagnetic Waves

25. ■ A helium-neon laser emits 632.8-nm-wavelength light. What's the frequency of this light, in vacuum?

26. ■ Find the wavelength ranges for (a) AM radio, with frequencies from 540 kHz to 1600 kHz, and (b) FM radio, with frequencies from 88 MHz to 108 MHz.

27. ■ If 546-nm laser light is passed through a frequency doubling device, what's its new wavelength?

28. ■■ Show, given the SI units of μ_0 and ϵ_0, that Equation 20.1 gives a result whose units are m/s.

29. ■■ Light travels at about $0.75c$ in water. What's the wavelength in water of light whose wavelength in vacuum is 624 nm?

30. ■■ The height of a typical AM radio station's tower is one-fourth of the radio wavelength. (a) How high should the tower be for a radio station broadcasting at 710 kHz? (b) If another station uses a higher frequency, should its tower be taller or shorter?

31. ■■ (a) Find the peak magnetic field in an electromagnetic wave whose peak electric field is 150 N/C. (b) Find the peak electric field in an electromagnetic wave whose peak magnetic field is 25 nT.

32. ■■ Solar radiation reaches Earth at a rate of about 1400 W/m². If all this energy were absorbed, what would be the average force exerted by radiation pressure on a square meter oriented at right angles to the sunlight?

33. ■■ A 5.0-mW laser beam has a circular cross section with diameter 0.60 mm. What is the maximum radiation pressure that this laser can exert on (a) a mirrored surface and (b) a black surface?

Section 20.2 The Electromagnetic Spectrum

34. ■ For electromagnetic waves with each of these frequencies, find the wavelength in vacuum. In what region of the electromagnetic spectrum does each lie? (a) 9.4×10^{12} Hz; (b) 6.2×10^{14} Hz; (c) 2.9×10^{15} Hz; (d) 5.0×10^{17} Hz.

35. ■ For electromagnetic waves with each of these wavelengths (in vacuum), find the frequency. In what region of the electromagnetic spectrum does each lie? (a) 2.0 km; (b) 2.0 m; (c) 2.0 mm; (d) 2.0 μm; (e) 2.0 nm.

36. ■ A wireless computer network uses microwaves at 5.0 GHz. What's the corresponding wavelength (in air, where λ is negligibly different than in vacuum)?

37. ■ Referring to Figure 20.6, find the range of frequencies in the microwave part of the electromagnetic spectrum.

38. ■ What is the frequency of the 12-cm microwaves used in a microwave oven?

39. ■ Antennas for radio transmission and reception are particularly effective if they're about a half-wavelength long. What's the appropriate length for an FM antenna, taking 100 MHz as a typical FM frequency? (You may have used such an antenna with your FM receiver.)

40. ■■ Suppose you set up a standing wave pattern using light at the red end of the visible spectrum. What's the distance between nodes in this standing wave?

Section 20.3 The Fundamental Speed c

41. ■ An electron is moving at $0.99c$ directly toward a light source. In the electron's reference frame, how fast is the light approaching?

42. ■■■ A boat moves at 6.00 m/s relative to the water. Find the boat's speed relative to shore when it's traveling (a) downstream and (b) upstream in a river with a 2.0-m/s current. (c) The boat travels 100 m downstream and then 100 m upstream, returning to its original point. Find the time for the round trip, and compare this time with the round-trip time if there were no current. (You can neglect relativity at these slow speeds.)

43. ■■■ For the boat and river of the preceding problem, compute the travel time for a round trip on a 100-m path perpendicular to the current, again assuming the boat returns to its starting point. Compare the result with that of the preceding problem. (Michelson's instrument sent light beams along two such perpendicular paths, one going with the Earth's velocity and one perpendicular to it.)

Section 20.4 Relativity of Time and Space

44. ■ A rod with a proper length of 10.0 m lies along the x-axis. If the rod is traveling in the $+x$-direction, find the rod's measured length when its speed is (a) $0.1c$; (b) $0.5c$; (c) $0.9c$; (d) $0.99c$.

45. ■■ A spaceship flies from Earth to Moon, a distance of 400,000 km. Find the elapsed time for this trip both on Earth and on a clock on board the spaceship when the spaceship's speed is (a) $0.01c$; (b) $0.3c$; (c) $0.75c$; (d) $0.99c$.

46. ■■ Our galaxy's diameter is about 10^5 light years. How much time does it take an electron moving at $0.9999c$ to cross the galaxy, measured in the reference frame of (a) the galaxy and (b) the electron?

47. ■■■ Suppose the light clock in Figure 20.11 is 1.5 m high and is moving at $0.5c$. (a) Find the elapsed time for the light's round trip as measured by (a) Sharon on the train and (b) Sam on the ground. (c) What distance does the light travel, as measured by Sam? (d) Use your answers to parts (b) and (c) to compute the speed of the light, as measured by Sam.

48. ■■ You want to gain a year on your earthbound peers, traveling fast enough on a round trip that when 10 years elapse for them, only 9 years elapse for you. What speed is required?

49. ■■ What speed will shorten the length of a round-trip journey by 1%, compared with its duration measured in an inertial reference frame?

50. ■■ In the future, speeding tickets for spaceships will be issued based on the spaceship's measured length. The Neptune 3000 spaceship has proper length 75 m. If the posted speed limit is $0.6c$, will the Neptune 3000 spaceship be ticketed if space police measure its length as 65 m?

51. ■■ A future spaceship makes the round-trip journey to the Moon in just 1 min, as measured by clocks on Earth. How much time elapses on a clock aboard the spaceship? (Use the average Earth-Moon distance listed in Appendix E.)

52. ■ Find the speed of a spacecraft if its clock, relative to clocks on Earth, runs at a rate (a) 90% as fast; (b) 60% as fast; (c) 10% as fast.

53. ■ A linear accelerator can accelerate subatomic particles to close to the speed of light. A proton travels through a 1-km section of the accelerator with a constant speed. Find the proton's speed if this section is only 1.0 m long in the proton's reference frame.

54. ■■ Sirius is a large blue star just 8.7 light years from Earth. Find the speed required for a one-way trip to Sirius that takes, as measured on board the spacecraft, (a) 40 years; (b) 20 years; (c) 5 years.

55. ■■ A subatomic particle has a 480-ns lifetime in its own rest frame. If it moves through the lab at $0.980c$, how far does it travel before decaying, as measured in the lab?

56. ■■ Spaceship A has proper length 100 m, and ship B has a proper length 50 m. They pass an observer on Earth, both moving in the same direction but with different speeds. The observer measures both ships' lengths to be 42 m. How fast is each moving?

57. ■■■ A relativistic train with proper length 80 m passes through a tunnel with a proper length 75 m. (a) If the train's speed is $0.45c$, is the train ever completely inside the tunnel? (b) At this speed, how much time elapses for a cross-country trip of 5000 km, as measured on the train?

58. ■■■ A spaceship travels at $0.40c$ relative to an observer. How much time does the spaceship's clock lose each day, compared with the observer's? Does the observer's clock gain this much time compared with the spaceship's? Explain.

Section 20.5 Relative Velocity and the Doppler Effect

59. ■ A spaceship approaching Earth at $0.73c$ fires a particle beam toward Earth, with speed $0.88c$ relative to the ship. At what speed does Earth receive the particles?

60. ■ Repeat the preceding problem if the ship is moving away from Earth at $0.73c$ but still fires the particle beam toward Earth.

61. ■■ Two spaceships approach Earth from opposite directions, each traveling at $0.7c$ relative to Earth. How fast is each moving, as measured by the other?

62. ■ An astronaut returning to Earth after a long trip is cruising earthward at 6.95×10^7 m/s. She launches a package toward home, at a speed of 1.50×10^7 m/s relative to her spacecraft. At what speed does the package reach Earth?

63. ■ A source emitting 540-nm light moves toward a receiver. Find the received wavelength for source speeds (a) $0.01c$; (b) $0.10c$; (c) $0.50c$; (d) $0.90c$; (e) $0.99c$.

64. ■ Repeat the preceding problem if the source moves away from the receiver.

65. ■■ A proton is moving at $0.25c$ to the right, as viewed in the laboratory. Find the velocity (relative to the lab) of another reference frame in which the proton is moving at $0.25c$ to the left.

66. ■■■ Three galaxies lie along a line. The outer two galaxies are receding at $0.25c$ relative to the middle one. (a) Determine the speed of each outer galaxy in the reference frame of the other outer galaxy. (b) If each galaxy emits visible light with wavelength 550 nm, at what wavelength does one of the outer galaxies receive light from the other two?

67. ■■ A light source moves toward a receiver with speed v. If the source emits light with a frequency f_0, what is the frequency of the received signals? *Hint:* Remember that $\lambda f = c$ for electromagnetic waves. Evaluate numerically for a source of 200-nm ultraviolet light moving toward an observer at $0.4c$.

68. ■■ Repeat the preceding problem if the source is moving away from the observer.

69. ■■ A space probe is leaving the solar system at $0.15c$, moving directly away from Earth. It emits radio pulses at the rate of 40 per minute. At what rate are the pulses received at Earth?

70. ■■ What relative speed between the sender and receiver is required to shift light from one end of the visible spectrum to the other (400 nm to 700 nm)?

71. ■■ Particle accelerators often employ colliding-beam technology, in which fast particles approach a target chamber from opposite directions. If each beam approaches the chamber at $0.990c$, what's each beam's speed relative to the other?

Section 20.6 Relativistic Momentum and Energy

72. ■■ Find the momentum, kinetic energy, and total energy of a proton with speed (a) $0.01c$; (b) $0.10c$; (c) $0.50c$; (d) $0.99c$.

73. ■■ Find the momentum, kinetic energy, and total energy of an electron with speed (a) $0.01c$; (b) $0.10c$; (c) $0.50c$; (d) $0.99c$.

74. ■■ A proton is accelerated from rest through a potential difference of 100 MV. Find the proton's kinetic energy and rest energy. What's the ratio of kinetic to rest energy?

75. ■■ A positron is the antiparticle of the electron, with the same mass but charge $+e$. Find the minimum energy required to produce an electron-positron pair.

76. BIO ■■ **PET Imaging.** Positron-emission tomography (PET) images body tissues by detecting gamma rays produced when electrons and positrons annihilate. (The positrons come from short-lived radioisotopes introduced into the patient.) What's the energy of such gamma rays? See the preceding problem.

77. ■■ A particle's kinetic energy is equal to its rest energy. What's its speed?

78. ■■ (a) Find the speed of an electron with a kinetic energy 4.80×10^{-13} J. (b) What's the speed of a proton with the same kinetic energy?

79. ■■■ (a) Find the speed required for the classical and relativistic values of (a) momentum and (b) kinetic energy to differ by 1%.

80. ■■■ An alpha particle (helium nucleus) has total energy 4.1 GeV. (a) What's its speed? (b) What's the speed of an electron with the same total energy?

81. ■■ A high-energy electron has $\gamma = 100$. Find (a) its momentum, (b) its total energy, and (c) the ratio of its total energy to its rest energy.

82. ■■ An atom moving at $0.890c$ has a momentum 1.15×10^{-16} kg·m/s. Find its mass.

83. ■■ A proton's total energy is ten times its rest energy. How fast is it moving?

General Problems

84. ■■ Find (a) the relativistic factor γ and (b) the speed of protons accelerated to energy 7 TeV in the Large Hadron Collider.

85. ■■ What power laser beam would be needed to suspend a 2.7-g ping-pong ball against Earth's gravity? The ball is painted black so it absorbs the entire beam.

86. ■■■ How much ice must melt at 0°C in order to gain 1 g of mass? What's the source of this mass? (The latent heat of ice melting at this temperature is 333 kJ/kg.)

87. ■■ A stretched spring has greater energy and therefore greater mass than an unstretched spring. What's the mass increase when you stretch a spring with $k = 480$ N/m by 20 cm?

88. ■■■ A spaceship heads away from Earth at $0.20c$. The ship beams 400-nm light toward Earth. The light strikes a mirror on Earth and is reflected back to the ship. What is the wavelength of light received by the ship? *Hint:* Find the wavelength of light received at Earth, and then consider that the mirror effectively retransmits light of that wavelength.

89. ■■ Imagine a "moving walkway" like you find in airports, except that this one goes super fast: $0.75c$, relative to the airport. Suppose you put an identical walkway on top of this one. How fast are passengers on the second walkway moving relative to the airport?

90. ■■■ You embark on a round-trip journey to a star 10 light years away, as measured in Earth's reference frame. You travel at a constant speed outbound, turn around rapidly, and travel back at constant speed. If the round trip takes 15 years by your clock, (a) how fast do you travel and (b) what's the round-trip time as measured on Earth?

91. ■■ Hydrogen atoms throughout the universe emit ultraviolet radiation with wavelength 121.6 nm. How fast must a galaxy be receding from us in order that we receive this radiation as visible light with wavelength 650 nm?

92. ■■ The most distant galaxies observed have redshifts of about 7, meaning the wavelength of light we receive is seven times longer than what was emitted at the galaxy. How fast is a redshift-7 galaxy moving, and is it receding or approaching?

93. ■■ Using the momentum and energy found in Example 20.11, show that Equation 20.14 gives the rest energy found in that example.

94. ■■■ The expression $(1 + x)^n$ can be approximated as $1 + nx$, provided $|x| \ll 1$. Here the exponent n does not have to be an integer. Use this approximation to show that Equation 20.11's expression for relativistic kinetic energy reduces to $K = \frac{1}{2}mv^2$ for $v \ll c$.

Answers to Chapter Questions

Answer to Chapter-Opening Question

Nearly all our information about the universe beyond Earth comes from electromagnetic waves, with astronomers now able to access a spectrum that ranges from radio waves (received by the radio telescopes shown in the chapter-opening photo) to gamma rays.

Answers to GOT IT? Questions

Section 20.1   (b) $1.05\,\mu$m < (d) 780 nm < (a) 550 nm < (c) 434 nm

Section 20.2   (c) 10^{14} Hz

Section 20.4   (a) Shannon measures a shorter time than Chris.

Section 20.5   (c) More than $0.6c$ but less than c

Section 20.6   (a) larger for the proton

21 Geometrical Optics

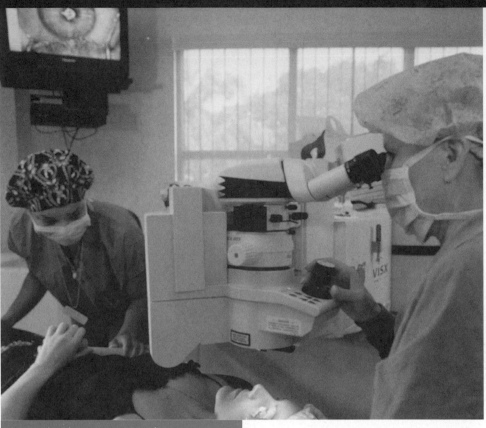

■ How does laser surgery provide permanent vision correction?

This chapter explores **geometrical optics**, which is based on the idea that light travels in straight lines called rays. We'll first consider reflection, and show how mirrors form images. Next we'll discuss refraction, the bending of light as it passes from one medium to another. Refraction governs image formation in lenses, and we'll develop methods for locating and describing those images. Finally, we'll consider optical instruments: microscopes, telescopes, and the human eye.

21.1 Reflection and Plane Mirrors

Most of the objects around us aren't sources of light, yet we can see them. Why? It's because they **reflect** light from sources like the Sun or electric lamps. Reflection from smooth, polished surfaces also lets us form images, as you'll see when we examine plane and curved mirrors.

Light Rays

Although light is a wave, we can often neglect its wave properties and consider that light generally travels in straight lines called **rays**. This is the approximation of **geometrical optics**, and it's valid when we consider only light's interactions with objects much larger than its wavelength. Given light's wavelength—less than a millionth of a meter—that's true in most everyday situations. Light's straight-line path bends when it goes from one medium to another, and it may follow curved paths in materials whose properties change with position. But as long as the wavelength is negligible, these situations are described by geometrical optics.

Reflection

Light rays can strike a surface at any angle. The **angle of incidence** is the angle between an incident ray and the normal (perpendicular) to the surface. Both experiment and electromagnetic theory show that the **angle of reflection** θ_r between the reflected ray and the normal is the same as the **angle of incidence** θ_i (Figure 21.1a):

$$\theta_r = \theta_i \qquad \text{(Law of reflection)} \qquad (21.1)$$

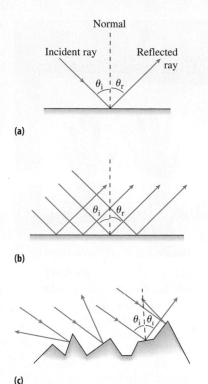

FIGURE 21.1 (a) Angles of reflection and incidence are equal. (b) In specular reflection, a smooth surface reflects a light beam undistorted. (c) A rough surface results in diffuse reflection.

✓**TIP**

Measurements of incident and reflected rays are always made with respect to the normal, not the surface.

When a beam of parallel light rays reflects from a smooth surface, all the rays go off in the same direction (Figure 21.1b). This is **specular reflection**, and it occurs with mirrors and other shiny surfaces. But most surfaces are rough, and as a result light reflects in essentially all directions. This is **diffuse reflection**, shown in Figure 21.1c. Because diffuse reflection sends light in all directions, you can see illuminated objects from any direction. Most objects reflect some wavelengths of light more than others, which is why you perceive colors. A red shirt, for example, reflects red light and absorbs most other colors.

An example of selective absorption of colors occurs in *chlorophyll*, the molecule in green plants that uses solar energy to produce carbohydrates from carbon dioxide and water. Chlorophyll mainly uses light in the blue to near ultraviolet (350 nm to 450 nm) and also red light (650 nm to 700 nm). The light in the middle of the visible spectrum is reflected, so plants appear green.

Images in Plane Mirrors

Throughout this chapter we'll use **ray tracing** to show how images are formed. Drawing any two rays from each point on an object suffices to locate the image of that point (Figure 21.2a). In practice, some special rays are easier to use. In Figure 21.2b we show two rays each from the top and bottom of a simple object, a lamppost. For each we use one ray that strikes perpendicular to the mirror, reflecting right back the way it came ($\theta_r = \theta_i = 0$). Another ray strikes at an angle, and it too obeys the law of reflection. Extend those rays on the other side of the mirror, and they meet in a point. When you look in the mirror, that's where the light *seems* to come from, so that's the image location. In Figure 21.2b, we've located images of the lamppost's top and bottom; the rest of the lamppost's image lies in between. There isn't really any light coming from behind the mirror, so this is a **virtual image**.

Figure 21.2b illustrates that the **image distance** d_i, the distance the image lies behind the mirror, is equal to the **object distance**, or the distance the object lies in front of the mirror: $d_i = d_o$. Also, the image and object heights are the same: $h_i = h_o$. We define **magnification**, M, as the ratio of image size to object size:

$$M = \frac{h_i}{h_o}$$

For reflection from a plane mirror, $M = 1$. With curved mirrors and lenses, we'll find less trivial relations between object and image distances and sizes, and magnifications that can be greater than or less than 1.

When you look at yourself in the mirror, you see an image your same size, standing equidistant from you on the other side of the mirror. That virtual person is an exact copy of

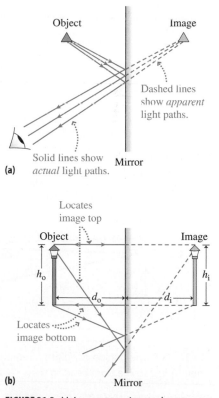

FIGURE 21.2 Using ray tracing to locate an image in a plane mirror.

FIGURE 21.3 The palm of a right hand faces the mirror. So does the image's palm. That makes the image look like a left hand, but it's still the image of a right hand.

you, but appears reversed. You may think of this as a left-right reversal, but because both you and your image are facing the mirror, what the mirror actually does is to reverse things front to back, along the axis perpendicular to the mirror. Figure 21.3 shows the result.

CONCEPTUAL EXAMPLE 21.1 **Point of View**

Referring again to Figure 21.2, does the size and position of an object's image in a plane mirror depend on the position from which you view it? That is, is the image location the same for different observers looking into the mirror?

SOLVE Figure 21.2a shows the triangle's image as viewed by an eye, as shown. Move the position of the eye and redraw the ray diagram, as shown in Figure 21.4. The reflected rays still converge in the same place, showing that the image hasn't moved. That's why it's not necessary to include an observer's eye in a ray diagram, such as the one in Figure 21.2b.

REFLECT Try for yourself moving the eye in Figure 21.2a to different locations and drawing ray diagrams for each case. You'll see that the image is in the same place for different viewers.

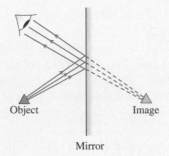

Object   Image

Mirror

FIGURE 21.4 Viewing from a different location does not change the image.

EXAMPLE 21.2 **A Big Enough Mirror?**

You stand in front of a mirror that's attached to a vertical wall. What's the minimum height of the mirror needed for you to see your body's entire height h?

ORGANIZE AND PLAN To see your whole body, rays from the top of your head and the bottom of your feet must reflect to your eye.

Known: Height $= h$.

SOLVE To solve the problem, break your height h into two segments h_1 and h_2 as shown in Figure 21.5, with h_1 from floor to eye level and h_2

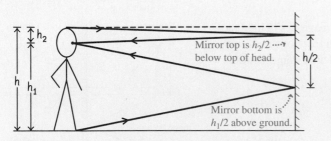

Mirror top is $h_2/2$ ····•
below top of head.

Mirror bottom is·
$h_1/2$ above ground.

FIGURE 21.5 The person of height h requires a mirror of height $h/2$ to see from head to toe.

from eye level to the top of your head, so $h = h_1 + h_2$. By the law of reflection, the bottom of the mirror need only be $h_1/2$ above the ground, as shown. Similarly, the top of the mirror can be $h_2/2$ below the top of your head. The total mirror height is then $h_1/2 + h_2/2 = (h_1 + h_2)/2$. But $h = h_1 + h_2$, so the total mirror height is $h/2$. The mirror needs to be only half your height.

REFLECT Although the minimum height is $h/2$, a mirror that size has to be placed carefully to match the diagram shown. Most "full-length" mirrors are much longer than half the average person's height, so placement isn't crucial and the tallest people can see their full images.

MAKING THE CONNECTION In this example, does it matter how close you stand to the mirror? Will you see more or less of yourself if you step back?

ANSWER The analysis and diagram in this example didn't depend on your distance from the mirror. Stepping back changes all the angles, but the minimum mirror height remains $h/2$.

21.2 Spherical Mirrors

You've probably seen mirrors in which your image is enlarged, and you may have viewed your reduced image reflected in a shiny ball. Changing the image size requires a curved mirror, the most common being a **spherical mirror**, generally made from a small portion of a sphere. Figure 21.6 shows that spherical mirrors can be either **concave** or **convex**. We'll determine images for both types.

Concave Mirrors

Concave mirrors are called converging mirrors because, as Figure 21.7a shows, parallel rays striking the mirror converge toward a **focal point**. Ideally, all parallel rays should meet right at the focus, but for spherical mirrors there's some deviation that worsens for rays arriving farther from the mirror's principal axis, as shown in Figure 21.7b. This poor focusing is called **spherical aberration**. It can be reduced by making the mirror only a small section of the sphere. Parabolic-shaped mirrors eliminate spherical aberration but are harder to make. They're used in critical applications such as telescopes.

Figure 21.7a also shows that the focal point lies halfway between the center of curvature and the mirror surface. The **focal length** f is defined as the distance from the mirror to the focal point, so this means that a mirror with curvature radius R has $f = R/2$. You can see how this works by ray tracing—essentially reproducing the drawings in Figure 21.7. Use a compass to draw an arc representing the mirror's surface. Then draw parallel rays that strike the mirror's surface, as shown in Figure 21.7. At each point where an incoming ray strikes the mirror, draw the normal, directed toward the center of curvature. Then draw the reflected ray such that $\theta_r = \theta_i$. You should see the pattern emerge as in Figures 21.7a and b: Rays close to the principal axis focus well, but rays farther out do not.

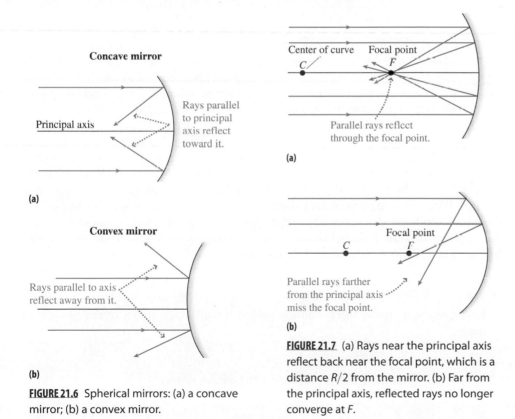

FIGURE 21.6 Spherical mirrors: (a) a concave mirror; (b) a convex mirror.

FIGURE 21.7 (a) Rays near the principal axis reflect back near the focal point, which is a distance $R/2$ from the mirror. (b) Far from the principal axis, reflected rays no longer converge at F.

Image Formation by Concave Mirrors

Parallel rays striking a concave mirror reflect through the focal point. You can use this fact, along with the law of reflection, to trace reflected rays that meet at the image point. As shown in Figure 21.8, you can select from four special rays:

1. Draw a ray parallel to the principal axis. The reflected ray goes through the focal point.
2. Draw a ray directly through the focal point. By symmetry, this ray reflects parallel to the central axis.
3. Draw a ray that strikes the mirror on the principal axis. By the law of reflection, the reflected ray makes the same angle on the opposite side of the principal axis ($\theta_r = \theta_i$).
4. Draw a ray through the center of curvature. This ray strikes the mirror normal to its surface and thus reflects straight back along the same path as the incoming ray.

To locate an image, it's sufficient to consider any two of these four rays, and typically we use the first two. Figure 21.9 uses these rays to locate images. Both reflected rays converge at the image point, either in reality for real images or when extended through the mirror for virtual images. This method gives the location, size, and orientation (upright or inverted) of the image.

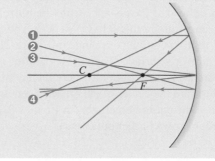

FIGURE 21.8 Four special rays that you can use to construct images formed in a concave mirror.

Knowing that parallel rays reflect through the focal point, you can use ray tracing to see how concave mirrors form images. Tactics 21.1 outlines the procedure for drawing representative rays from an object. In Figure 21.9 we illustrate how to use this procedure to form the image of a representative object, an arrow, with the arrow's bottom placed on the principal axis. Using the arrow allows you to see the image's location, size, and orientation (upright or inverted, relative to the object). The type of image depends on the location of the object relative to the focal point. There are three cases: $d_o > 2f$, $2f > d_o > f$, and $d_o < f$. Table 21.1 summarizes the results for these cases. In the first two cases (Figures 21.9a and 21.9b), light rays pass through the image location, so the eye perceives light that actually comes from the image. That makes these images **real images**. With concave mirrors, real images occur whenever the object lies beyond the focal point. In Figure 21.9c the object lies closer to the mirror than the focal point, and the light rays diverge. It looks like they're coming from a point behind the mirror, so we have a virtual image—just as with a plane mirror.

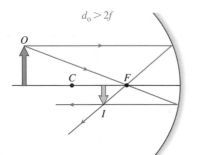

(a) Real, inverted, reduced image

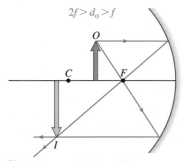

(b) Real, inverted, enlarged image

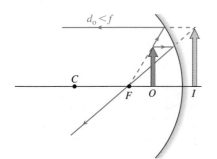

(c) Virtual, upright, enlarged image

FIGURE 21.9 Ray tracing for images formed by a concave mirror. In each case *O* designates the object and *I* the image.

✓**TIP**

Real images are really there. You can project a real image onto a screen or sheet of paper.

TABLE 21.1 Image Formation with Concave Mirrors: Three Cases

Object location	Image orientation	Image size	Real or virtual?
$d_o > 2f$	Inverted	Reduced	Real
$2f > d_o > f$	Inverted	Enlarged	Real
$d_o < f$	Upright	Enlarged	Virtual

What happens at the transitions between the three cases, $d_o = 2f$ and $d_o = f$? When $d_o = 2f$, the image is inverted and real, and this case marks the transition between a reduced and enlarged image. Thus the image is inverted, real, and the same size as the object. When $d_o = f$, a ray diagram helps show what happens (Figure 21.10a). In this case, all the reflected rays are parallel. They don't converge to form any image, real or virtual. Flashlights, car headlights, and searchlights exploit this situation by putting a light source at the focal point of a curved mirror to produce a parallel beam (Figure 21.10b).

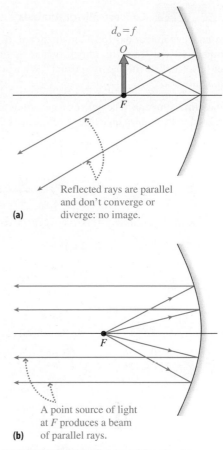

(a)

(b)

FIGURE 21.10 (a) With an object at the focus, there's no image. Instead, there emerge parallel rays that can't be focused. (b) A point source of light placed at the focal point produces a parallel beam of rays.

CONCEPTUAL EXAMPLE 21.3   Vanity Mirror

You want a handheld mirror that will show your face enlarged. How should the mirror be designed?

SOLVE A concave mirror produces an enlarged real image when $2f > d_o > f$. In this case, however, the image is inverted and behind the object. That's not what you want.

An enlarged virtual image results when $d_o < f$. That's what you want, because the virtual image is on the other side of the mirror, just as you're used to seeing, and it's also upright. For this to work, your face must be between the focal point and the mirror. You probably don't want to hold the mirror too close; a comfortable distance might be 30 cm. That means you want a focal length greater than 30 cm, giving a curvature radius $R = 2f$ greater than 60 cm.

REFLECT No matter how large the mirror is, you'll want to stand close to the principal axis to minimize spherical aberration.

Concave Mirrors: Quantitative

Using ray diagrams for concave mirrors, we'll develop some relationships among the focal length f, the object and image distances d_o and d_i, and the heights h_o and h_i. Figure 21.11a identifies a pair of similar triangles, showing that

$$\frac{-h_i}{h_o} = \frac{d_i}{d_o}$$

Here we've taken h_i as negative, denoting an inverted image; $-h_i$ is then the (positive) length of the triangle's vertical side. Using two other similar triangles (Figure 21.11b) gives

$$\frac{-h_i}{h_o} = \frac{f}{d_o - f}$$

Equating the two expressions for $-h_i/h_o$ and rearranging gives the **mirror equation**:

$$\frac{1}{f} = \frac{1}{d_o} + \frac{1}{d_i} \qquad \text{(Mirror equation; SI unit: m}^{-1}) \qquad (21.2)$$

If you know a mirror's focal length, Equation 21.2 lets you find the image distance for any object distance you choose.

The mirror's magnification M is the ratio of image height h_i to object height h_o, as for the plane mirror in Section 21.1. Using our first pair of similar triangles,

$$M = \frac{h_i}{h_o} = -\frac{d_i}{d_o} \qquad \text{(Magnification; SI unit: dimensionless)} \qquad (21.3)$$

Together, Equations 21.2 and 21.3 completely determine reflection from concave mirrors. They give the same results as carefully drawn ray diagrams, if you follow these sign conventions:

- Focal length f is always positive for concave mirrors, negative for convex mirrors.
- Object distance d_o is always positive.
- Image distance d_i is positive for a real image on the same side of the mirror as the object, and negative when the image is a virtual image on the opposite side of the mirror from the object.
- Object size h_o is always taken to be positive.
- Image size h_i is positive when the image is upright and negative when it's inverted.
- Magnification M is positive when the image is upright and negative when it's inverted.

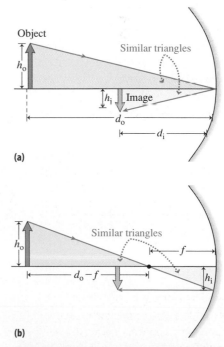

(a)

(b)

FIGURE 21.11 Using similar triangles to relate object and image distances and heights.

EXAMPLE 21.4 **Concave Mirror Analysis**

A concave mirror has focal length 20 cm. If you place a 3.0-cm-tall object 60 cm from the mirror, where is the image? Is it upright or inverted? Is it real or virtual? What are the image size and magnification?

ORGANIZE AND PLAN The unknown quantities are d_i, h_i, and M. They're related to the known quantities by Equations 21.2 and 21.3:

$$\frac{1}{f} = \frac{1}{d_o} + \frac{1}{d_i}$$

$$M = \frac{h_i}{h_o} = -\frac{d_i}{d_o}$$

Known: $f = 20\,\text{cm}$; $h_o = 3.0\,\text{cm}$; $d_o = 60\,\text{cm}$.

SOLVE Solving Equation 21.2 for d_i,

$$\frac{1}{d_i} = \frac{1}{f} - \frac{1}{d_o} = \frac{d_o - f}{f d_o}$$

Thus

$$d_i = \frac{f d_o}{d_o - f} = \frac{(20\,\text{cm})(60\,\text{cm})}{60\,\text{cm} - 20\,\text{cm}} = 30\,\text{cm}$$

Then by Equation 21.3,

$$M = -\frac{d_i}{d_o} = -\frac{30\,\text{cm}}{60\,\text{cm}} = -0.5$$

That is, the image is inverted and half as large as the object. This result is confirmed using Equation 21.3 to find h_i:

$$h_i = M h_o = (-0.5)(3\,\text{cm}) = -1.5\,\text{cm}$$

REFLECT The numbers in this example correspond to the scale in the ray diagram in Figure 21.9a. You should always be able to reconcile numerical results from Equations 21.2 and 21.3 with carefully drawn ray diagrams.

MAKING THE CONNECTION Using the same mirror and object, describe the image when the object is 30 cm from the mirror.

ANSWER This change has the effect of swapping d_o and d_i in Equation 21.2, so $d_i = 60\,\text{cm}$. This gives magnification $M = -2$ and image height $h_i = -6.0\,\text{cm}$. In effect, you've just swapped the object and image in the ray diagram.

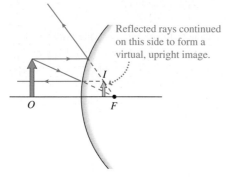

Reflected rays continued on this side to form a virtual, upright image.

FIGURE 21.12 Image formation in a convex mirror. The image is always virtual, upright, and reduced in size.

Convex Mirrors

Convex mirrors are diverging mirrors, because parallel rays that reflect from the mirror subsequently diverge. That makes real images impossible. Convex mirrors always form virtual images on the other side of the mirror, as shown in Figure 21.12. Despite the different geometry, the rays shown here follow the rules outlined in Tactics 21.1. The difference is that the convex mirror's focal point is on the opposite side of the mirror. Therefore, rays don't actually go through the focal point, but reflect as though they were coming from that point.

Equations 21.2 and 21.3 also hold for convex mirrors, with the added sign convention that the focal length f is *negative*, denoting a focal point on the opposite side of the mirror. Apply those equations and you'll find that d_i is always negative and its absolute value is less than d_o. That, in turn, gives a positive magnification. So we get a reduced, upright virtual image on the opposite side of the mirror. A convex mirror never forms a real image.

A convex mirror's curved shape makes for a larger field of view, which makes concave mirrors useful as the side-view mirrors on cars. The reduced image gives the illusion that cars behind you are farther away than they are. That's the reason for the warning label: "Objects in mirror are closer than they appear."

✓ TIP

Check out your reflection in both sides of a spoon. One side is a concave mirror, the other convex.

Reviewing New Concepts

- The law of reflection governs the direction of reflected rays.
- Plane mirrors produce virtual images, with magnification 1.
- Concave mirrors produce a variety of images, real and virtual, with different magnifications depending on the object's location relative to the focal point.
- Convex mirrors produce virtual images, reduced in size.

GOT IT? Section 21.2   If an object is placed 20 cm from a concave spherical mirror with 12-cm focal length, the resulting image is (a) reduced, inverted, real; (b) reduced, inverted, virtual; (c) enlarged, inverted, real; (d) enlarged, inverted, virtual; (e) enlarged, upright, real; (f) reduced, upright, real.

21.3 Refraction and Dispersion

So far we've considered light changing direction by reflection. Here you'll study **refraction**, the bending of light rays passing from one medium to another. Figure 21.13 shows refraction at an air-water interface. Note that the incoming light ray is partially reflected and partially transmitted into the water. Both reflection and refraction generally occur at interfaces between transparent media.

Why Refraction?

Refraction occurs because light travels slower in transparent materials than in vacuum. In water, for example, the speed of light is about three-fourths of c; in most glass, it's about two-thirds of c. In air it's almost, but not quite, c. We define the **index of refraction**, n, as the ratio of the speed of light c in vacuum to its speed in a given medium:

$$n = \frac{c}{v} \quad \text{(Index of refraction; SI unit: dimensionless)} \quad (21.4)$$

Table 21.2 lists the refractive index for various substances; it's greater than 1 for all but vacuum because light travels more slowly in materials. In a medium with refractive index n, Equation 21.4 gives $v = c/n$. But as you saw in Chapter 11, $v = \lambda f$ for any periodic wave. The frequency f counts the number of wave crests passing a given point per unit time; since waves don't appear or disappear, f is independent of the medium. Then light's wavelength in a medium of refractive index n is also smaller by $1/n$: $\lambda = \lambda_0/n$, where λ_0 is the wavelength in vacuum.

Figure 21.14 shows how this speed and wavelength change results in light's changing direction. Here we show wave crests on either side of the interface between two media.

TABLE 21.2 Selected Values of Index of Refraction for Yellow Sodium Light ($\lambda = 589$ nm)

Substance	Index of refraction n	Substance	Index of refraction n	Substance	Index of refraction n
Solids		*Glasses* (typical values)		*Liquids* (20°C)	
Ice	1.31	Crown	1.52	Water	1.33
Quartz	1.54	Flint (light)	1.58	Ethanol	1.36
Rock salt	1.54	Flint (medium)	1.62	Glycerin	1.47
Zircon	1.92	Flint (dense)	1.66	Oil (light)	1.52
Diamond	2.42	Lanthanum flint	1.80		

Convex mirrors help drivers emerge from narrow alleys or driveways onto busy streets. The wide field of view gets around obstacles such as utility poles and shrubbery. These mirrors are especially popular in Japan, where urban streets are often small and crowded.

FIGURE 21.13 Refraction of a laser beam passing from air into water.

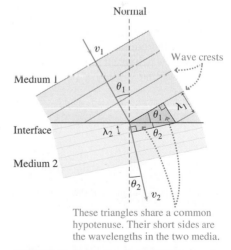

These triangles share a common hypotenuse. Their short sides are the wavelengths in the two media.

FIGURE 21.14 Refraction occurs because of differing wave speeds and hence wavelengths in different media.

Wave crests have to be continuous across the interface, while their speed and wavelength change. Under those conditions, Figure 21.14 shows that the direction of wave propagation will change. In our figure, medium 2 has the slower speed, and that causes the propagation direction to bend *toward* the normal to the interface.

Figure 21.14 shows a pair of triangles with a common hypotenuse. In each medium, that hypotenuse has length $\lambda/\sin\theta$, where λ and θ are the wavelength and angle between the light's propagation direction and the normal in that medium. Equating expressions for the common hypotenuse in the two media gives $\lambda_1/\sin\theta_1 = \lambda_2/\sin\theta_2$. But we've seen that $\lambda_1 = \lambda_0/n_1$ and $\lambda_2 = \lambda_0/n_2$. Using these results in our equation for the hypotenuse and canceling the common vacuum wavelength λ_0 gives

$$n_1 \sin\theta_1 = n_2 \sin\theta_2 \quad \text{(Snell's law; SI unit: dimensionless)} \quad (21.5)$$

Equation 21.5 is **Snell's law**, after the Dutch mathematician Willebrord Snell (1580–1626), who deduced it experimentally by studying refraction in water and glass. Notice that as with reflection, incident and refracted angles are measured with respect to the normal. Figure 21.15 shows these angles, along with the angle of the partial reflection.

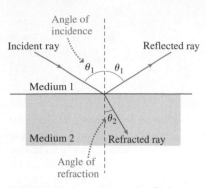

FIGURE 21.15 Quantities in Snell's law.

Through the Window

A beam of light in air enters a glass window pane ($n = 1.52$) at $40°$ to the normal. Describe the path of the light as it passes through the glass and emerges on the other side.

ORGANIZE AND PLAN Snell's law governs the refraction at each surface: $n_1 \sin\theta_1 = n_2 \sin\theta_2$. Refraction occurs when the light enters the glass and again when it leaves, because each time the refractive index n changes. Snell's law governs both refraction processes.

Known: n_1 (air) $= 1.00$; n_2 (glass) $= 1.52$; θ_1 (air) $= 40°$.

SOLVE Solving Snell's law for the refracted angle in glass,

$$\sin\theta_2 = \frac{n_1 \sin\theta_1}{n_2} = \frac{(1.00)(\sin 40°)}{1.52} = 0.423$$

so

$$\theta_2 = \sin^{-1}(0.423) = 25.0°$$

The window's surfaces are parallel, so the light makes this same $25°$ angle with the normal at the far surface (Figure 21.16). It then re-enters air, which is again medium 1: n_1 (air) $= 1.00$. Applying Snell's law a second time, with $\theta_2 = 25.0°$,

$$\sin\theta_1 = \frac{n_2 \sin\theta_2}{n_1} = \frac{(1.52)(\sin 25.0°)}{1.00} = 0.642$$

so

$$\theta_1 = \sin^{-1}(0.642) = 40°$$

The light leaves the glass at the same angle it entered.

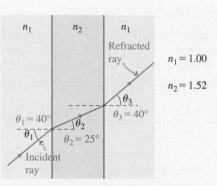

FIGURE 21.16 Snell's law applied to light going through a pane of glass.

REFLECT Note that the ray leaving the glass is parallel to the incident ray but displaced slightly. The displacement depends on the thickness of the glass. Light coming in from the right would have followed the same path, but in reverse.

MAKING THE CONNECTION What's the largest angle the light ray can make with the normal in the glass?

ANSWER This occurs when the incident ray approaches an angle of $90°$, so it's nearly skimming the surface. Applying Snell's law with $\theta_1 \approx 90°$ gives $\theta_2 \approx 41.1°$.

Light bends toward the normal when entering a material with a higher index of refraction, and away from the normal when entering a medium with a lower index.

Effects of Refraction

Refraction causes apparent changes in positions of objects viewed through different media. Consider viewing a fish underwater, as shown in Figure 21.17. Because of refraction, light from the fish doesn't go straight to your eye, so the position you perceive isn't the fish's actual position. A more subtle example is refraction in Earth's atmosphere, where the refractive index increases from 1 in the vacuum of space down to about 1.00029 in the denser air at the surface. This varying refractive index causes a gradual bend in the light's path. As a result, you can see the Sun just before sunrise and after sunset, when its actual position is just below the horizon. Mirages occur for a similar reason, when air's refractive index varies with temperature just above hot surfaces like roads or deserts.

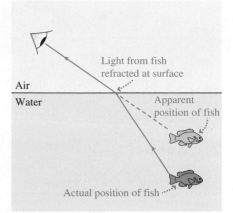

FIGURE 21.17 Observing a fish underwater.

EXAMPLE 21.6   Lost Keys!

You've dropped your keys into a 2.0-m-deep swimming pool. How far down do they appear to be?

ORGANIZE AND PLAN Figure 21.18 shows a light ray coming up from the keys at the pool bottom. Upon entering air, light refracts away from the normal, as given by Snell's law: $n_1 \sin \theta_1 = n_2 \sin \theta_2$. The keys appear at the apparent depth shown, corresponding to a straight light path to their actual horizontal position. Analyzing the two triangles shown gives the apparent depth d_{apparent} in terms of the actual depth d.

Known: n_1 (air) $= 1.00$; n_2 (water, from Table 21.1) $= 1.33$; $d = 2.0$ m.

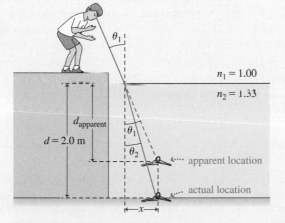

FIGURE 21.18 The keys appear closer to the surface than they actually are.

SOLVE For the tall right triangle with sides x and d, $x = d \tan \theta_2$. For the shorter triangle with sides x and d_{apparent}, $x = d_{\text{apparent}} \tan \theta_1$. Equating these expressions for x yields

$$d_{\text{apparent}} = d \frac{\tan \theta_2}{\tan \theta_1}$$

If you're looking almost straight down, the angles are small, and for small angles $\tan \theta \approx \sin \theta$. Then to a good approximation,

$$d_{\text{apparent}} = d \frac{\sin \theta_2}{\sin \theta_1}$$

But from Snell's law,

$$\frac{\sin \theta_2}{\sin \theta_1} = \frac{n_1}{n_2}$$

so

$$d_{\text{apparent}} \approx d \frac{n_1}{n_2} = (2.0 \text{ m}) \left(\frac{1.00}{1.33} \right) = 1.5 \text{ m}$$

REFLECT The apparent depth is three-fourths of the actual depth—an effect that would be more dramatic in glass, with a higher index of refraction. Try looking straight down into a block of glass, and the table below appears closer than it really is.

MAKING THE CONNECTION   Why is it important that you're looking almost directly downward in this example?

ANSWER You'd still see a change in depth if you were off to one side, but the computation in this example depends on the small-angle approximation $\tan \theta \approx \sin \theta$.

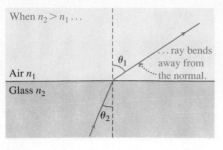

When $n_2 > n_1$ ...

...ray bends away from the normal.

θ_1

Air n_1

Glass n_2

θ_2

(a)

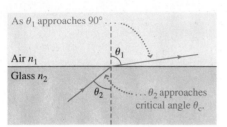

As θ_1 approaches 90° ...

θ_1

Air n_1

Glass n_2

θ_2

...θ_2 approaches critical angle θ_c.

(b)

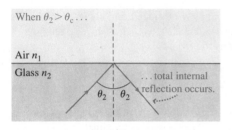

When $\theta_2 > \theta_c$ ...

Air n_1

Glass n_2

...total internal reflection occurs.

θ_2 θ_2

(c)

FIGURE 21.19 Total internal reflection.

Light propogating in a fiber-optic cable undergoes total internal reflection at the outer surface of the cable.

FIGURE 21.20 Total internal reflection in an optical fiber.

Total Internal Reflection

You've seen that light passing from glass to air bends away from the normal, as in Figure 21.19a. Increase the incidence angle θ_2, and the refracted angle θ_1 approaches 90° (Figure 21.19b). The **critical angle** θ_c is the incidence angle for which $\theta_1 = 90°$. For incidence angles $\theta_2 > \theta_c$, there's no transmission from glass to air. Instead, **total internal reflection** occurs, as shown in Figure 21.19c. Here we've considered glass and air, but total internal reflection can occur whenever light propagating through a medium with higher refractive index reaches the interface with a lower-index medium.

Snell's law gives the critical angle for light going from a medium with refractive index n_2 to a medium with index n_1. Figure 21.19b shows that θ_2 reaches the critical angle when θ_1 reaches 90°. Therefore, by Snell's law, $n_1 \sin 90° = n_2 \sin \theta_c$. Since $\sin 90° = 1$, $\sin \theta_c = n_1/n_2$, or

$$\theta_c = \sin^{-1}\left(\frac{n_1}{n_2}\right) \quad \text{(Critical angle)} \qquad (21.6)$$

For example, for light going from common glass ($n = 1.52$) into air, the critical angle is

$$\theta_c = \sin^{-1}\left(\frac{1}{1.52}\right) = 41.1°$$

Thus any incidence angle greater than 41.1° results in total internal reflection.

Diamonds have a high refractive index (Table 21.2), making total internal reflection more likely. A finished diamond's structure allows for multiple internal reflections and significant refraction when light emerges from the stone. The result is the characteristic sparkling, as you see light refracted from many points simultaneously. Artificial diamonds made from zircon have a refractive index somewhat lower than diamond but larger than glass.

An important application of total internal reflection is fiber-optic communication. Figure 21.20 shows that light signals can propagate along fiber-optic cables by total internal reflection at the interface between the glass fiber and the so-called cladding that surrounds it. Optical fibers are used extensively in telecommunications, especially telephone, television, and broadband lines that carry Internet data. Most use infrared light.

Dispersion

You've surely observed **dispersion**—the breaking up of white light into the colors of the visible spectrum, commonly using a prism. Dispersion results from slight variations in refractive index with wavelength (Figure 21.21), with shorter wavelengths generally having slightly higher indices. In a prism, incoming white light refracts twice—once on entering the glass and once on leaving (Figure 21.22). Each refraction bends shorter-wavelength light more, resulting in the separation of colors, with violet bent the most and red the least. The angular separation depends on the prism's shape and refractive index, as well as the angle of the incoming white light.

APPLICATION **Fiber Optics in Medicine**

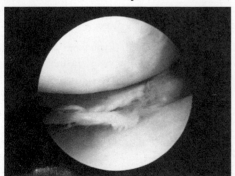

Optical fibers make it possible to peer inside the body with minimal invasiveness. Physicians examine the upper or lower digestive tract for tumors and other disorders by inserting a fiber-optic cable with a camera attached. Fiber-optic systems guide surgeons performing arthroscopic surgery on joints (such as the torn knee meniscus shown in the photo) and laparoscopic surgery in the abdomen; the fiber cables are very thin, and therefore the surgeon makes only a few small cuts rather than a long incision that would harm healthy tissue and take much longer to heal.

Figure 21.23 shows how rainbows result from a combination of refraction, reflection, and dispersion in raindrops. Light refracts upon entering a drop, undergoes total internal reflection at the back of the drop, and refracts again on leaving the drop. Water's refractive index and the drop geometry result in a clustering of outgoing rays deflected through approximately 42°. Absent dispersion, the rainbow would appear as a bright but colorless arc, centered on a line from the Sun to the viewer, and 42° wide. Dispersion results in slightly different angular deflection for different colors, with violet deflected more than red. That makes the rainbow a colored arc, with red on top and violet on the bottom. The fainter secondary rainbow results when light undergoes two reflections within the drop, and its colors are reversed.

✓ **TIP**

Falling raindrops are very nearly spherical, due to surface tension, not the "teardrop" shape often depicted.

GOT IT? Section 21.3   Light going from glass into water (a) bends toward the normal; (b) bends away from the normal; (c) isn't bent at all.

21.4 Thin Lenses

Lenses have been used for centuries as magnifiers and to correct vision. In 1609 Galileo put lenses together to make telescopes, thereby revolutionizing astronomy. Here you'll see how single lenses form images. Then in Section 21.5 we'll consider combinations of lenses used in microscopes and telescopes. Finally, in Section 21.6 we'll discuss human vision, which relies on the same optical properties as manufactured lenses.

How Lenses Work

A common lens is the **spherical lens**, shown in Figure 21.24a. This is a **biconvex lens**, since both lens surfaces bow outward from the middle. A **plano-convex lens** is flat on one side and convex on the other. A **convex lens** can be either biconvex or plano-convex.

Refraction occurs both when light enters a lens and when it leaves (Figures 21.24b and 21.24c). The combined effect of the two refractions is to direct parallel rays toward a **focal point** *F* (Figure 21.24d). As with mirrors in Section 21.2, focusing is never perfect and gets worse for rays far from the principal axis (spherical aberration). The ray tracing in Figure 21.24d neglects the details of refraction at each surface and shows instead a single bend in the center of the lens. This is the **thin-lens approximation**, valid when the lens is thin and so the two refracting surfaces are close.

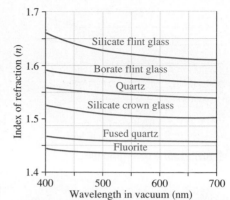

FIGURE 21.21 Refractive indices as functions of wavelength.

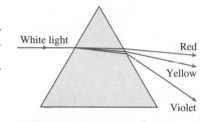

FIGURE 21.22 Dispersion of white light by a prism.

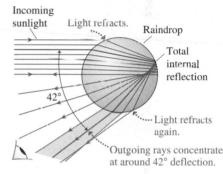

FIGURE 21.23 Formation of a rainbow.

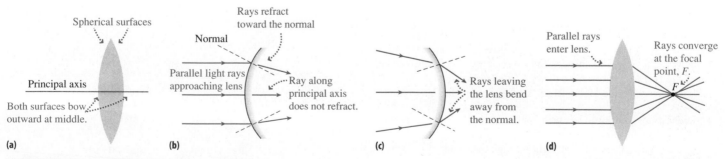

FIGURE 21.24 Details of refraction in a converging lens.

Concave lens

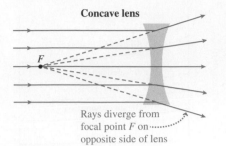

FIGURE 21.25 A diverging lens.

Figure 21.25 shows refraction by a **concave lens**, which bows inward. Here parallel rays diverge from a focal point, rather than converging toward it. Their refractive properties make convex lenses **converging lenses**, while concave lenses are **diverging lenses**.

Image Formation by Convex Lenses

We'll explore image formation in lenses using ray tracing, as we did for mirrors. One important difference between lenses and mirrors is that lenses are transparent. Light can pass either way, so there's a focal point on each side of the lens. For thin lenses the distance to the focal point is the same on each side, even if the curvatures of the two lens surfaces aren't the same.

Tactics 21.2 outlines a procedure for tracing two representative rays, with results as shown in Figure 21.27. As with reflection from concave mirrors in Section 21.2, the type of image formed by a convex lens depends on the location of the object relative to the focal point. Figure 21.27 illustrates three cases: $d_o > 2f$, $2f > d_o > f$, and $d_o < f$. A summary of the results for the three cases is given in Table 21.3. The image can be either upright or inverted, and enlarged or reduced. With the object beyond the focal point, the image is real and on the opposite side of the lens from the object. Since light is present at a real image, it can be projected on a screen—as with movies and computer projectors. But with the object inside the focal point, refracted rays don't converge and the image is virtual, appearing on the same side of the lens. To see the virtual image, you hold the lens between you and the object. That's just what you do with a simple magnifying lens; looking through it, you see an enlarged, upright, virtual image.

Lenses don't have to be made of glass. In Einstein's general relativity, gravity itself bends light. Space in the vicinity of massive objects therefore acts like a lens. Massive galaxy clusters become cosmic telescopes, allowing us to see objects so distant our earthbound telescopes alone couldn't detect them. Other lensing galaxies form multiple or distorted images of distant quasars. Closer to home, so-called "microlensing" allows us to detect unseen objects by their effect on light as they pass in front of more distant stars.

✓**TIP**

Real images form only on the opposite side of the lens and are always inverted.

TABLE 21.3 Summary of Results for the Three Cases

Object location	Image orientation	Image size	Real or virtual?
$d_o > 2f$	Inverted	Reduced	Real
$2f > d_o > f$	Inverted	Enlarged	Real
$d_o < f$	Upright	Enlarged	Virtual

TACTICS 21.2 **Image Formation by Convex Lenses**

Parallel rays striking a convex lens are directed through the focal point, while rays passing through the lens center are not deflected because the two refractions have the opposite effect. You can use these facts to trace two rays (Figure 21.26) that determine the image point.

1. Draw a ray parallel to the principal axis. The refracted ray goes through the focal point on the other side of the lens.
2. Draw the undeflected ray straight through the center of the lens to the other side.

Figure 21.27 shows how these rays are drawn, and how both converge at the image point. The image obtained using this method gives the location, size, and orientation of the image.

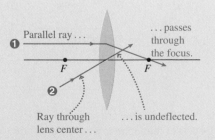

FIGURE 21.26 Two special rays for locating images formed by thin lenses.

cont'd.

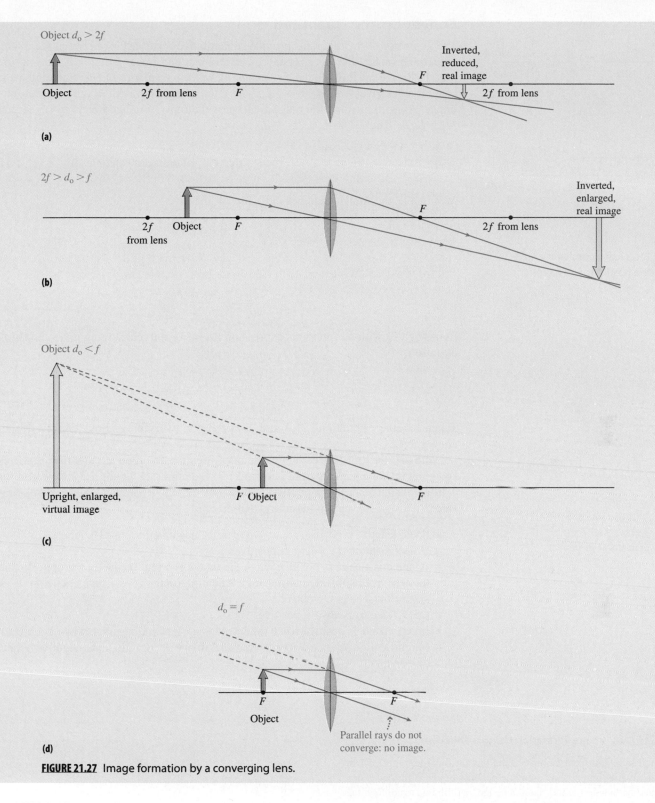

FIGURE 21.27 Image formation by a converging lens.

What happens at the transition points between the cases $d_o = 2f$ and $d_o = f$? When $d_o = 2f$, the image is inverted and real, and this case marks the transition between a reduced and an enlarged image. Thus the image is inverted, real, and the same size as the object. When $d_o = f$, Figure 21.27d shows what happens: the refracted rays emerge parallel. They don't converge to form any image, real or virtual.

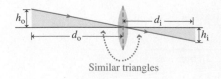

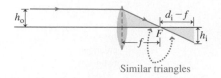

Similar triangles

(a)

Similar triangles

(b)

FIGURE 21.28 Using similar triangles to relate the positions and heights.

Convex Lenses: Quantitative

The geometry of ray diagrams provides quantitative relationships between focal length f, object and image distances d_o and d_i, and object and image heights h_o and h_i. In Figure 21.28a, the ray through the lens center makes two similar triangles. Equating the ratios of their sides gives $-h_i/d_i = h_o/d_o$, where the minus sign with h_i indicates an inverted image. As with mirrors, the magnification is the ratio of image height to object height: $M = h_i/h_o$. Thus $M = h_i/h_o = -d_i/d_o$, or

$$M = -\frac{d_i}{d_o} \quad \text{(Magnification; SI unit: dimensionless)} \quad (21.7)$$

Meanwhile, the similar triangles in Figure 21.28b give

$$\frac{-h_i}{d_i - f} = \frac{h_o}{f}$$

which rearranges to

$$\frac{h_i}{h_o} = -\frac{d_i - f}{f}$$

Equating this with our previous expression for h_i/h_o and rearranging gives the **thin-lens equation**.

$$\frac{1}{f} = \frac{1}{d_o} + \frac{1}{d_i} \quad \text{(Thin-lens equation; SI unit: m}^{-1}) \quad (21.8)$$

Taken together, Equations 21.7 and 21.8 tell you everything about images formed by convex lenses.

Note that the thin-lens equation and magnification equation are identical to the corresponding mirror equations from Section 21.2 (Equations 21.2 and 21.3). However, there's a different sign convention, because light passes through lenses but is reflected by mirrors. Here's the sign convention for convex lenses:

- Focal length f is positive for convex lenses, negative for concave lenses.
- Object distance d_o is always positive.
- Image distance d_i is positive when the image is a real image on the opposite side of the lens as the object, and negative when the image is a virtual image on the same side of the lens as the object.
- Object size h_o is always positive.
- Image size h_i is positive when the image is upright and negative when it's inverted.
- Magnification M is positive when the image is upright and negative when it's inverted.

These agree with the ray diagrams in Figure 21.27.

EXAMPLE 21.7 **Image Formation Using a Thin Lens**

A pencil is 60 cm from a convex lens having focal length 22.5 cm. Draw a ray diagram to show how the pencil's image is formed. Then compute the location of the pencil's image, and find its magnification.

ORGANIZE AND PLAN Our sketch is shown in Figure 21.29. Note that this case is similar to the one illustrated in Figure 21.27a. The image is real, inverted, and reduced in size. The image location follows from the thin-lens equation (Equation 21.8),

$$\frac{1}{f} = \frac{1}{d_o} + \frac{1}{d_i}$$

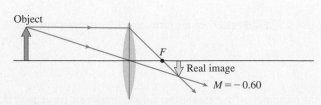

FIGURE 21.29 Our sketch for Example 21.7.

cont'd.

and the magnification from Equation 21.7,

$$M = -\frac{d_i}{d_o}$$

Known: $d_o = 60$ cm; $f = 22.5$ cm.

SOLVE Solving the thin-lens equation for the image distance d_i,

$$d_i = \frac{1}{\frac{1}{f} - \frac{1}{d_o}} = \frac{f d_o}{d_o - f} = \frac{(22.5\text{ cm})(60\text{ cm})}{60\text{ cm} - 22.5\text{ cm}} = 36\text{ cm}$$

By the sign convention for convex lenses, this means the image is 36 cm from the lens, on the opposite side from the object. With $d_i = 36$ cm, the magnification is

$$M = -\frac{36\text{ cm}}{60\text{ cm}} = -0.60$$

The magnification is negative, so the image is inverted, and it's 0.60 (three-fifths) as large as the object pencil.

REFLECT The computed values for the image distance and magnification agree with the ray diagram.

MAKING THE CONNECTION Where should you place the pencil in this example to produce a real image exactly the same size as the object?

ANSWER A real image always has a negative magnification, so $M = -1$ in this case. This requires $d_i = d_o = 2f = 45$ cm.

EXAMPLE 21.8 **Magnifying Glass**

Using the lens in the preceding example ($f = 22.5$ cm), how would you produce an upright, double-sized image? Draw a ray diagram for this situation.

ORGANIZE AND PLAN An upright image is always virtual. That's what you see when you use the lens as a magnifying glass. The desired magnification is $M = +2$ for an upright, double-sized image. Using $M = +2$ in the magnification and thin-lens equations allows you to find the object distance and image distance.

Known: $f = 22.5$ cm, $M = +2$.

SOLVE The magnification equation gives $M = -d_i/d_o = +2$, or $d_i = -2d_o$. Substituting into the thin-lens equation,

$$\frac{1}{f} = \frac{1}{d_o} + \frac{1}{d_i} = \frac{1}{d_o} + \frac{1}{-2d_o} = \frac{1}{2d_o}$$

With $f = 22.5$ cm, this gives d_o:

$$d_o = \frac{f}{2} = \frac{22.5\text{ cm}}{2} = 11.25\text{ cm}$$

Thus, the object has to be placed closer to the lens than the focal point. This produces an image at $d_i = -2d_o = -22.5$ cm. By the sign convention, this means the image is 22.5 cm from the lens, on the

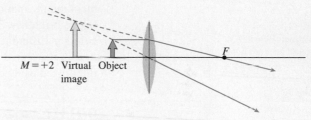

FIGURE 21.30 Our sketch for Example 21.8.

same side as the object. Placing the object at $d_o = 11.25$ cm with a focal length $f = 22.5$ cm allows you to complete the ray diagram (Figure 21.30).

REFLECT Once again, the ray diagram illustrates that the computed numbers are correct and that the image is virtual.

MAKING THE CONNECTION For larger magnification, should you move the lens closer to or farther from the object?

ANSWER Using the magnification equation or ray diagram, you can show that you need to move the lens farther, but keep it within the focal length.

Concave Lenses

Figure 21.25 showed that light diverges from concave lenses; therefore, they can't form real images. As Figure 21.31 demonstrates, concave lenses always form reduced virtual images. Ray tracing works the same way for concave and convex lenses, except that instead of converging to a focal point, parallel rays diverge from the opposite focal point. Otherwise, the rules of ray tracing and image formation are the same as you've learned.

The magnification and thin-lens equations (Equations 21.7 and 21.8) still hold for concave lenses, but with an important difference in the sign convention. Because the focal point is on the opposite side of the lens, *the focal length is negative for a concave (diverging) lens.* With a negative focal length, you can follow through Equations 21.7 and 21.8 to

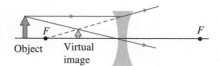

FIGURE 21.31 Image formation by a diverging lens.

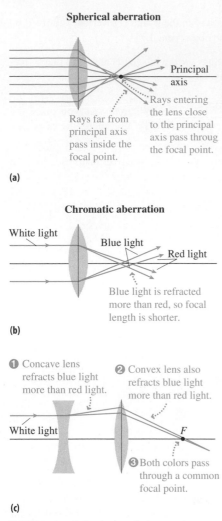

Spherical aberration

Rays far from principal axis pass inside the focal point.

Principal axis

Rays entering the lens close to the principal axis pass through the focal point.

(a)

Chromatic aberration

White light

Blue light

Red light

Blue light is refracted more than red, so focal length is shorter.

(b)

❶ Concave lens refracts blue light more than red light.

❷ Convex lens also refracts blue light more than red light.

White light

F

❸ Both colors pass through a common focal point.

(c)

FIGURE 21.32 Spherical and chromatic aberration.

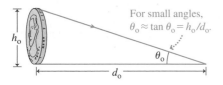

For small angles, $\theta_0 \approx \tan \theta_0 = h_0/d_0$.

h_0

θ_0

d_0

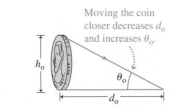

Moving the coin closer decreases d_0 and increases θ_0.

h_0

θ_0

d_0

FIGURE 21.33 Viewing a small object from different distances changes its angular size.

see that the quantitative results agree with ray tracing. With the object distance d_0 still positive, a negative value for f guarantees that d_i is negative, so the image always appears on the same side of the lens as the object and is virtual. Equation 21.8 also guarantees that the absolute value of d_i is smaller than that of f, so the virtual image lies inside the focal length. The absolute value of d_i is smaller than that of d_0, so the magnification is positive and less than 1. In a concave lens you always see an upright, virtual, reduced image, regardless of the object's placement.

Are there practical uses of diverging lenses? Later in this chapter we'll show how both converging and diverging lenses are used in microscopes and telescopes (Section 21.5) and in correcting vision (Section 21.6).

✓ **TIP**

A concave lens never forms a real image.

Lens Aberrations

Lenses suffer two important distortions, which affect microscopes and refracting telescopes. We've mentioned spherical aberration, which causes imperfect focusing in both mirrors and lenses (Figure 21.32a). Rays farthest from the axis are most affected, so blocking the outer edges of the lenses minimizes spherical aberration, but that reduces the image brightness. A second distortion is **chromatic aberration**, the focusing of different colors at different points (Figure 21.32b). This is a manifestation of dispersion, the variation in refractive index with wavelength (recall Section 21.3). Chromatic aberration is minimized using combinations of lenses (Figure 21.32c).

GOT IT? Section 21.4   In order to produce a magnified, real image with a convex lens, the object has to be placed at a distance (a) $d_0 > 2f$; (b) $2f > d_0 > f$; (c) $d_0 < f$.

21.5 Microscopes and Telescopes

In Section 21.4 (Example 21.8), you saw how a single convex lens can be used as a magnifier. Here we'll consider single-lens magnification in more detail. Then we'll show how to improve the magnification using two lenses to form a microscope. Finally, we'll consider telescopes, which image distant objects.

Magnification and the Near Point

We've defined the term *magnification* as the ratio of image size to object size. In viewing magnified images through one or more lenses, it's more appropriate to consider magnification as the ratio of the angular size of the magnified image to that of the naked-eye image. This is the **angular magnification**, designated m.

Hold a coin at arm's length and slowly bring it closer. As the coin approaches your eye, it grows larger in your field of view—and therefore it subtends a larger angle (Figure 21.33). For an object of size h_0 viewed at distance d_0, the angle θ_0 is given by $\tan \theta_0 = h_0/d_0$. For small objects, the angle θ_0 is small, and to good approximation $\theta_0 \approx \tan \theta_0$, with θ_0 in radians. Therefore, a good approximation is $\theta_0 = h_0/d_0$.

As you move the coin closer to your eye, d_0 decreases while h_0 remains constant. Therefore, the angle θ_0 increases, and the angular size of the coin increases. But if you move the coin too close, your eye can't focus on it. The angular size of the coin continues to grow, but that's not helpful because you can't see it clearly. For that reason we'll measure θ_0 when d_0 reaches the closest point your eye can focus, called the **near point** (symbol N). With $N = d_0$, the largest angular size you can view is $\theta_0 = h_0/N$. A typical near point for a young adult is about 25 cm, and this distance tends to grow as people age and the lenses in their eyes become less flexible. (We'll discuss human vision in Section 21.6.) For a penny with $h_o = 1.9\,\text{cm}$, a near point $N = 25\,\text{cm}$ gives a maximum angular size

$$\theta_o = \frac{h_o}{N} = \frac{1.9\,\text{cm}}{25\,\text{cm}} = 0.076\,\text{rad} \approx 4.4°$$

✓ **TIP**

The relation $\theta \approx \tan\theta$ works only with θ in radians, where $\theta \ll 1$.

Single-Lens Magnification

Section 21.4 showed that a convex lens works as a magnifying glass if the object is closer to the lens than the focal distance (Figure 21.34a). Figure 21.34b shows the relationship among viewing angle θ, object height, and object distance:

$$\theta \approx \tan\theta = \frac{h_i}{d_i} = \frac{h_o}{d_o}$$

Therefore, the angular magnification m—the ratio of magnified angle θ to naked-eye angle θ_o—is

$$m = \frac{\theta}{\theta_o} = \frac{h_o/d_o}{h_o/N} = \frac{N}{d_o}$$

From the thin-lens equation,

$$\frac{1}{d_o} = \frac{1}{f} - \frac{1}{d_i}$$

Making this substitution for $1/d_o$, the angular magnification becomes

$$m = N\left(\frac{1}{f} - \frac{1}{d_i}\right) \qquad \text{(Single-lens angular magnification; SI unit: dimensionless)} \qquad (21.9)$$

For you to see the image, it must lie somewhere between the near point and infinity. When it's at the near point $(d_i = -N)$,

$$m = 1 + \frac{N}{f} \qquad \text{(Maximum magnification)}$$

When the image is at infinity $(d_i = -\infty)$, $1/d_i = 0$, and

$$m = \frac{N}{f} \qquad \text{(Minimum magnification)}$$

In practice, the magnification is somewhere between these extremes, depending on where you find it comfortable to hold the magnifying glass and view the image. Equation 21.9 tells you that for greater magnification you should use a shorter focal length lens. For example, suppose you're viewing a flea using a magnifier with $f = 3.0\,\text{cm}$. If your near point is at $N = 25\,\text{cm}$, then the maximum magnification is

$$m = 1 + \frac{N}{f} = 1 + \frac{25\,\text{cm}}{3.0\,\text{cm}} = 9.3$$

You can increase the flea's apparent size by almost a factor of 10.

✓ **TIP**

You can use any distance units for the near-point distance and focal length, provided they're the same for both. The units must cancel in order to give the dimensionless magnification.

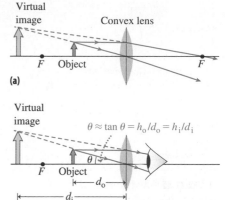

(a)

(b)

FIGURE 21.34 Magnification by a single lens.

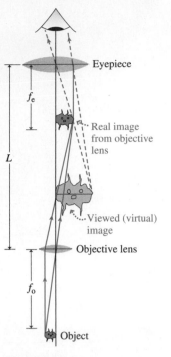

Eyepiece

f_e

Real image from objective lens

L

Viewed (virtual) image

Objective lens

f_o

Object

FIGURE 21.35 The compound microscope.

The Compound Microscope

Your near-point distance N and the magnifier's focal length f limit the magnification achievable with a single lens. The **compound microscope** gets around this limit by using two convex lenses. The first, the **objective lens**, forms a real image of the object. That intermediate real image is viewed through a second lens, the **eyepiece**. You look into the eyepiece and see a virtual image of the real image formed by the objective, as illustrated in Figure 21.35.

For a substantially magnified real image, the object should be placed just beyond the objective's focal point ($d_o \approx f_o$). The resulting intermediate image—which serves as the object for the eyepiece—should then fall just within the eyepiece's focal distance, thus maximizing the eyepiece magnification. The fact that the intermediate image lies near the eyepiece's focal point can be used to find an approximate expression for the microscope's overall magnification. The intermediate image is real, with magnification M_o given by Equation 21.7: $M_o = -d_i/d_o$. But with the intermediate image just within the eyepiece's focal distance, $d_i \approx L - f_e$, where L is the distance between the lenses. Therefore,

$$M_o = -\frac{d_i}{d_o} \approx -\frac{L - f_e}{f_o}$$

The eyepiece then acts as a simple magnifier, so by Equation 21.9, with $d_i \approx -\infty$, $m_e = N/f_e$. The overall magnification is the product of the two successive magnifications, or

$$M = M_o m_e = -\left(\frac{L - f_e}{f_o}\right)\left(\frac{N}{f_e}\right)$$

or

$$M = -\frac{N(L - f_e)}{f_o f_e} \quad \text{(Compound microscope; SI unit: dimensionless)} \quad (21.10)$$

Although Equation 21.10 is an approximation, the product $f_o f_e$ in the denominator suggests that microscopes with high magnification should have lenses with short focal lengths. Also note the minus sign in Equation 21.10, showing that the image is inverted relative to the object. We use M in Equation 21.10 because this magnification is the ratio of the image size to the actual object size.

EXAMPLE 21.9 **Compound Microscope Design**

Your microscope's objective and eyepiece lenses have focal lengths 3.0 cm and 1.0 cm, respectively. How far apart should the lenses be mounted to provide 100× magnification? (Assume a 25-cm near-point distance.)

ORGANIZE AND PLAN Equation 21.10 gives the magnification of the compound microscope:

$$M = -\frac{N(L - f_e)}{f_o f_e}$$

All the quantities are known except L, the distance between lenses, so you can solve for L. The magnification of a compound microscope is negative, so $M = -100$.

Known: $f_o = 3.0$ cm; $f_e = 1.0$ cm; $M = -100$; $N = 25$ cm.

SOLVE Solving for L gives

$$L = -\frac{f_o f_e M}{N} + f_e = -\frac{(3.0 \text{ cm})(1.0 \text{ cm})(-100)}{25 \text{ cm}} + 1 \text{ cm} = 13 \text{ cm}$$

cont'd.

REFLECT $L = 13$ cm is a reasonable spacing for a desktop microscope. Notice that the magnification is more than 10 times better than the value 9.3 found in the text using the 3-cm lens as a single magnifier. The compound microscope provides much greater magnification.

MAKING THE CONNECTION Assuming the same spacing L, how much would the magnification be improved if the objective lens were replaced by one with 2-cm focal length?

ANSWER The magnification becomes $M = -150$. Microscopes commonly have several objective lenses of different focal lengths, to provide user-selected magnifications.

Refracting Telescopes

Refracting telescopes use lenses to image distant objects. The concept is the same as with microscopes: Use an objective lens to form an intermediate image, then an eyepiece to make an enlarged, virtual image of the intermediate image. Figure 21.36 shows a ray diagram for a refracting telescope. We'll find an approximate expression for the magnification using the same notation as for the microscope, with θ_0 the angular size of the object to the naked eye and θ the angular size viewed through the telescope.

Rays from distant objects are nearly parallel, so the intermediate image forms essentially at the focus of the objective lens. As in the microscope, we want the intermediate image near the eyepiece's focal point for maximum magnification. So the two focal points very nearly coincide, as shown in Figure 21.36. The figure identifies the angles θ_0 and θ; since they're small, we can approximate their tangents by the angles themselves. Then $\theta_0 \approx h_1/f_0$ and $\theta \approx h_1/f_e$, where h_1 is the height of the intermediate image. This height is common to our expressions for both angles, so the angular magnification becomes approximately

$$m = -\frac{f_0}{f_e} \quad \text{(Telescope magnification; SI unit: dimensionless)} \quad (21.11)$$

Equation 21.11 shows that the focal length of the eyepiece should be much smaller than that of the objective for maximum magnification. Again, the negative sign means that the final image is inverted. In contrast to the microscope, the telescope's magnification is angular (m)—an increase in the angular extent of the object. Telescopes certainly don't enlarge the actual sizes of ships or galaxies! The design in Figure 21.36 is the **Keplerian telescope**, named for Johannes Kepler. (Remember Kepler's laws of planetary motion from Chapter 9.) The **Galilean telescope** uses a concave (diverging) lens as the eyepiece. This inverts the image again, resulting in an upright final image. Refracting telescopes are used for terrestrial viewing, in applications from nautical "spyglasses" to telephoto camera lenses. But they're rarely used in contemporary astronomy, for reasons you'll now see.

Reflecting Telescopes

The most important characteristic of an astronomical telescope is the amount of light it gathers, which depends on the area of its light-gathering element. It's difficult to manufacture large lenses, and difficult to mount them because that has to be done at their edges to avoid blocking light. It's much easier to make large mirrors, supported from behind. For these reasons, astronomers today use **reflecting telescopes**, whose light-gathering elements are mirrors. While the largest refracting telescope ever built has a 1-m-diameter lens, today's largest reflectors exceed 10 m—for over 100 times the light-gathering capability of the largest refractor. Manufacturing such a mirror is a feat of amazing precision, since the entire mirror surface must match its design shape to within a wavelength of light.

Unlike bathroom mirrors, telescope mirrors are silvered on the front surface, so light doesn't pass through glass and suffer chromatic aberration. And if they're parabolic, they also eliminate spherical aberration. The most pristine imaging uses a detector placed right at the mirror's focal point (Figure 21.37a), so the light encounters only the primary mirror. Other arrangements use secondary mirrors and often optical fibers to deliver light to spectrographs or other instruments (Figure 21.37b) or, in smaller amateur telescopes, to an

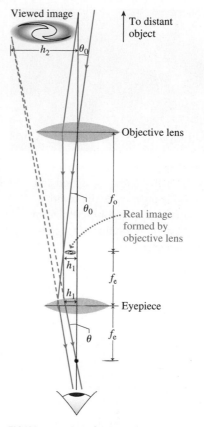

FIGURE 21.36 Ray diagram for a refracting telescope.

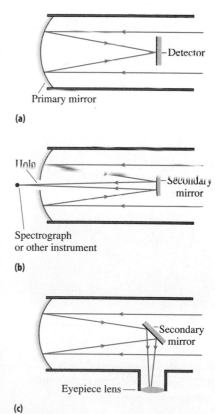

FIGURE 21.37 Reflecting telescopes.

FIGURE 21.38 Artist's conception of the Giant Magellan Telescope, scheduled for completion in 2016. The seven telescope segments combine to form the equivalent resolution of a single 21-m-diameter mirror.

eyepiece (Figure 21.37c). The largest telescopes have segmented mirrors (Figure 21.38) whose individual segments can be steered and reshaped to optimize focusing. So-called adaptive optics technology does this in real time to compensate for varying distortion in Earth's atmosphere.

Another way to eliminate atmospheric distortion is with space-based telescopes. The 2.4-m-diameter Hubble Space Telescope, although not large by ground-based standards, nevertheless produces spectacular images. That's despite a manufacturing error that left its primary mirror with significant aberration—an error rectified when astronauts installed additional corrective optics. Hubble's successor, the James Webb Space Telescope, is scheduled for launch in 2013. Its 6.5-m-diameter mirror is too large to fit inside a rocket, so it will use a folding design that unfurls once it's in space. The Webb telescope is optimized for infrared observations and will be deployed in a solar orbit 1.5 million km sunward of Earth, where the superposition of Earth's and Sun's gravity gives it an orbital period of exactly 1 year.

Reviewing New Concepts

- Light refracts when it enters a medium with a different index of refraction.
- Lenses use the property of refraction to form real and virtual images.
- Combinations of lenses are used in microscopes and refracting telescopes to enlarge near and distant objects.
- In a microscope or refracting telescope, an objective lens forms an intermediate real image, and the eyepiece is used to view a virtual image of the intermediate image.

GOT IT? Section 21.5 A backyard (Keplerian) telescope gives you an image of the Moon magnified eight times. If the objective has a focal length of 16 cm, the focal length of the eyepiece is (a) 32 cm; (b) 8 cm; (c) 4 cm; (d) 2 cm.

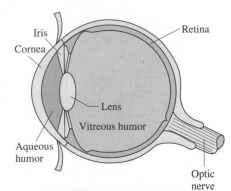

(a)

21.6 The Eye and Vision

Here we consider a most remarkable optical instrument—the human eye. We'll first describe the eye's anatomy and optics, which explain basic vision functions. Then we'll discuss common vision disorders—nearsightedness, farsightedness, and astigmatism—and how corrective lenses restore good vision.

Anatomy and Optics of the Eye

Figure 21.39a depicts the eye's anatomy. Incoming light passes through the **pupil**, an opening in the **iris**. The iris adjusts the size of the pupil to compensate for changing light conditions, with the opening narrowing when it's brighter. Light refracts in the **cornea** and **lens**, which together converge light to form a real image on the **retina** (Figure 21.40). Retinal receptor cells—**rods** and **cones** (Figure 21.39b)—sense the image and send information to the brain through the **optic nerve**.

For you to see a clear image, the eye has to focus light from whatever you're viewing directly onto the retina. Applying the thin-lens equation

$$\frac{1}{f} = \frac{1}{d_o} + \frac{1}{d_i}$$

presents a challenge, because you'd like to see clearly over a wide range of object distances d_o. Because the lens-to-retina distance d_i is fixed, the eye must adjust its focal length to compensate for different object distances. This adjustment is called **accommodation** and is done by contracting and relaxing the **ciliary muscle** that surrounds the flexible lens. To view distant objects, the muscle is relaxed, making the lens

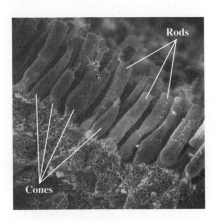

(b)

FIGURE 21.39 (a) Anatomy of the human eye. (b) Scanning electron micrograph showing retinal rods and cones.

relatively flat and its focal length relatively large. Viewing nearby objects requires a shorter focal length, so the ciliary muscle contracts to make the lens bulge. In Section 21.5 we introduced the *near point*, the shortest object distance for which the eye can produce a clear image. The near point depends on how short your eye can make its focal length. As people grow older, the lens becomes less flexible. Shorter focal lengths are no longer possible, and the near point distance increases. This condition is **presbyopia**.

Nearsightedness

At the other extreme, the eye has a **far point**—the largest distance for which clear images can form on the retina. In a healthy eye, the far point is essentially at infinity. You can see distant buildings and mountains, even details on the Moon. Your vision of distant objects is limited only by their size.

If the eye itself is too long, however, images of distant objects fall in front of the retina (Figure 21.41a). The far point is no longer at infinity, and objects farther than the far point appear out of focus. This common condition is **myopia**, or nearsightedness. Myopia can be corrected by glasses or contact lenses, as shown in Figure 21.41b. A diverging (concave) lens is used, causing a slight divergence of light rays before they enter the eye. The net effect of this divergence and convergence is to put the image right on the retina, restoring clear vision.

Lasers can reshape the cornea to correct myopia. (See the chapter-opening photograph.) The current technique is laser-assisted in situ keratomileusis (LASIK), in which the outer layer of cornea is cut and flapped aside, and a laser then ablates and reshapes the underlying corneal tissue. The idea of LASIK is the same as corrective lenses: increase the focal length, so that images of distant objects fall on the retina. LASIK is also used for hyperopia, although with more difficulty because the outer edge of the cornea must be reduced in thickness.

Farsightedness

The opposite of myopia is **hyperopia**, or farsightedness. Here the eye is too short, and images of nearby objects would focus behind the retina (Figure 21.42a). Converging lenses in glasses or contacts correct this problem, as shown in Figure 21.42b. Converging lenses also help people with presbyopia see close-up, effectively reducing the eye's near point.

Refractive Power and the Lensmaker's Equation

Opticians prescribe eyeglasses or contact lenses with specific focal lengths to correct a patient's degree for nearsightedness or farsightedness. Rather than specifying the actual focal length, however, opticians use **refractive power**, defined as

$$P_{\text{refractive}} = \frac{1}{f}$$

The units for refractive power are **diopters**, with 1 diopter $= 1 \text{ m}^{-1}$. Thus, a converging lens with focal length 25 cm has refractive power

$$P_{\text{refractive}} = \frac{1}{f} = \frac{1}{0.25 \text{ m}} = 4.0 \text{ m}^{-1} = 4.0 \text{ diopters}$$

Diverging lenses used to correct nearsightedness have negative focal lengths, so their refractive powers are negative. Thus, a lens with focal length $f = -25$ cm has a refractive power -4 diopters.

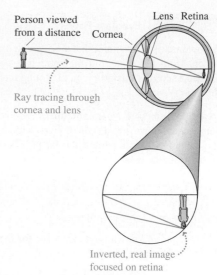

Person viewed from a distance Cornea Lens Retina

Ray tracing through cornea and lens

Inverted, real image focused on retina

FIGURE 21.40 Refraction occurs in both cornea and lens, forming an image on the retina.

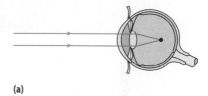

(a)

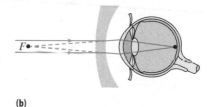

(b)

FIGURE 21.41 (a) In a nearsighted person, the image forms in front of the retina. (b) Correcting nearsightedness requires a diverging lens, which has a focal point *F*.

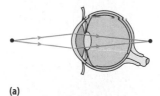

(a)

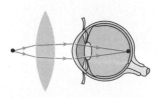

(b)

FIGURE 21.42 (a) In a farsighted person, nearby objects focus behind the retina. (b) Correcting farsightedness requires a converging lens.

EXAMPLE 21.10 **Eyeglass Power**

An optometrist determines that a nearsighted patient's far point is only 44 cm from his eye. Find the refractive power of eyeglass lenses needed to correct this problem, assuming the glasses are 2.0 cm from the eyes.

ORGANIZE AND PLAN Nearsightedness requires corrective lenses that form virtual images of distant objects, with the images no farther than the far point. In this patient the far point is 44 cm from the eye, or 42 cm from the lens. The thin-lens equation

$$\frac{1}{f} = \frac{1}{d_o} + \frac{1}{d_i}$$

relates the lens's focal length, the object distance, and image distance. With d_o and d_i known, you can solve for the focal length f and then determine the refractive power using $P_{refractive} = 1/f$.

Known: $d_o = \infty$ (distant objects); $d_i = -42$ cm (negative since it's a virtual image).

SOLVE Solving for the focal length,

$$\frac{1}{f} = \frac{1}{d_o} + \frac{1}{d_i} = \frac{1}{\infty} + \frac{1}{-42 \text{ cm}} = -\frac{1}{42 \text{ cm}}$$

Therefore, $f = -42$ cm. The refractive power is

$$P_{refractive} = \frac{1}{f} = \frac{1}{-0.42 \text{ m}} = -2.4 \text{ diopters}$$

REFLECT This is a reasonable prescription for a moderately nearsighted person. More severely nearsighted patients have shorter far-point distances and correspondingly larger diopter corrective lenses.

MAKING THE CONNECTION Contact lens boxes are stamped with the lenses' refractive power. You're nearsighted and your roommate is farsighted. You've mixed up your contact lenses; one box says -3.5, the other $+2.0$. Which lenses are yours, and who has the more serious vision defect?

ANSWER Diverging lenses—negative refractive power—correct nearsightedness, so yours are the -3.5-diopter lenses. Since the magnitude of your diopter measure is greater, you've got the weaker eyes.

Contact Lenses

For the patient in Example 21.10, how does the prescription change for contact lenses?

SOLVE The difference is that the 2.0-cm gap between eyeglasses and eye is gone. This changes the image distance to $d_i = -44$ cm, giving a contact lens focal length of -44 cm. That means a refractive power

$$P_{refractive} = \frac{1}{f} = \frac{1}{-0.44 \text{ m}} = -2.3 \text{ diopters}$$

REFLECT That's not much of a difference. Most contacts come in 0.25-diopter increments, so a –2.25-diopter lens would do.

A lens's focal length depends on its shape and refractive index n. For a given lens shape, a higher refractive index means more refraction, for both converging and diverging lenses. For lenses with spherical surfaces (Figure 21.43), opticians use the lens-maker's equation to determine the appropriate curvature:

$$P_{refractive} = \frac{1}{f} = (n - 1)\left(\frac{1}{R_1} - \frac{1}{R_2}\right) \qquad \text{(Lens-maker's equation;} \quad \text{SI unit: } m^{-1}[= \text{diopter}]) \qquad (21.12)$$

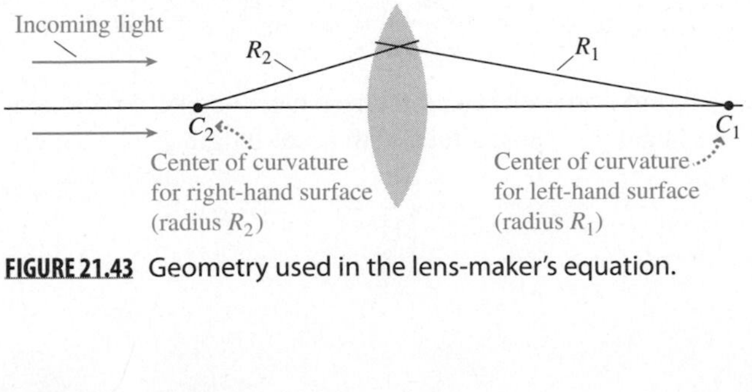

FIGURE 21.43 Geometry used in the lens-maker's equation.

As with the thin-lens equation, it's important to observe a sign convention for the curvature radii R_1 and R_2:

> **Sign convention for curvature radii:** The curvature radius is negative when the center of curvature C is on the side of the incoming light, and positive when it's on the opposite side.

The lens-maker's equation shows why the radii of curvature need not be the same on each side of a lens. The equation also works with one surface flat, with the corresponding radius taken as infinity.

For example, consider a lens of common glass ($n = 1.52$), convex on both sides as in Figure 21.43, with a curvature radius 0.50 m on each side. Then by the sign convention, $R_1 = 0.50$ m and $R_2 = -0.50$ m. The refractive power of this lens is

$$\frac{1}{f} = (n - 1)\left(\frac{1}{R_1} - \frac{1}{R_2}\right) = (1.52 - 1)\left(\frac{1}{0.50 \text{ m}} - \frac{1}{-0.50 \text{ m}}\right) = 2.08 \text{ m}^{-1}$$
$$= 2.08 \text{ diopters}$$

Then the focal length is $f = 1/P_{refractive} = 0.48$ m. The focal length and refractive power come out positive, as you know they should for a convex lens. A concave lens with the same curvature radii would have $P_{refractive} = -2.08$ diopters, because the sign convention reverses the sign on both radii.

Astigmatism

Astigmatism results when the lens is not symmetric. For example, think of a basketball deformed into an oblong shape like a football. That means different focal lengths for rays entering the lens at different points, resulting in blurred vision. Corrective lenses for astigmatism are cylindrical rather than spherical. Many people have a combination of myopia and astigmatism, a condition that requires negative corrective lenses with nonspherical elements.

Chapter 21 in Context

This chapter introduced *geometrical optics*, an approximation that considers light to travel in straight rays. You learned to use *ray tracing* to see how mirrors and lenses form images. You learned about *real images*, where light rays converge, and *virtual images*, where rays only appear to converge. You saw how light *refracts* when it passes from one medium into another, and how *dispersion* results when refraction differs for different wavelengths. Refraction in lenses leads to image formation and explains the design of optical instruments. These include single-lens *magnifiers*, *microscopes*, *telescopes*, and the human eye itself.

Looking Ahead In Chapter 22, the final chapter on optics, we'll consider effects associated with light's wave properties: interference, diffraction, and polarization. You'll see some practical applications, as well as limitations the wave nature of light imposes on our ability to image small or distant objects. The properties of light continue to be important in the study of atomic, nuclear, and quantum phenomena, subjects of Chapters 23–26. Much of this "modern physics" involves a refined understanding of the nature of light and its interaction with matter.

Reflection and Plane Mirrors

(Section 21.1) Geometrical optics is the approximation that light travels generally in straight lines, called **rays**. **Diffuse reflection** occurs when light rays striking a rough surface are reflected in many different directions. Rays striking a smooth surface like a mirror exhibit **specular reflection**. Light always reflects at an angle equal to the incidence angle.

Law of reflection: $\theta_i = \theta_r$

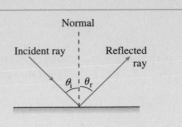

Spherical Mirrors

(Section 21.2) Ray tracing locates images formed by mirrors and lenses. Plane mirrors produce **virtual images**. Spherical mirrors can be either **concave** or **convex**. **Concave mirrors** form real or virtual images, depending on the object distance and focal length. **Convex mirrors** always form virtual images on the other side of the mirror.

Mirror equation: $\dfrac{1}{f} = \dfrac{1}{d_o} + \dfrac{1}{d_i}$

Mirror magnification: $M = \dfrac{h_i}{h_o} = -\dfrac{d_i}{d_o}$

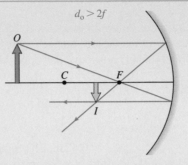

Refraction and Dispersion

(Section 21.3) Light **refracts** when it passes from one medium to another, traveling slower in a medium with a higher **index of refraction**. **Snell's law** describes refraction quantitatively. **Total internal reflection** occurs when light propagating in a medium with lower refractive index is incident on a medium with higher index, provided the incidence angle is greater than the **critical angle**. **Dispersion,** as when white light is sent through a prism, results from different wavelengths of light having different refractive indexes.

Index of refraction: $n = \dfrac{c}{v}$

Snell's law: $n_1 \sin \theta_1 = n_2 \sin \theta_2$

Critical angle: $\theta_c = \sin^{-1}\left(\dfrac{n_1}{n_2}\right)$

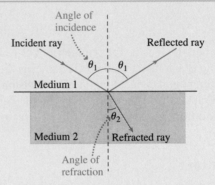

Thin Lenses

(Section 21.4) Convex (converging) lenses form real or virtual images, depending on the object distance and focal length. Parallel rays striking a convex lens are directed through the focal point. For a concave (diverging) lens, parallel rays diverge from the opposite focal point, so concave lenses always form virtual images.

Thin-lens equation: $\dfrac{1}{f} = \dfrac{1}{d_o} + \dfrac{1}{d_i}$

Lens magnification: $M = -\dfrac{d_i}{d_o}$

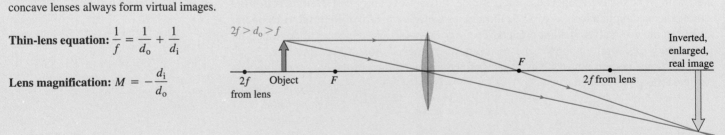

Microscopes and Telescopes

(Section 21.5) A single-lens magnifier uses a convex lens to form a virtual image. **Compound microscopes** and **refracting telescopes** use paired lenses (objective and eyepiece), with the objective forming an intermediate real image and the eyepiece forming a virtual image of the intermediate image. **Reflecting telescopes** replace the objective lens by a parabolic mirror, which allows much greater light-gathering power and eliminates aberrations.

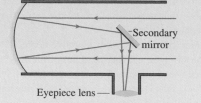

Secondary mirror

Eyepiece lens

Single-lens magnification (angular magnification): $m = N\left(\dfrac{1}{f} - \dfrac{1}{d_i}\right)$

Compound microscope magnification: $M = -\dfrac{N(L - f_e)}{f_o f_e}$

Telescope magnification (angular magnification): $m = -\dfrac{f_o}{f_e}$

The Eye and Vision

(Section 21.6) In the eye, incoming light passes through the **pupil** and is refracted by the **cornea** and **lens**, which comprise a converging lens system that forms a real image on the **retina**. Common refractive disorders are myopia (nearsightedness), hyperopia (farsightedness), and astigmatism.

Refractive power: $P_{refractive} = \dfrac{1}{f}$

Lens-maker's equation: $P_{refractive} = \dfrac{1}{f} = (n - 1)\left(\dfrac{1}{R_1} - \dfrac{1}{R_2}\right)$

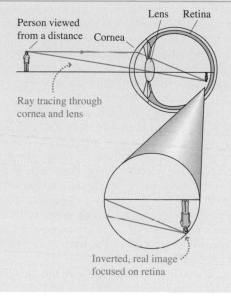

Person viewed from a distance   Cornea   Lens   Retina

Ray tracing through cornea and lens

Inverted, real image focused on retina

NOTE: Problem difficulty is labeled as ■ straightforward to ■■■ challenging. Problems labeled BIO are of biological or medical interest.

Conceptual Questions

1. Explain the difference between specular and diffuse reflection.
2. If you stood between two parallel vertical mirrors, what would you see?
3. Can a single flat mirror ever produce a real image?
4. Two vertical mirrors are mounted at right angles. Show that this combination allows you to see yourself as others see you.
5. How does a blue wall appear if it's illuminated with blue light? With red light?
6. How can you determine the focal length of a concave mirror? Of a convex mirror?
7. Draw carefully a circular arc to represent the surface of a spherical concave mirror. Drawing a ray diagram with incident rays parallel to the principal axis, show that spherical aberration isn't a problem for rays close to the principal axis but becomes more problematic the farther the incoming rays are from the axis.
8. Draw carefully a parabola to represent a section through a parabolic mirror. Drawing a ray diagram, show that all incoming parallel rays focus at the same point.
9. Why is the order of the colors in the secondary rainbow reversed from the primary rainbow?
10. A glass fish tank is filled with water. Is it possible for light originating within the water to undergo total internal reflection at the water-glass interface? Can light that travels through the water and glass suffer total internal reflection at the glass-air interface?
11. Using a convex lens, can you ever get a virtual image with magnification 1? Can you get a virtual image with magnification less than 1?
12. How would the refractive properties of a glass convex lens change if you used it under water?

13. BIO **Animal eyes.** The lens in the eyes of many animals, including humans, has a refractive index that varies from the center to the edge of the lens—a feature that helps correct for spherical aberration. In order to achieve that correction, should the refractive index be greater at the center or the edge?

14. Does a nearsighted person need positive or negative corrective lenses? Which should be stronger, eyeglasses or contact lenses?

15. After age 40, many people need bifocals. For nearsighted people, the main part of the bifocal lens is negative, with a small portion (usually near the bottom) also negative, but one or two diopters weaker than the main lens. What's the purpose of this design?

Multiple-Choice Problems

16. You have a concave mirror with focal length 25 cm. How far from the mirror should you place an object to get a real image 50 cm from the mirror? (a) 25 cm; (b) 35 cm; (c) 50 cm; (d) 75 cm.

17. An object is 80 cm in front of a concave mirror with 50-cm focal length. The image is magnified by (a) 1.3; (b) 1.6; (c) 2.0; (d) 2.7.

18. An object is 0.92 m from a concave mirror, and a real image forms 0.36 m from the mirror. What's the mirror's focal length? (a) 0.18 m; (b) 0.22 m; (c) 0.26 m; (d) 0.30 m.

19. You have a handheld mirror that gives an enlarged virtual image. If want your face to appear doubled in size when you hold the mirror 30 cm away, what should be the mirror's focal length? (a) 20 cm; (b) 40 cm; (c) 60 cm; (d) 80 cm.

20. The speed of light in a dense glass is measured at 1.56×10^8 m/s. The refractive index of this glass is (a) 1.56; (b) 1.92; (c) 2.24; (d) 3.69.

21. Light propagating in air strikes the surface of a lake. The reflected ray makes a 62° angle with the normal. What angle does the refracted ray in the water make with the normal? (a) 42°; (b) 48°; (c) 54°; (d) 58°.

22. A diver shines a flashlight upward toward the water surface. What's the maximum angle the light beam can make with the surface normal and still emerge from the water? (a) 30.0°; (b) 41.7°; (c) 48.8°; (d) 60.0°.

23. A lightbulb placed 30 cm from a convex lens with focal length 18 cm forms an image magnified by a factor of (a) 0.67; (b) 1.0; (c) 1.5; (d) 1.7.

24. You have a convex lens with focal length 15 cm. In order to form an upright image of newspaper type magnified by a factor of 2.5, how far should you hold the lens from the newspaper? (a) 6 cm; (b) 9 cm; (c) 12 cm; (d) 25 cm.

25. The focal lengths of a telescope's objective and eyepiece lenses are 100 cm and 4.0 cm, respectively. The telescope forms an image with angular magnification (a) 50; (b) 25; (c) 10; (d) 5.

Problems

Section 21.1 Reflection and Plane Mirrors

26. ■ A 1.60-m-tall student stands before a vertical mirror. She can just see her whole body, from head to toe. (a) How tall is the mirror? (b) If the student's eyes are 10 cm below the top of her head, how far is the bottom of the mirror above the ground?

27. ■■ You're standing 3.0 m from a plane mirror. If you want to photograph your image, at what distance should you focus the camera's lens?

28. ■■ A laser is mounted on a table 1.06 m above the ground, pointed at a vertical mirror a horizontal distance 3.50 m away. The laser beam strikes the mirror 1.62 m above the ground. Behind the laser is a wall, 2.80 m from the aperture where the laser light emerges. At what height does the reflected beam strike the wall?

29. ■■ A dog sitting 2.0 m from a vertical mirror thinks its image is another dog. (a) How far is the image from the dog? (b) To investigate, the dog walks directly toward the mirror at 0.25 m/s. At what rate does the dog approach its image?

30. ■ Figure P21.30 shows two mirrors at right angles. Light strikes one mirror at 45°, as shown. (a) Continue the ray diagram and show that after striking both mirrors the outgoing ray is parallel to the incoming ray. (b) What is the distance between the incoming and outgoing rays?

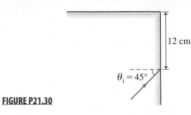

12 cm

$\theta_i = 45°$

FIGURE P21.30

31. ■■ Repeat the preceding problem for an incidence angle of 30°.

32. ■■ You're standing on the dock, holding a flashlight 3.0 m above the surface of a lake. A boat is 20 m away. You want to shine your flashlight so the light reflected from the water strikes the boat's bow, 1.2 m above the water line. Where should you aim the flashlight?

Section 21.2 Spherical Mirrors

33. ■ A concave mirror has focal length 35 cm. Where should you place an object to get a real image (a) 50 cm from the mirror and (b) 1.0 m from the mirror?

34. ■ An object is 66.0 cm from a concave mirror, and a real image forms 48.0 cm from the mirror. Find (a) the mirror's focal length and (b) the image magnification.

35. ■■ Sketch ray diagrams, including object and image, for a concave mirror when (a) $M = -1$; (b) $M = -0.5$; (c) $M = -3$; (d) $M = +1.5$.

36. ■ (a) Draw the ray diagram for a concave mirror with an object 100 cm from the mirror and an inverted image 25 cm from the mirror. (b) What's the mirror's focal length?

37. ■■ Draw the ray diagram for a convex mirror with an object 100 cm from the mirror and an upright image 25 cm on the other side of the mirror.

38. ■ You place a small lightbulb 70 cm from a concave mirror and find a real image 20 cm from the mirror. Where should you place the bulb so its real image is 70 cm from the mirror?

39. ■ An object is 125 cm in front of a concave mirror with focal length 75.0 cm. (a) Where is the image? (b) Is it real or virtual? (c) What's the magnification?

40. ■ A small lightbulb 2.0 m from a concave mirror gives a real image 6.0 cm from the mirror. (a) Find the mirror's focal length. (b) Where will the image be if the bulb is moved to 4.0 m from the mirror?

41. ■ A small lightbulb is 35 cm in front of a concave mirror with focal length 21 cm. What are the location, magnification, and orientation (upright or inverted) of the bulb's image?

42. ■■ You have a handheld mirror that gives a virtual image of your face, enlarged by a factor of 1.5, when you hold the mirror 30 cm away. What's the mirror's focal length?

43. ■■ A concave mirror has focal length 50 cm. You want this mirror to form a real image 1.2 times larger than an object. (a) Where should you place the object? (b) Draw a ray diagram to locate the image. (c) Where should you place the same object so there's a virtual image 1.2 times larger than the object?

44. BIO ■■ **Dentist's mirror.** A dentist's mirror produces an upright image of a tooth, magnified three times. (a) Is the mirror concave or convex? (b) If the dentist holds the mirror 2.0 cm from the tooth to achieve this magnification, what's the mirror's focal length?

45. ■■ Looking into a 6.0-cm-diameter reflective ball, you see your face half size. (a) Is your image upright or inverted? (b) Where is the image, relative to the center of the ball? (c) How far is your face from the surface of the ball?

46. ■■ A car's outside rear-view mirror is convex, with focal length −1.0 m. In the mirror you see a truck that's actually 3.5 m tall and 10.0 m behind you. What are its apparent height and location?

47. ■■ A spherical mirror is silvered on both sides, so it's concave on one side and convex on the other. When an object is 60 cm from one side, a real image forms 30 cm from that side. (a) Find the curvature radius. (b) Describe the image if the same object is held 60 cm from the other side of the mirror.

48. ■■■ Consider the illustrations in Figures 21.7a and 21.7b. Prove using geometry that reflected rays reach the focal point $f = R/2$ in the limit as the incoming rays approach the principal axis. *Hint:* Consider the triangle formed by the radius of curvature, principal axis, and reflected ray, and use the law of sines.

Section 21.3 Refraction and Dispersion

49. ■ For which material in Table 21.2 does light travel most slowly? What is that speed?

50. ■ The speed of light in a new type of plastic is measured at 2.11×10^8 m/s. What is its refractive index?

51. ■ Find the speed of yellow light in (a) ice; (b) quartz; (c) ethanol; (d) diamond.

52. ■ A park ranger's flashlight beam strikes a lake surface. The reflected ray makes a 55° angle with the normal. What angle does the refracted ray make with the normal in the water?

53. ■■ A rectangular glass block ($n = 1.53$) lies submerged at the bottom of a tank of water (Figure P21.53). A light beam enters the water from the air above, making a 20° angle to the normal in air. Determine the angle the light ray makes with the normal in (a) the water and (b) the glass.

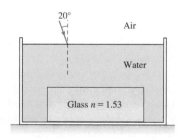

FIGURE P21.53

54. ■■ You're looking straight down into a 2.35-m-deep swimming pool. What's the apparent depth?

55. ■■ In winter a lake has a 0.25-m-thick ice layer over 1.10 m of water. A light beam from above strikes a spot on the ice, making a 30° angle with the normal. How far horizontally from that spot does the light ray strike the lake bottom?

56. ■■ A diver shines a flashlight upward toward the water surface. Describe what happens when the flashlight beam strikes the water's surface with these angles, measured from the normal: (a) 20°; (b) 40°; (c) 60°.

57. ■■ You're standing in waist-deep water with your eyes 0.52 m above the surface. A fish is swimming 0.65 m below the surface. Your line of sight to the fish makes a 45° angle with the water's surface. How far (horizontally) from you is the fish?

58. ■■ Looking directly downward from a boat, you see a shark that appears to be 5.0 m below the water's surface. What's the shark's actual depth?

59. ■■ What's the critical angle for light going from diamond into air? How does this value help explain diamond's appearance?

60. ■■ Find the speed of violet light ($\lambda = 400$ nm) and red light ($\lambda = 700$ nm) in silicate flint glass. By what percentage do the speeds differ?

61. ■■■ Light in air strikes a glass pane with $n = 1.50$, making a 35° angle with respect to the normal. (a) What angle does the light make with the normal in the glass? (b) The glass is 2.0 mm thick. As shown in Example 21.5, the ray emerging from the glass is parallel to the ray that entered the glass. Find the perpendicular distance between those parallel rays.

62. ■■■ A prism is in the shape of an equilateral triangle. An incoming beam of white light strikes the prism at 78° with respect to the normal, as shown in Figure P21.62. The refractive indices for far red (700 nm) and far violet (400 nm) are 1.62 and 1.65, respectively. Find the angles at which the two colors emerge on the other side of the prism, and compute the angular spread of the light. *Note:* For simplicity, Figure P21.62 shows a single ray of one color.

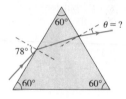

FIGURE P21.62

Section 21.4 Thin Lenses

63. ■ For a convex lens with focal length 50 cm, draw ray diagrams to show image formation for an object (a) 150 cm; (b) 80 cm; (c) 35 cm from the lens.

64. ■ Repeat the preceding problem for a concave lens with $f = -50$ cm.

65. ■ For a convex lens draw ray diagrams for the following cases: (a) $M = -0.5$; (b) $M = -1.0$; (c) $M = -2.0$.

66. ■■ A convex lens forms an image of a 12-cm-tall soda can. The image is 2.0 cm tall and appears 8.2 cm from the lens. (a) Where is the can? (b) What is the lens's focal length?

67. ■■ You're using a 20-cm-focal-length convex lens to image a small lightbulb. (a) Where should you place the bulb to get a half-size real image? Where is that image? (b) Where should you place the bulb to get a double-size real image? Where is that image?

68. ■■ A small lightbulb is 1.00 m from a screen. If you have a convex lens with 20-cm focal length, where are the two lens locations that will project an image of the lightbulb onto the screen? What's the magnification in each case?

69. ■■ A movie projector uses a single lens to project a real image on a screen 6.0 m from the lens. Each frame of the movie film is 3.0 cm tall, and the image is 1.20 m tall. (a) Should the lens be concave or convex? (b) Should the film be upright or inverted in the projector? (c) How far should the film be from the lens?

70. ■■ You're using a concave lens with $f = -5.4$ cm to read 4.0-mm-high newspaper type. How high do the type characters appear if you hold the lens (a) 1 cm; (b) 3 cm; (c) 10 cm from the newspaper?

71. ■■ Repeat the preceding problem for a convex lens with $f = +5.4$ cm, considering lens-to-newspaper distances (a) 1 cm; (b) 3 cm; (c) 5 cm.

72. ■■■ A small lightbulb is fixed to a meterstick's 0-cm mark, and at the 100-cm mark is a concave mirror with focal length 10 cm. At the 50-cm mark is a convex lens with a focal length of 20 cm. Describe the final image formed by the mirror.

73. ■■ You're using a convex lens with $f = 11.0$ cm to study an insect. How far from the insect should you hold the lens to get an upright image magnified by a factor of 1.8?

74. ■■■ You're using a 16-cm-focal-length convex lens as a magnifying glass, with $M = +3.0$. How far and in what direction (relative to the object) must you move the lens to change the magnification to -3.0?

Section 21.5 Microscopes and Telescopes

75. ■ A 4.5-cm-focal-length lens is used as a magnifying glass. (a) What's the maximum angular magnification possible for a person with near point at 25 cm? (b) Repeat for an older person with $N = 75$ cm.

76. ■ A magnifying glass has a maximum angular magnification 10.2, for a person with near point at 30 cm. What's the lens's focal length?

77. ■■ A person with near point at 25 cm views a 0.75-mm-long flea. (a) What is the flea's angular size (a) when viewed at the near point and (b) when viewed through a magnifying glass that enlarges the image 10 times?

78. ■■ A compound microscope has a 0.85-cm-focal-length objective and a 1.45-cm focal length eyepiece. (a) If the lenses are 11 cm apart, what's the microscope's magnification? Assume a 25-cm near point. (b) What should be the focal length of a new eyepiece to increase the magnification by 20%?

79. ■■ A compound microscope has magnification $M = -125$ and a 1.50-cm-focal-length objective located 11.7 cm from the eyepiece. The user has a 30-cm near point. What's the eyepiece's focal length?

80. ■ A telescope uses two convex lenses: an objective with $f = 120$ cm and an eyepiece with $f = 6.0$ cm. (a) What's the angular magnification? (b) Is the image upright or inverted?

81. ■■ Draw the ray diagram for the Galilean telescope, showing that the final image is upright.

82. ■■ (a) Explain how you would make a telescope from two convex lenses with focal lengths 1.2 cm and 24 cm. (b) What would be the telescope's angular magnification?

83. ■■■ (a) Explain why the length of a telescope using two convex lenses should be approximately equal to the sum of the lenses' focal lengths. (b) A telescope with a magnification of -40 has two lenses separated by 95 cm. Find the focal lengths of the two lenses.

Section 21.6 The Eye and Vision

84. ■ A convex lens ($n = 1.52$) has curvature radius 30 cm on both sides. (a) Find its focal length and refractive power. (b) Repeat for a lens that's concave, with 30-cm curvature on both sides.

85. ■■ A thin lens ($n = 1.58$) is convex on both sides, with curvature radii 40 cm on one side and 20 cm on the other. (a) Find the focal length. (b) Why doesn't your answer depend on which side faces the incoming light?

86. ■■ Draw a diagram like that in Figure 21.41 for a lens that's concave on both sides. Explain how the sign convention leads to a negative focal length and negative refractive power.

87. If a person's far point is at 39 cm, what should be the diopter power of corrective eyeglasses? Assume that the eyeglasses are 1.5 cm from the eye.

88. ■■ A patient is prescribed -4.5-diopter contact lenses. Where is this patient's far point?

89. ■■ Professor Deadwood has to hold a book at arm's length (65 cm from his eye) to read clearly. Find the refractive power of glasses that will reduce that distance to a more comfortable 30 cm.

90. ■■ A $+2.75$-diopter contact lens is prescribed for a patient, letting her see clearly objects as close as 25 cm. Where is her near point?

91. ■■■ A **plano-convex** lens is flat on one side and convex on the other. (a) Find a general expression for the refractive power of a plano-convex lens in terms of its refractive index n and curvature radius R on the convex side. (b) If $n = 1.65$, find the curvature radius needed for a $+5.0$-diopter lens.

92. ■■■ A **meniscus** lens is concave on one side and convex on the other. (a) Find the refractive power of a meniscus lens with $n = 1.60$ and curvature radius 30 cm on the convex side and 40 cm on the concave side. (b) Repeat for a similar lens but with the two curvature radii reversed.

General Problems

93. ■■■ Optical scientists are developing materials whose refractive indices change when an electric field is applied. One application may be zoom lenses in cell phone cameras. Suppose such a lens initially has index n_1 and focal length f_1. Find an expression for the refractive index n_2 required to triple the focal length.

94. ■■ By what factor is the image magnified when an object is placed 1.5 focal lengths from a converging lens? Is the image upright or inverted?

95. ■■ A lightbulb is 56 cm from a convex lens, and its image appears on a screen 31 cm on the other side of the lens. (a) What's the lens's focal length? (b) By how much is the image enlarged or reduced?

96. **BIO** ■■ **Squid vision.** The giant squid's dinner-plate size eye is the largest in the animal kingdom, and boasts a lens some 12 cm in diameter. That lens size, along with a light-emitting organ just below the eye, allows the squid to see in the dim ocean depths. If the retina lies 27 cm behind the lens, and the lens has refractive index 1.52, what should be the lens's curvature radius (assuming it has the same magnitude on both sides) to focus distant objects? Assume the lens is surrounded by fluid with water's refractive index—1.33—on both sides, and use a generalization of the lens-maker's equation for the case when the surrounding medium isn't air: $P_{\text{refractive}} = \dfrac{1}{f} = \left(\dfrac{n_{\text{lens}}}{n_{\text{surroundings}}} - 1\right)\left(\dfrac{1}{R_1} - \dfrac{1}{R_2}\right)$. (Actually, the lens's refractive index varies from the center to the edge of the lens—which the squid has in common with many other animals.)

97. ■■ An object is 6.0 cm from the surface of a reflecting ball, and its image appears three-quarter size. Find (a) the ball's diameter and (b) the image location.

98. ■■ A 3.0-cm-long pin is viewed through a concave lens with focal length -6.2 cm. (a) Where should you place the lens so the pin's image is 1.8 cm long? (b) Where is the image located? (c) Draw a ray diagram for this situation.

99. ■■ A candle is 84 cm from a screen. A converging lens placed between candle and screen produces a sharp image on the screen when the lens is in either of two positions. In one position, the magnification is 2 and in the other it's $1/2$. (a) Find the distance from lens to candle in each case. (b) Identify each image as either virtual or real. (c) What is the lens's focal length?

100. ■■ A 300-power compound microscope has a 4.5-mm-focal-length objective lens. If the distance from eyepiece to objective is 10 cm, what should be the focal length of the eyepiece?

101. ■ ■ ■ A camera can normally focus as close as 60 cm, but additional lenses mounted in front of the main lens can provide close-up capability. What type and power of auxiliary lens will allow the camera to focus as close as 20 cm? The distance between the lenses is negligible.

102. ■ ■ ■ Two vertical plane mirrors are placed with a 72° angle between them. You stand a nail upright, on the line bisecting the mirrors. How many images of the nail appear? Draw a ray diagram showing the image formation.

103. BIO ■ ■ ■ **Contact lenses.** A contact lens is in the shape of a convex meniscus (see Problem 92). The inner surface is curved to fit the eye, with curvature radius 7.80 mm. The lens is made from plastic with $n = 1.56$. If the lens is to have a focal length of 44.4 cm, what should be the curvature radius of its outer surface?

Answers to Chapter Questions

Answer to Chapter-Opening Question
Laser surgery reshapes the cornea, changing its refractive power in order to focus images on the eye's retina.

Answers to GOT IT? Questions
Section 21.2   (c) Enlarged, inverted, real
Section 21.3   (b) Bends away from the normal
Section 21.4   (b) $2f > d_o > f$
Section 21.5   (d) 2 cm

■ How does light's wave nature help explain why Blu-ray discs hold full-length high-definition movies, which require more than five times the information that fits on an ordinary DVD?

When light interacts with objects comparable in size to its wavelength, we can no longer ignore light's wave nature. Then we need to consider the phenomena of interference and diffraction. Here we'll see how interference enables very sensitive measurements and how diffraction lets us separate light by wavelengths. At the same time we'll see how diffraction limits our ability to image small or distant objects. We'll also explore the polarization of light, the phenomenon behind the LCD displays in our cell phones, iPods, and TVs, and which provides scientists information on everything from the structure of rocks to the distant cosmos.

22.1 Interference

In Chapter 20 you found that light is an electromagnetic wave. As described for other waves in Section 11.2, light undergoes **interference** when two or more waves meet in the same place. Here's a quick review; see also Figures 11.5 and 11.6:

■ Interference is governed by the **superposition principle**: When two or more waves interfere, the net wave has a displacement equal to the sum of the individual wave displacements.

■ When crests meet crests and troughs meet troughs, the result is **constructive interference**, giving a wave with larger amplitude than either individual wave.

■ When crests and troughs meet, **destructive interference** occurs and the waves tend to cancel.

Figure 22.1 shows a naïve attempt to observe interference of light; try this and you'll find the flashlight beams go right through each other with no obvious interference. Why didn't this work, when a similar experiment with two sources of water waves or sound waves (as in Figure 11.4) shows a pattern of constructive and destructive interference?

There are two reasons. First, a stable interference pattern requires that the sources be **coherent**, meaning they have exactly the same wavelength and maintain the same phase relation. But your flashlights produce white light—a mix of all visible wavelengths (Section 20.2)—and there's no specific phase relation between waves from the two flashlights. Although white-light interference is observed under certain conditions, it's easier to see interference when sources are **monochromatic**—producing light of a single wavelength. And light from two distinct sources won't work; we have to split light from a single source to produce two coherent beams. We achieved coherence with sound waves by driving two loudspeakers with the same electrical signal—but that's harder to do with light.

The second reason light interference is difficult to observe is the short wavelength—400 nm to 700 nm for visible light. You can easily detect interference of water waves with centimeter-scale wavelengths, or sound waves with wavelengths from centimeters to meters. But light's wavelength is smaller than 10^{-6} m. To see interference, you'll have to look closely! In Chapter 21 you got away with the approximation of geometrical optics, which ignores interference, because you weren't looking closely enough that the wave nature of light was evident.

Beams pass through one another.

FIGURE 22.1 Incoherent light beams pass through each other, with no obvious interference.

✓**TIP**

Remember that audible sound has wavelengths ranging from 1.7 cm to 17 m, while visible light ranges from 400 nm to 700 nm.

The Michelson Interferometer

Section 20.3 described Albert A. Michelson's attempt to detect variations in the speed of light due to Earth's motion through the ether. The apparatus he used is the **Michelson interferometer**, diagramed in Figure 22.2. Although Michelson invented the device for his ether experiments, Michelson interferometers are widely used even today for precision measurements.

The interferometer begins with a monochromatic light source. Then comes the key element: a partially silvered mirror. It's silvered just enough that about half the incident light passes through and half is reflected. Thus it's called a **beam splitter**. Light that passes through the beam splitter follows a straight path to mirror M_1. Light reflected by the beam splitter goes to mirror M_2. Mirrors M_1 and M_2 reflect light back to the beam splitter. Again, half the returning light in each beam is reflected and half passes through. The light heading to the observer is therefore a recombination of the two beams after they've traveled the different paths. If the paths are identical, they interfere constructively. But if one beam goes just half a wavelength farther, they're out of phase and interfere destructively. Because the beams spread slightly, light actually travels many paths of slightly different lengths, resulting in an interference pattern with alternating bands of constructive and destructive interference (also shown in Figure 22.2). This picture is beautiful evidence of light waves interfering!

Recall that Michelson was attempting to determine Earth's motion relative to the ether, the supposed medium for light waves. With Earth moving at 30 km/s as it orbits the Sun, he reasoned there would be an "ether wind" of this speed blowing past Earth—analogous to the wind you feel when you bicycle through still air. Depending on the orientation of his interferometer, light traveling on the two paths would be affected differently by the ether wind. Michelson's experiment consisted of watching the interference pattern as he rotated the apparatus. He expected the pattern to shift, due to the small changes in the speed of light caused by the changing direction of the ether wind. His apparatus was more than sensitive

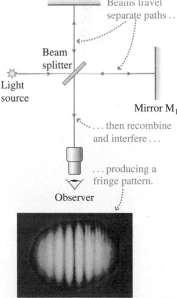

FIGURE 22.2 Schematic diagram of the Michelson interferometer and photograph of the resulting interference pattern.

enough to detect such changes, but they were never seen. Michelson repeated the experiment throughout the year, as the direction of Earth's orbital motion changed, but there was never any shift in the interference pattern. As discussed in Chapter 20, Einstein's 1905 theory of special relativity resolved this conundrum by declaring the ether a fiction and the speed of light in vacuum the same in all inertial reference frames.

CONCEPTUAL EXAMPLE 22.1   **Interferometric Measurement**

The Michelson interferometer provides an exquisitely precise instrument for measuring small distance changes. Suppose mirror M_2 moves slightly, shifting the interference pattern so a bright line moves to where the adjacent dark line was. How far did the mirror move? Your interferometer uses monochromatic 640-nm red light.

SOLVE The bright line results when the two beams return in phase, interfering constructively. To change to the adjacent dark line, the beam returning from M_2 has to be out of phase with the beam from M_1. Therefore, light on the M_2 leg has to travel an extra distance $\lambda/2$. Because the light makes a round trip to M_2, the mirror itself only has to move by $\lambda/4$ (Figure 22.3). With the 640-nm light, that's a mere 160 nm.

REFLECT It doesn't matter which way M_2 moves; moving $\lambda/4$ in either direction changes constructive interference to destructive. An optical interferometer is a great way to detect small changes. Here that change is in position; in Michelson's experiment it was an expected change in travel time due to differences in the speed of light. Either way, interferometry provides exquisite precision.

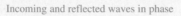
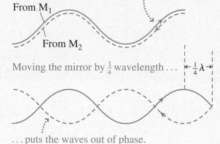

FIGURE 22.3 Changing from constructive to destructive interference requires a shift of half a wavelength, so it's necessary to move the mirror by only $\lambda/4$.

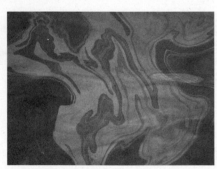

Interference fringes in an oil film on water

(a)

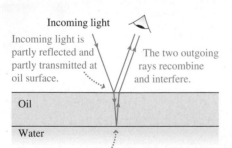

(b)

FIGURE 22.4 (a) Interference in an oil film illuminated with white light. The film's thickness varies substantially over the surface, resulting in the irregular pattern seen here. (b) Illustrating thin-film interference for an oil film on water.

Thin-Film Interference

It's a rainy day, and you see bands of color from a puddle in the oil-smeared parking lot. This is **thin-film interference**, resulting from a film of oil floating atop the puddle. The interference is between light waves reflected from the top of the oil and waves reflected from the oil-water boundary below (Figure 22.4). When the film's thickness is just right, you get constructive interference. Why do you see different colors? Color corresponds to wavelength, and as the oil thickness or viewing angle varies, the conditions at different points are right for constructive interference with different colors.

Quantitative analysis of thin-film interference requires a deeper understanding of reflection at boundaries between materials. Figure 22.5 illustrates the two rules that govern such reflection:

1. When light traveling in a medium with lower refractive index reflects from a boundary with a medium of higher index, the reflected wave is 180° out of phase with the incident wave.
2. When light traveling in a medium with higher refractive index reflects from a boundary with a medium of lower index, the reflected wave is in phase with the incident wave.

These behaviors are analogous to the reflection of waves on strings when the string end is, respectively, fixed or free, as in Figure 11.8.

With these rules, you can analyze thin-film interference like that shown in Figure 22.4. Table 21.2 gives water's refractive index as 1.33, and the index for oil is higher—typically around 1.5. Therefore, light reflecting from the top surface of the oil undergoes a 180° phase shift, while light reflecting from the oil-water boundary below suffers no phase shift. Suppose the oil thickness is $\lambda/4$ (Figure 22.6a). Some light is reflected from the top of the oil, while some enters the oil and, of that, some is reflected at the oil-water interface. Some of that light emerges from the top surface having traveled a round-trip distance $\lambda/2$ relative to the wave reflected from the top surface. This extra distance, combined with the 180° phase shift at the top interface, is just right to bring the two waves back into phase, resulting in *constructive interference*. If the oil thickness is $\lambda/2$, on the other hand, the round-trip distance in the oil is λ. The 180° phase shift then puts the two waves out of phase, resulting in *destructive interference* (Figure 22.6b).

There are two subtle points here. First, the wavelength λ we're talking about is the wavelength *in the oil*. As you learned in Chapter 20, that's given by $\lambda_{oil} = \lambda_0/n_{oil}$, where λ_0 is the wavelength in vacuum. The corrected wavelength λ is the one that determines the oil thickness required for constructive or destructive interference. Second, light must be incident essentially normal to the surface if our criteria for the film thickness are to hold. If light comes in at an angle, then the round-trip path will be longer than twice the thickness, and it's that longer path that gets compared with the wavelength. That's why our constructive and destructive interference conditions are met at different viewing angles for different wavelengths.

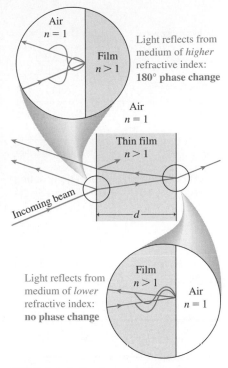

FIGURE 22.5 Reflection results in a phase shift or no phase shift, depending on the relative refractive indices of the two media.

EXAMPLE 22.2   **Oily Interference**

Red light ($\lambda_0 = 650$ nm) is incident normally on a thin oil film ($n = 1.52$) on water. (a) What minimum oil thickness gives constructive interference? (b) What other thicknesses will also result in constructive interference?

ORGANIZE AND PLAN You've just seen that constructive interference results when the oil thickness is $\lambda/4$. The wavelength of light in oil is given by $\lambda = \lambda_0/n$.

Known: $\lambda_0 = 650$ nm, $n = 1.52$.

SOLVE (a) The minimum oil thickness is $d = \lambda/4$, so

$$d = \frac{\lambda}{4} = \frac{\lambda_0}{4n} = \frac{(650 \text{ nm})}{4(1.52)} = 107 \text{ nm}$$

(b) Add a whole wavelength to the travel distance—a half-wavelength to the film thickness—and you won't change the phase. Therefore, a film thickness $\lambda/4 + \lambda/2 = 3\lambda/4$ will also work:

$$d = \frac{3\lambda}{4} = \frac{3\lambda_0}{4n} = \frac{3(650 \text{ nm})}{4(1.52)} = 321 \text{ nm}$$

Every additional $\lambda/2$ in thickness also gives constructive interference; thus $d = \lambda/4, 3\lambda/4, 5\lambda/4$, and so on will all work.

REFLECT The 107-nm minimum thickness isn't unreasonable because oil on water spreads into a thin layer. A thickness of $\lambda/2 = 214$ nm gives destructive interference for this wavelength. However, with white light you're likely to see constructive interference for some other color at that thickness. So you see a bright array of colors from an oil film, whenever the film thickness or viewing angle varies.

MAKING THE CONNECTION   Is the minimum oil thickness for constructive interference of 430-nm violet light larger or smaller than for red light? By how much?

ANSWER The shorter wavelength of violet light implies a thinner film. Going through the same procedure with the violet wavelength gives a minimum thickness of just 71 nm.

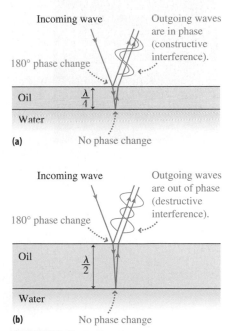

FIGURE 22.6 Thin-film interference, resulting in (a) constructive and (b) destructive interference.

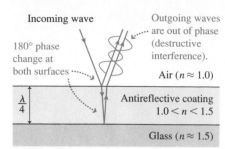

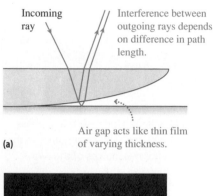

FIGURE 22.7 How an antireflection coating works.

Antireflection Coatings

Thin-film interference is pretty, as in oil slicks and soap bubbles, but is it useful? Yes: Destructive interference in thin films minimizes reflection and therefore maximizes transmission of light into the material behind the film. **Antireflection coatings** use this effect to maximize light collection in camera lenses and solar photovoltaic cells, and they eliminate glare from eyeglasses and telescope optics. Figure 22.7 shows an antireflection coating applied to glass. The coating's refractive index is higher than air's but lower than that of glass. That gives a 180° phase change at each surface, so we want the total path lengths to differ by $\lambda/2$ for destructive interference—giving a film thickness $\lambda/4$. A detailed analysis based on electromagnetic theory shows that a coating index equal to $n_{coating} = \sqrt{n_{glass}}$ gives exactly zero reflection. But that's only true at the one wavelength where the film thickness is $\lambda/4$, so actual antireflection coatings produce significant but not total reduction in reflection across the visible spectrum.

Newton's Rings

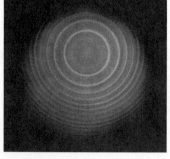

(a)

(b)

FIGURE 22.8 (a) A lens on a flat glass plate; (b) Newton's rings arise from the difference in path lengths for the two rays.

Placing a curved piece of glass on flat glass gives an interference pattern called **Newton's rings**, illustrated in Figure 22.8. Here the air gap between glass surfaces acts like a thin film of varying thickness. Interference between light reflected from the curved and flat surfaces produces the alternating pattern of constructive and destructive interference shown in Figure 22.8b. You can see Newton's rings clearly using white light, so this is one example of interference that doesn't require monochromatic light. Analysis of Newton's rings can provide precision measurements of the shapes and radii of lenses.

✓ TIP

If the bottom glass isn't flat, or the top glass isn't spherical or some other symmetric shape, Newton's rings appear irregular, not circular.

Isaac Newton believed that light consisted of particles, not waves, and he struggled to find a satisfactory explanation of the rings that bear his name. The controversy between the particle and wave models of light continued after Newton's time. Although interference phenomena solidly confirm the wave model, we'll see in later chapters how quantum physics blurs the distinction between waves and particles.

GOT IT? Section 22.1   Two light waves of the same wavelength are out of phase. Which of the following phase differences will result in destructive interference when the two waves meet? (a) $\lambda/2$; (b) 2λ; (c) $3\lambda/2$; (d) 3λ.

22.2 Double-Slit Interference

In 1801 the English physician Thomas Young (1773–1829) performed an experiment that provided definitive interference-based evidence for the wave nature of light. Young's experiment is conceptually simple, and with today's lasers it's easy to perform in undergraduate physics labs.

The Experiment

A modern Young's experiment uses monochromatic laser light and a pair of narrow, closely spaced parallel slits, as shown in Figure 22.9a. Young himself used sunlight, and split his beam with a thin card, but the net effect was similar. Figure 22.9a shows that each slit acts as a source of light waves, which spread in a circular pattern like ripples on a pond. This is an example of *Huygens' principle*, which we'll explore in Section 22.3.

The spreading light fills the region beyond the slits, but light from each slit generally travels a different distance to reach a given point, so light from the two slits generally meets out of phase. One exception is the mid-line of the apparatus, where light travels

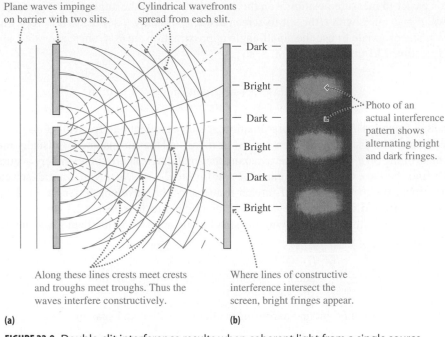

Plane waves impinge on barrier with two slits.

Cylindrical wavefronts spread from each slit.

Photo of an actual interference pattern shows alternating bright and dark fringes.

Along these lines crests meet crests and troughs meet troughs. Thus the waves interfere constructively.

Where lines of constructive interference intersect the screen, bright fringes appear.

(a)

(b)

FIGURE 22.9 Double-slit interference results when coherent light from a single source passes through closely spaced slits.

the same distance from each slit and is therefore in phase—resulting in constructive interference. A screen placed beyond the slits therefore is illuminated brightly at its midpoint. There's also constructive interference when the path difference is a full wavelength, resulting in other bright regions on the screen. Between are dark regions, corresponding to points where light paths from the two slits differ by a half wavelength, resulting in destructive interference. Figure 22.9b indicates the regions of constructive and destructive interference and shows how they produce a pattern of alternating light and dark bands called **interference fringes** on the screen. You might recall Conceptual Example 11.2's exploration of "dead spots" in the sound produced by a pair of loudspeakers. The speakers were analogous to the slits of our light-based experiment, and the "dead spots" were points of destructive interference. We showed an analogous interference pattern for water waves in Figure 11.4.

Quantitative Analysis

Figure 22.10 shows how the difference Δr in path lengths from the two slits to a point P is given by $\Delta r = d \sin \theta$, where d is the slit spacing and θ is the angle to P measured from the centerline. We'll get a bright fringe whenever Δr is a whole number of wavelengths—the condition for constructive interference. That is, when $\Delta r = n\lambda$, with n an integer. Equating our two expressions for Δr,

$$n\lambda = d \sin \theta \; (n = 0, 1, 2, \ldots) \quad \text{(Condition for bright fringes; SI unit: m)} \quad (22.1)$$

The integer n is the **order** of the interference maximum. The central fringe corresponds to $n = 0$; on either side are first-order fringes with $n = 1$, then second-order $n = 2$ fringes, and so on.

✓**TIP**

Don't confuse the interference order n with the refractive index n. We won't use the two symbols in the same context.

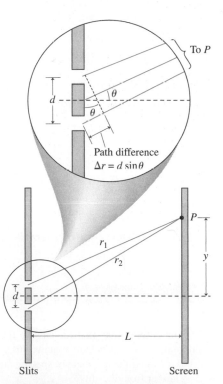

Path difference $\Delta r = d \sin \theta$

FIGURE 22.10 Geometry for double-slit interference. In the blowup you can see that for $L \gg d$, paths to P are nearly parallel and differ by distance $d \sin \theta$.

It's easier to measure position y on the screen rather than the angle θ. As Figure 22.10 shows, $y = L \tan\theta$, with L the slit-to-screen distance. Usually $y \ll L$, so θ is very small. Then it's appropriate to use the small-angle approximation $\sin\theta \approx \tan\theta$, so $y \approx L\sin\theta$. But Equation 22.1 gives $\sin\theta = n\lambda/d$, so

$$y = \frac{n\lambda L}{d} \qquad \text{(Positions of bright fringes; SI unit: m)} \qquad (22.2)$$

Although Equation 22.2 is an approximation, it's an excellent one for typical double-slit experiments. The distance L is often a meter or more, and the slit spacing d is usually much less than 1 mm. So $\theta \sim 10^{-3}$ radians, making the small-angle approximation very accurate.

To find the dark fringes, you can still use Figure 22.10. Now you want destructive interference, so Δr must be a half-integer multiple of the wavelength: $\Delta r = n\lambda/2$, with n restricted to odd integers. The condition analogous to Equation 22.1 then becomes $n\lambda/2 = d\sin\theta$, with n odd. Applying the small-angle approximation then gives the dark-fringe positions:

$$y = \frac{n\lambda L}{2d} \, (n \text{ odd}) \qquad \text{(Positions of dark fringes; SI unit: m)} \qquad (22.3)$$

In this approximation bright and dark fringes alternate with equal spacing.

EXAMPLE 22.3 **Finding the Fringes**

A double-slit experiment uses 633-nm red laser light and slits separated by 0.125 mm. The interference pattern appears on a screen 2.57 m away. Find the positions of the first three orders of bright fringes.

ORGANIZE AND PLAN Figure 22.11 is our sketch. The distance from the central maximum to the nth-order bright fringe is given by Equation 22.2:

$$y = \frac{n\lambda L}{d}$$

The first three fringes correspond to $n = 1, 2,$ and 3.

Known: $\lambda = 633$ nm; $d = 0.125$ mm; $L = 2.57$ m.

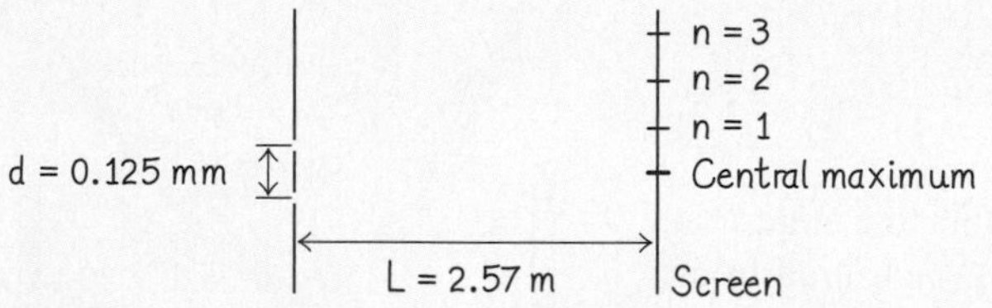

$$
\begin{array}{c}
\hline
\quad | \text{ --- } n = 3 \\
\quad | \text{ --- } n = 2 \\
\quad | \text{ --- } n = 1 \\
d = 0.125 \text{ mm } \updownarrow| \quad \text{ --- Central maximum} \\
\longleftarrow L = 2.57 \text{ m} \longrightarrow| \quad \text{Screen}
\end{array}
$$

FIGURE 22.11 Our sketch for Example 22.3.

SOLVE The first-order bright fringe corresponds to $n = 1$, so its position is

$$y_1 = \frac{n\lambda L}{d} = \frac{(1)(633 \times 10^{-9}\,\text{m})(2.57\,\text{m})}{1.25 \times 10^{-4}\,\text{m}} = 0.0130\,\text{m} = 1.30\,\text{cm}$$

Similar calculations give $y_2 = 2.60$ cm for second order, and $y_3 = 3.90$ cm for third order, all measured from the central maximum.

REFLECT Once we found the first-order fringe, we didn't really need Equation 22.2 for the higher-order fringes, since they're evenly spaced in the small-angle approximation. With 1.3 cm between fringes, the interference pattern is easily seen on the screen.

MAKING THE CONNECTION In this example, how far is the first dark fringe from the central maximum?

ANSWER The distance is just half the distance to the first bright fringe, or 0.65 cm. You could also find that using Equation 22.3, with $n = 1$.

CONCEPTUAL EXAMPLE 22.4 **Color and Spacing**

How would the interference pattern in Example 22.3 change with the following alterations: (a) decreasing the distance between slits; (b) retaining the original slit spacing but using green light instead of red; (c) retaining the original slit spacing and red laser but moving the screen farther from the slits?

SOLVE We've begun by sketching a sample two-slit interference pattern for reference (Figure 22.12). The remainder of Figure 22.12 shows solutions for the three cases.

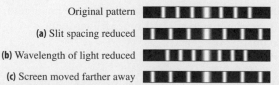

Original pattern

(a) Slit spacing reduced

(b) Wavelength of light reduced

(c) Screen moved farther away

FIGURE 22.12 Diffraction patterns for the cases described in Conceptual Example 22.4.

cont'd.

(a) By Equation 22.2, decreasing the slit spacing d increases the distance y for any order. The interference pattern spreads wider, with a larger distance between bright fringes.

(b) Green light has a shorter wavelength than red. The distance between bright fringes is smaller, by the wavelength ratio.

(c) Moving the screen away also widens the pattern, since $y \propto L$. You can think of the constructive interference lines extending outward in a wedge of fixed angular width; the farther the screen, the wider the wedge.

REFLECT For a given light source and fixed slits, the easiest way to enlarge the interference pattern is to move the screen. Move it too far, though, and the pattern gets dim and fuzzy.

EXAMPLE 22.5   **Interferometric Measurement, Again**

The tiny distance between slits can be difficult to measure. But you can use the interference pattern to find the spacing. Suppose you illuminate the slits with 589-nm yellow sodium light and observe a 4.50-cm bright-fringe spacing on a screen 1.75 m away. What's the slit spacing?

ORGANIZE AND PLAN Equation 22.2 gives the positions of bright fringes: $y = n\lambda L/d$. The distance between successive fringes is constant and equal to the first-order position: $y = \lambda L/d$. We can then solve for the slit spacing d.

Known: $\lambda = 589$ nm; $y = 4.50$ cm; $L = 1.75$ m.

SOLVE Solving for the slit spacing,

$$d = \frac{\lambda L}{y} = \frac{(589 \times 10^{-9}\,\text{m})(1.75\,\text{m})}{4.50 \times 10^{-2}\,\text{m}} = 2.29 \times 10^{-5}\,\text{m} = 22.9\,\mu\text{m}$$

REFLECT Compare this with the spacing in Example 22.3. Now the screen is closer and the wavelength smaller, yet the distance between fringes is much larger than in Example 22.3. That implies a much smaller slit spacing, as you've found. You'd have a hard time measuring this spacing directly without a microscope.

MAKING THE CONNECTION In this example, what's the angle θ for the first-order maximum? Is the small-angle approximation justified?

ANSWER Using $\tan\theta = y/L$ gives $\theta = 0.0257$ rad $= 1.5°$, so the small-angle approximation is justified. At this level there's virtually no difference between θ (in radians), $\sin\theta$, and $\tan\theta$.

Reviewing New Concepts

- When light waves interfere, constructive interference leads to an increase in brightness and destructive interference to a decrease in brightness.
- When waves interfere in phase, constructive interference occurs; when they're out of phase, interference is destructive.
- The double-slit experiment uses the interference of coherent waves from two slits to create an interference pattern of alternating bright and dark fringes.

GOT IT? Section 22.2 The three fringe patterns shown occur with the same double-slit arrangement and screen. The light source is white, and the three patterns are obtained using red, yellow, and blue filters. Which pattern corresponds to which color?

22.3 Diffraction

Diffraction is the bending of waves as they pass by an obstacle or through an opening. Diffraction occurs with mechanical as well as electromagnetic waves, and we'll use a water wave example to introduce the basic principle behind diffraction.

✓ **TIP**

Don't confuse *diffraction* with *refraction*. Diffraction occurs in a single medium; refraction (Chapter 21) occurs when waves pass from one medium to another.

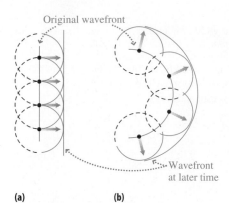

Original wavefront

Wavefront at later time

(a)   (b)

FIGURE 22.13 Each point on a wavefront acts like a source of circular "wavelets." This explains the advance of (a) straight and (b) circular wavefronts.

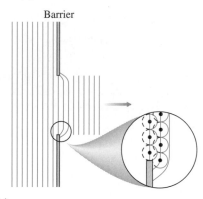

Barrier

(a)

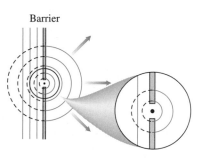

Barrier

(b)

FIGURE 22.14 Waves passing through gaps show (a) negligible diffraction when the gap is large compared with the wavelength and (b) significant diffraction when the gap is small.

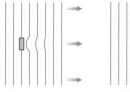

(a) Barrier width similar to wavelength   (b) Barrier much wider than wavelength

FIGURE 22.15 (a) Waves readily diffract around smaller objects, but (b) larger objects cast shadows where waves don't reach. "Small" and "large" are compared with the wavelength.

Huygens' Principle and Diffraction

Drop a pebble in a pond and circular waves spread outward. That's because you disturbed the medium at a single point. You can think of a continuous wavefront as a whole series of such disturbances, each generating circular waves. Normally, interference between the individual waves results in the wavefront simply propagating forward, as Figure 22.13 suggests. This idea that each point on a wavefront acts like a source of circular waves is **Huygens' principle**, after the Dutch physicist Christian Huygens (1629–1695), who first suggested that light is a wave. We invoked Huygens' principle in double-slit interference when we treated each of the narrow slits as a source of circular waves.

Figure 22.14a shows what happens when waves pass through a gap in a barrier. Near the edges there's no adjacent wavefront, so the circular Huygens' "wavelets" spread sideways beyond the barrier, causing the wavefront to bend; this is diffraction. In Figure 22.14a diffraction isn't very significant, because the gap is much wider than the wavelength. But in Figure 22.14b the gap is small compared with the wavelength, and diffraction results in circular waves beyond the gap. As Figure 22.14 suggests, diffraction always occurs when waves pass through a gap or aperture, but it's significant only when the gap is comparable in size to or smaller than the wavelength.

Suppose you're in a classroom while others are talking in the hallway around the corner from the door. You can hear them easily, because doors and halls are comparable in width to the wavelength of spoken sound—about 1 m for 300-Hz sound. That means sound waves diffract readily when passing through the doorway. But you can't see your friends in the hallway. That's because visible light's wavelength is less than 10^{-6} m, and so diffraction of light is insignificant.

 TIP

The wavelength of a 300-Hz sound wave—a typical audible frequency—is just over 1 m. But light's wavelength is less than 1 μm.

Figure 22.15 shows waves interacting with solid objects—the opposite of the gap situation in Figure 22.14. Again, diffraction occurs at the object's edges, and if the object is comparable in size to the wavelength, diffraction "fills in" the region behind the object. But if the object is much larger, there's a "shadow" behind the object, where waves can't penetrate. As a consequence, it's impossible to image objects whose size is significantly less than the wavelength of the light being used.

CONCEPTUAL EXAMPLE 22.6   AM versus FM

Recall from Chapter 20 that AM radio uses frequencies ranging from 540 kHz to 1600 kHz, while FM uses 88–108 MHz. Which type of radio is likely to be diffracted around obstacles such as hills and buildings?

SOLVE Waves diffract significantly around objects only when the object size is comparable to or smaller than the wavelength. Radio waves are electromagnetic, with wavelength $\lambda = c/f$. Using this equation gives the wavelength ranges of AM and FM radio:

$$\text{AM: 188 m to 556 m}$$

$$\text{FM: 2.8 m to 3.4 m}$$

The answer is clear. AM radio waves have much greater wavelength than FM, so AM diffracts more readily around hills and buildings, which are typically tens to hundreds of meters in size. But these obstacles cast substantial "shadows" for FM signals, which can block or weaken reception for nearby listeners.

REFLECT It's worse for broadcast TV, which uses still higher frequencies. That's why you may need a direct line of sight from your TV antenna to a station's broadcasting antenna if you get your TV over the air rather than through cable.

Diffraction Gratings

Diffraction is essential in two-slit interference, since diffraction at each slit results in the interfering circular wavefronts. In fact, two-slit interference is sometimes called two-slit diffraction. Interference also arises with multiple slits, and here the term diffraction is generally used.

A **diffraction grating** is a device designed for multiple-slit interference. The grating usually consists of a transparent material inscribed with a great many close, uniformly spaced parallel lines. The spaces between the lines function as slits, and light passing through interferes as in a two-slit system. But now there are typically hundreds or thousands of slits. The geometrical argument in Figure 22.10 still holds, giving the same condition for the formation of bright fringes:

$$n\lambda = d\sin\theta \ (n = 0, 1, 2, \ldots) \quad \text{(Bright fringes from diffraction grating; SI unit: m)} \quad (22.4)$$

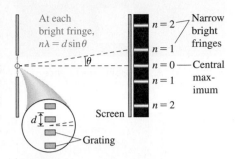

FIGURE 22.16 Interference pattern from a diffraction grating.

But because the bright fringes involve light from multiple slits, they're brighter and narrower than those from two slits. Between the bright fringes are darker areas, punctuated by weaker fringes resulting from interference that's partially constructive. If you increase the number of slits, the principal bright fringes become sharper and narrower, and the areas in between grow darker (Figure 22.16). With the large number of slits in a typical diffraction grating, you see just very narrow and very bright fringes at the angles given by Equation 22.4.

Because the condition in Equation 22.4 depends on wavelength, a diffraction grating sends different wavelengths out at different angles, thus separating light into its constituent colors. Diffraction gratings are therefore used in **spectroscopy**, which studies physical processes by analyzing the wavelengths of light emitted or absorbed. Each type of atom and molecule emits and absorbs a characteristic spectrum of wavelengths that identify it—whether it's in the lab or from a galaxy a billion light years distant. In Chapter 24 you'll see that atomic spectra played a crucial role in the development of atomic models.

In a grating spectrometer (Figure 22.17) light passes through an entrance slit and then a diffraction grating. Beyond the grating is a detector that records the amount of light incident at each angle. The grating spacing then gives precise values of wavelengths present. Those wavelengths can be compared with known standards to determine the composition of the light source. Deviations from standard wavelengths result from Doppler shift and allow astronomers, for example, to determine motions of distant astrophysical objects.

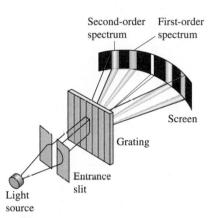

FIGURE 22.17 Essential elements of a grating spectrometer. An electronic detector is normally used in place of the screen.

EXAMPLE 22.7 | **Hydrogen's Spectrum**

When hydrogen is heated or subject to an electric discharge, it emits light with three specific visible wavelengths: 434 nm (violet), 486 nm (blue-green), and 656 nm (red). This light is sent through a diffraction grating with line spacing $d = 1.20\ \mu m$. At what angles do you see the first-order fringes? Do you see any second-order fringes, and if so, at what angles?

ORGANIZE AND PLAN The angles can all be computed using Equation 22.4: $n\lambda = d\sin\theta$.

Known: $\lambda_{violet} = 434\ nm$, $\lambda_{blue-green} = 486\ nm$, $\lambda_{red} = 656\ nm$; $d = 1.20\ \mu m$.

SOLVE Solving Equation 22.4 for $\sin\theta$ gives $\sin\theta = n\lambda/d$, so $\theta = \sin^{-1}(n\lambda/d)$. The shortest wavelength gives the smallest angle. The first-order angles are:

Violet: $\theta_{violet} = \sin^{-1}\left(\dfrac{n\lambda_{violet}}{d}\right)$

$= \sin^{-1}\left(\dfrac{434 \times 10^{-9}\ m}{1.20 \times 10^{-6}\ m}\right) = 21.2°$

Blue-green: $\theta_{blue-green} = \sin^{-1}\left(\dfrac{n\lambda_{blue-green}}{d}\right)$

$= \sin^{-1}\left(\dfrac{486 \times 10^{-9}\ m}{1.20 \times 10^{-6}\ m}\right) = 23.9°$

Red: $\theta_{red} = \sin^{-1}\left(\dfrac{n\lambda_{red}}{d}\right) = \sin^{-1}\left(\dfrac{656 \times 10^{-9}\ m}{1.20 \times 10^{-6}\ m}\right) = 33.1°$

Similar calculations with $n = 2$ give 46.3° and 54.1° for the violet and blue-green, respectively. But the calculation for the red wavelength gives $\theta_{red} = \sin^{-1} 1.09$. Since the sine of an angle must be 1 or less, this means there is no second-order fringe for the red wavelength.

REFLECT The angular spacing between the three first-order lines is large enough that they can be distinguished easily. Furthermore, there's no overlap between first- and second-order lines, which could cause confusion. (See Problem 52.)

MAKING THE CONNECTION Are any third-order hydrogen lines visible with this grating?

ANSWER No. With $n = 3$, not even the shortest wavelength gives $\sin\theta \leq 1$.

FIGURE 22.18 A single-slit interference pattern.

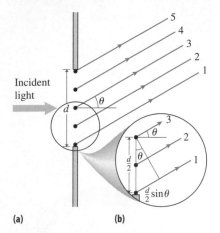

FIGURE 22.19 Geometry for single-slit diffraction.

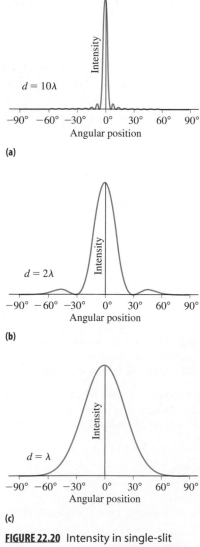

FIGURE 22.20 Intensity in single-slit diffraction, for three different slit widths.

Reflection Gratings

The gratings we've described are *transmission gratings*, because light goes through them. *Reflection gratings* are similar, but consist of a shiny surface ruled with parallel lines. You've probably noticed colors when light reflects from the underside of a CD or DVD, as shown in the photo on the first page of this chapter. That's a result of diffraction from the closely spaced tracks—about 1 μm apart—that store information on the disk. As you look at different regions on the disk, you're viewing light reflected at slightly different angles. Each angle corresponds to constructive interference for a different color, as in a transmission grating.

Single-Slit Diffraction

Figure 22.14a showed that light passing through a gap many wavelengths wide suffers essentially no significant diffraction, while Figure 22.14b showed that a narrow gap with width $d \ll \lambda$ produces circular wavefronts. But what if the gap width is comparable to the wavelength? Then the gap behaves something like a multiple-slit system, with each point in the gap becoming a source of circular waves. The resulting waves subsequently interfere to produce what's called **single-slit diffraction**, shown in Figure 22.18.

Figure 22.19 shows the geometry of single-slit diffraction. In this case it's easier to locate the *dark* interference fringes, where *destructive* interference occurs. Consider two points in the slit, separated by half the slit width (Figure 22.19b). Huygens' principle says that each point acts as a new source of circular waves. For light from these two sources to interfere destructively, the minimum path difference $(d/2) \sin \theta$ shown in Figure 22.19b is half a wavelength: $(d/2) \sin \theta = \lambda/2$, or $\lambda = d \sin \theta$. But every pair of points within the slit that are half a slit width apart meets the same criteria. Therefore, the relation $\lambda = d \sin \theta$ is true in general for the first dark interference fringe.

For the second-order fringe, consider pairs of points that are $d/4$ apart. Light from such pairs interferes destructively if the path difference, now $(d/4) \sin \theta$, is half a wavelength: $(d/4) \sin \theta = \lambda/2$, or $2\lambda = d \sin \theta$. Again, this holds for all point pairs $d/4$ apart, so it's a general expression for the nth-order dark fringe. Continue for successive orders, and you find

$$n\lambda = d \sin \theta \ (n = 1, 2, \dots) \qquad \text{(Dark fringes from a single slit; SI unit: m)} \qquad (22.5)$$

There's no $n = 0$ dark fringe, because there's still a central intensity maximum in this interference pattern.

Finding the intensity in single-slit diffraction requires a calculation involving the phases of the electric fields in light waves from all parts of the slit. Figure 22.20 shows the result for several ratios of slit width to wavelength. Note that the diffraction pattern broadens as the slit narrows—or, equivalently, as the wavelength increases. We'll see next how this effect limits our ability to image distant or small objects.

The Diffraction Limit

So far we've considered diffraction by a rectangular slit. Similar analysis applies to other shapes, including circular apertures that admit light to telescopes, microscopes, and our eyes. Here the diffraction pattern consists of concentric circular rings surrounding a disk-shaped central maximum (Figure 22.21). Equation 22.5 isn't quite right in circular geometry; instead, the angular position of the first-order dark ring for diffraction through a circular aperture of diameter D is given by

$$\sin \theta = 1.22 \frac{\lambda}{D}$$

Diffraction "smears out" the light that enters our eyes, telescopes, and microscopes, limiting our ability to form clear images or to resolve closely spaced objects. If the central maxima in the diffraction patterns of light from two distinct objects overlap, then we won't be able to distinguish them. The same applies to light from two points on the

Animal Iridescence

Some animals are equipped with natural reflection gratings. Bird feathers and butterfly wings, for example, have narrow, closely spaced ridges that reflect different colors at different angles. This effect is called **iridescence**. Some insects and shellfish also exhibit iridescence.

FIGURE 22.21 Diffraction pattern produced by a circular aperture.

same object; in that case we won't be able to image the object clearly. Figure 22.22b shows that we can just barely resolve two light sources if the peak of one central maximum lies on the first minimum of the other. This is the **Rayleigh criterion**, named for the English physicist John William Strutt, Lord Rayleigh (1842–1919). The Rayleigh criterion is generally important only when we're dealing with very small angles; then $\sin \theta \approx \theta$ in our expression for the first-order dark fringe. Since that angle θ is the separation between the central maximum and the first-order dark fringe, it's also the minimum angular separation between two objects that can be resolved through an aperture of diameter D. That is,

$$\theta_{\min} = 1.22 \frac{\lambda}{D} \quad \text{(Rayleigh criterion)} \qquad (22.6)$$

The resulting diffraction patterns from a pair of point sources are shown in Figure 22.22.

✓**TIP**

Because $\theta_{\min}$ is an approximation for $\sin \theta_{\min}$, the angle here is in radians.

The Rayleigh criterion sets the **diffraction limit**—the minimum angular separation you can resolve (Figure 22.23). Watch a car approaching at night from far away, and at first you see just a single light. That's because the headlights' angular separation is less than $\theta_{\min}$, so your eyes can't distinguish them. As the car approaches, the angle increases and eventually you see two distinct headlights. Astronomers trying to image distant objects face the same problem: angular sizes are often smaller than $\theta_{\min}$. Using a telescope with larger diameter D helps. The diffraction limit is one reason we can't directly image distant stars, although atmospheric distortion is usually a greater problem for ground-based telescopes. Hubble and other space-based telescopes, however, are diffraction limited. Radio wavelengths are much longer, so according to Rayleigh's criterion, radio telescopes should have worse resolution than optical telescopes. However, it's easier to build huge radio

Fully resolved

(a)

Barely resolved
(Rayleigh criterion)

(b)

Unresolved

Summed light intensity

Individual diffraction patterns

(c)

FIGURE 22.22 Two light sources can be resolved only if the central maxima of their diffraction patterns don't overlap too much. This sequence illustrates how resolution of a pair of light sources changes as their angular separation changes.

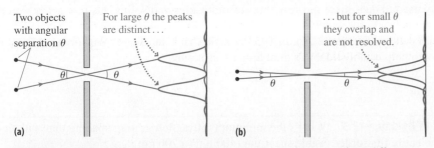

Two objects with angular separation θ

For large θ the peaks are distinct . . .

. . . but for small θ they overlap and are not resolved.

(a)

(b)

FIGURE 22.23 Distant light sources at different angular positions produce diffraction patterns whose central peaks have the same angular separation θ as the sources.

EXAMPLE 22.8 **Pixelation**

Your laptop screen has pixels separated by 0.20 mm. If your eyes' pupil diameter is 3.0 mm, what's the closest your eyes could be to the screen before you notice individual pixels emitting 650-nm red light?

ORGANIZE AND PLAN The minimum distance results when the viewing angle satisfies Rayleigh's criterion: $\theta_{min} = 1.22\lambda/D$, with D your pupil diameter. Figure 22.24 shows that the angular separation between pixels is approximately $\theta = s/r$, with s the pixel spacing and r the viewing distance. Equating this quantity to θ_{min} in Equation 22.6 lets you find the viewing distance.

Known: $\lambda = 650$ nm; $D = 3.0$ mm, pixel spacing $s = 0.20$ mm.

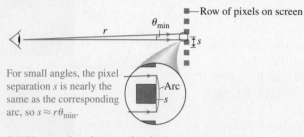

For small angles, the pixel separation s is nearly the same as the corresponding arc, so $s \approx r\theta_{min}$.

FIGURE 22.24 Resolving individual screen pixels.

SOLVE The Rayleigh criterion gives θ_{min}:

$$\theta_{min} = 1.22\frac{\lambda}{D} = (1.22)\left(\frac{650 \times 10^{-9} \text{ m}}{3.0 \times 10^{-3} \text{ m}}\right) = 2.64 \times 10^{-4} \text{ rad}$$

Then the screen distance is

$$r_{min} = \frac{s}{\theta_{min}} = \frac{2.0 \times 10^{-4} \text{ m}}{2.64 \times 10^{-4} \text{ rad}} = 0.76 \text{ m} = 76 \text{ cm}$$

REFLECT This answer is reasonable—try it yourself! You might need to get a little closer to see the pixels, and if you're farsighted you could have trouble focusing. The distance calculated in this example is the theoretical minimum, assuming otherwise perfect optics.

MAKING THE CONNECTION Are blue pixels easier or harder to resolve than red ones?

ANSWER A shorter wavelength gives a smaller λ_{min}, so you could resolve blue pixels from a greater distance.

telescopes. Radio telescope arrays consist of multiple dish antennas, some separated by thousands of kilometers, making huge effective values of D in Equation 22.6. For that reason radio-telescope images often resolve more detail than optical images.

Rayleigh's criterion also helps explain why some animals have better vision than others. Birds of prey flying at great heights need excellent eyesight to resolve small animals on the ground. They have large pupils, which Equation 22.6 shows results in smaller θ_{min}, giving their eyes greater resolving power.

Sometimes the diffraction limit is helpful. Move too close to your computer or TV screen, and your eyes resolve the individual pixels. That may distract you from the larger picture. At a more comfortable distance the angular spacing between pixels is less than Rayleigh's θ_{min}, so they blend together in a seemingly continuous picture.

X-ray Diffraction

Wilhelm Röntgen discovered x rays in 1895. (There's more on x rays in Chapter 23.) In 1912 the German physicist Max von Laue proposed that the regularly spaced atoms in a solid should act like a diffraction grating for x rays. Von Laue recognized that x-ray wavelengths (~1 nm) are comparable to the atomic spacing in solids, and he knew that diffraction is most obvious when the grating size is comparable to the wavelength. Figure 22.25 shows the principle behind x-ray diffraction. Today x-ray diffraction is widely used to determine the structures of molecules and solids. An important historical example is DNA, for which the English chemist Rosalind Franklin first obtained the diffraction pattern (Figure 22.26). In 1953 Francis Crick and James Watson used Franklin's results to help establish DNA's double-helix structure.

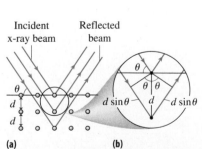

FIGURE 22.25 (a) X rays reflect off planes of atoms in a crystalline solid. (b) Constructive interference occurs when the extra distance $2d \sin \theta$ is an integer multiple of the x-ray wavelength.

GOT IT? Section 22.3 What's the minimum diffraction grating spacing that will let you see the entire first-order visible spectrum (400 nm to 700 nm)?

22.4 Polarization and Scattering

You may own a pair of polarized sunglasses that cut glare from roads and other surfaces. Polarization is another wave property of light. Here we'll explore the basic physics of polarization and consider some of its myriad applications.

Polarization of Light

Recall that electromagnetic waves are transverse, with electric and magnetic fields perpendicular to each other and to the propagation direction, as in Figure 20.3. The wave in Figure 20.3 is **linearly polarized**, meaning that its electric field lies along a fixed axis. So does the magnetic field, but since the two fields are always perpendicular, it's sufficient to mention only the electric field when discussing polarization.

Light from most sources is **nonpolarized**, consisting of many electromagnetic waves with random polarization axes. A polarizing filter, like the lens in your sunglasses, contains long organic polymers oriented linearly. Figure 22.27 shows that the molecules block electric field components parallel to their long dimension. Thus the horizontally oriented molecules pass only vertically polarized light—giving unpolarized light a definite polarization. Light tends to reflect from horizontal surfaces with a horizontal polarization, and thus your sunglasses are oriented as in Figure 22.27 to block such glare.

The electric field direction that passes through a polarizer defines its **transmission axis**. Using two successive polarizers with perpendicular transmission axes blocks all light, since the first polarizer yields light with polarization that can't get through the second polarizer (Figure 22.28a). But if the axes are parallel, then all the light that gets through the first one also gets through the second (Figure 22.28b). Rotating one polarizer relative to the other lets an intermediate amount of light through (Figure 22.28c).

✓ **TIP**

Experiment with two pairs of sunglasses. If they block light when oriented at right angles, they're both polarized. Otherwise, one or both aren't.

Recall that electric and magnetic fields carry energy proportional to the *square* of the field strength—and since E and B are proportional in an electromagnetic wave, that makes the wave's intensity (power per unit area) proportional to the square of its electric field. Sending polarized light through a polarizer passes only the field component parallel to the polarizer's transmission axis. If θ is the angle between the wave's electric field and the transmission axis, then the transmitted electric field is reduced by a factor $\cos\theta$. Since the intensity I is proportional to the square of the field,

$$I = I_0 \cos^2\theta \qquad \text{(Malus's law; SI unit: W/m}^2\text{)} \qquad (22.7)$$

Here I_0 is the intensity incident on the polarizer and I the intensity of the light that gets through. For polarizers with parallel axes, $\theta = 0$ and $\cos^2\theta = 1$—and thus light passes through with undiminished intensity. But crossed polarizers have $\theta = 90°$ and $\cos^2\theta = 0$,

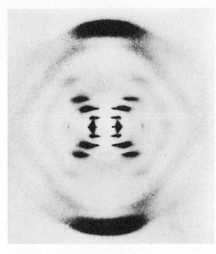

FIGURE 22.26 X-ray diffraction patterns formed by DNA. This is the historical photo by Rosalind Franklin.

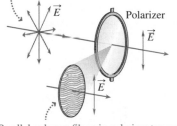

Nonpolarized incoming light — $\vec{E}$ field is oriented in all directions.

Polarizer

$\vec{E}$

$\vec{E}$

$\vec{E}$

Parallel polymer fibers in polarizer transmit only component of $\vec{E}$ perpendicular to fibers.

FIGURE 22.27 Polarization limits the electric field of the electromagnetic wave to a single axis.

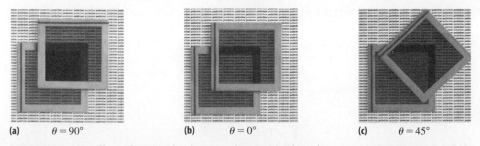

(a) $\theta = 90°$ (b) $\theta = 0°$ (c) $\theta = 45°$

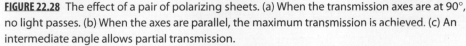

FIGURE 22.28 The effect of a pair of polarizing sheets. (a) When the transmission axes are at 90°, no light passes. (b) When the axes are parallel, the maximum transmission is achieved. (c) An intermediate angle allows partial transmission.

so no light gets through. Equation 22.7 is named for the French scientist Etienne-Louis Malus, who discovered the relation long before its simple explanation was obvious.

If you view nonpolarized light through polarized sunglasses, there's still an intensity reduction. Nonpolarized light contains random polarizations, so θ in Malus's law varies randomly from 0 to 90°. Thus $\cos^2 \theta$ varies from 0 to 1, with average value 1/2. Thus polarized sunglasses cut the unpolarized intensity in half.

Sunlight's interaction with Earth's atmosphere causes a slight degree of polarization. You can confirm this by holding your polarized sunglasses up to the sky and rotating them. You'll notice a slight change in intensity, indicating slightly polarized light. Bees sense this polarization, and use it to help navigate.

EXAMPLE 22.9 **Intensity versus Angle**

What angle between two polarizers will result in (a) one-half and (b) one-fourth of the light incident on the second polarizer getting through?

ORGANIZE AND PLAN Malus's law gives the transmitted intensity: $I = I_0 \cos^2 \theta$. Thus the ratio of transmitted to incident intensity is $I/I_0 = \cos^2 \theta$.

SOLVE For $I/I_0 = 1/2$, $\cos^2 \theta = 1/2$. Therefore, $\cos \theta = \sqrt{1/2} \approx 0.707$, and $\theta = \cos^{-1} 0.707 = 45°$.
Similarly, for $I/I_0 = 1/4$, $\cos^2 \theta = 1/4$, so $\cos \theta = \sqrt{1/4} = 1/2$. Then $\theta = \cos^{-1}(1/2) = 60°$.

REFLECT The transmitted intensity drops as the angle increases from 0 to 90°. However, this example shows that the variation isn't linear.

MAKING THE CONNECTION At what angle is the transmitted intensity down to 10 percent of the intensity incident on the second polarizer?

ANSWER Following the same procedure gives an angle of 71.6°.

CONCEPTUAL EXAMPLE 22.10 **Three Polarizers**

Two crossed polarizers don't let light pass. But insert another polarizer between the crossed pair, and some light gets through (Figure 22.29). Why? What's the intensity emerging from the last polarizer, assuming initially unpolarized incident light?

SOLVE Suppose the angle between the first and middle polarizers' transmission axes is θ. As long as $\theta \neq 90°$, some light passes through the middle polarizer, carrying the orientation θ of that polarizer. The angle between the middle and last polarizers' orientations is $90° - \theta$, so light polarized by the middle polarizer can pass through the last one.

We've seen that intensity is halved when unpolarized light passes through a single polarizer. So if I_0 is the intensity incident on the first polarizer, $I_1 = I_0/2$ is what emerges. By Malus's law, the intensity after the middle polarizer is $I_2 = I_1 \cos^2 \theta = (I_0/2) \cos^2 \theta$. Then for light emerging from the final polarizer,

$$I_3 = I_2 \cos^2 (90° - \theta) = \frac{I_0}{2} \cos^2 \theta \cos^2 (90° - \theta)$$

Since $\cos (90° - \theta) = \sin \theta$ and $\sin 2\theta = 2 \sin \theta \cos \theta$, this expression simplifies to

$$I_3 = \frac{I_0}{8} \sin^2 2\theta$$

REFLECT The transmitted intensity is a maximum when $\theta = 45°$. At either extreme (0° or 90°), the system becomes essentially just a pair of crossed polarizers, with zero transmitted intensity.

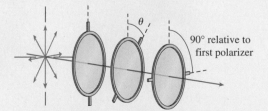

FIGURE 22.29 A sequence of three polarizers.

Liquid Crystal Displays

A common application of crossed polarizers is the liquid crystal display (LCD), used in everything from calculators to watches to cell phones, iPods, computer displays, and TVs. Figure 22.30 shows the basic LCD operation. Unpolarized light first passes through a single polarizer. The resulting polarized light then goes through the liquid crystal cell. A potential difference applied to the liquid crystal causes light to pass through the cell with its polarization unchanged. The light then encounters the second polarizer, oriented at 90° to the first, so it's blocked. The result is a dark spot on the display.

With the potential difference turned off, molecules in the liquid crystal rotate the polarization by 90°, so it's aligned with the second polarizer. The result is a bright spot on the display. Simple displays like those in your calculator turn on or off a few liquid crystal cells configured to make the various digits. Your high-definition TV, on the other hand, has millions of pixels, each a miniature version of Figure 22.30. You can confirm the polarized nature of an LCD device by rotating polarized sunglasses in front of the display.

Atmospheric Scattering of Light

Sunlight is white—a mixture of all visible wavelengths. So why does the sky appear blue, but reddens near sunrise and sunset, while clouds are white? The answer lies in the scattering of light from particles in the air. That scattering depends strongly on the particles' sizes. Water droplets that make up clouds are much larger than light's wavelength, so they scatter all visible wavelengths almost equally. That's why clouds appear white.

It's very different when the scattering particles are much smaller than the light's wavelength. This is the case with nitrogen (N_2) and oxygen (O_2) molecules that make up some 99% of dry air. The electric field of light waves interacts with the valence electrons in these molecules, resulting in a weak scattering that depends strongly on wavelength—with the intensity of scattered light proportional to $1/\lambda^4$. Therefore, the shorter wavelengths—blue and violet—are preferentially scattered (Figure 22.31). This effect is known as **Rayleigh scattering**. During most of the day, you look up and see blue light that's been scattered out of the direct beam of sunlight. Around sunrise or sunset, though, light has to travel through more atmosphere to reach you. More of the blue has been scattered away from your line of sight, so you see predominantly orange and red.

Scattering of sunlight plays a significant role in climate change. *Sulfate aerosols* are particles formed from sulfur emissions associated with burning coal and volcanic eruptions. These strongly scatter incoming sunlight, and therefore exert a cooling effect on Earth's surface. Although this effect is significant, it's not enough to counter global warming resulting from carbon dioxide emissions—also largely from burning coal. Some have suggested that we might intentionally inject sulfate aerosols into the upper atmosphere to counteract global warming, but most scientists consider such "geoengineering" schemes risky and uncertain.

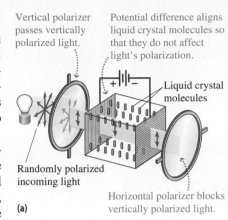

Vertical polarizer passes vertically polarized light.

Potential difference aligns liquid crystal molecules so that they do not affect light's polarization.

Liquid crystal molecules

Randomly polarized incoming light

Horizontal polarizer blocks vertically polarized light.

(a)

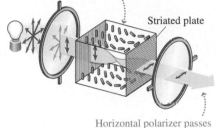

When potential difference is turned off, liquid crystal molecules align with striated plates and rotate the light's polarization.

Striated plate

Horizontal polarizer passes horizontally polarized light.

(b)

FIGURE 22.30 Polarization plays a central role in the operation of a liquid crystal display. Multiple units like the one shown—millions in a TV or computer screen—produce the individual pixels on an LCD screen.

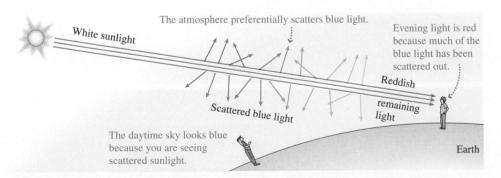

The atmosphere preferentially scatters blue light.

White sunlight

Evening light is red because much of the blue light has been scattered out.

Reddish remaining light

Scattered blue light

The daytime sky looks blue because you are seeing scattered sunlight.

Earth

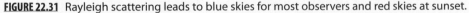

FIGURE 22.31 Rayleigh scattering leads to blue skies for most observers and red skies at sunset.

GOT IT? Section 22.4   Rank in decreasing order the intensity of light passed by each pair of polarizers, assuming the same incident intensity.

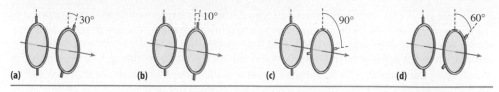

(a)   (b)   (c)   (d)

Chapter 22 in Context

This chapter considered *interference, diffraction*, and *polarization*—phenomena that result from the *wave nature* of light. Light waves exhibit both *constructive* and *destructive interference*—like the sound and water waves of Chapter 11. Light also diffracts, or bends around obstacles or when it passes through small openings. Sometimes diffraction and interference both occur, as in two-slit interference. Interference and diffraction are generally significant only when light interacts with objects comparable to or smaller than the wavelength. Another wave property is polarization, which describes the direction in which a light wave's electric field oscillates. Together, interference, diffraction, and polarization vividly confirm the wave nature of light.

Looking Ahead   Light's wave nature was well established by the end of the 19th century. In Chapter 23, however, you'll see that the early 1900s brought growing evidence that light also exhibits particle-like behavior. The apparent dual nature of light puzzled physicists. You'll see in subsequent chapters how this particle-wave duality led to a deeper understanding not only of light but also of atoms, subatomic particles, and fundamental forces.

Interference

(Section 22.1) Interference of light waves can be **constructive interference**, which results in brighter light, or **destructive interference**, where the two waves cancel to produce a dark region. The Michelson interferometer uses interference of light to make exquisitely precise measurements of variations in distance and/or travel time for light. **Thin films** produce interference between light waves reflecting off their top and bottom surfaces. Reflection at an interface with a material of higher refractive index results in a 180° phase change.

Corrected wavelength with thin-film interference: $\lambda = \dfrac{\lambda_0}{n}$

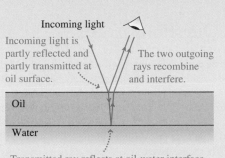

Incoming light

Incoming light is partly reflected and partly transmitted at oil surface.

The two outgoing rays recombine and interfere.

Oil

Water

Transmitted ray reflects at oil-water interface.

Double-Slit Interference

(Section 22.2) In **Young's double-slit experiment**, light passing through two slits interferes constructively and destructively, creating an interference pattern of alternating light and dark fringes on a distant screen.

Condition for bright fringes: $n\lambda = d \sin\theta \quad (n = 0, 1, 2, \ldots)$

Distance of bright fringe from central maximum: $y = \dfrac{n\lambda L}{d}$

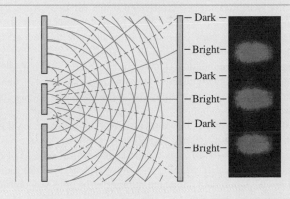

— Dark —
— Bright—
— Dark —
— Bright—
— Dark —
— Bright—

Diffraction

(Section 22.3) **Diffraction** is the bending of waves as they pass an obstacle or through an opening. Diffraction is more significant with obstacles or holes comparable in size to or smaller than the light's wavelength. A **diffraction grating** is a multiple-slit system that separates light precisely into its constituent wavelengths. Even **single-slit diffraction** produces a pattern of light and dark, as waves from different parts of the slit interfere. Regularly spaced atoms in a solid act as a diffraction grating for x rays.

Bright fringes from diffraction grating: $n\lambda = d \sin\theta$
$(n = 0, 1, 2, \ldots)$

Dark fringes from a single slit: $n\lambda = d \sin\theta \quad (n = 1, 2, \ldots)$

Rayleigh criterion: $\theta_{\min} = 1.22\dfrac{\lambda}{D}$

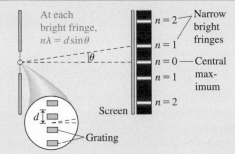

At each bright fringe, $n\lambda = d\sin\theta$

$n = 2$ — Narrow bright fringes
$n = 1$
$n = 0$ — Central maximum
$n = 1$
$n = 2$

Screen
Grating

Polarization and Scattering

(Section 22.4) Light is **polarized** when its electric field oscillates along a single axis. **Polarizers** pass light with a specific polarization; **Malus's law** describes the intensity of light passing through a polarizer as a function of its polarization direction. **Atmospheric scattering** of light is dominated by Rayleigh scattering, which favors scattering of shorter wavelengths.

Malus's law: $I = I_0 \cos^2\theta$

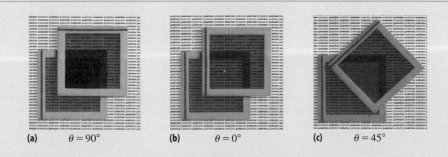

(a) $\theta = 90°$ (b) $\theta = 0°$ (c) $\theta = 45°$

Conceptual Questions

1. Why is interference easier to observe with sound than with light?
2. An antireflection coating needs its thickness matched to a particular wavelength of light. How might you make a coating that minimizes reflection throughout the visible spectrum, 400 nm to 700 nm?
3. How easy would it be to coat a military airplane's wings so they don't reflect 8-GHz radar waves?
4. Light with wavelength λ in a medium with refractive index n_1 strikes a thin film with index n_2 coating a material with index n_3. For each of the following cases, what's the minimum film thickness that will maximize the reflected light? (a) $n_1 > n_2 > n_3$; (b) $n_1 > n_3 > n_2$; (c) $n_2 > n_3 > n_1$; (d) $n_2 > n_1 > n_3$.
5. The center spot in Newton's rings, formed where the two glass surfaces are in contact, is always dark. Why?
6. Explain how you could do a double-slit experiment with infrared or ultraviolet light. If you used the same apparatus, how would the interference pattern for ultraviolet differ from that made with infrared?
7. In a double-slit experiment, how is the fringe pattern affected if you triple the slit-to-screen distance?
8. A double-slit interference pattern stretches from one side of a screen to the other. If you increase the wavelength of the light used in the experiment, does the number of bright lines on the screen increase or decrease?
9. Why is it easier for AM radio signals to get around obstacles like buildings or hills?
10. You're observing a spectrum produced by a diffraction grating of spacing d. How will the appearance of the spectrum alter if you change to a grating of size (a) $0.6d$; (b) $1.2d$.
11. What will you see if you look at white light through a diffraction grating?
12. Your eyes can barely resolve two green lights some distance away. Will you be able to resolve two lights in the same locations if their colors are (a) blue; (b) red?
13. Describe two tests you can perform to tell whether your sunglasses are polarized.
14. Rank in increasing order the amount of atmospheric scattering experienced by these colors: yellow, green, orange, violet.

Multiple-Choice Problems

15. Red light ($\lambda = 630$ nm) is incident on an oil film ($n = 1.50$) on a puddle of water. What minimum oil thickness will result in no reflection? (a) 630 nm; (b) 420 nm; (c) 315 nm; (d) 210 nm.
16. An antireflection coating ($n = 1.42$) of thickness 82 nm is on a plastic eyeglass lens ($n = 1.68$). What visible wavelength will be least reflected? (a) 466 nm; (b) 510 nm; (c) 551 nm; (d) 617 nm.
17. A soap film ($n = 1.33$) with a thickness of 120 nm forms a bubble, with air on each side. What visible wavelength is most reflected from this bubble? (a) 420 nm; (b) 480 nm; (c) 560 nm; (d) 640 nm.
18. Light with wavelength 560 nm is sent through a pair of slits 1.0 mm apart. On a screen 1.8 m away, bright-line fringes are separated by (a) 0.10 mm; (b) 0.50 mm; (c) 1.0 mm; (d) 1.5 mm.
19. Red laser light ($\lambda = 632.8$ nm) passes through a pair of slits, producing an interference pattern on a screen 1.80 m away. If the distance from the central maximum to the third-order bright fringe on the screen is 1.10 cm, what's the slit spacing? (a) 0.20 mm; (b) 0.31 mm; (c) 0.42 mm; (d) 0.53 mm.
20. Light of unknown wavelength passes through a pair of slits 0.10 mm apart, producing interference on a screen 1.40 m away. If the distance between the central maximum and the first dark fringe of the interference pattern is 3.50 mm, what's the wavelength? (a) 650 nm; (b) 600 nm; (c) 550 nm; (d) 500 nm.
21. Monochromatic light with wavelength 480 nm passes through a diffraction grating with spacing 1.4 μm. The angle at which a first-order bright line is observed is (a) 10°; (b) 20°; (c) 30°; (d) 45°.
22. What's the smallest diffraction grating spacing that enables you to see the entire visible spectrum? (a) 400 nm; (b) 700 nm; (c) 1000 nm; (d) 1400 nm.
23. Two polarizing sheets have their transmission axes at 25° to each other. The portion of light from the first polarizer that makes it through the second is (a) 0.60; (b) 0.72; (c) 0.82; (d) 0.91.
24. A third polarizer is placed between a pair of crossed polarizers. Find the fraction of light coming from the first polarizer that makes it through the third if the angle between the transmission axes of the first two is 45°. (a) 0.50; (b) 0.25; (c) 0.12; (d) 0.

Problems

Section 22.1 Interference

For problems involving reflection, assume incoming light is normal to the surface unless indicated otherwise.

25. ■ Visible light is incident on an oil film ($n = 1.51$) coating a puddle of water. What minimum oil thickness will maximize the reflection of each of these colors: (a) blue (460 nm); (b) yellow (580 nm); (c) red (640 nm)?
26. ■ For what minimum oil thickness in the preceding problem will you see no reflection for the red light?
27. ■ ■ You pour some oil on water and, looking straight down, see a reflection of 440-nm violet light. What will you see if you (a) double and (b) triple the oil thickness?
28. BIO ■ **Reflection from eyeglasses.** An antireflection coating ($n = 1.38$) of thickness 105 nm coats a plastic eyeglass lens ($n = 1.64$). For what visible wavelength will reflection be best minimized?
29. ■ ■ The soap film comprising a bubble is 102 nm thick and has refractive index 1.33. What visible wavelength is best reflected from this bubble? Assume air both inside and outside the bubble and a viewing angle normal to the bubble surface.
30. ■ ■ You're observing Newton's rings with 520-nm light. What minimum air gap produces a bright ring?
31. BIO ■ ■ **Protecting your eyes.** When illuminated by white light, a pair of reflective sunglasses appears blue, peaking at 480 nm. How thick is the reflective coating if its refractive index is 1.36?
32. BIO ■ **Avoiding UV.** Plastic with $n = 1.42$ is to be coated on sunglasses so eye-damaging near ultraviolet light (UVA; $\lambda = 380$ nm) will be reflected. What thickness coating should you use?

33. ■■ A Michelson interferometer uses 589-nm yellow sodium light. How far must one mirror be moved for the interference pattern to shift by 10 fringes?

34. ■■■ The Michelson interferometer is often used to measure the refractive index of gases. A transparent cell, initially evacuated, is placed in one arm of the interferometer and illuminated with monochromatic light with $\lambda = 540$ nm. (a) Explain why the interference fringes shift when air is let into the cell. (b) If the cell is 6.0 cm long, by how many bright fringes does the pattern shift when the cell fills with air ($n = 1.00029$) at 20°C and $P = 1.0$ atm?

Section 22.2 Double-Slit Interference

35. ■■ Yellow light with wavelength 589 nm passes through a pair of slits separated by 0.56 mm. What is the separation between bright-line fringes on a screen (a) 2.0 m; (b) 3.0 m away?

36. ■■ Violet light with $\lambda = 440$ nm strikes a double slit and creates an interference pattern on a distant screen. At the position of the third-order bright fringe, how much farther is this fringe from the more distant slit than from the closer one? Does it matter how far away the screen is?

37. ■■ Green light with $\lambda = 540$ nm strikes a pair of slits separated by 0.085 mm. (a) Find the angular deviation of each of the first three fringes, relative to the central maximum. (b) Is the small-angle approximation justified?

38. ■■ Orange light with $\lambda = 615$ nm strikes a pair of slits separated by 0.580 mm. On a screen 1.20 m away, what's the distance between (a) the two second-order bright fringes; (b) the central maximum and one of the fourth dark fringes; (c) the third bright fringe on one side of the central maximum and the third dark fringe on the other?

39. ■■ Monochromatic light of unknown wavelength passes through two slits separated by 0.12 mm, producing an interference pattern on a screen 1.55 m away. If the distance between the central maximum and the first dark fringe is 3.40 mm, what's the wavelength of the light?

40. ■■ A double-slit system with slit separation 0.340 mm is used to determine the wavelength of light passed by a filter. The resulting bright fringes are 2.58 mm apart on a screen 1.50 m away. (a) What's the wavelength? (b) What color is the filter?

41. ■■ White light shines through a far-red filter and the resulting 680-nm light passes through a double slit and produces 17 bright fringes across a distant screen. If the red filter is replaced by a violet one ($\lambda = 440$ nm), how many bright fringes will be on the screen?

42. ■■■ You're doing a double-slit experiment with two different colors. You notice that the second-order bright fringe using 680-nm red light is in the same place as the third-order bright fringe of the other color. (a) What's the wavelength of that color? (b) What color is it?

43. ■■ You have a white light source and two different filters: $\lambda = 480$ nm and $\lambda = 630$ nm. The first-order bright fringe from a double slit is 9.0 mm from the central maximum when you use the shorter wavelength. What's the corresponding distance for the longer wavelength?

44. ■■ Young's double-slit experiment can be done with waves throughout the electromagnetic spectrum. Suppose microwaves ($f = 100$ GHz) are sent through a pair of slits 25 cm apart, cut into a metal sheet. If you have a microwave detector 10 m on the other side of the slits, how far do you have to move the detector to go from maximum to minimum microwave intensity?

45. ■■ You're attempting a double-slit experiment with 10-nm x rays, using an x-ray detector 0.80 m from a pair of slits. If the detector is capable of resolving maxima separated by 0.15 mm, how close together must the slits be?

Section 22.3 Diffraction

46. ■■ A diffraction grating has 1.0-μm spacing. Find the first-order diffraction angles for the following wavelengths: (a) violet, 440 nm; (b) green, 540 nm; (c) red, 640 nm.

47. ■■ A diffraction grating has 1.25-μm spacing. (a) What wavelength is diffracted through a 26.1° angle? (b) Is this wavelength visible in second order? If so, at what angle?

48. ■■ (a) Find the smallest grating spacing that lets you see the entire visible spectrum. (b) What spacing lets you see the entire visible spectrum through second order?

49. ■■ You have a diffraction grating with spacing 1.40-μm that passes infrared, visible, and ultraviolet. Find the first-order diffraction angles of the following wavelengths: (a) ultraviolet, 300 nm; (b) visible, 550 nm; (c) infrared, 900 nm.

50. ■■ You're using a red laser with $\lambda = 632.8$ nm to illuminate a diffraction grating. (a) What spacing will allow you to see diffraction through fourth order? (b) If you can barely resolve the fourth-order line, at what angles do the first three orders appear?

51. ■■ A diffraction grating has a 1.0-μm spacing. What portion of the visible spectrum can be seen in (a) first order; (b) second order; (c) third order?

52. ■■■ You're using a diffraction grating to view the 400- to 700-nm visible spectrum. Suppose you can see the spectrum through third order. (a) Show that the first- and second-order spectra never overlap, regardless of the grating spacing. (b) Show that the second- and third-order spectra always overlap.

53. ■■ The bright yellow sodium line in the sodium spectrum is actually a pair of closely spaced lines at 589.0 nm and 589.6 nm. You observe the sodium spectrum using a diffraction grating with a spacing of 1300 nm. Find the angular separation between the two sodium lines (a) in first order; (b) in second order.

54. Monochromatic light with wavelength 560 nm passes through a single slit 2.50 μm wide and 2.00 m from a screen. Find the distance between the first- and second-order dark fringes on the screen.

55. ■■ In a single-slit diffraction pattern, the fourth-order minimum is 1.5° from the center. What's the slit width if the wavelength is (a) 450 nm and (b) 650 nm?

56. ■■ A 633-nm red laser is aimed through a 0.125-mm-diameter circular pinhole. In the resulting diffraction pattern, what's the angle from the central maximum to the first dark ring?

57. ■■ Monochromatic light $\lambda = 520$ nm is aimed through a pinhole, producing a diffraction pattern on a wall 1.75 m away. If the radius of the second dark ring is 2.50 mm, what's the pinhole's diameter?

58. ■■ Monochromatic light is aimed through a 0.24-mm-diameter circular hole. The resulting diffraction pattern appears on a screen, and its second bright ring has diameter 2.10 cm. How should the diameter of the hole be changed to increase the second bright ring's diameter to 3.00 cm?

59. ■■■ The brightest star in the sky is Sirius, which is actually a pair of stars 8.6 light years from Earth, separated from each other by about 3.0×10^{12} m. What minimum telescope diameter is needed to resolve the two stars with 550-nm light?

60. ■■■ When Venus is close to Earth it appears in a crescent phase. You can see this using a good pair of binoculars. (a) Use the astronomical data in Appendix E to estimate Venus's angular width when it's closest to Earth. (b) To see the crescent clearly, you need a resolution about half this angular width. What's the minimum diameter for your binocular lenses, assuming 550-nm light?

61. ■ ■ ■ The Hubble Space Telescope has a 2.4-m-diameter mirror. (a) What's Hubble's minimum angular resolution, using 550-nm light? (b) Hubble is used in the search for planets orbiting nearby stars. If a star is 10 light years distant, what's the minimum distance between planet and star if the two are to be resolved? Compare your answer with the distance from Earth to Sun.

62. ■ ■ The three principal visible spectral lines from hydrogen have wavelengths 434 nm, 486 nm, and 656 nm. The principal line of sodium is at 589 nm. A sodium lamp used to calibrate a diffraction grating shows the first-order sodium line at 25.0° from the central maximum. Find the angular positions of the three first-order hydrogen lines in hydrogen.

63. ■ ■ What's the spacing in a diffraction grating that produces a 10° separation between the two shortest-wavelength hydrogen lines in the preceding problem?

Section 22.4 Polarization and Scattering

64. ■ Unpolarized light passes through two polarizers. Find the fraction of light from the polarizer that gets through the second when the angle between their transmission axes is (a) 30°; (b) 45°; (c) 75°; (d) 90°.

65. ■ Two polarizing sheets have their transmission axes at 35°. What fraction of light from the first polarizer makes it through the second?

66. ■ ■ Polarized light passes through a polarizer, and only 10% gets through. Find the angle between the light's electric field and the polarizer's transmission axis.

67. ■ ■ A 10-mW unpolarized laser beam passes through two polarizers, whose transmission axes differ by 30°. What's the beam power (a) in the region between the two polarizers and (b) after the second polarizer?

68. ■ ■ A 5-mW laser beam has its electric field vertical. If it passes through a polarizer with its transmission axis at 60° to the vertical, what's the power of the transmitted beam?

69. ■ ■ ■ A polarizer is slipped between two crossed polarizers. Find the fraction of the light from the first polarizer that makes it through the last one if the angle between the transmission axes of the first two is (a) 30°; (b) 60°; (c) 89°.

70. ■ ■ Unpolarized light is passed through two polarizers. The light intensities emerging from the two are I_1 and I_2, respectively. Graph I_2/I_1 as a function of θ, where θ is the angle between the two transmission axes and varies from 0 to 180°.

71. ■ ■ ■ Three polarizers in sequence have transmission axes at 0°, 45°, and 90°. (a) What fraction of light from the first polarizer gets through the third? (b) How does the situation change if the first two polarizers are interchanged?

General Problems

72. ■ ■ The air gap in part of a Newton's rings apparatus varies from zero to 2.00 μm. How many bright rings can be observed in this region? Use an average visible wavelength of 550 nm.

73. ■ ■ A spy satellite is orbiting 300 km above Earth's surface. What diameter mirror is necessary for it to resolve the 15-cm-high numbers on a license plate with 550-nm light? Assume diffraction limits the optical resolution.

74. BIO ■ ■ **Antireflection glasses.** You're applying a 110-nm-thick antireflection coating to eyeglasses, hoping to limit reflection in the middle of the visible spectrum ($\lambda = 550$ nm). (a) What should be the coating's refractive index, assuming it's less than that of the lenses? (b) If instead you use a coating with $n = 1.42$, what's its minimum thickness?

75. ■ ■ ■ In a Newton's rings demonstration using 540-nm light, the distance from the center of the rings to the first bright ring is 0.30 mm. If the bottom piece of glass is flat, what's the curvature radius of the top piece?

76. ■ ■ The Hubble Space Telescope has a 2.4-m-diameter mirror; its successor, the James Webb Space Telescope, will have a 6.4-m mirror. Among the largest telescopes on the drawing board is the ground-based European Extra Large Telescope, whose mirror will measure 42 m across. Assuming diffraction is the limiting factor, could any or all of these telescopes resolve a star and its planet 75 light years from Earth, if the planet is the same distance from the star as Earth is from the Sun and the light used has 600-nm wavelength?

77. ■ ■ ■ The signal from a 103.9-MHz FM radio station reflects from two buildings 35 m apart, effectively producing two coherent sources of the same signal. You're driving at 60 km/h along a road parallel to the line connecting the two buildings and 400 m away. As you pass closest to the two sources, how often do you hear the signal fade?

78. BIO ■ ■ **Resolution of the eye.** You're viewing small objects from a distance of 2.0 m. What's the smallest object you can resolve under each of the following conditions, with light at the middle of the visible spectrum, $\lambda = 550$ nm? (a) The room is brightly lit, and your pupil diameter is 4.2 mm. (b) The lighting is dim, and your pupil has expanded to 9.5-mm diameter.

79. ■ ■ ■ Unpolarized light passes through a succession of three polarizers. Their orientations with respect to the vertical are 0°, 65°, and 45°, respectively. Find the fraction of the original light intensity remaining after each of the three polarizers.

80. ■ ■ You have a summer job with an aerospace firm providing instrumentation for a space probe to study the outer planets. The probe will carry an ultraviolet spectrometer with a 2.0-cm-wide grating ruled at 102 lines/mm. The spectrometer must be able to resolve spectral features only 1 mm apart in twelfth order, when observing at a wavelength of 155 nm. Does it meet these specifications?

81. ■ ■ You're investigating an oil spill for the state environmental protection agency. You have a sample of the slick on glass, and in the thinnest section of the slick you observe constructive interference with 580-nm light. The refractive indices of the oil and glass are 1.38 and 1.52, respectively. What do you report for the oil slick's thickness?

Answers to Chapter Questions

Answer to Chapter-Opening Question

The wave nature of light prevents us from resolving objects smaller than about the wavelength of light, and therefore the bits of information coded optically on a DVD or Blu-ray disc can't be too close. Ordinary DVDs use 650-nm red laser light, with bit spacing of 0.74 μm and corresponding information content of 4.7 gigabytes (GB). The invention of inexpensive blue lasers enabled Blu-ray technology, which uses 405-nm light and a 0.32-μm bit spacing. This and other improvements let Blu-ray discs store 25 GB of information, enough for high-definition movies.

Answers to GOT IT? Questions

Section 22.1   (a) $\lambda/2$ and (c) $3\lambda/2$
Section 22.2   (a) blue; (b) red; (c) yellow
Section 22.3   700 nm
Section 22.4   (b) > (a) > (d) > (c)

23 Modern Physics

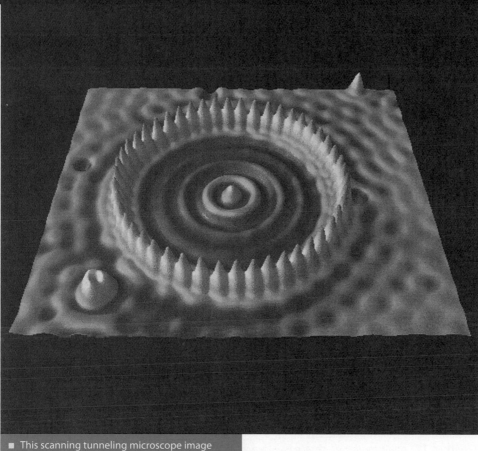

■ This scanning tunneling microscope image shows a "quantum corral" of 48 iron atoms on a copper surface. What unusual quantum phenomenon enables this type of spectroscopy?

This chapter introduces the key idea of quantum physics, namely, quantization of physical quantities. We'll stress the quantum nature of light and other electromagnetic radiation, whose energy comes in particle-like bundles called photons. Electromagnetic radiation emitted by hot objects provided early evidence of quantization. More direct evidence comes from experiments in which photons knock electrons out of metals (the photoelectric effect) or scatter, billiard-ball-like, from electrons (the Compton effect). Thus light, an electromagnetic wave, also exhibits particle properties. It's not only electromagnetic radiation that exhibits this wave-particle duality; so does matter itself. Matter particles exhibit wave properties, with wavelength related to their momentum. Wave-particle duality leads directly to the famous Heisenberg uncertainty principle. It also leads to practical devices, including several kinds of electron microscopes.

23.1 Quantization

Reach into a jar of pennies and grab a handful. Maybe you got 27 of them. Or maybe 32—but not 28.3 or $\sqrt{59}$. You're limited to integers: 26, 27, 28, and so on. That's the basic idea behind **quantization**. Something that's quantized has a smallest basic unit (its **quantum**)

531

that can't be divided, and you can have only integer multiples of that unit. You might argue that you could chop pennies into pieces. That's true, but then they wouldn't be pennies. And if you made the pieces ever smaller, you'd eventually reach the fundamental quanta of matter. Those fundamental quanta aren't coin-sized, but rather much too small to see. That's why quantization isn't obvious, and why the idea developed over centuries of accumulating experimental evidence. Here we'll outline that historical development and introduce some quantized physical quantities. You'll see these ideas developed further throughout Chapters 23–26, because quantization is fundamental to physicists' understanding of matter and energy.

✓**TIP**

Quantum is singular; the plural is **quanta**.

Early History

The idea of quantization dates to ancient times. In the fifth century BCE, the Greek philosopher Democritus, together with his teacher Leucippus, proposed the existence of small particles called $\alpha\tau o\mu o\sigma$ (atomos), meaning indivisible. What we call atoms today are divisible, but Democritus had nevertheless grasped the idea of quantization. Also in the fifth century BCE the Greek philosopher Empedocles developed the idea of four fundamental types of atoms—earth, air, water, and fire. This early atomic theory held that all materials are combinations of these four basic "elements." This four-element theory survived in various forms for centuries. Although this theory may not seem very scientific by today's standards, its early genesis shows that the notion of quantization runs deep though Western culture.

Chemical Evidence

Great advances in chemistry in the 18th and early 19th centuries sharpened scientists' understanding of quantization. The work of Charles, Boyle, and Gay-Lussac on ideal gases (Chapter 12) gave evidence that gases were atomic in nature. Around 1800, the French chemist Joseph Proust developed his *law of definite proportions*: In a given chemical reaction the proportion by mass of each participating element is always the same. For example, formation of water always involves an 8:1 mass ratio of oxygen to hydrogen.

About the same time, the English chemist John Dalton proposed the first modern atomic theory, with all atoms of a particular type—hydrogen, oxygen, nitrogen, and so on—being identical. The idea of identical atomic masses helps to explain the law of definite proportions. In 1811 the Italian chemist Amedeo Avogadro built on the new atomic theory to explain that atoms could join to form molecules, with the same constant proportions of atoms for each molecular type (O_2, CO_2, H_2O, etc.).

In the 1830s Michael Faraday (who also developed Faraday's law of induction; Chapter 19) experimented with **electrolysis**, the breaking of molecules using electric current. Faraday's results were consistent with the ideas of Dalton and Avogadro. In the electrolysis of water, for example, the volume of hydrogen produced is always twice that of oxygen. By performing electrolysis on different materials, Faraday established the chemical composition of many known molecules.

Quantized Mass and Electric Charge

Faraday's electrolysis experiments reinforced the idea that matter consists of atoms. Furthermore, Faraday found that the mass of a particular element released in electrolysis depends on the total electric charge supplied (current × time)—thus hinting at charge quantization. In 1897, J. J. Thomson used the motion of electrons in electric and magnetic fields to show that the charge-to-mass ratio for electrons is more than 1000 times larger than that of singly ionized hydrogen atoms (Section 18.3). Because hydrogen atoms are electrically neutral, the electron's charge must be exactly opposite that of the hydrogen ion. So Thomson's results suggest that the electron's mass is much smaller—by a factor of about 1800—than the hydrogen ion's mass. You'll see in Chapter 24 that this has significant implications for atomic structure.

In 1913 Robert Millikan's experiments firmly established charge quantization (Section 15.5). Knowing the elementary charge e and the charge-to-mass ratio e/m then gave the electron's quantized mass. Our modern value for that mass is about 9.11×10^{-31} kg, more than 1800 times smaller than the mass of a hydrogen atom.

Ionized atoms always have net charges that are integer multiples of the elementary charge $e \approx 1.602 \times 10^{-19}$ C. This strongly suggests that *all* charges (both positive and negative) found in atoms are multiples of e, and that charge is quantized in that base unit. The wide assortment of subatomic particles discovered during the last century all have charges that are integer multiples of e, or else they're neutral. Modern particle theory shows that many particles, including protons and neutrons but not electrons, are composites consisting of three **quarks**, which carry charges $\pm 2e/3$ and $\pm e/3$. The three quarks that form a proton have charges $2e/3$, $2e/3$, and $-e/3$, for a net charge $+e$. The quarks that form a neutron have charges $2e/3$, $-e/3$, and $-e/3$, giving net charge 0. However, individual free quarks are never observed, so for most practical purposes you can assume that charge is quantized in units of $\pm e$. We'll discuss quarks and other fundamental particles in Chapter 26.

Line Spectra

You've probably noticed the characteristic yellow glow of sodium-vapor streetlights, and the garish red from neon signs is unmistakable. Each type of atom—sodium, neon, and all the others—emits a **characteristic spectrum** of colors, unique to that element. The emissions are easily separated by wavelength using a diffraction grating (Section 22.3). Figure 23.1 shows some atomic spectra. Observation of an unfamiliar spectral line pattern in light from the Sun's atmosphere led to the discovery of helium—an element identified on the Sun before it was found on Earth.

Physicists in the late 19th century struggled to understand the wavelength patterns in atomic spectra. What they were missing is a connection between wavelength and energy, which we'll describe in Section 23.2. The line spectra in Figure 23.1, with emissions at specific wavelengths, therefore show that atoms emit energy only in certain discrete amounts—suggesting that energies within atoms are quantized. In Chapter 24 you'll see how physicists in the early 20th century used this information, along with mass and charge quantization, to develop models of the atom.

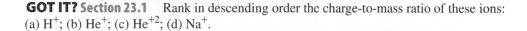

GOT IT? Section 23.1 Rank in descending order the charge-to-mass ratio of these ions: (a) H^+; (b) He^+; (c) He^{+2}; (d) Na^+.

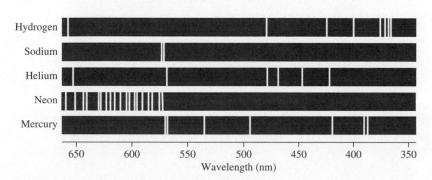

FIGURE 23.1 Emission spectra of selected elements.

23.2 Blackbody Radiation and Planck's Constant

The patterns of discrete spectral lines in Figure 23.1 are unique to different elements, and result when those elements are in the rarefied gaseous state. In contrast, a **continuous spectrum** contains the full range of visible wavelengths and beyond into the infrared and ultraviolet. This type of spectrum results from a hot, glowing object, such as an ordinary incandescent lightbulb. Although the emission isn't equally strong in all the visible wavelengths, there's enough of each color that the light appears white. Accordingly, a continuous spectrum is sometimes called a white-light spectrum.

A continuous spectrum results from **blackbody radiation**, introduced in Section 13.4. The thermal agitation of closely spaced atoms and molecules in a hot object produces electromagnetic radiation with a continuous range of wavelengths. Besides the lightbulb filament, other familiar examples of blackbody radiation include the glowing burner of an electric stove, the hot coals of a wood fire, as well as the Sun and Earth.

Experimental Studies of Blackbody Radiation

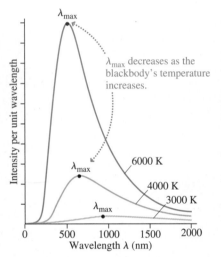

FIGURE 23.2 Intensity per unit wavelength emitted by blackbodies at different temperatures. At higher temperatures more radiation is emitted, and the "Wien peak" λ_{max} shifts to shorter wavelengths.

Physicists studying blackbody radiation in the late 19th century found a link between a blackbody's temperature and the distribution of emitted wavelengths. That lightbulb filament, at about 3000 K, emits what we perceive as white light. The stove burner's red-orange glow indicates its cooler temperature, no more than 1000 K. The Sun, around 5800 K, emits whiter light than the lightbulb, while Earth, at some 300 K, emits almost exclusively infrared. Figure 23.2 shows the relative amount of radiation at various wavelengths for some selected blackbody temperatures. Notice how the peak shifts to lower wavelengths as the temperature increases. The relationship between the peak wavelength λ_{max} and temperature T is given by **Wien's law**, named for the German physicist Wilhelm Wien (1864–1928), who proposed it in 1893:

$$\lambda_{max} T = 2.898 \times 10^{-3} \, \text{m} \cdot \text{K} \quad \text{(Wien's law; SI unit: m} \cdot \text{K)} \quad (23.1)$$

Although we speak of the "peak wavelength," we emphasize that λ_{max} in Equation 23.1 is the wavelength where the intensity *per unit wavelength interval* reaches its peak. We could equally well plot the intensity *per unit frequency interval*, and we'd get a peak at a different wavelength. A more meaningful quantity is the median wavelength, below (and above) which half the power is radiated:

$$\lambda_{med} T = 4.107 \times 10^{-3} \, \text{m} \cdot \text{K} \quad \text{(Median wavelength; SI unit: m} \cdot \text{K)} \quad (23.2)$$

Figure 23.2 also shows that increasing temperature results in more radiation at all wavelengths. Therefore, the total power emitted increases with temperature. As you saw in Chapter 13, power radiated by an object with surface area A at temperature T is given by the **Stefan-Boltzmann law**:

$$P = \epsilon \sigma A T^4 \quad \text{(Stefan-Boltzmann law; SI unit: W)} \quad (23.3)$$

Here the emissivity ϵ measures how efficiently the object absorbs and emits radiation. A perfect blackbody—so named because it absorbs all radiation incident on it and thus appears black—has emissivity 1. A perfectly reflective surface, in contrast, has $\epsilon = 0$. The quantity σ is the Stefan-Boltzmann constant, $\sigma = 5.67 \times 10^{-8} \, \text{W}/(\text{m}^2 \cdot \text{K}^4)$.

✓ **TIP**

The Stefan-Boltzmann law requires absolute temperatures (kelvins in SI).

EXAMPLE 23.1 **Stars to Ears**

Use the median wavelengths listed to find each object's surface temperature, assuming it's a perfect blackbody: (a) a star with $\lambda_{med} = 572$ nm; (b) an electric stove burner, $\lambda_{med} = 4.43\ \mu$m; (c) an eardrum with $\lambda_{med} = 13.21\ \mu$m.

ORGANIZE AND PLAN Equation 23.2 relates the median wavelength and surface temperature: $\lambda_{med}T = 4.107 \times 10^{-3}$ m·K. We're given λ_{med}, so we can find the temperature T.

SOLVE (a) For the star with $\lambda_{med} = 572$ nm, the temperature is

$$T = \frac{4.107 \times 10^{-3}\ \text{m·K}}{\lambda_{med}} = \frac{4.107 \times 10^{-3}\ \text{m·K}}{572 \times 10^{-9}\ \text{m}} = 7180\ \text{K}$$

(b) For the stove burner,

$$T = \frac{4.107 \times 10^{-3}\ \text{m·K}}{\lambda_{med}} = \frac{4.107 \times 10^{-3}\ \text{m·K}}{4.43 \times 10^{-6}\ \text{m}} = 927\ \text{K}$$

(c) For the ear,

$$T = \frac{4.107 \times 10^{-3}\ \text{m·K}}{\lambda_{med}} = \frac{4.107 \times 10^{-3}\ \text{m·K}}{13.21 \times 10^{-6}\ \text{m}} = 310.9\ \text{K}$$

REFLECT That ear temperature is 37.75°C, indicating a slight fever. (Normal body temperature is 37.0°C.) This example shows how radiation can serve as a "thermometer" for objects that radiate as blackbodies. In practice, thermometers using this technique typically measure two points on the blackbody curve, rather than the Wien peak or λ_{med}. The range of applications is vast: Infrared-sensing ear thermometers are quicker than oral fever thermometers, while astronomical spectroscopy gives the temperature of distant stars.

MAKING THE CONNECTION How does that star's median wavelength compare with the Sun's?

ANSWER We noted that the Sun is a 5800-K blackbody; then Equation 23.2 gives $\lambda_{med} = 708$ nm. This is very near the visible-infrared boundary, while the hotter star's 572-nm median wavelength is squarely in the visible.

CONCEPTUAL EXAMPLE 23.2 **Star Colors**

Look carefully at the stars in the night sky, and you'll notice that some appear bluer, others redder. What does this tell you?

SOLVE A hot object's color is a rough indicator of the wavelength distribution of the radiation it emits. Bluer corresponds to shorter median or Wien-peak wavelength and therefore to higher temperature. Red, in contrast, means more radiation at longer wavelengths and therefore lower temperature. Because the Sun appears yellow-white, it serves as a good comparison. Stars that appear distinctly red are cooler than the Sun; bluish stars are hotter.

REFLECT The Sun's 708-nm median wavelength suggests it emits about half its radiation in the infrared and, because there's a little ultraviolet, somewhat less than half in the visible. Nevertheless, the Sun is quite bright in visible light. So color gives only a relative indication of temperature.

EXAMPLE 23.3 **Sun's Power**

Find the total power the Sun emits, considering it to be a spherical blackbody with radius $R = 6.96 \times 10^8$ m. Compare your answer with humankind's energy consumption rate, about 4.2×10^{20} J per year.

ORGANIZE AND PLAN The total radiated power follows from the Stefan-Boltzmann law, with $\epsilon = 1$ for a blackbody: $P = \sigma A T^4$. For the spherical Sun, $A = 4\pi R^2$. We've noted that the Sun's surface temperature is about 5800 K.

Known: $\sigma = 5.67 \times 10^{-8}$ W/(m²·K⁴); $R = 6.96 \times 10^8$ m; $T = 5800$ K.

SOLVE Inserting these values into the Stefan-Boltzmann law,

$$P = \sigma A T^4 = (5.67 \times 10^{-8}\ \text{W/(m}^2\text{·K}^4)) (4\pi (6.96 \times 10^8\ \text{m})^2) (5800\ \text{K})^4 = 3.9 \times 10^{26}\ \text{W}$$

With 1 W = 1 J/s, this is equivalent to $(3.9 \times 10^{26}\ \text{J/s})(3.15 \times 10^7\ \text{s/year}) = 1.2 \times 10^{34}$ J/year. That's more than 10^{13} times humankind's energy consumption rate!

REFLECT Only a tiny fraction of the Sun's energy reaches Earth—but that's still some 10,000 times the power we humans consume.

MAKING THE CONNECTION Suppose the Sun maintained the same energy output but expanded slightly. What would happen to its surface temperature?

ANSWER By the Stefan-Boltzmann law, a constant power P with larger area A implies a lower temperature. The Sun's surface would cool.

Stars evolve, and they change color as their energy output and radius vary. A star with the mass of our Sun appears about the same through most of its 10-billion-year life. At the end, though, the Sun will evolve into a huge, cool red giant star, and eventually it will shrink to a hot white dwarf. By observing color and spectral distribution, astronomers can determine a star's approximate mass and age.

Planck's Quantum Theory

Wien's law and the Stefan-Boltzmann law were first deduced experimentally. In 1884 Boltzmann derived the Stefan-Boltzmann law from classical thermodynamics. However, similar attempts to understand blackbody curves (Figure 23.2) and Wien's law did not yield theoretical explanations.

In 1900 the German physicist Max Planck (1858–1947) found a solution. Planck modeled the atoms in the blackbody as simple harmonic oscillators, and he predicted how a blackbody would produce radiation from its ensemble of atomic oscillators. Planck found that the only way to reproduce the actual blackbody curves in Figure 23.2 was to alter the equation describing those curves. A profound implication of Planck's work is that energies of atomic vibrations are quantized, with energy E proportional to oscillation frequency f. Mathematically,

$$E = hf \qquad \text{(Planck's energy quantum; SI unit: J)} \qquad (23.4)$$

Here h is **Planck's constant**, with approximate value $h = 6.626 \times 10^{-34}$ J·s. Planck's constant h is a fundamental constant of nature. With his quantization condition, Planck successfully reproduced the experimental results in Figure 23.2. He also was able to derive Wien's law, providing a theoretical basis for blackbody radiation.

Planck's results were surprising, and physicists were unsure how to interpret them. Classically, there should be no restriction on oscillation energies, but quantization seemed necessary to explain blackbody spectra. The tiny size of Planck's constant, 6.626×10^{-34} J·s in SI, indicates that a single quantum of energy is quite small. That's why quantum effects aren't obvious in our macroscopic world and weren't obvious to Planck and other physicists around 1900. Example 23.4 makes this point.

EXAMPLE 23.4 **Quantizing a Mass-Spring System**

A mass-spring system consists of a 1.0-kg mass on a spring with spring constant $k = 120$ N/m, oscillating with amplitude 14 cm. Assuming its energy is quantized according to Planck's formula, compare the size of its energy quantum with its actual energy.

ORGANIZE AND PLAN Equation 23.4 gives the energy quantum: $E = hf$. We studied the mass-spring oscillator in Chapter 7, where Equation 7.3 gave the angular frequency of the oscillations: $\omega = \sqrt{k/m}$. The frequency f (cycles per second, or Hz) is then $f = \omega/2\pi$. The system's energy is equal to the spring's potential energy at maximum displacement: $U = \frac{1}{2}kx^2$.

Known: $h = 6.626 \times 10^{-34}$ J·s; $m = 1.0$ kg; $k = 120$ N/m; $x = 14$ cm (maximum displacement).

SOLVE The oscillation frequency is

$$f = \frac{\omega}{2\pi} = \frac{1}{2\pi}\sqrt{\frac{k}{m}} = \frac{1}{2\pi}\sqrt{\frac{120\,\text{N/m}}{1.0\,\text{kg}}} = 1.74\,\text{Hz}$$

giving an energy quantum $E = hf = (6.626 \times 10^{-34}\,\text{J·s})(1.74\,\text{Hz}) = 1.2 \times 10^{-33}$ J. The actual energy is $\frac{1}{2}kx^2 = \frac{1}{2}(120\,\text{N/m})(0.14\,\text{m})^2 = 1.2$ J.

REFLECT The oscillator's energy is 33 orders of magnitude greater than the energy quantum! No wonder we don't notice quantization in macroscopic systems.

MAKING THE CONNECTION Compare the energy quantum and actual energy for an electron on a molecular "spring" with $k = 0.98$ N/m and oscillation amplitude 0.67 nm.

ANSWER A similar calculation gives $hf = 1.09 \times 10^{-19}$ J, and an actual energy that's twice this. Here the energy quantum hf is significant, and the system's energy is clearly an integer multiple of this quantum. The value of hf here is typical of atomic systems, which is why physicists often measure atomic energies in electron volts (1 eV $= 1.6 \times 10^{-19}$ J). Here $hf = 0.68$ eV.

One notable attempt to understand blackbody curves without using Planck's quantum was made by the English physicists John William Strutt (Lord Rayleigh) and Sir James Jeans in 1905. They also considered emission by harmonic oscillators but without quantization of energy. Their result, known as the Rayleigh-Jeans law, is illustrated in Figure 23.3. The Rayleigh-Jeans law agrees with the experimental results (and Planck's theory) at long wavelengths but deviates substantially at shorter wavelengths. Physicists refer to this discrepancy at short wavelengths as the **ultraviolet catastrophe**. The catastrophe shows that we really must accept Planck's quantum as an aspect of physical reality. In the next section we'll consider additional experiments that extend energy quantization to electromagnetic radiation itself.

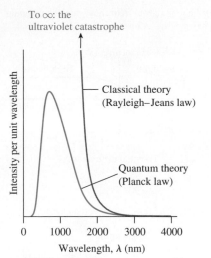

FIGURE 23.3 Intensity per unit wavelength for a 5000-K blackbody, showing the difference between the classical and quantum theories.

Reviewing New Concepts

- Many physical quantities are quantized, such as the mass and charge of fundamental particles.
- Blackbody radiation is described by Wien's law (or, alternatively, the median wavelength) and the Stefan-Boltzmann law.
- Planck explained blackbody observations by introducing his constant h, with the implication that atomic oscillations in solids have energies quantized according to $E = hf$.

23.3 Photons

In the 1880s, Heinrich Hertz noticed that ultraviolet light striking a metal surface in vacuum ejected electrons from the metal. This **photoelectric effect** would later prove to be important in expanding Planck's quantization idea.

Photoelectric Experiments

Hertz's former assistant Phillip Lenard explored the photoelectric effect further in the period 1898–1902. Figure 23.4 shows a typical photoelectric experiment. Incident light strikes a metal surface, the *cathode*, inside an evacuated tube. Electrons—called **photoelectrons**—are emitted from the cathode and can travel to another metal electrode, the *anode*. A variable potential difference ΔV and ammeter connect anode and cathode externally to form a complete circuit. The ammeter reads the circuit current, which is equal to the **photocurrent** carried by electrons flowing from cathode to anode. With virtually no resistance in external circuit, the potential difference between anode and cathode is essentially equal to ΔV. Adjusting ΔV then changes the photocurrent. A positive ΔV—anode positive relative to cathode—attracts photoelectrons to the anode, maximizing the photocurrent. With ΔV negative, electrons are repelled from the anode and the photocurrent is lower.

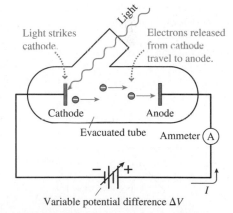

FIGURE 23.4 Apparatus for studying the photoelectric effect.

✓ **TIP**

The light striking the cathode need not be visible; it can also be ultraviolet or infrared.

For light of a given frequency, stopping potential is independent of light intensity.

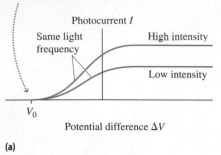

(a)

Stopping potential does depend on frequency . . .

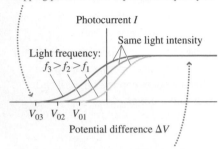

. . . but maximum photocurrent depends only on rate at which photons strike cathode (light intensity).

(b)

FIGURE 23.5 (a) Photocurrent versus potential difference for light with fixed frequency (or wavelength) but different intensities, showing that stopping potential is independent of intensity. (b) Photoelectric data showing that stopping potential does vary with light frequency (or wavelength).

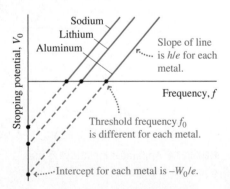

FIGURE 23.6 Stopping potential versus frequency for three different metals.

Figure 23.5a shows typical results using monochromatic light. At a fixed ΔV, you can see that increasing the light intensity increases the photocurrent, meaning more electrons ejected from the cathode. Changing the polarity retards the electron flow, decreasing the photocurrent. The photocurrent reaches zero when $\Delta V = -V_0$, where V_0 is called the **stopping potential**. Note from Figure 23.5a that the stopping potential is independent of light intensity. However, the stopping potential does depend on the frequency or wavelength of the light (Figure 23.5b). A systematic study of the stopping potential V_0 as a function of light frequency f leads to the relationship shown in Figure 23.6. It's evident that stopping potential increases linearly with frequency. There's a minimum frequency f_0, called the **threshold frequency**, for emission of photoelectrons; as Figure 23.6 shows, the threshold is different for different metals.

✓**TIP**

Remember that frequency and wavelength are inversely related ($f = c/\lambda$). Higher frequencies correspond to shorter wavelengths.

Classical physics can account for some of the observed effects. Classically, photoelectrons could be ejected when the oscillating electric field in the electromagnetic wave interacts with electrons in the metal. Another possibility is that light heats the metal surface, with thermal energy responsible for electron emission. Either of these explanations is consistent with the relationship between light intensity and photocurrent (Figure 23.5a).

However, Figure 23.5a also presents problems for classical physics. Why should the stopping potential be independent of light intensity? Shouldn't a more intense wave produce more energetic photoelectrons, requiring a higher stopping potential? And why should the light frequency matter, as it clearly does in Figures 23.5b and 23.6? Further, photoelectron emission takes place virtually instantaneously when the light is turned on, regardless of intensity or frequency. Classically, low-intensity light should need more time to give electrons enough energy to escape the metal. Together, these problems made it difficult for physicists to explain the photoelectric effect.

Einstein Explains the Photoelectric Effect

In 1905, Einstein explained the photoelectric results by extending Planck's quantum idea to include light itself. For Planck, an oscillating atom's energy comes in multiples of a basic quantum: $E = hf$, where f is the oscillation frequency. Einstein proposed that the same relation holds for the energy of electromagnetic waves. Because electromagnetic wave frequency and wavelength are related by $f = c/\lambda$, we can write

$$E = hf = \frac{hc}{\lambda} \quad \text{(Quantum of electromagnetic wave energy; SI unit: J)} \quad (23.5)$$

Einstein's quantization condition implies that electromagnetic wave energy comes in particle-like "bundles." Physicists in the 1920s coined the term **photons** (literally, "light particles") for these bundles.

EXAMPLE 23.5 **Dangerous Rays!**

Find the energy quantum associated with 650-nm red light and with 300-nm ultraviolet-B (UVB).

ORGANIZE AND PLAN In terms of wavelength, the energy quantum is $E = hc/\lambda$ (Equation 23.5).

Known: $h = 6.626 \times 10^{-34}$ J·s; $c = 3.00 \times 10^8$ m/s; $\lambda = 650$ nm (red) and 300 nm (UVB).

cont'd.

SOLVE For the 650-nm red wavelength:

$$E = \frac{hc}{\lambda} = \frac{(6.626 \times 10^{-34}\,\text{J}\cdot\text{s})(3.00 \times 10^8\,\text{m/s})}{650 \times 10^{-9}\,\text{m}} = 3.06 \times 10^{-19}\,\text{J}$$

or 1.91 eV. A similar calculation for the 300-nm UVB yields 6.62×10^{-19} J, or 4.14 eV.

REFLECT Visible light isn't harmful, but the higher-energy photons in ultraviolet radiation can damage DNA in skin cells, which may lead to cancer. Even more energetic UVC radiation ($\lambda < 280$ nm) is, fortunately, blocked by ozone high in Earth's atmosphere.

MAKING THE CONNECTION Microwave radiation can have deleterious health effects, but can its photons directly damage DNA?

ANSWER No. Microwaves have much longer wavelength than visible light. Since visible photons aren't energetic enough to damage DNA, neither are microwave photons.

Einstein's proposal straightforwardly explains the photoelectric effect. A single photon from the incident light ejects one electron from the metal surface. The reason for a threshold frequency f_0 is that the electron needs a minimum energy hf_0 to escape the metal. If the photon's energy exceeds hf_0, then the excess energy becomes kinetic energy of the photoelectron. Thus the photoelectron's maximum kinetic energy is $K_{max} = hf - hf_0$, or $K_{max} = hf - W_0$, where $W_0 = hf_0$ is the **work function** of the metal, equal to the minimum energy that can dislodge an electron. Table 23.1 lists work functions for several metals.

A negative anode will slow the electron, sapping its kinetic energy. Recall from Section 16.2 that potential difference involves energy per unit charge, so the electron (charge $-e$) loses kinetic energy equal to eV moving against a negative potential difference of magnitude V. Electrons stop completely, cutting off the photocurrent, when eV is equal to the maximum electron kinetic energy K_{max}. The corresponding V is the stopping potential, so $K_{max} = eV_0$. Combining our results yields **Einstein's equation**, which relates the photoelectric quantities:

$$eV_0 = K_{max} = hf - W_0 \quad \text{(Einstein's equation; SI unit: J)} \quad (23.6)$$

Equation 23.6 expresses a linear relationship between stopping potential and frequency, in agreement with the experimental results in Figure 23.6. According to Einstein, the photoelectrons' kinetic energy should depend only on the light's frequency, not its intensity. Because one photon ejects one electron, we get photoelectrons even at low light intensities. What does depend on intensity is the *number* of photoelectrons ejected in a given time, and thus the photocurrent does depend on intensity.

Consider the experimental results in Figure 23.6, showing stopping potential V_0 as a function of frequency f. According to Einstein's equation, this function is

$$V_0 = \frac{hf}{e} - \frac{W_0}{e}$$

You should recognize this as the equation of a straight line, with slope h/e and intercept $-W_0/e$. Knowing e, then, we can get the work functions from the intercepts in the experimental graphs. More fundamentally, the *slope* of the graphs determines Planck's constant h. Photoelectric data from Lenard and others gave a value for h that agreed reasonably well with the value Planck found he needed to explain blackbody radiation. Historically, Einstein's interpretation of the photoelectric effect gave strong support to the idea of quantization. In Chapter 24 you'll see how the Planck-Einstein quantum idea led to quantized energy states in atoms—the foundation of modern atomic theory.

TABLE 23.1 Work Functions for Selected Metals

Metal	Work function W_0 (J)
Sodium	3.78×10^{-19}
Lithium	3.78×10^{-19}
Aluminum	6.73×10^{-19}
Lead	6.81×10^{-19}
Zinc	6.91×10^{-19}
Copper	7.18×10^{-19}
Silver	7.43×10^{-19}
Iron	7.48×10^{-19}
Platinum	9.04×10^{-19}

Photovoltaic Solar Energy

Photovoltaic (PV) cells use the energy of solar photons to generate electricity. Instead of a metal cathode, PV cells use a semiconductor junction that's engineered to capture solar photons efficiently. Some cells layer different semiconductors in order to absorb energy in different wavelength ranges throughout the solar spectrum. Given that solar energy is incident on Earth at some 10,000 times humankind's energy consumption rate (Example 23.3), photovoltaic systems have great promise for clean energy generation.

CONCEPTUAL EXAMPLE 23.6 **Threshold Wavelength**

Is there a threshold wavelength corresponding to each metal's threshold frequency? What's the physical meaning of a threshold wavelength?

SOLVE For electromagnetic waves, frequency and wavelength are related by $\lambda = c/f$. The threshold frequency is $f_0 = W_0/h$, so the threshold wavelength is

$$\lambda_0 = \frac{c}{f_0} = \frac{c}{W_0/h}$$

or

$$\lambda_0 = \frac{hc}{W_0}$$

Threshold frequency is the *minimum* frequency of light required for the photoelectric effect. Since wavelength and frequency are inversely related, the threshold wavelength is the *maximum* wavelength for the photoelectric effect. All shorter wavelengths produce photoelectrons, but longer wavelengths don't.

REFLECT This is a real effect, and a striking one. If you have a visible light source and colored filters, you can observe a photoelectric effect for shorter wavelengths but not longer ones. But only in certain metals; for many metals the threshold wavelength lies in the ultraviolet and no visible wavelength will dislodge photoelectrons.

EXAMPLE 23.7 **Photoelectric Effect in Sodium**

(a) What are the threshold frequency and wavelength for sodium? (b) What's the stopping potential for sodium using 400-nm violet light?

ORGANIZE AND PLAN From Conceptual Example 23.6, the threshold frequency is $f_0 = W_0/h$ and the threshold wavelength is $\lambda_0 = c/f_0$. For a metal with a given work function, the stopping potential follows from Einstein's equation:

$$eV_0 = K_{max} = hf - W_0$$

For sodium, Table 23.1 gives $W_0 = 3.78 \times 10^{-19}$ J.

Known: $W_0 = 3.78 \times 10^{-19}$ J; $c = 3.00 \times 10^8$ m/s; $\lambda = 400$ nm; $h = 6.626 \times 10^{-34}$ J·s.

SOLVE (a) The threshold frequency for sodium is

$$f_0 = \frac{W_0}{h} = \frac{3.78 \times 10^{-19} \text{ J}}{6.626 \times 10^{-34} \text{ J·s}} = 5.70 \times 10^{14} \text{ Hz}$$

giving a threshold wavelength

$$\lambda_0 = \frac{c}{f_0} = \frac{3.00 \times 10^8 \text{ m/s}}{5.70 \times 10^{14} \text{ Hz}} = 5.26 \times 10^{-7} \text{ m} = 526 \text{ nm}$$

Photons with wavelengths longer than 526 nm can't eject photoelectrons from sodium.

(b) For a photon wavelength 400 nm, the stopping potential follows from Einstein's equation:

$$V_0 = \frac{hf}{e} - \frac{W_0}{e} = \frac{hc}{e\lambda} - \frac{W_0}{e}$$

Numerically, that's

$$V_0 = \frac{hc}{e\lambda} - \frac{W_0}{e}$$
$$= \frac{(6.626 \times 10^{-34} \text{ J·s})(3.00 \times 10^8 \text{ m/s})}{(1.60 \times 10^{-19} \text{ C})(4.00 \times 10^{-7} \text{ m})} - \frac{3.78 \times 10^{-19} \text{ J}}{1.60 \times 10^{-19} \text{ C}}$$
$$= 0.74 \text{ V}$$

REFLECT The stopping potential is low. That's because the light's wavelength isn't far below the threshold wavelength, so the photoelectrons don't have much kinetic energy and are easily stopped.

MAKING THE CONNECTION Does the stopping potential increase, decrease, or remain the same if the light's wavelength changes to 460 nm? What happens with 560-nm light?

ANSWER Stopping potential decreases for 460-nm light. The 560-nm light is above the threshold wavelength, so it doesn't produce photoelectrons.

Photon Properties

Einstein's interpretation of the photoelectric effect provided convincing evidence for the quantum nature of light. Both that work and his relativity theory (Chapter 20) were published during Einstein's "magic year," 1905. What you learned about light in Chapter 20 provides additional insights into the nature of photons. First, recall the relationship between energy and momentum for light: $E = pc$. (In Chapter 20 we used U rather than E for energy, to avoid confusion with E for the electric field; here we'll revert to E for energy.) Combining $E = pc$ with energy quantization (Equation 23.5, $E = hf = hc/\lambda$) gives

$$\lambda = \frac{h}{p} \quad \text{(Photon wavelength-momentum relation; SI unit: m)} \quad (23.7)$$

Next, consider the relativistic relationship energy-momentum relationship: $E_0 = mc^2 = \sqrt{E^2 - p^2c^2}$, where m is a particle's mass and E_0 its rest energy. Using $E = pc$ for photons, the photon rest energy becomes

$$E_0 = mc^2 = \sqrt{E^2 - p^2c^2} = \sqrt{E^2 - E^2} = 0$$

Therefore:

Photons have zero mass.

The fact that a photon is massless is connected to the fact that it travels at the speed of light c. To see how this works, consider the relativistic relation between energy and speed (Equation 20.14):

$$E = \gamma mc^2 = \frac{mc^2}{\sqrt{1 - v^2/c^2}}$$

As speed approaches c, the denominator $\sqrt{1 - v^2/c^2}$ approaches zero. For a particle to have finite energy, this means that the numerator must approach zero, too. Thus, a particle that travels at the speed of light has zero mass.

The photon is massless, yet it carries momentum and energy. This might seem strange in terms of classical physics, but it's one of the principal results of special relativity. In the remainder of this section we'll discuss other experiments that can only be understood using the fact that light consists of the massless particles we call photons.

Compton Effect

In the early 1920s, American physicist Arthur Compton (1892–1962) noticed that x rays scattered from solids had longer wavelengths than the incident x rays—a result appropriately named the **Compton effect**. There was no way to understand this result classically, so Compton proposed a solution suggested in Figure 23.7. Because x rays are electromagnetic waves, Compton argued that their energy should be quantized in units of hf, just like visible light. He therefore pictured a billiard-ball-like collision between a photon and an electron in the solid, as shown in Figure 23.7.

As we did for elastic collisions in Chapter 6, Compton worked out the dynamics of the collision based on conservation of momentum and energy. Rather than using classical momentum ($p = mv$) and energy ($K = \frac{1}{2}mv^2$), Compton used relativistic expressions. Equation 23.7 shows that a photon's momentum is inversely proportional to its wavelength: $p = h/\lambda$. For the electron, with mass m, Compton used the relativistic energy-momentum relation (Equation 20.15): $E^2 = p^2c^2 + m^2c^4$. He then applied conservation of momentum and energy, assuming the electron was initially essentially at rest. Because momentum is a vector, the analysis is lengthy; the result is

$$\Delta\lambda = \lambda - \lambda_0 = \frac{h}{mc}(1 - \cos\theta) \quad \text{(Compton effect; SI unit: m)} \quad (23.8)$$

Here θ is the photon's scattering angle, relative to its original motion, as shown in Figure 23.7. The photon wavelength is λ_0 before scattering and λ after. Note that Equation 23.8 predicts a longer wavelength for the scattered photon, which is just what Compton observed. This makes sense, because the photon gives up some of its energy to the electron during the collision. Because $E = hc/\lambda$, a lower energy implies a longer wavelength.

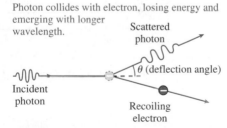

Photon collides with electron, losing energy and emerging with longer wavelength.

FIGURE 23.7 The Compton effect.

The quantity h/mc has dimensions of length and is called the **Compton wavelength** λ_C. Using the numerical values of Planck's constant, the speed of light, and the electron mass, the electron's Compton wavelength is

$$\lambda_C = \frac{h}{mc} = 2.43 \times 10^{-12}\,\text{m} = 2.43\,\text{pm}$$

Knowing the Compton wavelength makes it easy to apply Equation 23.8 to relate the photon wavelengths before and after scattering. The electron's Compton wavelength is much smaller than the wavelength of visible light, which is why the Compton effect is most easily observed with x rays. Other particles, like the proton, are more massive and therefore have even shorter Compton wavelengths. Compton's work showed that Einstein's quantization extended to the highest-frequency electromagnetic waves. Also, it provided a convincing example of electromagnetic radiation behaving just as particles do.

One application of the Compton effect is to diagnose calcium loss in bones (osteoporosis). The intensity of Compton-scattered x-ray or gamma-ray photons depends on the number of scattering electrons, which is proportional to bone density. A low intensity of scattered photons is therefore a sign of low bone density.

EXAMPLE 23.8   **Compton Effect in Graphite**

Compton used graphite in many of his early experiments. Suppose x rays with wavelength of 6.25 pm are incident upon a graphite block. What will be the wavelengths of the x rays scattered at angles of (a) 45° and (b) 90°?

ORGANIZE AND PLAN   Equation 23.8 gives the change in wavelength

$$\Delta\lambda = \lambda - \lambda_0 = \frac{h}{mc}(1 - \cos\theta)$$

Then the final wavelength is

$$\lambda = \lambda_0 + \frac{h}{mc}(1 - \cos\theta)$$

Known: Compton wavelength $\lambda_C = h/mc = 2.43\,\text{pm}$; $\lambda_0 = 6.25\,\text{pm}$; $\theta = 45°$ or $90°$.

SOLVE   (a) For 45° scattering,

$$\lambda = \lambda_0 + \frac{h}{mc}(1 - \cos\theta)$$

$$= 6.25\,\text{pm} + (2.43\,\text{pm})(1 - \cos 45°) = 6.96\,\text{pm}$$

This is longer than the incident wavelength, as expected.

(b) For 90°, a similar calculation gives $\lambda = 8.68\,\text{pm}$. Scattering through this larger angle results in a larger increase in the photon's wavelength. That's because the collision is more "head-on" and transfers more of the photon's energy to the electron.

REFLECT   At 90°, $\cos\theta = 0$, and the wavelength shift is equal to the Compton wavelength. If the x rays bounce back, then $\theta > 90°$ and $\cos\theta$ is negative, so the wavelength shift is even larger. Note in these calculations that you can keep all wavelengths in picometers (pm) without having to convert to meters.

MAKING THE CONNECTION   What's the largest possible wavelength shift in Compton scattering?

ANSWER   The largest shift occurs when photons undergo head-on collisions, scattering directly backward at $\theta = 180°$. Then the wavelength shift is $2\lambda_C = 4.86\,\text{pm}$.

CONCEPTUAL EXAMPLE 23.9   **X ray or Visible?**

Why is the Compton effect normally observed using x rays? Why not use visible light, which is easier to produce and detect?

SOLVE   We just saw that the largest possible wavelength shift is twice the Compton wavelength: $\Delta\lambda_{max} = 2\lambda_C = 4.86\,\text{pm} = 4.86 \times 10^{-12}$ m. Equation 23.8 shows that this shift is independent of the incident wavelength. But the wavelength of visible light—hundreds of nanometers—is some 10^5 times greater. Thus the Compton shift in visible light is only about one part in 10^5, and would be extremely difficult to detect.

REFLECT   This analysis suggests that x-ray wavelengths much longer than the Compton wavelength aren't changed significantly on scattering from a solid. Wavelengths on the order of 100 pm—many Compton wavelengths—are used in x-ray diffraction (Section 22.3), so the Compton effect isn't usually observed together with x-ray diffraction.

Pair Production and Annihilation

Einstein's $E_0 = mc^2$ suggests that energy can manifest itself either as the rest energy of massive particles or in massless photons. In classical physics, energy takes different forms such as kinetic, thermal, and gravitational. In relativity, $E_0 = mc^2$ suggests another energy transformation, between the rest energy of massive particles or massless photons. Indeed, it's possible to transform photon energy into rest energy, creating new particles in the process. The low-mass electron is the easiest of the familiar particles to create in this way. But photons are neutral and electrons are charged, so creating a single electron would violate charge conservation. Instead, photon energy can make an **electron-positron pair**, where a positron is the so-called antiparticle to the electron: a particle that's identical in mass but with charge $+e$ in contrast to the electron's $-e$. In such **pair production**, the minimum photon energy needed corresponds to the total mass of both particles, in this case $2m$, where m is the electron (and positron) mass:

$$E = 2mc^2 = 2(9.11 \times 10^{-31}\,\text{kg})(3.00 \times 10^8\,\text{m/s})^2 = 1.64 \times 10^{-13}\,\text{J}$$

That's the minimum photon energy for electron-positron pair production. With $E = hc/\lambda$, this corresponds to a photon wavelength

$$\lambda = \frac{hc}{E} = \frac{(6.626 \times 10^{-34}\,\text{J}\cdot\text{s})(3.00 \times 10^8\,\text{m/s})}{1.64 \times 10^{-13}\,\text{J}} = 1.21 \times 10^{-12}\,\text{m}$$

which is the longest wavelength that can initiate pair production.

The reverse process is **pair annihilation**. When an electron and positron meet, they disappear and are replaced by two photons. Two photons are required in order to conserve momentum. If the electron and positron are at essentially rest, then the two photons are ejected in opposite directions with equal energies. The total available energy is the sum of the electron and positron rest energies, or $2mc^2$, giving each photon an energy

$$E = mc^2 = (9.11 \times 10^{-31}\,\text{kg})(3.00 \times 10^8\,\text{m/s})^2 = 8.20 \times 10^{-14}\,\text{J}$$

or 511 kilo-electron-volts (keV). Detection of 511-keV photons, particularly from astrophysical sources, is a sure sign of electron-positron annihilation.

✓ **TIP**

In pair production or annihilation, both momentum and energy must be conserved.

Reviewing New Concepts

You've now seen a number of ways in which photons interact with matter:

- The photoelectric effect, where photons knock electrons free from a metal surface;
- The Compton effect, in photons scattered from electrons, losing energy and increasing their wavelength;
- Pair production, in which photon energy creates a particle-antiparticle pair;
- X-ray diffraction, in which x-ray photons reflect from crystal planes in a solid, resulting in a characteristic interference pattern (Section 22.3);
- Photons absorbed by an atom, leaving the atom in a higher energy state. You'll see more on this process in Chapter 24.

GOT IT? Section 23.3   Rank in decreasing order the energy of one quantum of each of the following kinds of radiation: (a) visible; (b) ultraviolet; (c) radio; (d) x ray; (e) infrared; (f) gamma ray.

APPLICATION

Positron Emission Tomography (PET)

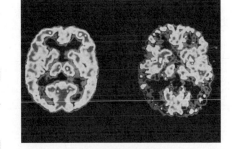

In the medical diagnostic technique known as PET scanning, patients are administered common substances (H_2O, O_2, etc.) "tagged" with a radioactive version of a constituent element. The radioactive atoms decay by emitting positrons, which annihilate with electrons in the surrounding tissue. Detectors register the resulting pairs of 511-keV photons and localize the annihilation site. Computers use inputs from multiple detectors to image the body's interior. Unlike other medical imaging techniques, PET can reveal active processes in the brain and other organs and is therefore used for studies of organ function as well as medical diagnosis.

23.4 Wave-Particle Duality

By the early 1920s, it was clear that light exhibits behaviors characteristic of both waves and particles. Physicists refer to this "split personality" as **wave-particle duality**. In Chapter 22 you saw that interference, diffraction, and polarization can only be explained by treating light as a wave. In this chapter you've seen how the photoelectric and Compton effects show light behaving as particles, called photons. Physicists have come to accept this unusual duality, and recognize that a full description of light involves both wave and particle aspects.

De Broglie Waves

In 1924, the French physicist Louis Victor de Broglie (1892–1987) proposed that if light shows both wave and particle behaviors, then so should matter. Quantitatively, he suggested a relation between a particle's momentum and the associated wavelength. Equation 23.7 gives a photon's wavelength in terms of its momentum: $\lambda = h/p$. De Broglie thought that the same relation should describe waves associated with matter, and defined the **de Broglie wavelength** of a particle with momentum p as $\lambda = h/p$. For particles with mass m and speed $v \ll c$, momentum is $p = mv$. Therefore, de Broglie's formula gives

$$\lambda = \frac{h}{mv} \qquad \text{(de Broglie wavelength, nonrelativistic particle; SI unit: m)} \qquad (23.9)$$

Particle wavelengths cover an enormous range, depending on mass and speed. For an electron (mass 9.11×10^{-31} kg) moving at 4.3×10^6 m/s, $\lambda = 1.7 \times 10^{-10}$ m—about the size of an atom. For a 45-g golf ball moving at 60 m/s, λ is an unimaginably tiny 2.5×10^{-34} m. You've seen that wave effects are apparent only when waves interact with systems whose size is comparable to or smaller than the wavelength. Thus we expect electrons to exhibit wave behavior at atomic scales, but we'll never observe wave behavior in golf balls.

Evidence for Matter Waves

Soon after de Broglie's proposal, American physicists Clinton Davisson and Lester Germer began looking for wave effects in electrons. The electron wavelength we just calculated is comparable to the atomic spacing in crystals, and Davisson and Germer reasoned that electrons should undergo diffraction from solids, similar to the x-ray diffraction we discussed in Section 22.3. They accelerated electrons through a potential difference $\Delta V = 54$ V; from Chapter 16 you know that gives them kinetic energy $K = e\Delta V$ and, using $K = \frac{1}{2}mv^2$, you can show that the corresponding speed is $v = 4.3 \times 10^6$ m/s—the speed we used in our wavelength calculation above. In their 1927 experiment, Davisson and Germer used a nickel target, and found a diffraction pattern corresponding to a wavelength very nearly equal to 1.7×10^{-10} m = 0.17 nm, which we calculated using de Broglie's formula. Thus the Davisson-Germer experiment confirmed the wave behavior of particles. Since that first experiment, many others have demonstrated the wave nature of particles. Diffraction of neutrons (Figure 23.8) is widely used in crystallography, while beams of electrons and even atoms exhibit double-slit interference just as light does.

The experiments we've described here demonstrate clearly that particles exhibit the same wave properties as light. This makes wave-particle duality universal, applying to both matter and light. Light consists of electromagnetic waves, but in some experiments it exhibits particle properties. Similarly, matter particles sometimes exhibit wave properties. In principle, any moving particle has a wavelength, although for macroscopic objects—like the golf ball discussed above—it's invariably too small to notice. Wave-particle duality is a well-established and fundamental property of nature.

Quantum Mechanics

The modern theory of **quantum mechanics** builds on the wave properties of matter. Two equivalent forms of the theory were developed in the 1920s by the Austrian physicist

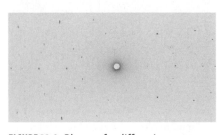

FIGURE 23.8 Photo of a diffraction pattern formed when a beam of neutrons passes through solid crystals. The diffraction pattern is the same as the one formed using a beam of x rays having the same wavelength as the neutrons.

Erwin Schrödinger (1887–1961) and the German Werner Heisenberg (1901–1976). Schrödinger's theory describes each particle using a wave function, Ψ, which depends on position and time. Like any wave, the quantum wave function Ψ isn't localized to a point in space—and neither, therefore, is the corresponding particle. What the wave function tells us is the *probability* of finding the particle at a particular point. Specifically, that probability is proportional to Ψ^2, the square of the wave function evaluated at that point.

Quantum mechanics is strange! It replaces the certainty of classical predictions with probabilities. Consider, for example, a double-slit experiment with electrons. Figure 23.9 shows a sequence of photos as the interference pattern gradually develops, one electron at a time. When a single electron heads toward the slits, it's impossible to predict where it will end up on the screen. All quantum mechanics can tell us is the probabilities that the electron will end up in particular spots, with the highest probabilities corresponding to the brightest interference fringes.

The two-slit experiment with electrons (or photons) provides deeper insights into wave-particle duality. As it stands, the interference pattern shows clearly that electrons (and light) have wave properties. With particles, you wouldn't get the pattern of multiple interference fringes; instead, you might expect particles to accumulate in two regions opposite the two slits. So here's a question: Which slit does an individual electron go through? That's a question that explicitly assumes the electron behaves like a particle. Try to answer this question, perhaps by putting electron detectors at the back of each slit, and a remarkable thing happens: The interference pattern disappears! Remove the detectors, so you can't tell what's happening at the slits, and the interference pattern returns. The strange conclusion is this: If we treat matter (or light) as if it consists of particles, then we'll observe particle-like behavior. That's what happens when we try to catch an electron (or a photon) going through one slit or the other. It's also what happens when we observe photon-electron collisions in the photoelectric and Compton effects. But leave matter (or light) alone, and it exhibits wave behavior—like the interference that occurs in the two-slit system. And don't think that perhaps several electrons (or photons) are somehow interacting to produce the interference pattern. Turn down the electron beam (or light intensity) so there's only a single electron (or photon) in the apparatus at any time, and you still get the interference pattern after many electrons (or photons) have gone through.

So light and matter exhibit both wave and particle aspects. But there's no contradiction, because the two aspects can't manifest themselves simultaneously. Look for particle behavior, and you'll find it—in the process eliminating any evidence of waves. Look for wave behavior, and you'll find it—in the process losing information associated with particles, such as exact positions or which slit they go through. The Danish physicist Niels Bohr (another founder of quantum mechanics) used the term **complementarity** to describe wave-particle duality. Both wave and particle aspects, said Bohr, are needed for a full description of either matter or light. But you can never observe both aspects simultaneously. Thus there's no contradiction in the odd-sounding statement that matter and light are both wave and particle. Neither the wave nor the particle view is superior; rather, both are needed for complete understanding.

100 electrons: Distribution looks random

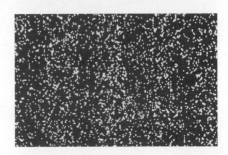

3000 electrons: Double-slit pattern appearing

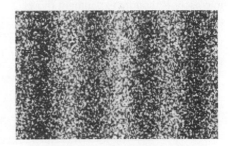

70,000 electrons: Double-slit pattern clear

FIGURE 23.9 Gradual appearance of the interference pattern in a two-slit experiment with electrons.

Heisenberg's Uncertainty Principle

With particles described by waves spread out over space, and with positions determined only by probabilities, there's some fuzziness inherent in quantum mechanics. In his work on quantum mechanics, Heisenberg managed to quantify that fuzziness. Suppose you want to determine a particle's position and velocity. To do so, you have to interact with it somehow. Maybe you look at it with visible light (or other electromagnetic radiation). This involves bouncing at least one photon off the particle, as illustrated in Figure 23.10a (next page). But the incoming photon has momentum $p = h/\lambda$. When the photon scatters off the particle, its momentum changes—and so, by conservation of momentum, does the particle's. So the act of observing disturbs the particle's motion—the very thing you were trying to measure. Because $p = h/\lambda$, you could reduce this disturbance by increasing the wavelength of the light you're using (Figure 23.10b, next page). But light can't be focused onto a spot much smaller than its wavelength, so that makes the position measurement less

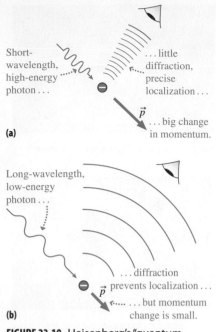

Short-wavelength, high-energy photon...

...little diffraction, precise localization...

$\vec{p}$

...big change in momentum.

(a)

Long-wavelength, low-energy photon...

...diffraction prevents localization...

$\vec{p}$

...but momentum change is small.

(b)

FIGURE 23.10 Heisenberg's "quantum microscope" thought experiment shows the origin of quantum uncertainty.

certain. The result is that there's a tradeoff: measure a particle's motion accurately, and you can't learn much about its position. Measure its position accurately, and you lose information about its momentum.

Heisenberg reasoned that the uncertainty Δp_x in the particle's momentum after the collision was approximately equal to the photon's momentum $p = h/\lambda$. (We're using just one component of momentum here, but the same is true for each component.) There's also an uncertainty Δx in the particle's position, with $\Delta x \approx \lambda$, because the light can't be focused to a spot smaller than about a wavelength. This implies that the product of the two uncertainties is

$$\Delta p_x \Delta x \approx \left(\frac{h}{\lambda}\right)(\lambda) = h$$

Using the full mathematics of quantum mechanics, Heisenberg found a more exact minimum uncertainty, giving the result known as **Heisenberg's uncertainty principle**:

$$\Delta p_x \Delta x \geq \frac{h}{4\pi} \qquad \text{(Heisenberg uncertainty principle; SI unit: J · s)} \qquad (23.10)$$

Equation 23.10 quantifies the tradeoff evident in Figure 23.10—that the more we know about a particle's position, the less we know about its momentum, and vice versa. This quantum uncertainty isn't like the uncertainties you encounter in everyday measurements—uncertainties that can be reduced using more precise instruments. Rather, quantum uncertainty is a fundamental fact of reality, limiting *in principle* how well you can measure position and momentum simultaneously. That absolute limit is given by the $=$ part of the $\geq$ sign. With less than perfect equipment, the minimum uncertainty is higher—the $>$ part of the symbol.

Another uncertainty principle involves measurement of energy E and the time t involved in making that measurement. The uncertainties in these quantities are related in the same way as those in momentum and position:

$$\Delta E \Delta t \geq \frac{h}{4\pi}$$

For example, many excited atomic states last only a short time; if $\Delta t = 10\,\text{ps}\ (10^{-11}\,\text{s})$, then the energy of the state must be uncertain by

$$\Delta E \geq \frac{h}{4\pi \Delta t} = \frac{6.626 \times 10^{-34}\,\text{J}\cdot\text{s}}{4\pi(10^{-11}\,\text{s})} = 5.3 \times 10^{-24}\,\text{J}$$

That may seem small, but it can be significant relative to the energy of an atomic state.

EXAMPLE 23.10 **Working the Uncertainty Principle**

An electron is moving at approximately 1000 m/s. Suppose you try to locate the electron with visible light ($\lambda = 550\,\text{nm}$), and this results in an uncertainty in the electron's position equal to the light wavelength. (a) Find the minimum uncertainty in the electron's momentum and the corresponding uncertainty in velocity. (b) Repeat using infrared light ($\lambda = 5500\,\text{nm}$) and ultraviolet light ($\lambda = 55\,\text{nm}$).

ORGANIZE AND PLAN The Heisenberg principle gives the uncertainty, with the minimum uncertainty following from

$$\Delta p_x \Delta x = \frac{h}{4\pi}$$

Thus, the minimum uncertainty in momentum is

$$\Delta p_x = \frac{h}{4\pi \Delta x}$$

Here the position uncertainty Δx is the light's wavelength λ. Momentum, mass, and velocity are related by $p = mv$.

Known: $\lambda = 550\,\text{nm}$; $m = 9.11 \times 10^{-31}\,\text{kg}$, $v = 1000\,\text{m/s}$; $h = 6.626 \times 10^{-34}\,\text{J}\cdot\text{s}$.

cont'd.

SOLVE (a) The minimum uncertainty in momentum is

$$\Delta p_x = \frac{h}{4\pi\Delta x} = \frac{6.626 \times 10^{-34}\,\text{J}\cdot\text{s}}{4\pi(550 \times 10^{-9}\,\text{m})} = 9.59 \times 10^{-29}\,\text{kg}\cdot\text{m/s}$$

Because $p = mv$, this corresponds to a velocity uncertainty $\Delta v = \Delta p/m = 9.59 \times 10^{-29}\,\text{kg}\cdot\text{m/s}/9.11 \times 10^{-31}\,\text{kg} = 105\,\text{m/s}$, or just over 10%.

(b) *Increasing* the wavelength (and Δx) by a factor of 10 *reduces* the momentum and velocity uncertainties by a factor of 10, while *reducing* the wavelength (and Δx) *increases* the uncertainties by a factor of 10—making the momentum and velocity uncertain by some 100%!

REFLECT Note the tradeoff: Using short-wavelength light involves high-energy photons—so energetic that they knock the electron completely off course.

MAKING THE CONNECTION How would the momentum and velocity uncertainties differ for a neutron, traveling at the same 1000 m/s and viewed with 550-nm light?

ANSWER The momentum uncertainty Δp_x doesn't involve the particle's mass, so the uncertainty Δp_x is the same as for the electron, $\Delta p_x = 9.59 \times 10^{-29}\,\text{kg}\cdot\text{m/s}$. However, the neutron has more mass, so its fractional uncertainty is smaller: $\Delta p_x/p_x = 5.7 \times 10^{-5}$.

CONCEPTUAL EXAMPLE 23.11   An Uncertain Golf Ball?

What's the relative momentum uncertainty for a 45-g golf ball moving at 60 m/s when viewed with a photon of 550-nm visible light?

SOLVE The minimum uncertainty Δp_x doesn't involve the properties of the golf ball, so the uncertainty Δp_x is the same as for the electron, $\Delta p_x = 9.59 \times 10^{-29}\,\text{kg}\cdot\text{m/s}$. However, the golf ball's actual momentum is much higher: $p = mv = 2.7\,\text{kg}\cdot\text{m/s}$. Therefore, the fractional uncertainty is much smaller:

$$\frac{\Delta p_x}{p_x} = \frac{9.59 \times 10^{-29}\,\text{kg}\cdot\text{m/s}}{2.7\,\text{kg}\cdot\text{m/s}} = 3.6 \times 10^{-29}$$

REFLECT That's too small to notice. The pan balance and photogate timer you use on the golf ball are subject to much larger errors than this. When you make measurements on macroscopic objects like golf balls, the uncertainty in Heisenberg's principle is much smaller than the uncertainty of your measuring instruments. That's why you don't notice the uncertainty principle in everyday life.

Modern Microscopy

You've seen how the wave properties of electrons are used in diffraction experiments. Those same wave properties are also used in several kinds of modern microscopes, which can image objects much smaller than the wavelength of visible light—the limit for conventional microscopes.

Equation 23.9 shows that we can choose a particle's wavelength by choosing its speed. **Electron microscopes** exploit this fact, imaging their subjects with beams of electrons. The electrons are accelerated from rest through a potential difference ΔV, gaining kinetic energy $K = \frac{1}{2}mv^2 = e\Delta V$. The resulting momentum is $p = mv = \sqrt{2mK} = \sqrt{2me\Delta V}$ so the electrons have wavelength

$$\lambda = \frac{h}{p} = \frac{h}{\sqrt{2me\Delta V}}$$

Using this equation, you can show that an electron wavelength of 0.1 nm—far below that of visible light—requires an easily achievable potential difference of 150 V.

There are two basic types of electron microscopes: the **transmission electron microscope (TEM)** and the **scanning electron microscope (SEM)**. The TEM, shown schematically in Figure 23.11, takes short-wavelength electrons through a series of

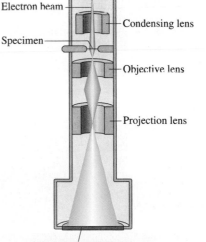

FIGURE 23.11 Schematic of a transmission electron microscope (TEM).

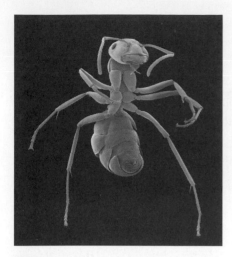

FIGURE 23.12 Image of an insect produced by an SEM.

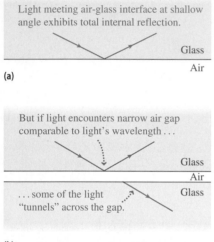

Light meeting air-glass interface at shallow angle exhibits total internal reflection.

Glass

Air

(a)

But if light encounters narrow air gap comparable to light's wavelength . . .

Glass

Air

Glass

. . . some of the light "tunnels" across the gap.

(b)

FIGURE 23.13 Tunneling, shown here for light, lets waves cross small gaps.

"lenses" and then *through* the specimen itself (hence *transmission*) to form an image on a detector. The lenses are actually magnets, and the magnetic force on the electrons (Section 18.1) is what focuses the beam. Aberrations in the lens system give an effective resolution typically no better than 0.5 nm, even with electrons of shorter wavelength. That's still a great improvement over the hundreds-of-nanometer resolution of optical microscopes.

The SEM forms images using electrons reflected from the specimen. The electron beam is scanned slowly over the surface, with the reflected electrons collected at a nearby anode. The positions and intensities of the reflected electrons are used to generate a three-dimensional image with resolution on the order of 1 nm (Figure 23.12). Since the SEM uses reflection, its specimens can be thicker objects, whereas the TEM requires thin specimens for transmission. However, SEM specimens must be coated with a conductor, such as gold, to make them reflect the electrons.

The **scanning tunneling microscope** (STM) takes advantage of **quantum tunneling**. Tunneling is a wave effect that can be demonstrated using visible light (Figure 23.13). Because electrons have wave properties, they too can tunnel across a small gap. That's what happens in the STM, shown schematically in Figure 23.14. As the STM's sharp tip passes over the sample surface, electrons tunnel through the gap between tip and sample, producing a current that is extremely sensitive to the gap spacing. A feedback mechanism moves the tip up and down to keep the current—and thus the gap—constant. Therefore, the tip's motion traces contours of the surface. The STM produces impressive images that can even show individual atoms, as with the "quantum corral" in this chapter's opening photo.

The **atomic force microscope** (AFM) is similar to the STM but more mechanical. As shown in Figure 23.15, the probe tip is attached to a tiny silicon cantilever, which bends as the tip moves up and down over the sample surface. Laser light reflects from the cantilever arm, and a photodiode detects the reflected light. The final result is an image like that shown in Figure 23.16.

These new forms of microscopy have a variety of applications in science and industry. In the biological sciences, STMs and AFMs probe the structures of amino acids, DNA, proteins, and cell clusters from organisms. Figure 23.17 shows an STM image of cellular DNA, with several turns of the double-helix structure clearly visible.

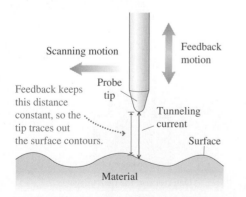

Scanning motion

Feedback motion

Feedback keeps this distance constant, so the tip traces out the surface contours.

Probe tip

Tunneling current

Surface

Material

FIGURE 23.14 Operation of a scanning tunneling microscope (STM).

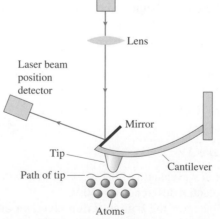

Laser

Lens

Laser beam position detector

Mirror

Tip

Path of tip

Cantilever

Atoms

FIGURE 23.15 Schematic of an atomic force microscope (AFM).

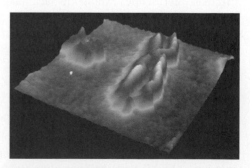

FIGURE 23.16 AFM image of a human X chromosome.

GOT IT? Section 23.4   What effect does increasing a particle's speed have on its de Broglie wavelength? The wavelength (a) increases; (b) decreases; (c) remains the same.

Chapter 23 in Context

This chapter's key idea is *quantization. Planck's constant h* is a fundamental constant of nature that determines the size of quanta. You saw plenty of experimental evidence suggesting that light energy is quantized in particle-like *photons*. Thus light exhibits both wave and particle aspects. Remarkably, as you learned later in the chapter, matter also exhibits both wave and particle aspects. Thus *wave-particle duality* is an essential feature of the world on the atomic scale. Bohr's *complementarity principle* declares that you cannot observe both wave and particle aspects in the same experiment, thus eliminating the seeming contradiction in wave-particle duality. You've seen experimental and practical uses for the wave properties of matter, some of them building on your study of wave optics in Chapter 22. Wave-particle duality is closely associated with the *Heisenberg uncertainty principle*, which shows that it is impossible to measure precisely both a particle's position and its momentum.

Looking Ahead The remaining chapters build upon quantum ideas you encountered in this chapter. Chapter 24 concerns the physics of atoms, which can only be understood in terms of quantized energy states. In Chapter 25 we'll focus on the nucleus at the heart of the atom. Then in Chapter 26 we'll turn to the world of subatomic particles—the individual particles that make up atoms, and the exotic "zoo" of short-lived particles resulting from high-energy interactions. Understanding these fundamental particles has long been a goal in physics, with considerable progress in the last half-century. Quantum ideas also show up in cosmology—the study of the universe as a whole.

FIGURE 23.17 This STM image shows cellular DNA magnified about 2 million times, with the double-helix structure clearly visible.

Quantization

(Section 23.1) **Quantization** means that a physical quantity has a smallest basic unit, an indivisible **quantum**. Early work in chemistry gave evidence for quantization of mass at the atomic level, while Thomson and Millikan added quantization of charge.

Electric charge is quantized in base units of $\pm e$.

Blackbody Radiation and Planck's Constant

(Section 23.2) **Blackbody radiation** is electromagnetic radiation emitted by hot objects. A variation of **Wien's law** relates the median wavelength λ_{med} and temperature T, where λ_{med} is the wavelength above and below which a blackbody emits half its power. The **Stefan-Boltzmann law** shows that the total power emitted by a blackbody increases rapidly with temperature. Planck was able to describe blackbody radiation, but his description introduced the new constant h and implied that the energies of atomic oscillators are quantized.

Wien's law: $\lambda_{max}T = 2.898 \times 10^{-3}\ \text{m} \cdot \text{K}$

Median wavelength: $\lambda_{med}T = 4.107 \times 10^{-3}\ \text{m} \cdot \text{K}$

Stefan-Boltzmann law: $P = \epsilon \sigma A T^4$

Planck's energy quantum: $E = hf$

Planck's constant: $h = 6.626 \times 10^{-34}\ \text{J} \cdot \text{s}$

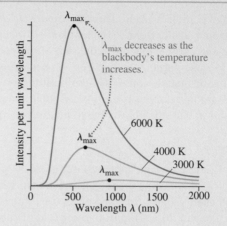

λ_{max} decreases as the blackbody's temperature increases.

Photons

(Section 23.3) In the **photoelectric effect**, light ejects electrons from a metal. The **threshold frequency** is the minimum needed for emission of photoelectrons from a particular metal used. Einstein explained the photoelectric effect by introducing **quantization of light energy**, suggesting that light consists of particle-like energy bundles now called **photons**. In the **Compton effect**, photons interact with electrons just like particles would, recoiling with lower energy and longer wavelength. Photons have zero mass and momentum inversely proportional to wavelength. **Pair production** converts photon energy into the rest energy of a particle-antiparticle pair. The reverse process is **pair annihilation**: When a particle and antiparticle meet, they disappear and are replaced by two photons.

Quantized photon energy: $E = hf = \dfrac{hc}{\lambda}$

Einstein's equation: $eV_0 = K_{max} = hf - W_0$

The **work function** W_0 is the energy required to eject an electron from a particular metal.

Photon momentum, energy, and wavelength: $p = \dfrac{E}{c} = \dfrac{h}{\lambda}$

Compton effect: $\Delta\lambda = \lambda - \lambda_0 = \dfrac{h}{mc}(1 - \cos\theta)$

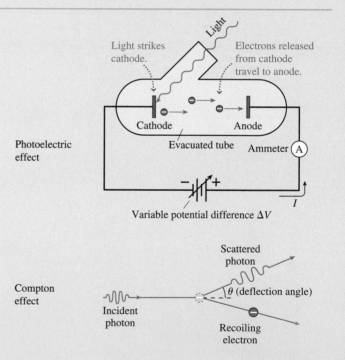

Photoelectric effect

Compton effect

Wave-Particle Duality

(Section 23.4) **Wave-particle duality** describes the fact that light exhibits both particle and wave behaviors. De Broglie suggested that wave-particle duality applies to matter, and electron diffraction experiments show electrons acting like waves. The **Heisenberg uncertainty principle** limits the precision with which a particle's momentum and position can be measured simultaneously.

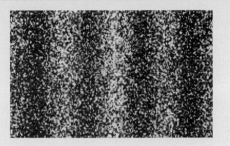

de Broglie wavelength: $\lambda = \dfrac{h}{p}$

Heisenberg uncertainty principle: $\Delta p_x \Delta x \geq \dfrac{h}{4\pi}$

NOTE: Problem difficulty is labeled as ■straightforward to ■■■challenging. Problems labeled BIO are of biological or medical interest.

Conceptual Questions

1. Why was helium first observed on the Sun?
2. Which has a higher temperature—a red-hot object or a white-hot one?
3. What is the "ultraviolet catastrophe"? In what sense is it a "catastrophe"?
4. A red laser and a blue laser emit the same number of photons each second. Which laser emits more power?
5. A red laser and a blue laser have the same power. Which (if either) emits more photons per second?
6. When a photon scatters from an electron initially at rest, why does the photon's wavelength increase?
7. What is a metal's "work function"? Why is it different for different metals?
8. Why is there a threshold frequency for photons in the photoelectric effect?
9. A particular metal ejects photoelectrons when illuminated with green light but not yellow. Do you expect that photoelectrons will be ejected from this metal with (a) orange light; (b) blue light?
10. From the standpoint of classical physics, why is it surprising that electrons are ejected almost immediately in a photoelectric experiment? How does quantum physics resolve this quandary?
11. An electron is traveling at a high but nonrelativistic speed. If its speed doubles, what happens to its de Broglie wavelength?
12. A rock is dropped from a tall building. Ignoring the effects of air resistance, how does its de Broglie wavelength vary as it falls?
13. It has been estimated that the proton's lifetime is at least 10^{36} years. Does that suggest we can know the proton's rest energy precisely, or only approximately?
14. Do baseball batters need to worry about the Heisenberg uncertainty principle when they swing at the ball?
15. Speculate on how the uncertainty principle affects the notion of objectivity in science.
16. Compare the relative advantages of the transmission electron microscope and the scanning electron microscope.

Multiple-Choice Problems

17. How many charge quanta are there in 20 mC? (a) 3.2×10^{16}; (b) 1.3×10^{17}; (c) 3.2×10^{17} (d) 1.3×10^{18}.
18. How many hydrogen atoms are there in a 1-g sample of pure hydrogen? (a) 6.0×10^{23}; (b) 6.0×10^{24}; (c) 6.0×10^{25}; (d) 6.0×10^{26}.
19. A blackbody radiates with median wavelength 2200 nm. What's its temperature? (a) 975 K; (b) 1050 K; (c) 1850 K; (d) 3020 K.
20. A blackbody heated to 900°C emits radiation at the rate of 850 W. If the blackbody's temperature is increased to 1200°C, the power becomes (a) 1130 W; (b) 1340 W; (c) 2110 W; (d) 2700 W.
21. The energy of one quantum of 12-GHz microwave radiation is (a) 8.0×10^{-22} J; (b) 4.0×10^{-23} J; (c) 4.0×10^{-24} J; (d) 8.0×10^{-24} J.
22. A 50-mW laser produces 532-nm green light. The number of photons it emits each second is (a) 1.3×10^{17}; (b) 1.6×10^{17}; (c) 2.7×10^{17}; (d) 2.7×10^{18}.
23. Silver's work function is 4.64 eV. The longest wavelength that will eject photoelectrons from silver is (a) 535 nm; (b) 400 nm; (c) 361 nm; (d) 267 nm.
24. What's the energy of a photon with wavelength 590 nm? (a) 1.1 eV; (b) 2.1 eV; (c) 4.2 eV; (d) 5.8 eV.
25. What's the momentum of a photon with wavelength 600 nm? (a) 3.2×10^{-26} kg·m/s; (b) 5.5×10^{-26} kg·m/s; (c) 7.5×10^{-26} kg·m/s; (d) 1.1×10^{-24} kg·m/s.
26. An x-ray photon with wavelength 2.98 pm scatters from a stationary electron, deflecting through a 55° angle. What's the wavelength of the scattered photon? (a) 3.06×10^{-3} nm; (b) 3.57×10^{-3} nm; (c) 4.02×10^{-3} nm; (d) 4.56×10^{-3} nm.
27. If they're all traveling at the same speed, which particle has the shortest de Broglie wavelength? (a) proton; (b) neutron; (c) electron; (d) alpha particle.
28. Electrons in an electron microscope must have de Broglie wavelengths of 0.45 nm. What's the electrons' speed? (a) 5.6×10^5 m/s; (b) 1.6×10^6 m/s; (c) 7.2×10^6 m/s; (d) very close to the speed of light.

Problems

Section 23.1 Quantization

29. ■ How many quanta of charge are present in (a) 1 C; (b) 1 μC?

30. ■ A toy balloon is filled with helium gas (He) at 1 atm and 20°C. If the volume of the balloon is 0.027 m³, how many atoms does it contain?

31. ■■ You have a piece of solid iron containing 2.50 moles of atoms. (a) What's its mass? (b) Suppose one atom in 10^{12} were missing one electron. What would be the net charge on the iron? Would you notice this if you touched the iron?

32. ■■ An antiproton is similar to a proton, but with charge $-e$. What are the charges on the three quarks that form an antiproton?

Section 23.2 Blackbody Radiation and Planck's Constant

33. ■ Radiation from a blackbody has its Wien peak at 1270 nm. What's the blackbody's temperature?

34. ■ A piece of iron is heated to 600°C. (a) What is the median wavelength of its blackbody radiation? (b) What temperature would give a median wavelength of half of what you found in part (a)?

35. ■ Find the surface temperature of a star whose Wien peak is in the near ultraviolet, with $\lambda = 390$ nm.

36. ■ A red giant star has surface temperature 3100 K. What's the median wavelength in its blackbody spectrum? Compare with the Sun (see Making the Connection in Example 23.1).

37. ■■ The tungsten lightbulb filament operates at about 3000 K (not far below tungsten's melting point). (a) What's the median wavelength it emits, and in what region of the spectrum does this lie? (b) If it's a 100-W bulb, what's the filament's surface area?

38. ■ A white dwarf star has surface temperature 20 kK. (a) Where's the Wien peak in this star's blackbody curve? (b) In what part of the electromagnetic spectrum is that peak?

39. ■ In the 1960s astronomers detected blackbody radiation with Wien peak at 1.06 mm, apparently coming from everywhere in space. What's the temperature of the radiation source? (This "cosmic microwave background" radiation was key to understanding the evolution of the universe.)

40. ■■ A blue supergiant star has surface temperature 30 kK and has total power output 100,000 times that of the Sun. (a) Where is the Wien peak of this star's blackbody curve? (b) Why does the star appear blue? (c) Given the Sun's 6.96×10^8 m radius, what's the radius of the supergiant?

41. ■■ An 800-K blackbody emits 450 W of radiation. At what temperature will the radiated power double to 900 W?

Section 23.3 Photons

42. ■ A photoelectric experiment uses 265-nm light and a silver target. Find (a) the maximum kinetic energy of the photoelectrons and (b) the stopping potential.

43. ■■ Find the threshold frequency and wavelength for production of photoelectrons from a lead target.

44. ■■ What's the longest wavelength that will produce photoelectrons from iron? In what part of the electromagnetic spectrum is this?

45. ■■ (a) What are the threshold frequency and wavelength for photons from a silver target? (b) What's the stopping potential for photoelectrons from silver under 135-nm ultraviolet light?

46. ■■ Suppose you want to produce photoelectrons with kinetic energy 1.50 eV from a copper target. What wavelength of electromagnetic radiation should you use?

47. ■■ A sodium target is illuminated with 442-nm violet light. Find (a) the stopping potential for the photoelectrons produced and (b) the photoelectrons' maximum speed.

48. ■■ Of the elements in Table 23.1, which can produce photoelectrons when illuminated with visible light?

49. ■■ Photoelectrons from a metal target have a 1.20-V stopping potential when the target is illuminated with 340-nm ultraviolet radiation. Find (a) the stopping potential for the same target under 260-nm radiation and (b) the work function for this metal.

50. ■■ A 1.0-mW laser with a 405-nm wavelength illuminates a sodium target. If 1 in 10^5 incident photons generates a photoelectron, what's the photocurrent?

51. ■■■ In an experiment to measure Planck's constant, a metal target is illuminated with ultraviolet light. When the wavelength is 300 nm, the measured stopping potential is 1.10 V. When the wavelength is changed to 200 nm, the stopping potential becomes 3.06 V. (a) What value does this experiment yield for Planck's constant? What's the percent error from the accepted value of h? (b) Using the experimental value h, find the threshold frequency and work function for this metal.

52. ■ An x-ray photon with wavelength 1.94 pm scatters from a stationary electron, deflecting through an angle of 105°. What's the scattered photon's wavelength?

53. ■ Find the energy of one quantum of microwave radiation with frequency (a) 1 GHz and (b) 300 GHz.

54. ■■ When perceiving 630-nm red light, your unaided eye can barely detect light at a threshold power around 2.5×10^{-15} W. At what rate are photons entering your eye at this level?

55. ■■ A semiconductor laser produces a continuous 10.0-W beam with wavelength 2.4 μm. Find (a) the energy of each photon and (b) the number of photons emitted each second.

56. ■■ (a) Find the wavelength of a 4.1-eV photon. In what part of the electromagnetic spectrum is this? Repeat for photons with energies (b) 2.3 eV and (c) 0.69 eV.

57. ■■ While sitting in direct sunlight, you absorb UVB radiation with average wavelength 310 nm. (a) Find the energy of an average UVB photon, and compare with an average visible photon ($\lambda = 550$ nm). (b) How many photons are in 1 J of energy at each wavelength?

58. ■■ A medical x-ray beam has wavelength 1.26 nm. (a) What's the energy of one photon in this beam? (b) How many photons are in a typical chest x-ray dose that delivers 0.021 J of energy?

59. ■■■ Example 23.3 showed that the Sun emits energy at the rate of about 3.9×10^{26} W. (a) Using the astronomical data in Appendix E, estimate the fraction of this energy that falls on Earth. (b) With the Sun directly overhead, what's the rate at which energy strikes each square meter of Earth's surface? (Ignore reflection and absorption by the atmosphere.) (c) Discuss the implications of your answer in part (b) for solar energy use.

60. ■■ A 639 keV x-ray photon scatters from an electron at rest, deflecting through 105°. What is the scattered photon's energy?

61. ■■ You're studying Compton scattering using x rays with wavelength equal to the Compton wavelength. At what scattering angle will you observe x rays with a wavelength twice that of the incident x rays?

62. ■■ A photon with a wavelength 4.50 pm scatters from an electron initially at rest. Find (a) the maximum kinetic energy for the scattered electron and (b) the wavelength of the scattered photon when the scattered electron has its maximum kinetic energy.

63. ■■ X rays with wavelength 5.0 pm scatter from electrons, recoiling with wavelength 8.2 pm. What's the scattering angle?

64. ■■ Compton scattering can occur from atoms as well as from electrons. If photons scatter from atoms of helium (He) gas, what's their maximum wavelength change? Should the photon energies be larger or smaller than those used in Compton scattering by electrons?

65. ■ ■ Your Compton-effect instrumentation can detect a 5.0% change in the photon's wavelength. What's the maximum initial photon wavelength in this case?

66. ■ ■ A hydrogen atom initially at rest emits a 434-nm photon. Find (a) the photon's energy and momentum, (b) the momentum of the atom after emitting the photon, and (c) the kinetic energy of the recoiling atom. Compare with the photon's energy.

67. ■ ■ (a) Find the energy needed to produce a proton-antiproton pair. (b) Find the energies of the two photons produced when a proton and antiproton at rest annihilate.

68. ■ ■ ■ In a colliding-beam experiment, an electron and positron traveling at the same speeds but opposite directions collide, producing two photons. If each of the photons has energy equal to three times the electron's rest energy, what were the electron and positron speeds?

Section 23.4 Wave-Particle Duality

69. ■ Find (a) the speed of an electron with de Broglie wavelength 1.0 nm and (b) the de Broglie wavelength of a proton with that speed.

70. ■ The electrons in an electron microscope have speed of 4.0 Mm/s. Find their de Broglie wavelength.

71. ■ An electron microscope calls for electrons with de Broglie wavelength 0.25 nm. What's the electrons' speed?

72. ■ ■ In an undergraduate version of the Davisson-Germer experiment, electrons are accelerated through several potential differences. For each potential difference, find the corresponding de Broglie wavelength and electron speed: (a) 12 V; (b) 30 V; (c) 60 V.

73. ■ ■ Find the kinetic energy of each of the following particles, each having de Broglie wavelength 0.30 nm: (a) electrons; (b) neutrons; (c) alpha particles.

74. ■ ■ Suppose you want to observe neutron diffraction in a crystal with 0.29-nm spacing between lattice planes. (a) Find the momentum and kinetic energy of neutrons having wavelength equal to this spacing. (b) If these neutrons constituted an ideal gas, what would be the gas temperature? Explain why these are called "cold" neutrons.

75. ■ ■ ■ What's the de Broglie wavelength of a nitrogen (N_2) molecule in air at room temperature (293 K)? How likely is it that you'll be able to observe diffraction effects in such molecules?

76. ■ ■ ■ A 25-g marble passes through a 2.0-cm-wide opening. What should be the marble's speed in order for it to experience a 0.1° diffraction angle through the opening? (Assume first-order single-slit diffraction.)

77. ■ ■ (a) What's the kinetic energy of an electron having de Broglie wavelength 400 nm? (b) Compare with the energy of a photon having the same wavelength.

78. ■ ■ If you want to measure the position of a 142-g baseball moving at 25 m/s to within 1.0 μm, what's the corresponding uncertainty in its speed? Do you think you can really measure the speed with that precision?

79. ■ ■ ■ An electron is confined to a box the size of a small atom, 0.10 nm across. (a) What's the uncertainty in the electron's momentum? (b) Suppose the momentum is just equal to the minimum uncertainty value you computed in part (a). What's the electron's energy? What wavelength photon would have the same energy?

80. ■ ■ The W^+ is an elementary particle with rest energy around 80 GeV, uncertain by about 2.1 GeV. Use Heisenberg's uncertainty principle to estimate this particle's minimum lifetime.

When a particle is moving with a relativistic speed, its de Broglie wavelength is still $\lambda = h/p$, where p is the relativistic momentum. Use this fact in the following problems.

81. ■ ■ Find the de Broglie wavelength of an electron moving at (a) 0.10c; (b) 0.50c; (c) 0.99c.

82. ■ ■ In a certain electron microscope, a de Broglie wavelength of 0.015 nm is required. Find the potential difference required to accelerate electrons to the proper speed.

83. ■ ■ ■ The Stanford linear accelerator can accelerate electrons to kinetic energies of 50 GeV. What's the de Broglie wavelength of these electrons? Compare with the diameter of a proton, about 2 fm.

84. ■ ■ ■ In his double-slit experiment with electrons, Jönsson used electrons with kinetic energy 50 keV. (a) Find the electrons' de Broglie wavelength. (b) The two slits were separated by 2.0 μm. On a screen 0.350 m from the slits, what was the distance between bright diffraction fringes?

General Problems

85. ■ ■ (a) Find the Compton wavelength of a proton. (b) Find the energy (in eV) of a gamma ray whose wavelength equals the proton's Compton wavelength.

86. ■ ■ Electrons in a photoelectric experiment emerge from an aluminum surface with a maximum kinetic energy of 1.3 eV. What is the wavelength of the incident radiation?

87. ■ ■ The most energetic cosmic rays ever detected are photons with energies approaching 10 J. Find the wavelength of such a photon.

88. ■ ■ A cosmic ray interacts for a mere 12 fs with its detector. Having just learned about the uncertainty principle, you realize that this interaction time establishes a precision limit on measurement of cosmic-ray energy. Find (a) the minimum uncertainty in that energy and (b) the fractional uncertainty for a 4.5-MeV cosmic ray.

89. ■ ■ An electron is initially at rest. What will be its kinetic energy after a 0.10-nm x-ray photon scatters from it at 90° to its original direction?

90. ■ ■ ■ A photocathode emits electrons with maximum kinetic energy 0.85 eV when illuminated with 430-nm violet light. (a) Will it eject electrons under 633-nm red light? (b) Find the threshold wavelength for this material.

91. BIO ■ ■ ■ **Human vision threshold.** (a) Estimate the number of photons per second emitted by a 100-W lightbulb, assuming a photon wavelength in the middle of the visible spectrum, 550 nm. (b) A person can just see this bulb from a distance of 800 m, with the pupil diameter dilated to 7.5 mm. How many photons per second are entering the pupil?

Answers to Chapter Questions

Answer to Chapter-Opening Question
It's quantum tunneling, the ability of particles to pass through a gap that classical physics says they don't have enough energy to overcome.

Answers to GOT IT? Questions
Section 23.1   (a) H^+ > (c) He^{+2} > (b) He^+ > (d) Na^+

Section 23.3   (f) gamma ray > (d) x-ray > (b) ultraviolet > (a) visible > (e) infrared > (c) radio

Section 23.4   (b) decreases

24 Atomic Physics

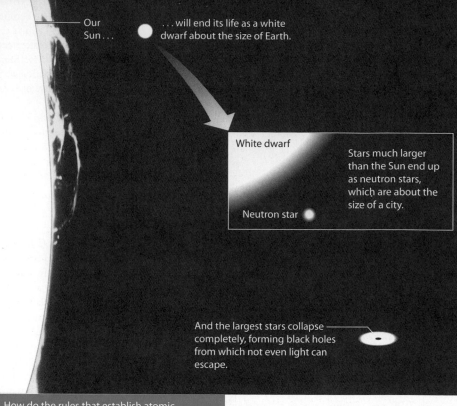

Our Sun...   ...will end its life as a white dwarf about the size of Earth.

White dwarf

Stars much larger than the Sun end up as neutron stars, which are about the size of a city.

Neutron star

And the largest stars collapse completely, forming black holes from which not even light can escape.

■ How do the rules that establish atomic structure also determine the fate of stars?

This chapter is about atoms. Early-20th-century experiments showed that nearly all the atom's mass is contained in a tiny, positively charged nucleus. Negative electrons surround the nucleus. But classical physics suggested that such a structure couldn't be stable, and furthermore could not explain atomic spectra. Niels Bohr then applied quantization to develop the first successful atomic model. A more complete quantum theory of atoms followed, including the introduction of four quantum numbers that completely describe atomic states. You'll see how this theory fully describes the hydrogen atom, and how it explains multi-electron atoms and the periodic table of the elements. We'll end with a discussion of atomic emissions, including x rays from transitions involving inner electrons and the stimulated emission that's essential to laser operation.

24.1 The Nuclear Atom

By 1910 physicists thoroughly understood classical dynamics (Chapters 2–11), thermodynamics (Chapters 12–14), and electromagnetism (Chapters 15–20). Even special relativity (Chapter 20) and the early foundations of quantum theory (Chapter 23) had been established.

One remaining quandary was the structure of the atom. Chemistry showed that a limited number of different atoms existed—hydrogen, carbon, nitrogen, oxygen, and so on. Their chemical properties were organized in a periodic table, similar to the table you know today. (We'll discuss the periodic table in Section 24.4.) You've learned how J.J. Thomson identified the electron as a negatively charged subatomic particle and measured its charge-to-mass ratio (Section 23.1). But physicists still didn't know the size and composition of the atom's positive part—required for atoms to be electrically neutral. Naturally, the positive and negative charges in atoms mutually attract, but it was unclear how they fit together to form a stable system.

"Plum-Pudding" Atoms

Thomson himself proposed the "plum-pudding" model (Figure 24.1), consisting of a positive lump ("pudding") that fills the atom's volume, with smaller electrons ("plums") moving about within the "pudding." The model made sense, based on the relative masses of the electrons and complete atoms. The lightest atom (hydrogen; 1.67×10^{-27} kg) is nearly 2000 times more massive than the electron (9.11×10^{-31} kg). But although Thomson tried, he could not give a satisfactory explanation of atomic spectra using his model.

Discovery of the Nucleus

A key step in understanding atomic structure occurred in 1909–1910. Ernest Rutherford, a New Zealand–born physicist, had studied radioactivity at Canada's McGill University. In 1907 he moved to the University of Manchester in England. There he tried using alpha particles—doubly ionized helium atoms emitted in some radioactive decays—as probes of atomic structure. Rutherford's collaborator Hans Geiger and student Ernest Marsden bombarded thin gold foils with alpha particles (Figure 24.2), hoping to isolate the interaction between an alpha particle and a single gold atom. Most alpha particles traveled through the gold foil with little deflection. But to the experimenters' surprise, some were deflected at large angles, even directly backward.

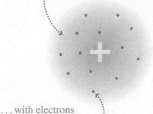

Thomson proposed that an atom consists of a diffuse cloud of positive charge . . .

. . . with electrons embedded in it like plums in a pudding.

FIGURE 24.1 Thomson's plum-pudding model of the atom.

✓**TIP**

Alpha particles—doubly ionized helium atoms—are missing both electrons. Hence, they carry positive charge $+2e$.

This result was remarkable: As Rutherford wrote, it was ". . . as if you fired a 15-inch naval shell at a piece of tissue paper and the shell came right back and hit you." As Figure 24.2 shows, it would be impossible for the spread-out positive "pudding" in Thomson's model to repel the alpha particle with enough force to deflect it significantly. However, Rutherford realized, if the atom's positive charge were confined to a small **nucleus**, then the force would be sufficient to make an alpha particle bounce back in a head-on collision, while passing through with little deflection when it did not have a close encounter with a

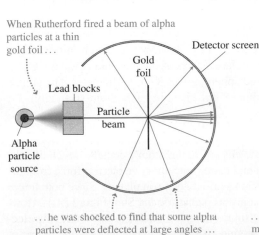

When Rutherford fired a beam of alpha particles at a thin gold foil . . .

Detector screen

Gold foil

Lead blocks

Particle beam

Alpha particle source

. . . he was shocked to find that some alpha particles were deflected at large angles . . .

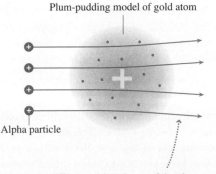

Plum-pudding model of gold atom

Alpha particle

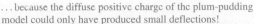

. . . because the diffuse positive charge of the plum-pudding model could only have produced small deflections!

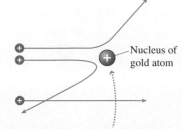

Nucleus of gold atom

So he proposed instead that the atom's positive charge is located in a tiny but massive nucleus that can deflect alpha particles sharply.

FIGURE 24.2 The Rutherford-Geiger-Marsden alpha particle scattering experiment. Most alphas suffer little deflection, but a few are deflected substantially. Using Rutherford's nuclear model, only those alphas that approach the nucleus closely scatter through large angles.

nucleus. Rutherford analyzed collisions between the gold nucleus and alpha particle using classical mechanics. His calculations agreed with the experimental results, provided the nucleus is a sphere with radius on the order of 10^{-15} m. The relatively large distances between nuclei (~10^{-10} m) explain why most alphas traverse the foil with little deflection, and the concentrated charge and mass in the tiny nucleus explain the rare backscattering events.

✓**TIP**

Nucleus is the singular; multiple ones are **nuclei**.

EXAMPLE 24.1 **Alpha Backscattering**

In some of Rutherford's experiments, the alpha particles (charge $+2e$, mass 6.64×10^{-27} kg) had initial kinetic energy 7.7 MeV. Assume an alpha particle with this energy is fired directly toward a pointlike gold nucleus with charge $+79e$. How close does the alpha get to the nucleus before reversing direction?

ORGANIZE AND PLAN Figure 24.3 is our sketch. This is a conservation-of-energy problem (Chapter 5). The alpha's total energy is the sum of kinetic and potential energy: $E = K + U$, with $K = \frac{1}{2}mv^2$. In this case the potential energy is electrical: $U = kq_1q_2/r$ (Equation 16.1), where r is the distance between the gold nucleus and alpha particle. Far from the nucleus, the alpha's energy is all kinetic. At closest approach it's instantaneously at rest so its energy is all potential. Equating the initial and closest-approach energies will give the unknown distance r.

Known: $K_0 = 7.7\,\text{MeV}$; $m = 6.64 \times 10^{-27}$ kg; $q_1 = 2e$; $q_2 = 79e$; $e = 1.60 \times 10^{-19}\,\text{C}$; $k = 8.99 \times 10^9\,\text{N}\cdot\text{m}^2/\text{C}^2$.

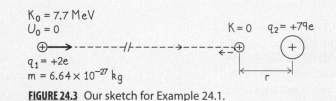

$K_0 = 7.7\,\text{MeV}$
$U_0 = 0$

$K = 0 \quad q_2 = +79e$

$q_1 = +2e$
$m = 6.64 \times 10^{-27}\,\text{kg}$

FIGURE 24.3 Our sketch for Example 24.1.

SOLVE Mass and charge are in SI units, so we need the energy in SI. Using $1\,\text{eV} = 1.60 \times 10^{-19}$ J,

$$K_0 = 7.7 \times 10^6\,\text{eV} \times \frac{1.60 \times 10^{-19}\,\text{J}}{1\,\text{eV}} = 1.23 \times 10^{-12}\,\text{J}$$

Equate the initial energy to the potential energy at the turning point: $K_0 = kq_1q_2/r$. Then solve for r:

$$r = \frac{kq_1q_2}{K_0} = \frac{k(2e)(79e)}{K_0} = \frac{158ke^2}{K_0}$$

$$r = \frac{158(8.99 \times 10^9\,\text{N}\cdot\text{m}^2/\text{C}^2)(1.60 \times 10^{-19}\,\text{C})^2}{1.23 \times 10^{-12}\,\text{J}}$$

$$= 2.96 \times 10^{-14}\,\text{m}$$

REFLECT In fact the gold nucleus is smaller than this, with radius about 7×10^{-15} m, and the alpha particle is even smaller. Thus we're justified in treating them as point particles.

MAKING THE CONNECTION How would the calculation change if the gold were replaced with a lighter element, such as aluminum?

ANSWER Aluminum has a smaller nuclear charge—only $13e$. Therefore, the alpha particle would get closer before reversing direction.

Although Geiger and Marsden performed the experiment, the effect is called **Rutherford scattering**, because it was Rutherford who interpreted the results. A startling implication of the Rutherford-Geiger-Marsden work is that even solids are mostly empty space. The nuclear radius is on the order of 1 fm (10^{-15} m), while the atomic radius is some 100,000 times larger. To appreciate the relative sizes of nucleus and atom, imagine a scale model with the nucleus a baseball-sized sphere, about 8 cm in diameter. The atom's diameter would be 100,000 times larger—about 8 km, spanning a medium-sized city!

The "Solar System" Atom

The tiny, massive, and positively charged nucleus sits at the atom's center—much like the Sun in the solar system. The analogy doesn't end there. The attractive electric force from the nucleus acts on negative electrons just as the Sun's gravity does on planets. Since both forces fall as $1/r^2$, electrons should orbit the nucleus just as planets do the Sun (Figure 24.4). However, there are problems with this classical "solar system" model. For one thing, the electron's motion is accelerated. As you saw in Section 20.1, accelerated charge is the source of electromagnetic radiation. So the electron should radiate continuously and quickly lose energy—so quickly, in fact, that atoms should collapse in a fraction of a second! Yet atoms are generally stable and ordinarily don't radiate energy. Thus the solar system model seems inconsistent with the laws of electromagnetism.

In the solar system model, the electron is held in "orbit" around the nucleus by the attractive electric force.

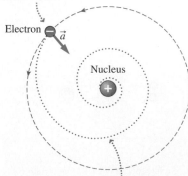

Electron

Nucleus

Because the electron is accelerated, the laws of electromagnetism suggest that it should radiate continuously and hence spiral into the nucleus — which it doesn't!

FIGURE 24.4 "Solar system" model of the hydrogen atom, with a single electron in circular orbit about the proton.

Another problem is that when atoms do radiate, they release brief bursts of radiation at particular wavelengths—the discrete spectral lines noted in Chapter 23. But the solar system model suggests continuous radiation. Furthermore, electromagnetism shows that the radiation frequency should be that of the orbital motion. But any orbit should be allowed, and so, therefore, should any frequency or wavelength of radiation. These quandaries did not last long, but it took a radical application of quantum ideas to resolve them.

GOT IT? Section 24.1   Alpha particles with the same energy are directed toward foils of gold and aluminum. In head-on collisions between alphas and nuclei, the alpha's distance of closest approach will be (a) smaller for the gold target; (b) smaller for the aluminum target; (c) the same for both the gold and aluminum target.

24.2 The Bohr Atom

Distinctive atomic spectra demanded an explanation, and at the same time clearly held clues to atomic structure. Physicists focused their attention on hydrogen, the simplest atom.

The Balmer Series

Scientists had known atomic spectra since the early 19th century, and they were intrigued by the patterns evident in arrangements of spectral lines—patterns most obvious for hydrogen (Figure 24.5). In 1885 the Swiss schoolteacher Johann Balmer (1825–1898) found that the visible wavelengths from hydrogen fit the expression

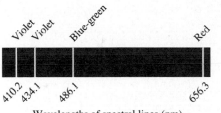

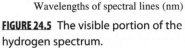

Wavelengths of spectral lines (nm)

FIGURE 24.5 The visible portion of the hydrogen spectrum.

$$\lambda = (364.56 \text{ nm})\frac{k^2}{k^2 - 4} \qquad \text{(Balmer formula)} \qquad (24.1)$$

where $k = 3, 4, 5$, and 6. You can check Balmer's results for yourself with the wavelengths in Figure 24.5. For example, Balmer's formula with $k = 3$ gives

$$\lambda = (364.56 \text{ nm})\frac{3^2}{3^2 - 4} = (364.56 \text{ nm})\frac{9}{5} = 656.2 \text{ nm}$$

corresponding to the red spectral line. (The wavelengths have been measured more accurately since Balmer's time, and those in Figure 24.5 are modern values.) The other visible lines correspond to $k = 4, 5$, and 6. Balmer predicted that spectral lines corresponding to higher integers might be observed in the ultraviolet. The first of these ($k = 7, \lambda = 397$ nm) had in fact already been found. The entire set of spectral lines, which includes all possible values of k, is called the **Balmer series**.

A more general form of Balmer's equation was found in 1890 by Swedish physicist Johannes Rydberg (1854–1919):

$$\frac{1}{\lambda} = R_{\text{H}}\left(\frac{1}{j^2} - \frac{1}{k^2}\right) \qquad (j < k) \qquad \text{(Rydberg equation; SI unit: m}^{-1}\text{)} \qquad (24.2)$$

Here j and k are integers, with $j < k$. $R_{\text{H}} = 1.097 \times 10^7 \text{ m}^{-1}$ is the **Rydberg constant** for hydrogen. With $j = 2$, the Rydberg equation becomes Balmer's formula. Spectral lines corresponding to other j values were discovered in the 20th century. We'll discuss their significance shortly.

✓**TIP**

The Rydberg constant can also be written $R_{\text{H}} = 0.01097 \text{ nm}^{-1}$, in which case $1/\lambda$ is in nm^{-1} and λ itself in nm. The restriction $j < k$ ensures positive wavelengths.

EXAMPLE 24.2   **Rydberg Equation and the Balmer Series**

Use the Rydberg equation to explore the wavelength pattern for the entire Balmer series.

ORGANIZE AND PLAN The Balmer series corresponds to the Rydberg equation (Equation 24.2)

$$\frac{1}{\lambda} = R_H\left(\frac{1}{j^2} - \frac{1}{k^2}\right)$$

with $j = 2$. With the restriction $j < k$, this means that k can be any integer larger than 2.

Known: $j = 2$; $k = 3, 4, 5, \ldots$; Rydberg constant $R_H = 0.01097 \text{ nm}^{-1}$.

SOLVE You already know that the k values 3, 4, 5, and 6 correspond to the four visible spectral lines in Figure 24.5, and $k = 7$ corresponds to the 397-nm ultraviolet line. Using the Rydberg equation, the wavelengths to four significant figures are:

$$k = 3 : \lambda = 656.3 \text{ nm (red)}$$
$$k = 4 : \lambda = 486.2 \text{ nm (blue-green)}$$
$$k = 5 : \lambda = 434.1 \text{ nm (violet)}$$

$$k = 6 : \lambda = 410.2 \text{ nm (violet)}$$
$$k = 7 : \lambda = 397.0 \text{ nm (ultraviolet)}$$

Note that as k increases, the spectral lines get closer. For large k, the term $1/k^2$ becomes very small and the spectrum appears almost continuous. The shortest possible wavelength in the series is for $k \to \infty$:

$$\frac{1}{\lambda} = R_H\left(\frac{1}{j^2} - \frac{1}{k^2}\right) = (0.01097 \text{ nm}^{-1})\left(\frac{1}{2^2}\right) = 0.0027425 \text{ nm}^{-1}$$

This corresponds to $\lambda = 364.6$ nm and is called the **series limit**. Thus, the entire Balmer series lies in the range from 364.6 nm to 656.3 nm.

REFLECT The pattern you've seen here, with larger gaps between spectral lines at longer wavelengths and smaller gaps toward a series limit ($k \to \infty$), occurs in other series with different values of j.

MAKING THE CONNECTION What are the possible values of k for a series with $j = 7$?

ANSWER k must be an integer larger than j. Therefore, the allowed values of k are 8, 9, 10, and so on to infinity.

Bohr's Theory

In 1912, the young Danish physicist Niels Bohr (1885–1962) visited Rutherford in England. Bohr found Rutherford's nuclear model convincing, and sought a theoretical framework that would also predict the observed spectral lines. Bohr's model was a mix of classical and quantum ideas and was built on four general assumptions:

Bohr's general assumptions:

1. The electron orbits the nucleus in a circular orbit. However, only certain specific orbits—and, correspondingly, electron energies—are allowed. These correspond to so-called **stationary states** of the atom.
2. Electromagnetic radiation is emitted only when an electron makes a transition from one stationary state to another. The difference in energy ΔE between states shows up as the energy of the emitted photon: $\Delta E = hf$.
3. Classical physics governs the electron orbits in stationary states, but the rules of classical dynamics do not apply to transitions between states.
4. The angular momentum L of each stationary state is quantized. Specifically, $L = \dfrac{nh}{2\pi}$, where $n = 1, 2, 3, \ldots$ is the **quantum number** characterizing a particular orbit.

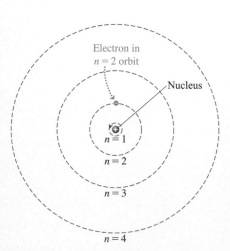

FIGURE 24.6 The first four orbits in the Bohr atom.

Figure 24.6 shows the Bohr model of hydrogen. The lowest orbit, with $n = 1$, is the **ground state**, with higher orbits ($n > 1$) corresponding to **excited states**.

Bohr's assumption 4 quantizes angular momentum, and that, along with assumption 3, leads directly to quantized electron energies. Classically, the orbiting electron has kinetic energy $K = \frac{1}{2}mv^2$, while the atom's potential energy U results from charges $+e$ (the nucleus) and $-e$ (the electron) separated by a distance r (Equation 16.1):

$$U = \frac{kq_1q_2}{r} = \frac{k(e)(-e)}{r} = -\frac{ke^2}{r}$$

The atom's total energy, which is conserved, is therefore $E = K + U$:

$$E = K + U = \frac{1}{2}mv^2 - \frac{ke^2}{r}$$

As detailed in Figure 24.7, Newton's second law for the orbiting electron leads to $mv^2 = ke^2/r$. Using this expression in the atom's total energy E gives

$$E = \frac{1}{2}\frac{ke^2}{r} - \frac{ke^2}{r}$$

or

$$E = -\frac{ke^2}{2r} \quad \text{(Energy of hydrogen atom, Bohr model; SI unit: J)} \quad (24.3)$$

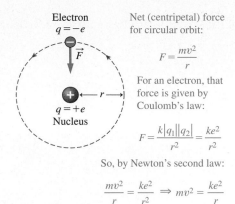

Electron
$q = -e$

$\vec{F}$

$q = +e$
Nucleus

Net (centripetal) force
for circular orbit:

$$F = \frac{mv^2}{r}$$

For an electron, that
force is given by
Coulomb's law:

$$F = \frac{k|q_1||q_2|}{r^2} = \frac{ke^2}{r^2}$$

So, by Newton's second law:

$$\frac{mv^2}{r} = \frac{ke^2}{r^2} \Rightarrow mv^2 = \frac{ke^2}{r}$$

FIGURE 24.7 Using the electron's motion to find the Bohr atom's energy.

Note that the total energy is negative. You might recall that a similar analysis of satellite orbits in Chapter 9 led to

$$E = -\frac{GM_E m}{2r}$$

where r is the radius of a circular orbit, M_E Earth's mass, and m the satellite's mass. The similarity with Equation 24.3 is no coincidence: Classical analysis of the electron orbit in hydrogen is completely analogous to circular orbits under gravity, because each orbit results from an inverse-square force.

So far we've used just classical physics. Now we invoke Bohr's fourth assumption to quantize the orbits. In Chapter 8 we defined angular momentum $L = I\omega$, where I is rotational inertia and ω is angular velocity. In hydrogen the electron is essentially a point particle, so for an electron with mass m and orbital radius r, $I = mr^2$ and $\omega = v/r$. Combining these expressions with Bohr's fourth assumption gives the quantized angular momentum:

$$L = \frac{nh}{2\pi} = (mr^2)\left(\frac{v}{r}\right)$$

Solving for orbital speed v,

$$v = \frac{nh}{2\pi mr}, \text{ or } v^2 = \frac{n^2h^2}{4\pi^2m^2r^2}$$

But from Figure 24.7, $v^2 = \frac{ke^2}{mr}$. Equating our two expressions for v^2,

$$v^2 = \frac{ke^2}{mr} = \frac{n^2h^2}{4\pi^2m^2r^2}$$

Then solving for the allowed orbital radii,

$$r = n^2\left(\frac{h^2}{4\pi^2mke^2}\right)$$

The quantity in parentheses has units of length and is called the **Bohr radius**, a_0. Numerically,

$$a_0 = \frac{h^2}{4\pi^2mke^2} = \frac{(6.626 \times 10^{-34}\,\text{J}\cdot\text{s})^2}{4\pi^2(9.11 \times 10^{-31}\,\text{kg})(8.99 \times 10^9\,\text{N}\cdot\text{m}^2/\text{C}^2)(1.60 \times 10^{-19}\,\text{C})^2}$$
$$= 5.29 \times 10^{-11}\,\text{m}$$

Amazingly, that's just about the size of an atom! In general, the radius of the hydrogen atom in the nth quantum state is

$$r = n^2 a_0 \quad n = 1, 2, 3, \ldots \text{(Radius of Bohr atom; SI unit: m)} \quad (24.4)$$

The corresponding energy E follows by substituting Equation 24.4 into Equation 23.3 for the energy:

$$E = -\frac{ke^2}{2r} = -\frac{ke^2}{2n^2 a_0} = -\frac{1}{n^2}\left(\frac{ke^2}{2a_0}\right)$$

Here the quantity in parentheses is a combination of constants with dimensions of energy, designated E_0. You can show that its value is $E_0 = 2.18 \times 10^{-18}$ J. Then the energy of the nth quantum state is

$$E = -\frac{E_0}{n^2} \quad n = 1, 2, 3, \ldots \text{(Energy of Bohr atom; SI unit: J)} \tag{24.5}$$

✓**TIP**

At the atomic scale it's often convenient to work with energies in electron volts. Then $E_0 = 13.6$ eV; using this value, the energies in Equation 24.5 come out in eV.

Reviewing New Concepts: The Bohr Model of Hydrogen

- The electron makes a circular orbit around the positive nucleus.
- Classical physics describes the orbital dynamics, but the orbits are quantized according to Bohr's assumptions.
- Each orbit is characterized by an integer quantum number n.
- The radius of the nth orbit is $r = n^2 a_0$, where $a_0 = 5.29 \times 10^{-11}$ m is the Bohr radius.
- The energy of the nth orbit is $E = -\dfrac{E_0}{n^2}$, where $E_0 = 2.18 \times 10^{-18}$ J or 13.6 eV.

CONCEPTUAL EXAMPLE 24.3 **Energy and Orbital Radius**

As the quantum number n for the hydrogen atom increases, how does this affect (a) the atom's radius and (b) its energy?

SOLVE (a) Equation 24.4, $r = n^2 a_0$, shows that the radius increases as the *square* of the quantum number n. As you saw in Figure 24.6, the orbital spacing therefore grows rapidly with increasing n.

(b) Equation 24.5 gives the energy: $E = -E_0/n^2$, with $E_0 = 13.6$ eV. The reciprocal relationship $(1/n^2)$ and negative sign make the energy a little trickier to analyze. As n increases, $1/n^2$ decreases. Because of the negative sign, however, the energy is actually increasing—approaching zero from below—as n increases. This is best illustrated by computing the first few energy levels:

$$n = 1 : E_1 = -\frac{E_0}{1^2} = -13.6 \text{ eV}$$

$$n = 2 : E_2 = -\frac{E_0}{2^2} = -3.40 \text{ eV}$$

$$n = 3 : E_3 = -\frac{E_0}{3^2} = -1.51 \text{ eV}$$

These values show that the hydrogen atom's energy increases with increasing n.

REFLECT Note that in the limit of large n the atom gets very large. These so-called "Rydberg atoms" have been observed for $n \approx 300$, making the atom's diameter some 5μm! For such an atom the total energy E is very close to 0, and it wouldn't take much energy to dislodge the electron altogether.

The Bohr Atom and the Hydrogen Spectrum

Any successful theory of the hydrogen atom has to account for the spectral lines seen in the Balmer series. In Bohr's model, the atom can transition from one quantum state to another. If it drops from a higher state to a lower one, energy lost by the atom shows up as a photon of energy hf (Bohr's assumption 2). A transition between two specific quantum states always produces a photon of the same specific energy and wavelength, which explains why only certain wavelengths are observed.

For example, suppose a hydrogen atom undergoes a transition from the $n = 3$ state to $n = 2$. Using the energies of those two states from Conceptual Example 24.3, the energy of the emitted photon can be computed:

$$E_{photon} = E_3 - E_2 = -1.51 \text{ eV} - (-3.40 \text{ eV}) = 1.89 \text{ eV}$$

or 3.03×10^{-19} J. With $E = hc/\lambda$ for a photon (Equation 23.4), this photon's wavelength is

$$\lambda = \frac{hc}{E_{photon}} = \frac{(6.626 \times 10^{-34} \text{ J} \cdot \text{s})(3.00 \times 10^8 \text{ m/s})}{3.03 \times 10^{-19} \text{ J}} = 6.56 \times 10^{-7} \text{ m} = 656 \text{ nm}$$

This corresponds exactly to the red spectral line in the Balmer series!

To see how other observed spectral lines relate to quantized energy levels in hydrogen, it's helpful to draw an **energy-level diagram** (Figure 24.8). The Balmer series consists of all the photons produced when a hydrogen atom makes a transition ending at the $n = 2$ level. You've already seen that the red spectral line corresponds to the $n = 3 \rightarrow 2$ transition. The other visible lines correspond to $n = 4 \rightarrow 2$ (blue-green), $n = 5 \rightarrow 2$ (violet), and $n = 6 \rightarrow 2$ (violet).

More generally, Bohr showed that his quantized energy levels account for all the photon wavelengths predicted by the Rydberg equation, with all allowed values of j and k. Suppose a hydrogen atom makes a transition from a higher level ($n = k$) to a lower level ($n = j$). Then from Equation 24.5, the atom emits a photon of energy

$$E_{photon} = E_k - E_j = -\frac{E_0}{k^2} - \left(\frac{E_0}{j^2}\right) = E_0 \left(\frac{1}{j^2} - \frac{1}{k^2}\right)$$

With $E_{photon} = hc/\lambda$, this gives

$$\frac{1}{\lambda} = \frac{E_{photon}}{hc} = \frac{E_0}{hc}\left(\frac{1}{j^2} - \frac{1}{k^2}\right)$$

This matches the Rydberg equation, provided $R_H = E_0/hc$. Evaluating the constant E_0/hc numerically,

$$\frac{E_0}{hc} = \frac{2.18 \times 10^{-18} \text{ J}}{(6.626 \times 10^{-34} \text{ J} \cdot \text{s})(3.00 \times 10^8 \text{ m/s})} = 1.097 \times 10^7 \text{ m}^{-1}$$

That's the Rydberg constant! Thus Bohr's theory of the hydrogen atom accounts for precisely the wavelengths predicted by the Rydberg equation.

Other Spectral Series in Hydrogen

You've seen that the Balmer series results from transitions ending at $n = 2$. Bohr's theory and the hydrogen energy-level diagram suggest there should be additional series, with different ending states. How about transitions ending at $n = 1$? They result in the **Lyman series** of spectral lines. Notice in Figure 24.8 that the arrows representing transitions to the $n = 1$ level are longer than those representing transitions to $n = 2$. Those longer arrows correspond to greater photon energies. Since the Balmer series extends through the visible and into the near ultraviolet, it's not surprising that the Lyman series corresponds to shorter-wavelength ultraviolet photons. Next, consider transitions ending at $n = 3$ (the **Paschen series**). The energy gap between this level and higher levels isn't so large—hence the shorter arrows. The lower-energy photons from these transitions will be in the infrared.

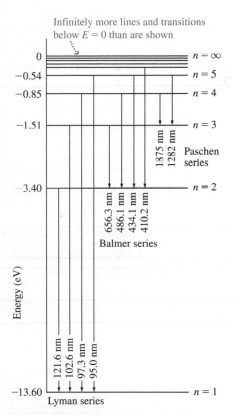

FIGURE 24.8 Energy-level diagram for the Bohr atom, showing the transitions responsible for parts of the first three series of spectral lines.

Spectral series outside the visible region are naturally harder to observe than the visible part of the Balmer series, and most weren't detected until after Bohr presented his atomic theory. Discovery of these other series (Table 24.1) provided substantial confirmation of Bohr's theory. In each case, the observed wavelengths matched the corresponding transitions between Bohr's energy levels. One characteristic of some of the most successful theories in physics is that they have not only accounted for observed phenomena but also predicted others that were observed later. The Bohr atom is a dramatic example.

TABLE 24.1 Hydrogen Spectral Series

Discoverer (year)	Wavelength range	$n = j$ (final state)	$n = k$ (initial state)
Lyman (1916)	Ultraviolet	1	>1
Balmer (1885)	Visible, near UV	2	>2
Paschen (1908)	Infrared	3	>3
Brackett (1922)	Infrared	4	>4
Pfund (1924)	Infrared	5	>5

EXAMPLE 24.4 **Range of the Lyman and Paschen Series**

Find the wavelength ranges of the Lyman and Paschen series.

ORGANIZE AND PLAN These series correspond to $j = 1$ (Lyman) and $j = 3$ (Paschen) in the Rydberg equation (Equation 24.2):

$$\frac{1}{\lambda} = R_H\left(\frac{1}{j^2} - \frac{1}{k^2}\right)$$

Then the allowed values of k range from $j + 1$ to infinity, giving a corresponding range in wavelengths.

Known: Rydberg constant $R_H = 0.01097\,\text{nm}^{-1}$.

SOLVE For the Lyman series, allowed k values range from 2 to ∞. The maximum wavelength is found using $k = 2$:

$$\frac{1}{\lambda} = R_H\left(\frac{1}{j^2} - \frac{1}{k^2}\right) = (0.01097\,\text{nm}^{-1})\left(\frac{1}{1^2} - \frac{1}{2^2}\right)$$

$$= 0.0082275\,\text{nm}^{-1}$$

giving

$$\lambda = \frac{1}{0.0082275\,\text{nm}^{-1}} = 122\,\text{nm}$$

The shortest wavelength—the Lyman series limit—results when $k \rightarrow \infty$:

$$\frac{1}{\lambda} = R_H\left(\frac{1}{j^2} - \frac{1}{k^2}\right) = (0.01097\,\text{nm}^{-1})\left(\frac{1}{1^2}\right) = 0.01097\,\text{nm}^{-1}$$

which gives $\lambda = 91.2$ nm. Thus the Lyman series ranges from 91.2 nm to 122 nm—entirely within the ultraviolet part of the spectrum.

For the Paschen series, we do similar calculations with $j = 3$, starting with $k = 4$ for the longest wavelength and $k = \infty$ for the series limit. The result is a wavelength range from 820 nm to 1875 nm, entirely in the infrared.

REFLECT The pattern continues to the other series. The wavelengths for the Brackett and Pfund series (described in Table 24.1) are even longer—farther into the infrared.

MAKING THE CONNECTION What's the series limit for the series with $j = 7$?

ANSWER Using the method of this example, the series limit is 4466 nm, or over 4 μm—well into the infrared.

Emission and Absorption Spectra

You may have seen the hydrogen Balmer series from a hydrogen gas discharge tube, somewhat akin to a common fluorescent lightbulb. When excited by an electric current, the hydrogen glows brightly, with the resulting color a blend of the four visible Balmer wavelengths. A handheld diffraction grating reveals the individual lines. This is an **emission spectrum**, so called because it results from atoms emitting photons as they make downward transitions. In contrast, the solar spectrum in Figure 24.9 is an **absorption spectrum**, an otherwise continuous spectrum missing certain wavelengths. Here diffuse gas

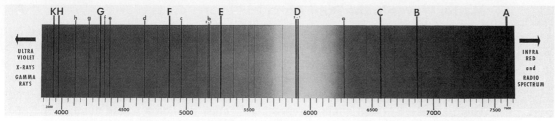

FIGURE 24.9 The solar spectrum, showing dark absorption lines resulting from removal of those wavelengths by the solar atmosphere.

(mostly hydrogen) in the Sun's atmosphere absorbs specific wavelengths from the continuous blackbody radiation emitted by the dense solar surface. Those wavelengths correspond to photons with the right energies to produce upward transitions among atomic states—the same wavelengths that result when downward transitions occur. Thus the absorption lines are, for hydrogen, at the Balmer and other series wavelengths. Since photons with these energies are removed from the continuous blackbody spectrum, we see dark spectral lines. Thus both emission and absorption spectra provide good confirmation of Bohr's theory.

✓ **TIP**

The solar spectrum also shows evidence of other elements, such as helium, which was discovered on the Sun before it was known on Earth (Chapter 22).

APPLICATION   **Galaxies and Intergalactic Space**

The absorption spectra of distant galaxies show patterns similar to those of the solar spectrum, with lines of such elements as hydrogen, helium, and sodium. However, their wavelengths are longer than in the solar spectrum. The longer wavelengths result from a Doppler shift (Chapter 20) and indicate that distant galaxies are receding from us at high speeds—evidence of an expanding universe. Additional absorption occurs in the intergalactic medium and provides an important probe of this extremely tenuous gas. The so-called "Lyman alpha forest" is a hodgepodge of absorption lines in the near ultraviolet and visible spectrum, resulting from absorption of the shortest wavelength hydrogen series line—the Lyman alpha spectral line with wavelength 91 nm—by clouds of intergalactic matter receding at high speeds. The Lyman alpha line is Doppler shifted, by different amounts, to longer wavelengths—hence the "forest" of lines.

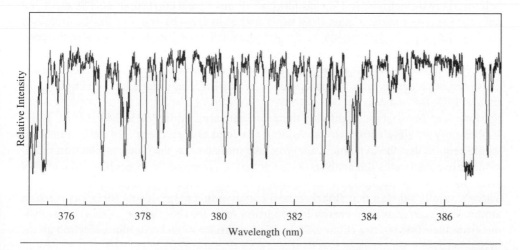

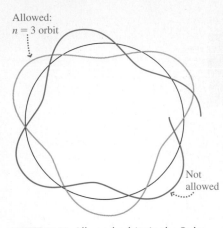

Allowed:
n = 3 orbit

Not
allowed

FIGURE 24.10 Allowed orbits in the Bohr model are those that fit an integer number of de Broglie wavelengths.

Matter Waves in the Bohr Atom

When he developed his particle-wave hypothesis (Section 23.4), de Broglie noticed immediately a connection to Bohr's quantized orbits. He reasoned that if electrons have wave properties, then an electron orbiting the nucleus could be modeled by a standing wave, as shown in Figure 24.10. De Broglie suggested that a fixed standing wave pattern—one that closes continuously back on itself—would correspond to a stable orbit. That means the orbital circumference ($2\pi r$) must include a whole number of wavelengths ($n\lambda$), so $n\lambda = 2\pi r$. Applying de Broglie's hypothesis $\lambda = h/p$, this equation can be rearranged to read $rp = nh/2\pi$. The quantity rp is the orbiting electron's angular momentum L, so $L = nh/2\pi$. But that's precisely Bohr's assumption about angular momentum quantization! So de Broglie's picture of the electron's orbit fitting a standing wave is equivalent to Bohr's angular momentum quantization. Given other evidence for matter waves (discussed in Chapter 23), this correspondence between electron wavelengths and the Bohr atom helped justify Bohr's hypothesis of quantized angular momentum. Thinking of electrons as waves then helped Erwin Schrödinger and others develop a more complete quantum theory of the hydrogen atom. You'll encounter aspects of that theory later in this chapter.

Assessment of the Bohr Model

You've seen how Bohr's quantum theory successfully predicted the spectrum of hydrogen and how it fit well with de Broglie's matter waves. But it wasn't perfect. First, Bohr's theory as we've presented it assumes the nucleus is at rest. But classical physics says the nucleus and electron should both orbit their common center of mass. All that's needed to correct for that effect, it turns out, is to replace the electron mass m with the *reduced mass* μ, defined as

$$\mu = \frac{m}{1 + m/M}$$

where M is the mass of the nucleus. Because $m/M \approx 1/1840$ for hydrogen, the correction is small. Nevertheless, it alters the Bohr radius and Rydberg constant and actually yields more accurate results than the simpler version of the theory we presented. The best way to see this correction is to compare the spectra of hydrogen and its isotope deuterium, whose nucleus has about twice the mass. The reduced-mass correction accurately predicts the slight differences in wavelength between hydrogen and deuterium spectra.

Bohr's atomic theory also works for other single-electron atoms, such as He^+. Neutral helium has two electrons, so He^+ has just one. In this case a correction must be made not only for the larger nuclear mass (four times that of hydrogen), but also for the increased nuclear charge ($+2e$ for helium). In general, the nucleus of atomic number Z in the periodic table has charge $+Ze$. If you use nuclear charge Ze in all the Bohr theory calculations, you'll find that the Rydberg constant is larger than hydrogen's by a factor of Z^2. This correction agrees with the observed spectra of He^+ and other single-electron atoms such as Li^{2+}, Be^{3+}, and so on.

However, this success highlights the principal shortcoming of Bohr's theory: It describes only single-electron atoms. Adding a second electron (e.g., for neutral helium) would require that the dynamical analysis take into account the electron-electron repulsion. This complicates the analysis so much that it's impossible to get Bohr-like energy levels for helium. Adding more electrons for the other elements compounds the difficulty. There's simply no way that Bohr's model can give us energy levels for multi-electron atoms. However, quantized energy levels clearly exist for those atoms, a fact that's obvious from their line spectra (Figure 23.1). We'll consider models for multi-electron atoms in Section 24.3.

EXAMPLE 24.5   **Spectrum of the Helium Ion**

Find the wavelength of the photon emitted when the electron in He$^+$ makes a transition from the $n = 3$ level to $n = 2$. Ignore the correction for reduced mass.

ORGANIZE AND PLAN The helium ion He$^+$ has a single electron, so Bohr's theory applies. The emitted wavelength follows from the Rydberg equation

$$\frac{1}{\lambda} = R\left(\frac{1}{j^2} - \frac{1}{k^2}\right)$$

where the Rydberg constant R is a factor of $Z^2 = 4$ greater than the hydrogen Rydberg R_H, as explained in the text. For this transition, $j = 2$ and $k = 3$.

Known: Hydrogen Rydberg constant $R_H = 0.01097\,\text{nm}^{-1}$; $j = 2$; $k = 3$.

SOLVE Writing the Rydberg constant as $R = 4R_H$, we have

$$\frac{1}{\lambda} = 4R_H\left(\frac{1}{j^2} - \frac{1}{k^2}\right) = 4(0.01097\,\text{nm}^{-1})\left(\frac{1}{2^2} - \frac{1}{3^2}\right)$$

$$= 0.0060944\,\text{nm}^{-1}$$

Thus

$$\lambda = \frac{1}{0.0060944\,\text{nm}^{-1}} = 164\,\text{nm}$$

REFLECT This is in the ultraviolet, and it's just one-fourth of 656 nm, the wavelength of the $n = 3 \rightarrow 2$ transition in hydrogen. Generally, you can expect larger energies (shorter wavelengths) associated with transitions in larger atoms. This will be important for x-ray emissions, discussed in Section 24.5.

MAKING THE CONNECTION Is there a transition in He$^+$ that produces a photon with wavelength close to 656 nm?

ANSWER Yes. The transition from $n = 6 \rightarrow 4$ gives exactly that wavelength.

GOT IT? Section 24.2 Rank in decreasing order the wavelengths of photons emitted when a hydrogen atom makes the following transitions: (a) $n = 4 \rightarrow 2$; (b) $n = 2 \rightarrow 1$; (c) $n = 4 \rightarrow 1$; (d) $n = 8 \rightarrow 3$.

24.3 Quantum Numbers and Atomic Spectra

Bohr's atomic theory explained hydrogen but not multi-electron atoms. However, Bohr's success led others to an improved quantum theory that applied to all atoms. In Chapter 23 we described how Erwin Schrödinger invoked matter waves in his more complete quantum theory, with all the properties of a particle contained in its wave function Ψ. This new theory would soon lead to a deeper understanding of hydrogen and other atoms.

More Quantum Numbers

Schrödinger's wave theory, applied to the hydrogen atom, predicts the same energy levels as the older Bohr theory: $E = -E_0/n^2$, where n—now called the **principal quantum number**—again takes integer values 1, 2, 3, and so on. The new quantum theory added two more quantum numbers: l, the **orbital quantum number**, and m_l, called the **magnetic orbital quantum number**. To describe the orbital state of an electron in hydrogen requires all three quantum numbers: n, l, and m_l.

The orbital quantum number l is related to the electron's orbital angular momentum, L. In classical physics this quantity could have any value, but in quantum mechanics its magnitude is quantized according to the rule

$$L = \sqrt{l(l + 1)}\,\frac{h}{2\pi} \quad \text{(Orbital angular momentum; SI unit: J·s)} \quad (24.6)$$

The quantum number l takes integer values 0, 1, 2, and so on, but with the restriction that $l < n$. Thus hydrogen's ground state ($n = 1$) requires $l = 0$. Excited states ($n > 1$) permit higher values of l. With $n = 3$, for example, then the allowed values of l are 0, 1, or 2. Note that this differs from the Bohr model, where orbital angular momentum was

z-component of
angular momentum

$$L_z = m_l \frac{h}{2\pi}$$

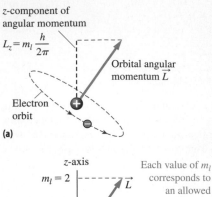

Orbital angular
momentum $\vec{L}$

Electron
orbit

(a)

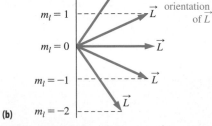

(b)

FIGURE 24.11 (a) The angular momentum vector is perpendicular to the electron's orbital plane. (b) With $l = 2$, the angular momentum vector has five possible spatial orientations, corresponding to the allowed values of m_l.

$L = nh/2\pi$. In fact, Equation 24.6 predicts that the orbital angular momentum is zero when $l = 0$, which isn't consistent with Bohr's circular orbits. We stress that Equation 24.6 is correct, and Bohr's planetlike orbits don't accurately describe atoms.

✓ **TIP**

Like other quantum numbers, l is dimensionless. Planck's constant h has dimensions of angular momentum, so Equation 24.6 gives the proper units.

The magnetic quantum number m_l is restricted to integers ranging from $-l$ to $+l$, including 0. For example, if the orbital quantum number is $l = 2$, then the possible values of m_l are $-2, -1, 0, 1$, and 2. Figure 24.11 shows the physical significance of m_l. As you learned in Section 8.9, the angular momentum vector is perpendicular to the orbit in a direction given by the right-hand rule (Figure 24.11a). But quantum mechanics admits only certain allowed values for that orientation—a feature known as **space quantization**. Figure 24.11b shows the allowed orientations of the angular momentum vector, which correspond to the different m_l values. In general, the relationship between m_l and the z-component of angular momentum L_z is

$$L_z = m_l \frac{h}{2\pi}$$

What axis are we supposed to use for space quantization? In fact, any axis will do, but space quantization is most significant when that axis is in the direction of a magnetic field—that's why m_l is the *magnetic* quantum number.

In addition to orbital angular momentum, the electron has intrinsic angular momentum called **spin**. Although it's tempting to think of the electron as a little spinning ball, we emphasize that spin is a purely quantum-mechanical effect. To be an electron is to have this intrinsic angular momentum; it's not something imposed on the electron by setting it into spinning motion. Every electron has spin angular momentum with the same magnitude S:

$$S = \sqrt{\frac{3}{4}} \frac{h}{2\pi}$$

Electron spin requires a fourth quantum number, the **magnetic spin quantum number** m_s. It's analogous to the magnetic orbital quantum number because it quantizes the orientation of the spin angular momentum vector:

$$S_z = m_s \frac{h}{2\pi}$$

similar to the relationship between L_z and m_l. For the electron there are two possible values of m_s: $-1/2$ and $+1/2$, which physicists call "spin down" and "spin up," respectively. The value of m_s is independent of the other three quantum numbers, so for a given combination of n, l, and m_l, the fourth quantum number m_s may be either $-1/2$ or $+1/2$. The state of an atomic electron is fully described by its four quantum numbers, as summarized in Table 24.2.

TABLE 24.2 Quantum Numbers for Atomic Electrons

Name	Symbol	Allowed values
Principal	n	$1, 2, 3, \ldots$
Orbital angular momentum	l	$0, 1, 2, \ldots, n-1$
Magnetic	m_l	$-l, \ldots, -2, -1, 0, 1, 2, \ldots, l$
Magnetic spin	m_s	$-1/2$ or $+1/2$

CONCEPTUAL EXAMPLE 24.6   **Quantum Numbers for $n = 3$**

What are the possible combinations of quantum numbers for an electron on the $n = 3$ level of hydrogen? How many combinations are there?

SOLVE For $n = 3$, l can assume values 0, 1, or 2. For each l, m_l ranges from $-l$ to $+l$. Finally, for each combination of n, l, and m_l, m_s is either $\frac{1}{2}$ or $-\frac{1}{2}$. The table below summarizes the resulting 18 possible combinations of the four quantum numbers.

Table of Quantum Numbers for $n = 3$

Quantum number l	Quantum number m_l	(n, l, m_l, m_s)
$l = 0$	$m_l = 0$	$(3, 0, 0, 1/2)$ or $(3, 0, 0, -1/2)$
$l = 1$	$m_l = -1$	$(3, 1, -1, 1/2)$ or $(3, 1, -1, -1/2)$
	$m_l = 0$	$(3, 1, 0, 1/2)$ or $(3, 1, 0, -1/2)$
	$m_l = 1$	$(3, 1, 1, 1/2)$ or $(3, 1, 1, -1/2)$
$l = 2$	$m_l = -2$	$(3, 2, -2, 1/2)$ or $(3, 2, -2, -1/2)$
	$m_l = -1$	$(3, 2, -1, 1/2)$ or $(3, 2, -1, -1/2)$
	$m_l = 0$	$(3, 2, 0, 1/2)$ or $(3, 2, 0, -1/2)$
	$m_l = 1$	$(3, 2, 1, 1/2)$ or $(3, 2, 1, -1/2)$
	$m_l = 2$	$(3, 2, 2, 1/2)$ or $(3, 2, 2, -1/2)$

REFLECT Even with the restriction $l < n$, the number of possible combinations grows rapidly with increasing n. Notice that there would be only two combinations of the four quantum numbers for $n = 1$, eight for $n = 2$ (all those with $l = 0$ or $l = 1$ in our table), and 18 for $n = 3$.

For an electron with principal quantum number n, there are n different values of the quantum number l, from 0 to $n - 1$. For each l there are $2l + 1$ values of m_l. As a result, for any quantum number n, there are n^2 combinations of n, l, and m_l. Including the spin quantum number doubles the number of combinations to $2n^2$. This rapidly growing number of combinations is important for the electronic structure of multi-electron atoms, which we'll consider in Section 24.4.

Spectroscopic Notation

Physicists use shorthand notation to describe atomic states, based on the quantum numbers n and l. The different l states are designated with the letters s, p, d, and f, for $l = 0, 1, 2$, and 3, respectively. (Historically, the letters stand for sharp, principal, diffuse, and fundamental, terms describing spectral lines in the early days of atomic spectroscopy.) In this notation, a state with $n = 3$ and $l = 2$ is a $3d$ state; a state with $n = 2$ and $l = 0$ is a $2s$ state.

The Zeeman Effect

Evidence for the orbital and magnetic quantum numbers is seen in the **Zeeman effect**, a splitting of spectral lines when atoms are in a strong magnetic field (Figure 24.12). This occurs because the energies for different m_l and m_s—normally almost the same for given n and l—take significantly different values in a magnetic field, as shown in Figure 24.13. This happens because, as you learned in Chapter 18, the orbiting electron constitutes a magnetic dipole, which has slightly different energy depending on its orientation in the magnetic field. Each different m_l state therefore has a different energy, giving rise to transitions with different energies and therefore different photon wavelengths, hence the spectral-line splitting.

Given the complexity of energy levels (Figure 24.13), why does the line only split into three lines? That follows from a so-called **selection rule**, which limits allowed transitions to those for which $m_l = 0$ or ±1. This and other selection rules are related to angular

No magnetic field: single spectral line

Strong magnetic field: spectral line splits into three

FIGURE 24.12 Zeeman splitting of one spectral line into three.

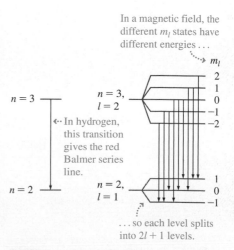

In a magnetic field, the different m_l states have different energies . . .

. . . so each level splits into $2l + 1$ levels.

FIGURE 24.13 Energy-level diagram for the Zeeman effect.

momentum conservation and can be derived theoretically in Schrödinger's quantum theory. Here's a summary of the selection rules that apply to atomic electron transitions:

- Δn = anything
- $\Delta l = \pm 1$
- Δm_l = 0 or ± 1
- Δm_s = 0 or ± 1 (that is, anything is allowed)

Consider, for example, the $n = 3$ to $n = 2$ transition in hydrogen, which produces the red line in the Balmer series. The orbital angular quantum number must change by 1 in this transition, as shown in Figure 24.13b, from $l = 2$ to $l = 1$ (a 3d to 2p transition). A transition from $l = 2$ to $l = 0$ is forbidden by the $l = \pm 1$ rule. The Zeeman effect depends on the m_l = 0 or ± 1 rule. Thus, an electron can go from $n = 3$, $m_l = 2$ to $n = 2$, $m_l = 1$, but not to $m_l = 0$ or $m_l = -1$. The transition arrows shown in Figure 24.13 are the only ones allowed by the selection rules.

A close examination of spectral lines even in the absence of an applied magnetic field reveals a very slight splitting, called **fine structure**. As you saw in Chapter 18, moving charge is the source of magnetic fields. As the electron orbits the nucleus, it "sees" a magnetic field resulting from the relative motion of nucleus and electron. The two spin states—up and down—have slightly different energies in this "internal" magnetic field, which explains fine-structure splitting.

Electron Clouds

Recall from Chapter 23 that quantum theory deals in probabilities, with Schrödinger's wave function giving only the probability of finding a particle. For atomic electrons, that translates into "clouds" that map out regions where we're likely to find an electron. Figure 24.14 shows electron probability distributions for different quantum states. Note that s states are spherically symmetric, while states with $l \neq 0$ have only axial symmetry. You can also see that p states with different m_l values have vastly different shapes. These probability clouds look nothing like Bohr's orbits!

GOT IT? **Section 24.3** Which of the following transitions are allowed in hydrogen? In each case the notation represents the quantum numbers (n, l, m_l, m_s) for an atom's state. (a) $(2, 1, 0, 1/2)$ to $(1, 0, 0, -1/2)$; (b) $(5, 2, 1, 1/2)$ to $(3, 2, 0, -1/2)$; (c) $(6, 2, -1, -1/2)$ to $(2, 1, 0, -1/2)$; (d) $(2, 0, 0, 1/2)$ to $(3, 1, -1, -1/2)$.

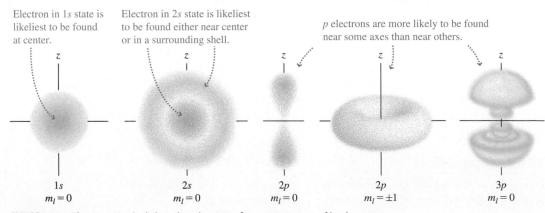

Electron in 1s state is likeliest to be found at center.

Electron in 2s state is likeliest to be found either near center or in a surrounding shell.

p electrons are more likely to be found near some axes than near others.

| 1s | 2s | 2p | 2p | 3p |
| $m_l = 0$ | $m_l = 0$ | $m_l = 0$ | $m_l = \pm 1$ | $m_l = 0$ |

FIGURE 24.14 Electron probability distributions for some states of hydrogen.

24.4 Multi-Electron Atoms and the Periodic Table

Hydrogen has one electron, making it the simplest atom. The theories of Bohr and subsequently Schrödinger gave physicists in the early 20th century a firm understanding of the hydrogen atom. Neither theory yields exact solutions for multi-electron atoms, but the quantum numbers and electron states we introduced for hydrogen serve as the basis for understanding multi-electron atoms.

The Pauli Exclusion Principle

An atom of an element with atomic number Z carries charge $+Ze$ in its nucleus. In the neutral atom, Z electrons surround the nucleus. But how are they distributed? As in hydrogen, each electron has a set of four quantum numbers. The atom, like other physical systems, normally assumes a state of lowest possible energy. From what you know of hydrogen, you might expect all the electrons to end up in the $n = 1$ state, since that state has the lowest energy. But that's not what happens. The reason it doesn't is articulated in the **Pauli exclusion principle**, named for the Austrian physicist Wolfgang Pauli (1900–1958), who suggested it in 1925. Pauli's principle is a fundamental feature of the quantum world that applies to electrons and many other common particles:

Pauli exclusion principle: No two electrons in an atom can be in the same quantum state. Because an electron's quantum state is specified by its four quantum numbers, that means no two electrons can have the same set of quantum numbers.

To see how the exclusion principle helps explain multi-electron atoms, we'll begin with helium, which has just two electrons. For lowest energy, the first electron should be at the $n = 1$ level. Then both l and m_l are zero. However, the spin quantum number m_s can be either $-1/2$ or $+1/2$. Because there are two choices, the second electron can also go into the $n = 1$ level, provided the two have opposite spins (opposite values of m_s). Thus, the two electrons in helium have quantum numbers (n, l, m_l, m_s) equal to $(1, 0, 0, -1/2)$ and $(1, 0, 0, +1/2)$.

Consider the next element, lithium, with atomic number 3. Lithium's first two electrons have quantum numbers $(1, 0, 0, -1/2)$ and $(1, 0, 0, +1/2)$, just like helium, because those are the lowest in energy. But the exclusion principle prohibits a third electron on the $n = 1$ level because $n = 1$ admits only two distinct sets of quantum numbers, corresponding to the two different spin states. Therefore, the third electron must have $n = 2$. This electron's second quantum number is $l = 0$, because that state has slightly lower energy than $l = 1$. Thus, the third electron in lithium has quantum numbers $(2, 0, 0, -1/2)$ or $(2, 0, 0, +1/2)$.

The same general ideas apply to larger atoms. Before going on, we'll introduce some of the notation used with multi-electron atoms. A **shell** is characterized by the principal quantum number n, and a **subshell** by the values of both n and l. Thus, you can speak of the $n = 3$ shell and the $3s$, $3p$, and $3d$ subshells. An atom's **electronic configuration** is a listing of each occupied subshell, with the number of electrons on that subshell as a superscript. The electronic configurations of the elements we've discussed to this point are $1s^1$ for hydrogen, $1s^2$ for helium, and $1s^2 2s^1$ for lithium. From now on, we'll use that shorthand, rather than listing all the sets of quantum numbers.

After lithium comes beryllium, with four electrons. Because there are two possible spin orientations, the exclusion principle allows a second electron in the $2s$ subshell, so beryllium's electronic configuration is $1s^2 2s^2$. Next comes boron, with five electrons. With no room left in the $2s$ subshell, the fifth electron goes into the $2p$ subshell, making boron's configuration $1s^2 2s^2 2p^1$.

Filling the Periodic Table

The strategy we've outlined for filling shells and subshells is the basis for understanding the electron configurations of other elements in the periodic table. As you've seen, it's important to know when a subshell is full, at which point the next electron goes into a

TABLE 24.3 Maximum Number of Electrons for Each Subshell

Orbital angular momentum quantum number l	Spectroscopic symbol for subshell	Allowed m_l values	Maximum number of electrons = (number of m_l values) $\times$ 2
0	s	0	2
1	p	$-1, 0, 1$	6
2	d	$-2, -1, 0, 1, 2$	10
3	f	$-3, -2, -1, 0, 1, 2, 3$	14

higher subshell. The number of electrons filling a subshell depends on the value of the quantum number l—or, in the language of subshells, whether it's an s, p, d, or f subshell. You've seen that an s subshell can contain two electrons. A p subshell has $l = 1$, with three possible values of m_l: $-1, 0$, and 1. But each m_l value can have two different m_s values ($-1/2$ and $+1/2$). Thus, a p subshell can hold $3 \times 2 = 6$ electrons. Table 24.3 summarizes the m_l values and electron capacities for each type of subshell. The maximum number of electrons increases with increasing l, and you can show that it's equal to $2(2l + 1)$.

For elements following boron ($1s^2 2s^2 2p^1$), electrons continue to fill the $2p$ subshell until it's full. That configuration ($1s^2 2s^2 2p^6$) is reached with a total of 10 electrons—the element neon. Next comes sodium, whose last electron must go into the $3s$ subshell. Thus, sodium's configuration is $1s^2 2s^2 2p^6 3s^1$.

Figure 24.15 shows electron configurations for the entire periodic table of the elements. Notice that the configuration for sodium is listed simply as $3s^1$, not the full $1s^2 2s^2 2p^6 3s^1$

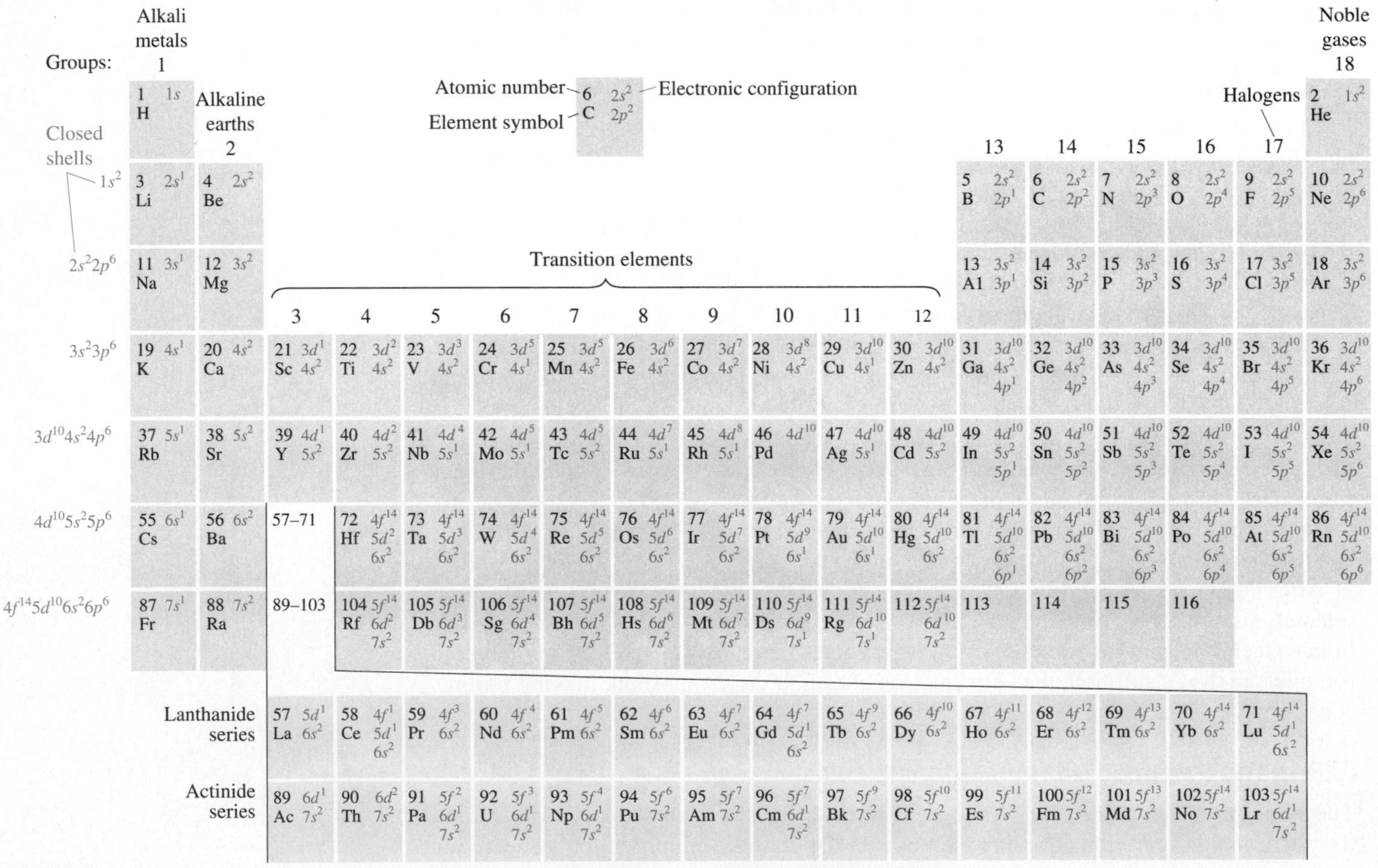

FIGURE 24.15 The periodic table, showing the ground-state electron configuration of each element.

that we just developed. That's because all elements beyond neon have full $1s$, $2s$, and $2p$ subshells, with remaining electrons in higher subshells. Therefore, there's no need to list the full electronic configuration, but only the highest filled and unfilled subshells. For convenience we've listed the highest filled subshells preceding each row of the periodic table, to the left of each row.

✓ **TIP**

The periodic table in Figure 24.15 lists the ground state configuration—that is, the state of lowest energy. When the atom is in an excited state, one or more electrons are in higher subshells.

Note that there are some irregularities in the order in which the subshells fill. After argon ($3s^23p^6$), you might expect that the last electron in the next element (potassium) would go into a $3d$ subshell. However, the energy of the $4s$ subshell is actually lower, so potassium has the configuration $3s^23p^64s^1$. Next is calcium ($3s^23p^64s^2$), followed by scandium ($3s^23p^63d^14s^2$). Then the $3d$ subshell continues to fill, but something unusual happens when it's nearly full. After nickel ($3s^23p^63d^84s^2$), the pattern indicates that the next element should add a ninth electron to the $3d$ subshell. However, it takes less energy to complete the $3d$ subshell by removing one electron from the $4s$ subshell, so the next element, copper, has $3s^23p^63d^{10}4s^1$. A number of similar anomalies occur in building up the larger elements.

Chemical Properties

The horizontal rows of the periodic table are called **periods**, which gives the table its name. The vertical columns are **groups**, numbered as shown in Figure 24.15. Elements within a group exhibit similar chemical properties, because they have the same number of electrons in their highest unfilled subshells; it's those outermost electrons that participate in chemical reactions. Some groups have names, such as **halogens, alkali metals,** and **noble gases.**

For example, the behavior of fluorine ($2s^22p^5$) is similar to that of chlorine ($3s^23p^5$); both are halogens. Both need one more electron to complete a p subshell. Completing a subshell comes with relatively little energy cost, so these elements easily gain an electron to form negative ions F^- and Cl^-. In contrast, elements in Group 1 such as sodium and potassium have a single s electron. It's energetically easy to lose that electron, forming positive ions Na^+ and K^+ and again leaving a filled outer shell. In chemistry, you learn that these positive and negative ions join to form common salts, including NaCl and KCl.

You've probably heard of *halogen lights*. They contain a tungsten filament, just like ordinary incandescent lightbulbs. The filament's high temperature continually vaporizes some tungsten. In an ordinary bulb, tungsten deposits on the glass, gradually blackening it and reducing light output. In a halogen bulb, the tungsten vapor combines with a halogen vapor (usually bromine), causing the tungsten to redeposit on the filament. This greatly increases the bulb's lifetime. Halogen lights have replaced incandescent bulbs in many applications, such as automobile headlights.

The **transition elements** in the periodic table consist of those groups that are in the process of filling d subshells. Transition metals often exhibit the magnetic properties of ferromagnetism and paramagnetism (see Section 18.6). That's because the d subshell electrons with different magnetic quantum numbers tend to have their spins aligned, resulting in a net magnetic moment.

The noble gases have entirely filled subshells. This makes it hard to gain or lose electrons, and as a result they tend not to bond with other atoms. That's why they're *inert*, meaning they don't readily undergo chemical reactions. As a result, they remain monatomic gases down to low temperatures. Helium has the lowest boiling point of any element, just 4.2 K at atmospheric pressure.

In both molecules, each carbon has four bonds.

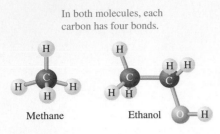

Methane   Ethanol

FIGURE 24.16 Carbon bonds in methane and ethanol.

Organic chemistry relies on the bonding properties of carbon. With configuration $2s^2 2p^2$, carbon needs four electrons to complete a subshell. Therefore, chemists draw carbon compounds showing four bonds between each carbon and its neighboring atoms, as in Figure 24.16. The varied three-dimensional geometry of carbon compounds comes from the fact that the unfilled subshell is a p subshell, with the spatial distributions shown in Figure 24.14. We won't pursue this topic here, but you'll likely encounter it in a course in organic chemistry or biology. Living organisms are carbon based and depend on this geometrical variety.

Electron configuration is also important in the semiconductors we discussed in Section 17.6. Silicon ($3s^2 3p^2$) is the basis of many semiconductor devices. "Doping" silicon with a small amount of impurities greatly changes its electrical properties. Using arsenic ($4s^2 4p^3$) introduces an extra electron that greatly increases the number of conduction electrons, thus raising the electrical conductivity. In contrast, doping with indium ($5s^2 5p^1$) creates an electron deficiency relative to silicon. This results in the formation of "holes," which act as positive charge carriers. Again, the result is an increase in electrical conductivity. By adjusting doping, engineers fine-tune the properties of the semiconductor devices that form the heart of modern microelectronics.

Russian chemist Dmitri Mendeleev (1834–1907) created the first periodic table in 1869, arranging the 60 or so known elements in order of atomic weight. Although Mendeleev noticed much of the group behavior we've described here, some elements seemed strangely out of order. That's because the order of atomic weight and atomic number isn't always the same. You've seen that the filling of subshells follows atomic number, so that's the proper way to order the periodic table. In Section 24.5 you'll learn how Mendeleev's table was corrected and our current understanding of atomic number was developed.

The Pauli exclusion principle isn't just about atoms. It applies to any configuration of multiple electrons as well as protons and neutrons, such as stars that have exhausted their nuclear fuel. With no thermal pressure to counter gravity, a star collapses to a fraction of its original size. In our Sun's case, the resulting *white dwarf* will cram most of the Sun's mass into the size of Earth. (See the illustration at the opening of this chapter.) Electrons get so close that the whole structure behaves like a giant atom, with no electrons in the same state. Therefore, most of the 10^{57} electrons end up in high-energy states, producing a pressure that sustains the star against further collapse. For stars with more than 1.4 times the Sun's mass, electron pressure isn't enough and the star collapses to a *neutron star*, where protons and electrons join to form neutrons. The exclusion principle now forces neutrons into higher states, again sustaining the star against collapse. But for stars with more than several times the Sun's mass, there's no force that can halt the collapse, and the result is a *black hole*.

CONCEPTUAL EXAMPLE 24.7   **Copper, Silver, and Gold**

Examine the electronic configuration of copper, silver, and gold. How does the observed behavior of these elements result from their configurations?

SOLVE These three elements are in the same group of the periodic table. Their electronic configurations are similar. Copper ($3d^{10} 4s^1$), silver ($4d^{10} 5s^1$), and gold ($5d^{10} 6s^1$) all have one s subshell electron outside filled subshells. As a result, this electron is very weakly bound. So it's nearly free and responds readily to an applied electric field. Thus these three elements are very good electrical conductors, with the three highest conductivities of any element (Table 17.3). This behavior results directly from the nearly free s subshell electron.

REFLECT These metals are also shiny. The conduction electrons on the metal's surface respond to the electric field in light, undergoing oscillation that reradiates the light—hence reflection. Selective absorption of some wavelengths results in the distinctive colors of copper and gold.

GOT IT? Section 24.4   Which of these elements do you consider the best candidate for bonding with potassium? (a) neon; (b) calcium; (c) iron; (d) iodine.

24.5 Radiation from Atoms

You've seen evidence for quantized atomic energy levels through visible, ultraviolet, and infrared spectra. Here we'll consider atomic x-ray emissions, whose explanation requires the shell model of Section 24.4. Then we'll explore lasers, showing how they use atomic transitions to produce intense, coherent light.

X rays

You've surely had dental x rays, and probably medical x rays as well. Your medical practitioner's x-ray source has the same basic design as the one German physicist Wilhelm Röntgen (1845–1923) used to discover x rays in 1895 (Figure 24.17). X rays result when electrons in an evacuated tube are accelerated through a high potential difference—on the order of kilovolts to megavolts—and then strike a target made from a high-melting-point metal like tungsten. The high-energy electrons stop quickly in the target. Some of their kinetic energy goes into heating, but much emerges as x-ray photons—another instance of accelerated charge producing electromagnetic radiation (Figure 24.18). Because the x rays result from slowing electrons, this form of x-ray production is called **bremsstrahlung**, German for "braking radiation."

Figure 24.19 graphs x-ray intensity versus wavelength for a molybdenum target. Note the continuous x-ray spectrum covering an order of magnitude in wavelength. However, there's a minimum wavelength λ_{min}, below which no x rays are produced. That's because the electrons' kinetic energy (Equation 16.6) is $K = eV_0$, where V_0 is the potential difference used to accelerate the electrons. Electrons lose part or all of their kinetic energy in hitting the target; that's why there's a continuous x-ray spectrum. The shortest wavelength in that spectrum (λ_{min}) results when an electron loses all its kinetic energy, and $K = eV_0$. The lost energy appears as a photon; since photon energy is $E = hc/\lambda$, we have

$$eV_0 = \frac{hc}{\lambda_{min}}$$

or

$$\lambda_{min} = \frac{hc}{eV_0} = \frac{hc}{e}\frac{1}{V_0}$$

The combination hc/e is

$$\frac{hc}{e} = \frac{(6.626 \times 10^{-34}\,\text{J}\cdot\text{s})(3.00 \times 10^8\,\text{m/s})}{1.60 \times 10^{-19}\,\text{C}} = 1.24 \times 10^{-6}\,\text{V}\cdot\text{m} = 1240\,\text{V}\cdot\text{nm}$$

Thus the minimum x-ray wavelength, as a function of accelerating potential V_0, is

$$\lambda_{min} = \frac{1240\,\text{V}\cdot\text{nm}}{V_0} \quad \text{(Duane-Hunt rule)} \tag{24.7}$$

Equation 24.7 is the **Duane-Hunt rule**, and its explanation in 1915 provided further confirmation that electromagnetic radiation is quantized.

In medicine, x rays are used not only for diagnostic purposes, but also to treat cancer. This application requires x rays of extremely short wavelengths and correspondingly high energies to penetrate the patient's skin and underlying tissue to reach the cancer. Depending on the cancer type, x-ray specialists use accelerating potentials as high as 25 MV.

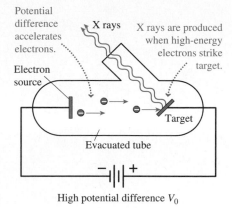

FIGURE 24.17 An x-ray tube.

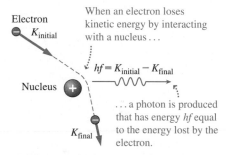

FIGURE 24.18 The bremsstrahlung process.

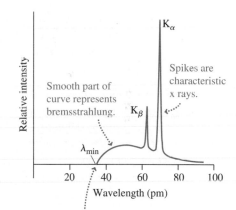

λ_{min} is the wavelength at which all the electron's kinetic energy is converted into a photon.

FIGURE 24.19 X-ray spectrum of molybdenum.

EXAMPLE 24.8 **Accelerating Potential**

What potential difference was used to accelerate the electrons that produced the x-ray spectra in Figure 24.19?

ORGANIZE AND PLAN The Duane-Hunt rule (Equation 24.7) relates the minimum wavelength and accelerating potential difference: $\lambda_{min} = 1240\,\text{V}\cdot\text{nm}/V_0$. The minimum wavelength can be read from the graph, so we can solve for V_0.

cont'd.

SOLVE On the graphs, the minimum wavelength appears to be about 3.6×10^{-2} nm = 0.036 nm. Therefore, the Duane-Hunt rule gives

$$V_0 = \frac{1240 \text{ V} \cdot \text{nm}}{\lambda_{\min}} = \frac{1240 \text{ V} \cdot \text{nm}}{0.036 \text{ nm}} = 3.4 \times 10^4 \text{ V} = 34 \text{ kV}$$

REFLECT We've rounded the answer to two significant figures, because it's difficult to read the graph with greater precision. A 34-kV potential difference is typical for medical diagnostic x rays.

MAKING THE CONNECTION What's the minimum wavelength of the high-energy x rays used for cancer treatment, produced by electrons accelerated through a 25-MV potential difference?

ANSWER Using the Duane-Hunt rule, $\lambda_{\min} = 5.0 \times 10^{-5}$ nm = 50 fm—an extremely short x-ray wavelength.

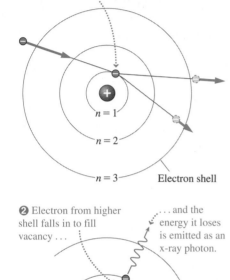

❶ High-energy incoming electron knocks electron out of inner shell.

$n = 1$

$n = 2$

$n = 3$ Electron shell

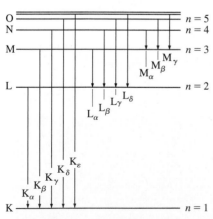

❷ Electron from higher shell falls in to fill vacancy . . .   . . . and the energy it loses is emitted as an x-ray photon.

Nucleus

FIGURE 24.20 Production of characteristic x rays. Actual target materials have many more shells, and the energy differences between the innermost shells are quite large—hence x-ray emission.

FIGURE 24.21 Energy-level diagram illustrating the production of characteristic x rays.

Characteristic X Rays

On top of its continuous spectrum, Figure 24.19 shows dramatic spikes in x-ray intensity at particular wavelengths. These are **characteristic x rays**, resulting from the process in Figure 24.20. Here a high-energy electron in the x-ray tube knocks an atomic electron out of the $n = 1$ shell, and another electron falls from $n = 2$ to take its place. Analogous processes involve electrons knocked out of $n = 2$ and replaced from the $n = 3$ shell, and so on. So there are a number of possible transitions, with different energies and wavelengths (Figure 24.21). Historically, the shells $n = 1, 2, 3, 4,$ and 5 were labeled K, L, M, N, and O. These letters label the x rays produced by a transition to that level, and the subscripts $\alpha, \beta, \gamma, \delta,$ and ε designate x rays of increasing energy from each shell. For example, a transition from the L shell to the K shell results in a K_α x ray, and a transition from the M to the K shell produces a K_β x ray.

In 1913–1914, the English physicist Harry Moseley (1887–1915) performed a systematic study of characteristic x rays. Moseley's results for K shell and L shell x rays are shown in Figure 24.22. The key feature of these **Moseley plots** is the linear relationship between atomic number and square root of the x-ray frequency for each type of x ray. Bohr's atomic theory explains this linear relationship. Recall from Section 24.2 that the Rydberg constant for atoms of atomic number Z increases as Z^2. An electron falling from the L shell to the K shell experiences an effective nuclear charge of $(Z - 1)e$, because only one electron lies interior to its orbit. Thus the energy E and frequency hf of the K_α x ray are (as in the Rydberg equation)

$$E = hf = (Z - 1)^2 E_0 \left(\frac{1}{1^2} - \frac{1}{2^2} \right) = \frac{3E_0}{4}(Z - 1)^2 \qquad (24.8)$$

Thus, the x-ray frequency should be proportional to $(Z - 1)^2$, and this is precisely what Moseley's data indicated. We stress that Bohr's atomic theory can't be applied in general to multi-electron atoms, but it works well in this case because outer shell electrons can be ignored in transitions involving only the inner shells. Moseley's work thus served as confirmation of quantized atomic energy levels and shell structure in multi-electron atoms. Further, Moseley's work closed the debate about whether the periodic table should be ordered by atomic number or atomic mass. By confirming that the nuclear charge is $+Ze$, Moseley showed that the correct ordering is by atomic number Z. This helped scientists better understand the ordering of groups in the periodic table (Section 24.4) in those cases in which the ordering by atomic number and atomic mass is different—for example, cobalt and nickel. Finally, Moseley found that there were gaps in his data where there were three elements still to be discovered: technetium ($Z = 43$), promethium ($Z = 61$), and rhenium ($Z = 75$).

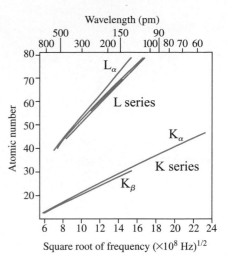

FIGURE 24.22 Moseley plots for characteristic x rays.

EXAMPLE 24.9   **Characteristic X-ray Wavelengths**

Predict the wavelength of the K_α x ray from a copper target. Compare the result with the Moseley plot in Figure 24.22.

ORGANIZE AND PLAN   A K_α x ray is produced by a transition from the L shell ($n = 2$) to the K shell ($n = 1$). As described in the text (Equation 24.8), the x-ray photon emitted in this process has energy

$$E = \frac{3E_0}{4}(Z - 1)^2$$

From the periodic table, $Z = 29$ for copper. As with any photon, energy and wavelength λ are related by $E = hc/\lambda$.

Known: $Z = 29$; $E_0 = 2.18 \times 10^{-18}$ J.

SOLVE   The x-ray energy is

$$E = \frac{3E_0}{4}(Z - 1)^2 = \frac{3(2.18 \times 10^{-18}\,\text{J})}{4}(29 - 1)^2$$

$$= 1.28 \times 10^{-15}\,\text{J}$$

giving a wavelength

$$\lambda = \frac{hc}{E} = \frac{(6.626 \times 10^{-34}\,\text{J} \cdot \text{s})(3.00 \times 10^8\,\text{m/s})}{1.28 \times 10^{-15}\,\text{J}}$$

$$= 1.55 \times 10^{-10}\,\text{m}$$

That's a 0.155-nm photon, with corresponding frequency

$$f = \frac{c}{\lambda} = \frac{3.00 \times 10^8\,\text{m/s}}{1.55 \times 10^{-10}\,\text{m}} = 1.94 \times 10^{18}\,\text{Hz}$$

The plots in Figure 24.22 show atomic number versus $\sqrt{f}$. The theoretical value here is $\sqrt{f} = \sqrt{1.94 \times 10^{18}\,\text{Hz}} = 1.39 \times 10^9\,\text{Hz}^{1/2}$, matching the plotted value for copper (Cu) in the K_α series.

REFLECT   Check other elements in the Moseley plot, and you'll find the same excellent agreement—good evidence of the atomic shell model's utility.

MAKING THE CONNECTION   Why is the frequency of the K_β x ray slightly larger than that of the K_α x ray for a given atomic number?

ANSWER   The K_β x ray comes from the $n = 3 \rightarrow 1$ transition—a slightly larger energy gap than the $n = 2 \rightarrow 1$ gap associated with the K_α x ray.

CONCEPTUAL EXAMPLE 24.10   **Characteristic X rays**

Figure 24.19 shows characteristic x rays for molybdenum ($Z = 42$). If instead you use a target made of tungsten ($Z = 74$), will the characteristic x rays appear at shorter or longer wavelengths?

SOLVE   According to the Moseley plots (Figure 24.22), x-ray wavelength decreases with increasing atomic number, for both the K series and L series. Therefore, the wavelengths should be shorter for x rays produced by tungsten, relative to those produced by molybdenum. This analysis is confirmed by a quantitative analysis using Equation 24.8, which predicts that the tungsten target produces a K_α x ray with energy 8.71×10^{-15} J, and hence a wavelength of about 23 pm—much shorter than the K_α x ray from molybdenum.

REFLECT   Note the tremendous range of characteristic wavelengths represented in the Moseley plot. This means you select characteristic x-ray wavelengths with your target choice.

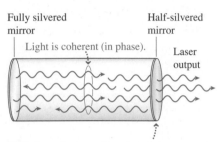

FIGURE 24.23 Stimulated emission.

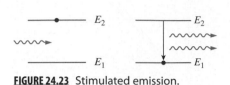

Fully silvered mirror   Half-silvered mirror

Light is coherent (in phase).

Laser output

Half-silvered mirror allows laser beam to emerge, while both mirrors reflect light to maintain stimulated emission.

FIGURE 24.24 Design of a laser.

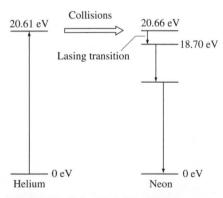

FIGURE 24.25 Energy levels in a helium-neon laser.

Lasers

It's hard to imagine the modern world without lasers. They scan our purchases at the supermarket checkout, read video information from our DVD and Blu-ray discs, reshape our corneas for better vision, machine and harden metal products, guide power tools, send our emails down optical fibers, and on and on.

In many lasers, light emission results from electronic transitions in atoms. So far we've considered only **spontaneous emission**, which occurs when an electron drops to a lower energy level without any assistance. Lasers, in contrast, use **stimulated emission**, illustrated in Figure 24.23. In stimulated emission, a photon passing an atom in an excited state triggers the emission. The emitted photon moves in the same direction and has the same phase as the original photon. That makes laser light coherent, a term we introduced in Chapter 22 to describe light sources in phase. The laser beam's coherence is one of its fundamental properties and is important in many laser applications. Stimulated emission in lasers takes place in a cavity bounded by two mirrors, as shown in Figure 24.24. Photons reflect back and forth between the mirrors, triggering more and more stimulated emission. One mirror is partially silvered, allowing some light to leak out; that's the laser's output.

✓ **TIP**

Laser is an acronym for Light Amplification by Stimulated Emission of Radiation.

There's another concept important in laser operation, relating to atomic energy levels. The laser's energy ultimately comes from electricity, other light sources, or even chemical reactions that "pump" atoms into excited states. Most excited states are short-lived and end in spontaneous emission. However, certain **metastable states** last for times on the order of 1 ms, compared with 1 ns for a typical excited state. Normally most atoms are in the ground state. But in a laser, energy is pumped into metastable states that last long enough to produce a **population inversion**, with more excited than ground state atoms. That makes stimulated emission much more likely than spontaneous emission.

The Helium-Neon Laser

Figure 24.25 shows energy levels important to the helium-neon laser, common in laboratory and commercial use. Most He-Ne lasers contain about 90% helium and 10% neon at around 0.01 atm of pressure. An electric field in the laser tube excites the helium atoms from the ground state to the excited state 20.61 eV higher, as shown. This is close to the 20.66-eV metastable state in neon. Excited helium atoms transfer their excess energy to neon through collisions, promoting neon atoms to the metastable state, where they remain long enough for stimulated emission. The emission occurs when neon falls from its 20.66-eV state to an 18.70-eV state. This energy difference corresponds to a 632.8-nm photon—the familiar red light of the helium-neon laser.

Other Lasers

Modern lasers often use semiconductors, rather than gases or other materials, as the so-called lasing medium. Energy levels in semiconductors can be engineered by doping with small concentrations of impurities (Sections 17.6 and 23.3), giving us lasers with most any desired wavelength. Early semiconductor lasers operated in infrared or red, but newer ones produce shorter wavelengths—a development that enabled Blu-ray discs that store high-definition movies (Section 22.3). Other semiconductor lasers are tunable, with output wavelength dependent on temperature.

Lasers are most commonly used in elective surgery (LASIK—see Section 21.6) to correct nearsightedness and astigmatism. However, lasers also cure serious conditions such as a torn retina. The outer layers of the retina can pull away from the underlying surface as the result of trauma or extreme nearsightedness. The laser energy fuses the two tissues, preventing blindness.

GOT IT? Section 24.5 For a given x-ray tube and fixed target, rank in descending order the wavelengths of the following x-rays: (a) K_α; (b) K_β; (c) L_α; (d) L_β.

Chapter 24 in Context

This chapter explored the structure of atoms, based largely on the quantum ideas from Chapter 23. *Rutherford scattering* implies a *nuclear atom*, with the atom's positive charge and most of its mass confined to a small *nucleus*. Building on Rutherford's work, Bohr developed a model of the hydrogen atom that explained the observed spectra. Although Bohr's model does not apply directly to multi-electron atoms, its *electron shells* are the basis for understanding those atoms. Evidence for shell structure is found in atomic emissions, especially *characteristic x rays*.

Looking Ahead In Chapter 25 we'll plunge into the heart of the atom—the nucleus. You'll learn that the nucleus is composed of protons and neutrons, bound by the strong force. Our study of the nucleus also covers radioactive decay and nuclear energy from fission and fusion. Chapter 26 focuses on the individual subatomic particles themselves, and explores theories that explain and classify the many different types of subatomic particles. You can consider Chapters 24–26 as a journey into the small, then smaller, then smallest scale that physics can study.

APPLICATION **Holography**

Laser light's coherence makes it possible to capture wavefronts along the path from an object. This gives a **hologram**—a pattern that, when illuminated, recreates a three-dimensional image of the object. Such images are quite realistic; in the photo, the man on the left is real, standing next to a holographic image. You've probably seen transmission holograms in museums or classroom demonstrations; you look through them to see the 3-D image. In contrast, the security device on your credit card is a reflection hologram. By using layered sheets that reflect different wavelengths, the reflection hologram produces an image when illuminated with white light.

CHAPTER 24 SUMMARY

The Nuclear Atom

(Section 24.1) In **Rutherford's nuclear model** of the atom, suggested by **alpha particle scattering**, the positive charge is confined to a small nucleus. But Rutherford couldn't explain atoms' stability and discrete line spectra.

Results of Rutherford scattering: Radius of a nucleus $\approx 10^{-15}$ m; radius of whole atom $\approx 10^{-10}$ m, or about 100,000 times larger.

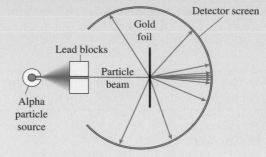

The Bohr Atom

(Section 24.2) **Bohr's model** mixes classical and quantum ideas to explain the hydrogen atom's stable orbits and emission spectrum. The atom has stable **quantum states** that correspond to specific allowed electron orbits and energies. Electromagnetic radiation is emitted only when an electron makes a transition from one state to another. Bohr showed that quantized energy levels in hydrogen account for all photon wavelengths predicted by the **Rydberg equation**, and thus for observed spectra.

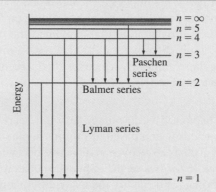

Rydberg equation: $\dfrac{1}{\lambda} = R_{\mathrm{H}}\left(\dfrac{1}{j^2} - \dfrac{1}{k^2}\right) (j < k)$

Radius of Bohr atom: $r = n^2 a_0, n = 1, 2, 3, \ldots$

Energy of Bohr atom: $E = -\dfrac{E_0}{n^2}, n = 1, 2, 3, \ldots$

Quantum Numbers and Atomic Spectra

(Section 24.3) The state of a hydrogen atom is described by four **quantum numbers**: n, l, m_l, and m_s. Transitions between states are governed by selection rules that tell how quantum numbers can change during transitions. The **Zeeman effect**, which occurs with light-emitting atoms in a strong magnetic field, provides evidence of orbital and magnetic quantum numbers.

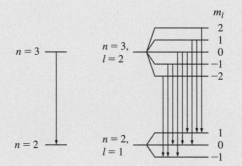

Orbital angular momentum: $L = \sqrt{l(l+1)}\,\dfrac{h}{2\pi}$

Multi-Electron Atoms and the Periodic Table

(Section 24.4) The **Pauli exclusion principle** states that no two electrons in an atom can be in the same quantum state. Electrons in a multi-electron atom are in shells (n level) and **subshells** (combination of n and l); levels fill from lowest energy to highest. The state of a multi-electron atom is described using spectroscopic notation. The **periodic table** lists elements in order of atomic number; elements fall into groups (table columns) that exhibit similar chemical behavior because they have the same number of electrons in their highest unfilled subshells.

Electronic configuration:

l	Subshell	m_l values	Max. # of electrons
0	s	0	2
1	p	$-1, 0, 1$	6
2	d	$-2, -1, 0, 1, 2$	10
3	f	$-3, -2, -1, 0, 1, 2, 3$	14

Radiation from Atoms

(Section 24.5) **X rays** are produced when high-energy electrons strike a target. The electrons lose some kinetic energy (**bremsstrahlung**) or they cause atomic transitions in atoms in the target (**characteristic x rays**). **Lasers** use **stimulated emission** to create coherent beams of electromagnetic radiation.

Duane-Hunt rule: $\lambda_{min} = \dfrac{1240 \text{ V} \cdot \text{nm}}{V_0}$

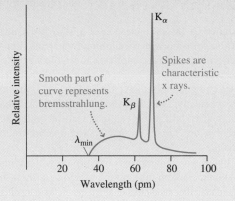

NOTE: Problem difficulty is labeled as ■ straightforward to ■■■ challenging. Problems labeled BIO are of biological or medical interest.

Conceptual Questions

1. How does Thomson's plum-pudding model fail to account for Rutherford scattering? How does the nuclear atom account for the backscattering events?
2. Would Rutherford scattering work if protons were used as projectiles instead of alpha particles?
3. When an electron in the Bohr atom jumps to a larger-radius orbit, does each of the following quantities increase, decrease, or remain the same? (a) Total energy; (b) kinetic energy; (c) potential energy; (d) electron speed.
4. Why doesn't Bohr's theory apply to the neutral helium atom?
5. Does the $n = 3$ to $n = 2$ transition produce a longer-wavelength photon in hydrogen or deuterium?
6. List successes and failures of Bohr's atomic theory.
7. What's the experimental evidence for each of the four quantum numbers that describes the state of a hydrogen atom?
8. Explain how to find the maximum number of electrons allowed on a given subshell.
9. Based on their electronic configurations, discuss the likely ionization states of sodium and chlorine. Explain why it's favorable for these atoms to combine in the form NaCl.
10. Why are the noble gases so stable? Why are they unlikely to form chemical compounds?
11. What's the physical significance of the quantum numbers (a) m_l and (b) m_s?
12. When sodium is dropped into water, a violent reaction occurs. Why?
13. Aluminum isn't in the same group as copper, silver, and gold, yet it's an excellent electrical conductor. Why?
14. Why do you suppose that some science fiction stories have used the concept of silicon-based life?
15. How do you think Moseley measured the wavelengths of the x rays he created in the lab? For each element he studied, how did he know which x rays were K_α, K_β, and so on?
16. If the potential difference in an x-ray tube is increased, how does this affect the wavelengths of (a) bremsstrahlung x rays and (b) characteristic x-rays?

Multiple-Choice Problems

17. A hydrogen atom makes a transition from the $n = 4$ level to $n = 2$. This results in the emission of a photon with wavelength (a) 410 nm; (b) 434 nm; (c) 486 nm; (d) 656 nm.
18. A ground-state hydrogen atom absorbs a photon and ends up in the $n = 3$ state. The photon's energy was (a) 2.18×10^{-18} J; (b) 1.94×10^{-18} J; (c) 1.24×10^{-18} J; (d) 2.42×10^{-19} J.
19. The ground-state energy of the helium ion He^+ is equal to the ground-state energy of hydrogen multiplied by (a) 4; (b) 2; (c) 1/2; (d) 1/4.
20. What transition in hydrogen corresponds to the emission of an infrared photon with $\lambda = 1282$ nm? (a) $n = 4$ to $n = 2$; (b) $n = 4$ to $n = 3$; (c) $n = 5$ to $n = 3$; (d) $n = 6$ to $n = 3$.
21. If a hydrogen atom makes a transition from the $n = 5$ level directly to $n = 1$, the emitted photon will be in what part of the spectrum? (a) Microwave; (b) infrared; (c) visible; (d) ultraviolet.
22. The spectroscopic notation for a hydrogen atom with an electron having quantum numbers $n = 3, l = 2, m_l = -1, m_s = -1/2$ is (a) $3s$; (b) $3p$; (c) $3d$; (d) $3f$.
23. Which of the following transitions is not allowed for an electron in hydrogen? (a) $4p$ to $1s$; (b) $3d$ to $2s$; (c) $3d$ to $4f$; (d) $2p$ to $4s$.
24. For a hydrogen atom in the $4d$ state, the orbital angular momentum is (a) $2(h/2\pi)$; (b) $\sqrt{6}(h/2\pi)$; (c) $\sqrt{12}(h/2\pi)$; (d) $\sqrt{20}(h/2\pi)$.
25. What's the shortest-wavelength x ray produced in a 150-kV x-ray tube? (a) 8.3 nm; (b) 0.83 nm; (c) 0.083 nm; (d) 8.3×10^{-3} nm.
26. Which of the following characteristic x rays has the longest wavelength? (a) K_α; (b) K_β; (c) L_α; (d) L_β.

Problems

Section 24.1 The Nuclear Atom

27. ■ In some Rutherford scattering experiments, the alpha particles had kinetic energies of 7.7 MeV. What's the speed of such an alpha particle?

28. ■■ Suppose a proton and an alpha particle each have kinetic energy 8.0 MeV. (a) Find the speed of each particle. (b) Find each particle's distance of closest approach in a head-on collision with a gold nucleus.

29. ■■ Find the distance of closest approach of a 7.7-MeV alpha particle in a head-on collision with an aluminum nucleus. Compare your answer with the closest approach to a gold nucleus, given in Example 24.1.

30. ■■ Alpha particles and gold nuclei have radii 1.9×10^{-15} m and 7.0×10^{-15} m, respectively. Find the kinetic energy of an alpha particle such that the two will just touch in a head-on collision. (At this energy Rutherford's point-particle approximation would begin to break down.)

31. ■■ Suppose the hydrogen atom consists of a proton with radius 1.2×10^{-15} m, with an electron orbiting the proton at a distance of 5.3×10^{-11} m. Take the atom to be a sphere with a radius equal to the electron's orbital radius. (a) What's the atom's volume? What's the volume of the nucleus? (b) What's the atom's density? Compare with that of water, 1000 kg/m^3. (c) What is the density of the nucleus? Compare with the atomic density you found in part (b).

Section 24.2 The Bohr Atom

32. ■ Find the wavelength of the photon emitted when hydrogen makes a transition (a) $n = 8 \rightarrow 5$; (b) $n = 6 \rightarrow 1$; (c) $n = 10 \rightarrow 2$.

33. ■ What approximate value of the quantum number n will make the hydrogen atom's energy (a) -1.36×10^{-20} J; (b) -2.69×10^{-21} J; (c) -1.29×10^{-21} J?

34. ■■ What's the series limit for (a) the Lyman series; (b) the Paschen series; (c) the Brackett series?

35. ■■ Find the range of wavelengths seen in the hydrogen emission spectrum in the Pfund series.

36. ■■ A hydrogen atom absorbs a photon with a wavelength $\lambda = 486$ nm. What are the atom's initial and final states?

37. ■■ A hydrogen atom is in the $n = 2$ state. (a) How much energy is required to remove the electron completely from the atom? (b) What's the wavelength of a photon with this energy?

38. ■■■ A hydrogen atom in the $n = 2$ state absorbs a photon with wavelength 656 nm. Find the change in (a) the atom's total energy; (b) the atom's potential energy; (c) the orbiting electron's kinetic energy.

39. ■■ (a) Use Bohr's theory to find the speed of an electron orbiting on the $n = 1$ level. Should relativistic calculations be used? (b) Repeat for $n = 2$.

40. ■■■ Find the wavelengths of photons emitted by Li^{2+} for the transitions (a) $n = 3 \rightarrow 2$ and (b) $n = 4 \rightarrow 2$.

41. ■■■ A hydrogen atom at rest makes a transition from the $n = 4$ state to $n = 2$. Find (a) the wavelength of the emitted photon and (b) the recoil speed of the atom.

42. ■■ For a hydrogen atom in the ground state, find the total energy, the potential energy, and the electron's kinetic energy. Verify that $E = K + U$.

43. ■■ An electron in a hydrogen atom orbits with radius of $4a_0$. (a) What's the change in the atom's total energy if the radius increases to $9a_0$? (b) Does this process correspond to the emission or absorption of a photon? What's the photon's wavelength?

44. ■■■ Find the de Broglie wavelengths of electrons in hydrogen's (a) $n = 1$ state; (b) $n = 2$ state; (c) $n = 10$ state.

45. ■■ Find the minimum quantum number needed to make a hydrogen atom at least 0.50 μm in diameter.

46. ■■ A hydrogen atom in the $n = 5$ state drops to the $n = 2$ state by undergoing two downward transitions. What are all possible combinations of the resulting photon wavelengths?

47. ■■■ A light source emits a continuous wavelength range from 400 nm to 1000 nm. If this light is incident on a gas of atomic hydrogen, find all the transitions that can occur due to photon absorption.

48. ■■■ A hydrogen atom is in the $n = 8$ state. (a) What's its electron's orbital radius? (b) Use your answer to find the electron's de Broglie wavelength. (c) From your answer to part (b), determine the electron's speed. (d) Compute the electron's speed using Bohr theory, and compare with your answer to part (c).

49. ■■ Find the ground-state energies of (a) He^+; (b) Li^{2+}.

50. ■■■ Consider the spectrum of He^+ resulting from downward transitions to the $n = 4$ level. (a) What wavelength results when initially $n = 6$? (b) To which transition in hydrogen does that wavelength correspond? (c) What transitions (initial n to final n) in He^+ produce the rest of the wavelengths seen in hydrogen's Balmer series?

Section 24.3 Quantum Numbers and Atomic Spectra

51. ■ Give spectroscopic notation for each of these electron states: (a) $n = 2, l = 0$; (b) $n = 4, l = 1$; (c) $n = 4, l = 3$; (d) $n = 3, l = 2$.

52. ■ For each of the following electronic states in hydrogen, find the magnitudes of the orbital and spin angular momenta: (a) $3s$; (b) $3p$; (c) $3d$.

53. ■■ Which of the following transitions are allowed for an electron in hydrogen? For each allowed transition, find the wavelength of the emitted photon: (a) $3p$ to $2s$; (b) $3d$ to $2s$; (c) $4f$ to $3d$; (d) $3p$ to $1s$.

54. ■■ For an electron at the $n = 4$ level in hydrogen, find all sets of allowed quantum numbers. How many different sets are there?

55. ■■ For an electron in the $3p$ state of hydrogen, calculate all allowed values of (a) orbital angular momentum; (b) z-component of orbital angular momentum; (c) z-component of spin angular momentum.

56. ■■■ A hydrogen atom has energy -1.36×10^{-19} J, and the electron can be in any one of 10 different quantum states. Find the spectroscopic notation for this atom.

57. ■■ By what factor does a hydrogen atom's angular momentum change when it makes the following transitions: (a) $3d$ to $2p$; (b) $3p$ to $4d$; (c) $4f$ to $3d$?

58. ■■■ In a Zeeman experiment done in a 2.0-T magnetic field, the atomic energy levels in hydrogen change by $(1.86 \times 10^{-23} \text{ J}) m_l$. When each Balmer series line is split into three, what's the difference in wavelength between the two outer lines? Is this easy to detect?

Section 24.4 Multi-Electron Atoms and the Periodic Table

59. ■ Write the ground-state electron configuration for each of the following atoms: (a) calcium; (b) iron; (c) gold; (d) uranium.

60. ■ Identify the atoms with ground-state electron configuration (a) $2s^2 2p^4$; (b) $3s^2 3p^6$; (c) $3d^{10} 4s^2 4p^1$; (d) $4f^{14} 5d^4 6s^2$.

61. ■ For argon in the ground state, list (a) the electronic configuration and (b) the set of four quantum numbers for each of the electrons.

62. ■■ Find the maximum number of electrons allowed on each of the following shells by writing out all the possible sets of quantum states for electrons on that shell: (a) $n = 2$; (b) $n = 3$; (c) $n = 4$.

63. ■■ What type of excess charge carriers (electrons or holes) are created when (a) silicon is doped with gallium and (b) germanium is doped with antimony?

Section 24.5 Radiation from Atoms

64. ■ What's the minimum wavelength of x rays produced in an x-ray tube with potential difference (a) 100 kV; (b) 1.0 MV?

65. ■ A dental x-ray source has minimum wavelength 0.031 nm. What's the potential difference in the x-ray tube?

66. ■ Old-fashioned tube televisions were a weak source of x rays, due to electrons stopping at the screen (leaded glass helped prevent x-ray exposure). For a TV tube operating at 30 kV, what minimum x-ray wavelength was produced?

67. ■ What's the shortest-wavelength photon that can be produced by the Stanford Linear Accelerator, where electrons are accelerated to energies of 50 GeV?

68. ■■ What wavelengths would Moseley have predicted for K_α x rays from the three missing elements, $Z = 43, 61,$ and 75?

69. ■■ Find the minimum potential difference required to produce K_α x rays in an x-ray tube using a cobalt target.

70. ■■ A K_α x ray is observed with wavelength 0.337 nm. Identify the target element.

71. ■ Helium-neon lasers can be made to produce green laser light with a wavelength of 532 nm. What's the gap between energy levels in neon responsible for this emission?

72. ■■ (a) How many photons per second are emitted from a 1.0-mW helium-neon laser with wavelength 632.8 nm? (b) If a laser contains 0.025 mole of neon, what fraction of the neon atoms participate in the lasing process each second?

73. ■■ A laser emits 4.50×10^{18} photons per second, using a transition from 2.33 eV above the ground state to the ground state. Find (a) the laser light's wavelength and (b) the laser's power output.

74. BIO ■■ **Laser surgery**. A laser used to repair a detached retina has wavelength of 532 nm. The beam comes in short pulses, each lasting 25 ms and carrying 1.0 J of energy. (a) What's the power of this beam while it's on? (b) How many photons are in each pulse?

General Problems

75. ■■ What transition in hydrogen corresponds to emission of an infrared photon with $\lambda = 1282$ nm?

76. ■■ For an electron on the $n = 6$ level of hydrogen, how many different sets of quantum numbers are there?

77. ■■ Find all the sets of allowed quantum numbers for an electron with the spectroscopic designation $5f$.

78. ■■ Determine the electronic configuration (n, l) for a hydrogen atom with energy -1.36×10^{-19} J and angular momentum of magnitude $\sqrt{12}h/2\pi$.

79. ■■■ A hydrogen atom is in the $3d$ state. (a) What are the possible angles its orbital angular momentum vector can make with a given axis? (b) What are the possible values for the component of angular momentum along that axis?

80. ■■■ Bohr's *correspondence principle* says that as quantized objects approach macroscopic size, quantum theory should reduce to the classical result. As one example, consider a hydrogen atom with a large value of n. (a) Find the orbital frequency of an electron at the $n = 100$ level. (b) Find the frequency of a photon emitted when an electron makes a transition from the $n = 101$ state to the $n = 100$ state, and compare with the frequency you computed in part (a). Are the discrete steps in frequency as noticeable as for lower-level transitions?

81. ■■ A solid-state laser made from lead-tin selenide has a lasing transition at a wavelength of $30 \,\mu$m. If its power output is 2.0 mW, how many transitions occur each second?

82. ■■■ Following the method outlined in the text for the K_α x rays, find a formula that predicts the wavelengths of K_β x rays. Use this formula to predict the wavelength of the K_β x ray for copper, and compare it with the experimental result shown in the Moseley plot in the text.

83. ■■■ For the L-series x rays, Moseley found he could predict the observed wavelengths by assuming that an electron falling to the L shell of an atom feels the equivalent of 7.4 electrons inside the L shell. (a) Use this information to find a formula that predicts the wavelengths of L_α x rays. (b) Use this formula to predict the wavelengths of an L_α x ray from tungsten, and compare with the experimental result shown in the Moseley plot in the text.

Answers to Chapter Questions

Answer to Chapter-Opening Question

The Pauli exclusion principle dictates atomic structure by prohibiting atomic electrons from occupying the same state. The exclusion principle also applies in the dense state of a collapsed star at the end of its life—and that requires many electrons to occupy high-energy states. These high-energy electrons create a pressure that sustains the star against further collapse.

Answers to GOT IT? Questions

Section 24.1   (b)
Section 24.2   (d) > (a) > (b) > (c)
Section 24.3   (a), (c), and (d)
Section 24.4   (d) Iodine
Section 24.5   (c) > (d) > (a) > (b)

Nuclear Physics

To Learn

By the end of this chapter you should be able to

- Explain the composition of the atomic nucleus.
- Distinguish elements and isotopes, atomic number and mass number.
- Describe the strong (nuclear) force that binds nuclei.
- Tell how the nuclear and electrical forces compete to determine stable nuclei.
- Explain the curve of binding energy as a function of mass number.
- Distinguish stable from radioactive isotopes.
- Describe alpha, beta, and gamma decay.
- Discuss the biological effects of radiation and units for measuring radiation dose.
- Explain radioactive half-life.
- Distinguish nuclear fission and fusion.
- Tell how nuclear power plants extract energy from fission.
- Describe challenges in developing controlled nuclear fusion.

■ How will these archeologists determine the ages of the bone fragments they're unearthing?

I n this chapter we focus on the atomic nucleus. We'll begin with nuclear structure and the forces that bind neutrons and protons to form the nucleus. For some nuclei those forces result in stable arrangements. Many other nuclei are unstable and undergo radioactive decay. You'll learn about the three main decays: alpha, beta, and gamma. We'll introduce radioactive half-life, which establishes a time scale for the decay of each radioactive material. Then we'll turn to fission and fusion, nuclear reactions that release large amounts of energy. Fusion powers our Sun and other stars, while fission supplies some 15% of humankind's electrical energy. But fission and fusion also threaten our very survival, through their use in nuclear weapons.

25.1 Nuclear Structure

Rutherford's work in the early 20th century established the nucleus as the tiny, massive atomic core. Today we know that nuclei are composed of neutrons and protons—collectively, **nucleons**. We classify nuclei by the numbers of each that they contain. Here we'll explore

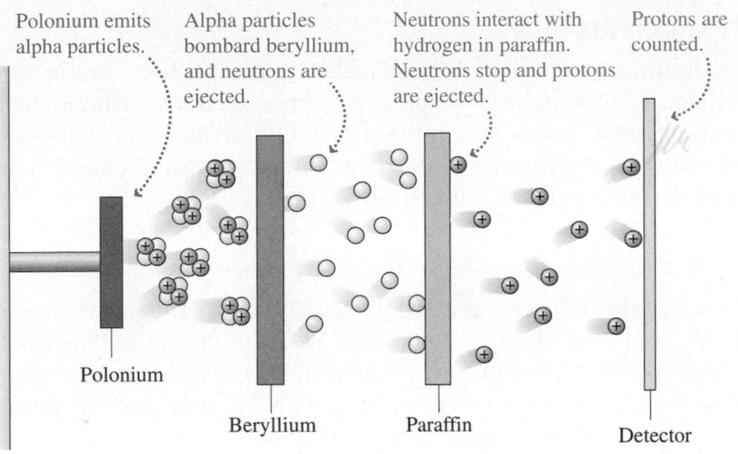

Polonium emits alpha particles.

Alpha particles bombard beryllium, and neutrons are ejected.

Neutrons interact with hydrogen in paraffin. Neutrons stop and protons are ejected.

Protons are counted.

Polonium

Beryllium

Paraffin

Detector

FIGURE 25.1 Discovery of the neutron involves (a) bombardment of beryllium foil with alpha particles, producing energetic, electrically neutral rays; (b) determination that the rays can pass through lead, showing they aren't photons; and (c) demonstration that the new particles can knock protons out of paraffin, showing they have approximately the same mass as protons.

the force that binds nucleons and show why some nuclei are stable but others are radioactive. These ideas serve as foundations for the concepts discussed later: radioactive decay, fission, and fusion.

Discovery of the Neutron

For years after the discovery of the nucleus and its immediate application in Bohr's atomic model (Chapter 24), relatively little progress occurred in understanding the nucleus. Charge and mass presented one quandary: Why, for example, does helium have four times the mass but only twice the charge of hydrogen? Some thought that helium might consist of four hydrogen nuclei, plus two electrons to bring the charge down to $+2e$. But Heisenberg's uncertainty principle shows that electrons confined to the small nucleus would have unrealistically high energies (see Problem 37).

In 1920 Rutherford suggested the presence of neutral particles—precisely the particles we now call **neutrons**—with about the same mass as the hydrogen nucleus, which he called a **proton**. Thus, an atomic nucleus could be a combination of protons and neutrons, with the protons supplying mass and positive charge, and neutrons additional mass. In the late 1920 and early 1930s, experiments involving alpha particle bombardment of beryllium led to the identification of Rutherford's neutron (Figure 25.1). These culminated in 1932, when the English physicist James Chadwick conclusively demonstrated that these experiments produced neutral particles with mass just slightly greater than the proton's. This earned Chadwick credit for the neutron's discovery. Table 25.1 gives modern values for the masses of the proton, neutron, and electron; note the small difference between proton and neutron masses.

✓**TIP**

The neutron's mass is about 0.14% larger than the proton's.

TABLE 25.1 Particle Charges and Masses

Particle	Electric charge	Mass (kg)	Mass (u)
Proton	$+e$	1.6726×10^{-27}	1.00728
Neutron	0	1.6749×10^{-27}	1.00866
Electron	$-e$	9.1094×10^{-31}	5.4858×10^{-4}

Nuclear Properties

Chadwick's discovery made clear that the nucleus is composed of protons and neutrons. Ordinary helium, for example, has two protons and two neutrons, giving it charge $+2e$ and mass about four times that of the proton. In general, a nucleus consists of Z protons (the **atomic number**) and N neutrons (the **neutron number**). Proton and neutron masses are similar, so the **mass number** A is defined as

$$A = Z + N \qquad \text{(Mass number; SI unit: dimensionless)} \qquad (25.1)$$

and is the total number of nucleons in the nucleus. Table 25.1 includes masses in **atomic mass units** (symbol u), with $1\ \text{u} = 1.66054 \times 10^{-27}\ \text{kg}$. Proton and neutron masses are each only slightly greater than 1 u, and electrons add very little mass, so an atom's mass (in u) is close to its mass number. For example, with 2 protons and 2 neutrons, helium's mass number is $A = 4$, and its actual mass is 4.0026 u.

✓ **TIP**

An alpha particle is a helium nucleus.

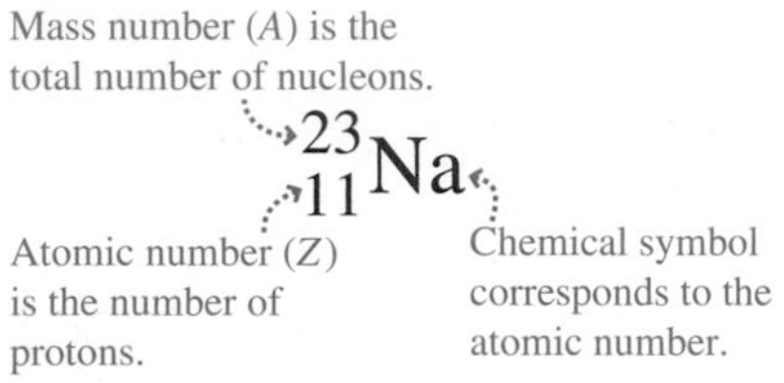

Mass number (A) is the total number of nucleons.

$$^{23}_{11}\text{Na}$$

Atomic number (Z) is the number of protons.

Chemical symbol corresponds to the atomic number.

FIGURE 25.2 Anatomy of a nuclear symbol.

Each combination of Z and N comprises a **nuclide**, designated symbolically in the form $^A_Z X$, as explained in Figure 25.2. However, the atomic number Z is unnecessary, because it corresponds to the chemical symbol (e.g., 1 for H, 6 for C, 92 for U). So we'll usually use only the chemical symbol and mass number: $^A X$. For example, in ^{23}Na the symbol Na tells you that it's the element sodium, meaning the atomic number $Z = 11$. Here $A = 23$, so by Equation 25.1 the neutron number is $N = A - Z = 23 - 11 = 12$. When speaking, we usually identify a nuclide by its name and mass number, as in "sodium-23." Sometimes you'll see nuclides written that way (sodium-23 or Na-23) instead of ^{23}Na.

Nuclides of a given element have the same atomic number Z but may have different numbers of neutrons and hence different mass numbers. These different nuclides are **isotopes** of that element. For example, ^{12}C, ^{13}C, and ^{14}C are all carbon isotopes. Some isotopes are stable; others are radioactive (more in Section 25.3). Most naturally occurring elements contain a mixture of stable isotopes. For example, about 99.98% of atmospheric oxygen is ^{16}O; nearly all the rest is ^{18}O. Carbon is about 99% ^{12}C and 1% ^{13}C. All elements have radioactive isotopes as well: ^{15}O is a short-lived isotope used in medical PET scans, while ^{14}C, formed in the atmosphere by reactions involving cosmic rays, is used in radiocarbon dating (Section 25.3). Appendix D lists some important isotopes.

Because atoms of different isotopes have the same number of protons and hence electrons, their chemical behavior is similar. However, the mass difference leads to subtle changes in rates of chemical reactions. The small mass difference, along with the dramatically different behavior of radioactive isotopes, makes isotopic analysis an important tool in many areas of science.

Isotope analysis can be used to study climate change. Carbon dioxide is increasing in Earth's atmosphere, leading to global warming. Meanwhile, atmospheric ^{13}C and ^{14}C are decreasing relative to ^{12}C. The diminishing portion of radioactive ^{14}C suggests that new carbon comes from a source that's been out of contact with the atmosphere for so long that all its ^{14}C has decayed. That points to an underground source. Plants selectively take up the lighter isotope ^{12}C, so the diminishing portion of ^{13}C suggests that the source is plant matter and not, for example, volcanoes. Taken together, change in the atmospheric mix of carbon isotopes is convincing evidence that carbon being added to the atmosphere comes from combustion of fossil fuels.

In another climate application, the lighter isotope ^{16}O evaporates preferentially, and the heavier ^{18}O precipitates more readily. The isotopic balance in both processes depends on temperature, and therefore the isotopic composition of polar ice cores demonstrates long-term climate trends.

Nuclear Shape and Size

Figure 25.3 shows the nucleus modeled as Z protons and N neutrons bound in a spherical shape. This is a crude, classical picture; quantum mechanics (Chapter 23) dictates that individual nucleons can't be localized precisely. Yet this simple model lets you visualize the nucleus and the interactions between protons and neutrons.

In our simple model, the nucleons act as hard spheres with roughly equal volumes. Therefore, the nuclear volume $V = \frac{4}{3}\pi r^3$ is proportional to the mass number A. That suggests that the nuclear radius should be proportional to $A^{1/3}$. Nuclear scattering experiments confirm this, giving the approximate relationship:

$$r = R_0 A^{1/3} \quad \text{(Nuclear radius; SI unit: m)} \quad (25.2)$$

where $R_0 = 1.2 \times 10^{-15}$ m $= 1.2$ fm. The slow $A^{1/3}$ dependence means that nuclear radii vary by less than a factor of 10. The smallest nucleus, ^{1}H, has $r = R_0 A^{1/3} = (1.2 \text{ fm})(1)^{1/3} = 1.2$ fm, while ^{238}U, the largest naturally occurring nucleus, has $r = R_0 A^{1/3} = (1.2 \text{ fm})(238)^{1/3} = 7.4$ fm. Thus, all nuclear radii are on the order of femtometers.

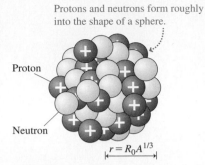

FIGURE 25.3 Model of the nucleus showing protons and neutrons packed tightly in the shape of a ball.

✓ **TIP**

The unit fm is a "femtometer" in SI. Physicists sometimes call this unit the "fermi," after nuclear physics pioneer Enrico Fermi.

EXAMPLE 25.1   Nuclear Density

Compute the density of the ^{238}U nucleus. Compare with the density of water, 1000 kg/m^3. Why the difference?

ORGANIZE AND PLAN Density ρ is mass per volume: $\rho = m/V$. The mass of a U-238 is close to 238 u (see Appendix D). We just found $r = 7.4$ fm for ^{238}U, so we can get the volume from $V = \frac{4}{3}\pi r^3$.

Known: $m = 238$ u; $r = 7.4$ fm.

SOLVE Converting the mass to SI,

$$m = 238 \text{ u} \times \frac{1.66054 \times 10^{-27} \text{ kg}}{1 \text{ u}} = 3.952 \times 10^{-25} \text{ kg}$$

Then the nuclear volume is

$$V = \frac{4}{3}\pi r^3 = \frac{4}{3}\pi (7.4 \times 10^{-15} \text{ m})^3 = 1.70 \times 10^{-42} \text{ m}^3$$

giving density

$$\rho = \frac{m}{V} = \frac{3.952 \times 10^{-25} \text{ kg}}{1.70 \times 10^{-42} \text{ m}^3} = 2.3 \times 10^{17} \text{ kg/m}^3$$

This is a factor of 2.3×10^{14} greater than water's density! That's because atoms are mostly empty space, while nuclei consist of closely packed nucleons.

REFLECT If you repeat this calculation for different nuclides, you'll find that nuclear density is essentially independent of mass number. That's consistent with our model in Figure 25.3, with its tightly packed nucleons.

MAKING THE CONNECTION   What's the mass number of a nuclide with half U-238's radius? Identify a stable nuclide with this mass number.

ANSWER Using $r = 3.7$ fm in $r = R_0 A^{1/3}$ gives $A = 29$ when rounded to an integer. A stable nuclide with $A = 29$ is ^{29}Si.

GOT IT? Section 25.1   Suppose you have two nuclides A and B, with B having twice the mass of A. The radius of nuclide A is r_A. What's the radius of nuclide B? (a) 1.26 r_A; (b) 1.41 r_A; (c) 2 r_A; (d) 2.82 r_A; (e) 4 r_A; (f) 8 r_A.

25.2 The Strong Force and Nuclear Stability

What holds the nucleus together? The protons repel via the electrostatic force, which should make them fly apart, while the neutral neutrons don't experience electrical attraction or repulsion. As you saw in Chapter 15, gravity is far too weak to play a role in binding the nucleus. So there must be some other force at work here. That's the **strong force**, also called the **nuclear force** in this context. Here are the main properties of the nuclear force:

- The nuclear force is always attractive, with almost equal strength for any nucleon pair: proton-proton, proton-neutron, or neutron-neutron.
- It's *really* strong. As Table 9.1 shows, the nuclear attraction between protons is some 100 times stronger than their electrostatic repulsion.
- Although very strong, it has very short range, becoming ineffective at distances greater than about 3 fm.

Because the nuclear force has such a short range, more distant protons don't feel significant nuclear attraction—but they still experience the longer-range electrical repulsion. Neutrons are therefore crucial in providing the "glue" that holds the nucleus together; they bind nucleons without introducing electrical repulsion. Figure 25.4 shows an important consequence for nuclear structure. The figure plots number of protons Z versus neutron number N for different nuclei, both stable and radioactive. Note that for smaller nuclei, the number of protons and neutrons is about the same, but larger nuclei have $N > Z$. We won't prove it, but electrostatic energy considerations from Section 16.1 show that the repulsive energy in the nucleus grows as Z^2. Thus, atoms with higher atomic numbers Z require more and more neutrons to bind the nucleus. But for very large nuclei that's still not enough, with the result that there are no stable nuclei for $Z > 83$—although some heavier elements, like uranium, have lifetimes measured in billions of years. The concept of half-life—shown in Figure 25.4—will be discussed in Section 25.4.

Binding Energy

If you look closely at atomic masses in Appendix D and particle masses in Table 25.1, you'll see that things don't quite add up. For example, an atom of ^{12}C has mass 12 u (that's exact, and serves to define the atomic mass unit u). This atom contains 6 protons, 6 neutrons, and 6 electrons. Using Table 25.1, you can compute the total mass:

$$6 \text{ protons: mass} = 6m_p = 6(1.00728 \text{ u}) = 6.04368 \text{ u}$$

$$6 \text{ neutrons: mass} = 6m_n = 6(1.00866 \text{ u}) = 6.05196 \text{ u}$$

$$6 \text{ electrons: mass} = 6m_e = 6(0.000548585 \text{ u}) = 0.00329 \text{ u}$$

$$\text{Total mass: } 12.09893 \text{ u}$$

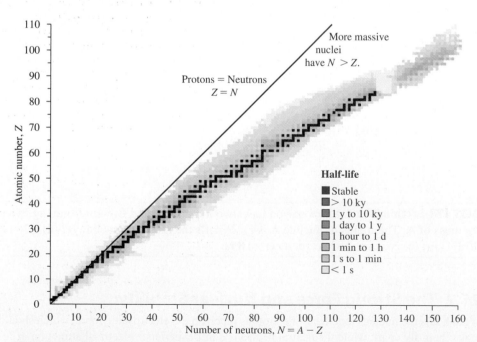

FIGURE 25.4 Chart of the nuclides, showing stable isotopes lying along the line of stability and radioactive isotopes coded by half-life.

So the atom's mass is *less* than the combined masses of its constituent particles. That's because energy is released as particles come together to form the bound atom—just as energy is released when you drop a rock to the ground, forming the bound system rock + Earth. Einstein's $E = mc^2$ (Chapter 20) says there's an associated decrease in mass, and that's reflected in the lower atomic mass. The mass difference Δm is the **mass defect**. Put differently, you'd have to supply the energy equivalent $(\Delta m)c^2$ in order to break the bonds that hold the atom and nucleus together; this is called the **binding energy**. Because electrons are loosely bound by the electrostatic force, the bulk of the atom's binding energy is in the nucleus.

A neutral atom has Z protons, Z electrons, and N neutrons. Therefore, the binding energy of an atom of isotope $^A X$ is

$$E_b = (\Delta m)c^2 = (Z m_p + N m_n + Z m_e - M(^A X))c^2 \quad \text{(Binding energy; SI unit: J)} \quad (25.3)$$

where M is the mass of the whole atom. When working with binding energies, it's convenient to express masses in u and energies in MeV. With

$$1\,\text{u} \cdot c^2 = (1.66054 \times 10^{-27}\,\text{kg})(2.9979 \times 10^8\,\text{m/s})^2$$
$$= 1.4924 \times 10^{-10}\,\text{J} = 931.5\,\text{MeV}$$

the binding energy of ^{12}C becomes

$$(\Delta m)c^2 = 0.09893\,\text{u} \cdot c^2 = (0.09893)(931.5\,\text{MeV}) = 92.2\,\text{MeV}$$

This is huge compared with electronic binding energies, which are typically on the order of electron-volts. That's a hint of the vast energy difference between nuclear and chemical reactions, the latter involving only the atom's outermost electrons.

EXAMPLE 25.2 **Binding Energy of Helium**

Use data in Appendix D to find the binding energy of the ^{4}He atom. Compare the result with the binding energy of ^{12}C, computed in the text.

ORGANIZE AND PLAN Equation 25.3 gives the binding energy: $E_b = (\Delta m)c^2 = (Z m_p + N m_n + Z m_e - M(^A X))c^2$. For the ^{4}He nucleus, $Z = 2$ and $N = 2$. Also needed is the atomic mass for the ^{4}He atom, from Appendix D: 4.0026 u.

Known: $M(^4\text{He}) = 4.0026\,\text{u};\ m_n = 1.00866\,\text{u};\ m_p = 1.00728\,\text{u};$ $m_e = 0.00054858\,\text{u};\ Z = 2;\ N = 2;\ 1\,\text{u} \cdot c^2 = 931.5\,\text{MeV}.$

SOLVE Using the known values in Equation 23.2,

$$E_b = ((2)(1.00728\,\text{u}) + (2)(1.00866\,\text{u}) + (2)(0.00054858\,\text{u})$$
$$- 4.0026\,\text{u})c^2 = 0.0304\,\text{u} \cdot c^2$$

Using $\text{u} \cdot c^2 = 931.5\,\text{MeV}$, this is 28.3 MeV.

REFLECT Our result is little less than one-third the binding energy of ^{12}C—not surprising, since ^{12}C contains three times as many protons, neutrons, and electrons as ^{4}He. Again, the total binding energy is vastly greater than the electron-volt-level energies of atomic electrons.

MAKING THE CONNECTION What's the binding energy of deuterium, whose nucleus contains one proton and one neutron?

ANSWER Repeating Example 25.2 gives 2.22 MeV. That's much less than half the binding energy of helium, for reasons we'll discuss shortly.

Binding Energy per Nucleon

Binding energies tend to grow, roughly in proportion to the number A of nucleons, as you saw in comparing ^{12}C and ^{4}He. However, there are deviations from this general pattern, revealed by computing the *binding energy per nucleon*, E_b/A. A plot of this quantity is the curve of binding energy, shown in Figure 25.5 (next page). Note that binding energy per nucleon generally increases as A increases from 1 to 56. The ^{56}Fe nucleus has the highest binding energy per nucleon and is therefore the most stable nuclide; for heavier nuclei the binding energy per nucleon gradually decreases. The shape of the binding-energy curve is important in explaining two nuclear-energy-releasing processes: fission and fusion. In

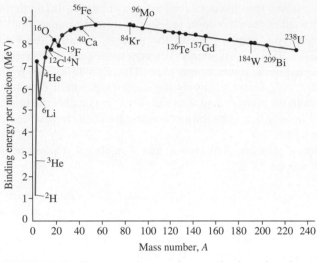

FIGURE 25.5 Binding energy per nucleon, with a broad peak around ^{56}Fe.

fission, a nucleus much heavier than iron splits into two lighter ones. The binding energy per nucleon of the fragments is greater than that of the original nucleus, so energy is released. **Fusion** occurs when nuclei much lighter than iron join to form a heavier one. Once again the increased binding energy per nucleon results in energy release. We'll discuss fission and fusion in Sections 25.5 and 25.6.

✓**TIP**

Energy is released in any process that leaves a final state with higher binding energy.

You may be puzzled that energy is *released* when the binding energy per nucleon *increases*. That's because binding energy measures the energy needed to separate the nucleons or, equivalently, the energy released when separate nucleons came together to form a nucleus. Equation 25.3 shows quantitatively how this works: The mass of nucleus ^{A}X is always lower than the total mass of its constituent protons and neutrons and so, by $E = mc^2$, that implies a lower energy. In Sections 25.5 and 25.6, you'll see how that released energy powers the Sun, supplies us with nuclear-generated electricity, and threatens us with the destructive potential of nuclear weapons.

EXAMPLE 25.3   **Iron and Uranium**

Use data in Appendix D to compute the binding energy per nucleon of ^{56}Fe and ^{235}U.

ORGANIZE AND PLAN From A and Z, $N = A - Z$. Then binding energy is computed as in the preceding example: $E_b = (\Delta m)c^2 = (Zm_p + Nm_n + Zm_e - M(^{A}X))c^2$. We'll divide by the number of nucleons A to get the binding energy per nucleon.

Known: $M(^{56}\text{Fe}) = 55.93494$ u; $M(^{235}\text{U}) = 235.043923$ u; $m_n = 1.00866$ u; $m_p = 1.00728$ u; $m_e = 0.00054858$ u; $Z = 26$, $A = 56(^{56}\text{Fe})$; $Z = 92$, $A = 235(^{235}\text{U})$; $1 \text{ u} \cdot c^2 = 931.5 \text{ MeV}$.

SOLVE For iron, the binding energy is

$$E_b = ((26)(1.00728 \text{ u}) + (30)(1.00866 \text{ u}) + (26)(0.00054858 \text{ u}) \\ - 55.93494 \text{ u})c^2 = 0.52840 \text{ u} \cdot c^2$$

Converting to MeV per nucleon,

$$E_b/A = \left(0.52840 \text{ u} \cdot c^2 \times \frac{931.5 \text{ MeV}}{\text{u} \cdot c^2} \right) \Big/ (56 \text{ nucleons})$$

$$= 8.79 \text{ MeV/nucleon}$$

Similar calculations for uranium give a total binding energy of 1784 MeV, or 7.59 MeV/nucleon. Both per nucleon values agree with Figure 25.5.

cont'd.

REFLECT The corresponding mass defect is significant: For ^{235}U, it's almost 2 u, nearly the equivalent of two missing nucleons! The mass defect is obvious in nuclei, which is why $E = mc^2$ is often mistakenly considered to be only about nuclear physics. There's also a mass defect when, for example, hydrogen and oxygen atoms join to form a water molecule. But it's the mass equivalent of a few electronvolts, not mega-electronvolts; thus it's so small compared with the molecular mass as to be unnoticeable.

MAKING THE CONNECTION Compare the binding energy per nucleon of ^{56}Fe and ^{55}Mn, an atom with almost the same number of nucleons.

ANSWER Repeating Example 25.3 gives 8.76 MeV/nucleon for ^{55}Mn. That's just slightly less than for ^{56}Fe, as you'd expect from Figure 25.5.

Stable Nuclei and Magic Numbers

You might notice "spikes" in the binding energy curve (Figure 25.5), such as the nuclides ^{4}He and ^{12}C. Each of these has an even number of protons and neutrons (they're "even-even" nuclei). There's some preference in nuclear bonding for proton-proton and neutron-neutron pairing, which explains the higher stability of even-even nuclei. There are many reasons for this, one being that protons and neutrons obey the Pauli exclusion principle and have half-integer spin, like electrons (Chapter 24). Thus, two protons (or neutrons) can "pair off" in the same quantum state except for opposite spins.

There are also many stable nuclei with even-odd or odd-even proton and neutron numbers. But odd-odd nuclei are rare, with only four stable nuclides: ^{2}H, ^{6}Li, ^{10}B, and ^{14}N. This illustrates the preference for pairing like nucleons.

The high stability of nuclides such as ^{4}He and ^{12}C might remind you of the stable inert gases in atomic physics (Chapter 23). There the preference for closed shells makes it difficult for inert gases to react chemically. A similar effect occurs in the nucleus, although there are no "Bohr orbits" for nucleons. However, in the quantum-mechanical nuclear **shell model**, there are nuclear shells that fill when the number of protons or neutrons is 2, 8, 20, 28, 50, 82, or 126. These so-called **magic numbers** further explain the spikes in Figure 25.5, as well as more subtle differences not evident there. Another consequence of magic numbers is that elements with those atomic numbers ($Z = 2, 8, 20$, and so on) tend to have more stable isotopes than other elements, particularly those with odd Z. For example, calcium ($Z = 20$) has six stable isotopes, but potassium ($Z = 19$) has two and scandium ($Z = 21$) has only one.

GOT IT? Section 25.2 Rank in order the binding energy per nucleon of the following nuclides: (a) ^{235}U; (b) ^{3}He; (c) ^{208}Pb; (d) ^{40}Ca; (e) ^{56}Fe.

Reviewing New Concepts

- The nucleus contains protons and neutrons, bound by the strong force—an attractive, short-range force between nucleons.
- Elements may have several isotopes, which differ only in the numbers of neutrons in their nuclei.
- There are only a few stable isotopes for each element; other isotopes are radioactive.
- Binding energy suggests the net strength of the attractive force holding a particular nucleus together, and binding energy per nucleon indicates the relative stability of different nuclei.

25.3 Radioactivity

Figure 25.4 shows many more unstable nuclei than stable ones. Unstable nuclei undergo radioactive decay. Here we'll explore radioactivity and introduce three principal decays: alpha, beta, and gamma. We'll quantify radioactive decay through the concept of half-life and then consider some applications of radioactivity.

Discovery of Radioactivity

In 1896 the French physicist Henri Becquerel (1852–1908) noticed that uranium salts placed near a covered photographic plate produced a foggy image on the plate. Pierre (1859–1906) and Marie Curie (1867–1934) soon found such **radioactivity** in other materials. Marie Curie is noted for discovering many radioactive isotopes and for finding a technique to isolate radium, a particularly useful radioactive source. In 1898 the young Ernest Rutherford found that the uranium source emitted two kinds of radiation: less penetrating **alpha radiation** and more penetrating **beta radiation**. By 1900 the French physicist Paul Villard had identified a third, the more penetrating **gamma radiation**. Gradually physicists came to understand the nature of these radiations:

- Alpha radiation consists of **alpha particles** (symbol α), which are helium-4 nuclei. Thus, they have mass 4.0015 u and charge $+2e$.
- Beta radiation consists of two types of **beta particles**: electrons (symbol β^-) and positrons (symbol β^+). The positron has the same mass as the electron but opposite charge, $+e$.
- Gamma radiation is electromagnetic radiation. Each gamma decay results in emission of a photon, called a **gamma ray** (symbol γ). Gamma-ray photons are on the high-energy (short-wavelength) end of the electromagnetic spectrum, with wavelengths less than 10^{-10} m.

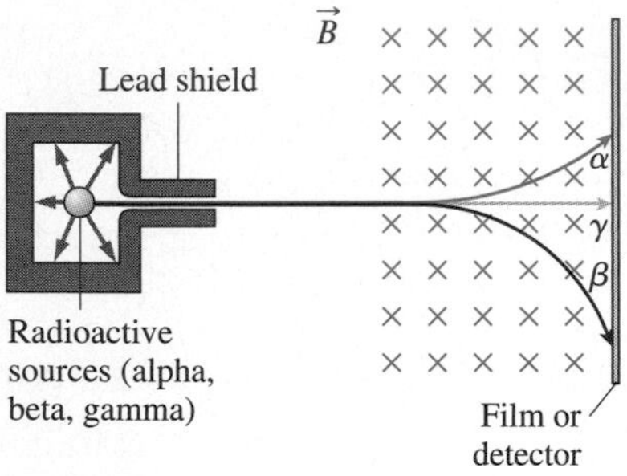

Figure 25.6 shows an experiment to reveal the nature of radioactive emissions. Recall from Chapter 19 that positive and negative particles curve in opposite directions when they pass through a magnetic field. Thus, alpha particles are deflected upward in the diagram and β^- particles downward. Photons are uncharged, so gamma radiation passes with no deflection. Positive beta particles (β^+) would be deflected in the same direction as alphas. However, β^+ particles have a much higher charge-to-mass ratio than alphas, which makes them easy to distinguish.

FIGURE 25.6 Distinguishing alpha, beta, and gamma radiation.

✓ TIP

In other contexts, it's common to see electrons and positrons represented by e^- and e^+, respectively. Using β^- and β^+ here reminds you that the source of the particles is beta decay of the nucleus.

Alpha Decay

Physicists refer to an original radioactive nucleus as the **parent nucleus** and the nucleus that remains after decay as the **daughter**. Since an alpha particle is $^4_2$He, the daughter in alpha decay has two fewer protons and two fewer neutrons than the parent, so its atomic number is smaller by two and its mass number by four. We write a radioactive decay as you would a chemical reaction, with the parent nucleus on the left of the reaction arrow and the daughter and other particles on the right. An example is alpha decay of radium-226, whose daughter has mass number $226 - 4 = 222$, and atomic number two less than radon's $Z = 90$, or 88. Element 88 is radon, so this decay is

$$^{226}\text{Ra} \rightarrow {}^{222}\text{Rn} + {}^4\text{He}$$

As with all forms of radioactivity, alpha decay is spontaneous, analogous to the spontaneous emission of a photon as an atom drops from a higher to a lower energy state (Chapter 24). Like the atomic process, alpha decay is possible when there's excess energy in the parent nucleus. That energy shows up as kinetic energy of the reaction products; in alpha decay it's typically several million electron-volts.

CONCEPTUAL EXAMPLE 25.4   **Alpha Decay**

Find the daughter nuclei when the ^{208}Po and ^{238}U undergo alpha decay. In each case, write the full reaction.

SOLVE The daughter following alpha decay has two fewer protons and two fewer neutrons than the parent, dropping its atomic number by 2 and mass number by 4. Polonium has $Z = 84$, so ^{208}Po has $Z = 82$ (lead, Pb), and is the lead isotope with mass number 204. Thus the reaction is

$$^{208}\text{Po} \rightarrow {}^{204}\text{Pb} + {}^{4}\text{He}$$

Similarly, the parent ^{238}U has $Z = 92$ and $A = 238$. So the daughter has $Z = 90$ (thorium, Th), and is the isotope with $A = 234$:

$$^{238}\text{U} \rightarrow {}^{234}\text{Th} + {}^{4}\text{He}$$

REFLECT An essential feature here is that the total number Z of protons and total number $A - Z$ of neutrons don't change. This implies conservation of electric charge, one of the fundamental conservation laws (Chapter 20). Mass, however, is *not* precisely conserved, because of the mass change $\Delta m = E/c^2$ associated with energy E released in the reaction.

Many radioactive nuclides undergo alpha decay. One reason is that ^{4}He is extremely stable, relative to other small nuclei—as evident in Figure 25.5. For this reason, alpha decay occurs rather than the nucleus losing, say, a single proton, ^{2}H, ^{3}He, or other small nucleus. Also, alpha decay occurs mostly in heavier nuclides. Recall that these need $N > Z$ to provide the necessary neutron "glue." Alpha decay—losing two protons and two neutrons—therefore lowers the number of protons Z by proportionately more, thus creating a more favorable ratio of neutrons to protons.

Beta Decay

In the most common form of beta decay, a nucleus emits an electron (β^-). Recall that the nucleus cannot contain electrons, so conservation of charge suggests that β^- decay involves a neutron changing into a proton. Therefore, the daughter nucleus has atomic number one *larger* than that of the parent. For example, the well-known beta decay of the isotope carbon-14 is

$$^{14}\text{C} \rightarrow {}^{14}\text{N} + \beta^-$$

Note that the mass number doesn't change, because the electron's mass is tiny relative to that of a nucleon. Charge is conserved here: The carbon nucleus has charge $+6e$, the nitrogen has charge $+7e$, and the electron's charge is $-e$.

CONCEPTUAL EXAMPLE 25.5   **Positron Emission**

Oxygen-15 is a radioisotope often used in positron emission tomography (PET scanning; see Chapter 23). This isotope decays by emitting a positron—a positive beta particle, β^+. Identify the product nucleus and write this positive beta decay symbolically.

SOLVE In β^- decay, like that of ^{14}C described above, the atomic number Z goes *up* one to compensate for the emitted *negative* charge. So in β^+ emission Z must drop by one, in this case from oxygen's $Z = 8$ to $Z = 7$. The positron (β^+) has the same mass as the electron, so again the mass number doesn't change. Thus the final nucleus has $Z = 7$ and $A = 15$. Element 7 is nitrogen, so the beta decay of ^{15}O is

$$^{15}\text{O} \rightarrow {}^{15}\text{N} + \beta^+$$

REFLECT ^{15}N is one of nitrogen's two stable isotopes, comprising just 0.37% of natural nitrogen. The rest—99.63%—is the common nitrogen-14. Positron emission is useful in medical imaging because the emitted positron soon annihilates with an electron, emitting two oppositely directed gamma-ray photons whose detection then localizes the emission site.

Detecting Neutrinos

Neutrinos interact only rarely with matter, making their detection difficult. Neutrino detectors must be huge in order to record a significant number of events. Neutrinos react occasionally with matter to produce beta particles, which can be detected directly or when they induce other observable reactions. Neutrino "telescopes" are often vast volumes of liquid surrounded by radiation detectors. They're built deep underground, sometimes in abandoned mines, for shielding from cosmic rays whose reactions would mask the neutrino signal. Shown here is a large pool of water surrounded by radiation detectors at the Super-Kamiokande neutrino observatory in Japan.

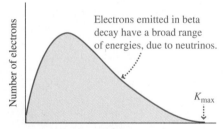

FIGURE 25.7 Energy spectrum in a typical beta decay. Emitted electrons have kinetic energies over a wide range, up to some maximum value K_{max} determined by the mass of the parent and daughter nuclei.

A third form of beta decay is **electron capture**, where the nucleus captures an inner-shell electron. This process converts a proton to a neutron, dropping the atomic number by one. That's the same result as in β^+ decay (Conceptual Example 25.5), so nuclides that undergo β^+ decay can generally also decay by electron capture. As in other beta decay processes, electron capture leaves the mass number unchanged.

Most radioactive nuclei undergo some form of beta decay, making it the most common radioactive decay. Elements generally have a few stable isotopes. Isotopes heavier than the stable ones tend to undergo β^- decay, while lighter isotopes tend to undergo β^+ decay and electron capture. Think about why this happens. For a particular element, the atomic number Z is fixed, and the neutron number N is different for each isotope. A relatively heavy isotope has a larger ratio N/Z than is needed for stability. A β^- decay increases Z and decreases N, producing a nuclide that is more likely to be stable with a smaller ratio N/Z. Similarly, β^+ decay and electron capture by lighter isotopes increase the N/Z ratio, again making the daughter nucleus more stable.

Beta Decay Energy and the Neutrino

You might expect electrons emitted in a particular β^- decay to have a specific kinetic energy associated with excess energy in the original radioactive nucleus. However, as Figure 25.7 shows, a particular nuclide produces electrons with a broad spectrum of energies, up to a maximum value. This odd behavior puzzled physicists, because it seemed to violate conservation of energy. In 1930 Wolfgang Pauli (of the exclusion principle; Chapter 24) suggested that some undetected particle must carry away energy, leaving the electron with less than the maximum possible kinetic energy. This unseen particle is a **neutrino**, meaning "small neutral particle." Its small mass and neutral charge make the neutrino extremely difficult to detect, and it wasn't found experimentally until 1956.

Physicists are still working to understand neutrinos more fully. For many years these particles were thought to have zero mass and to travel at the speed of light. Now physicists realize that neutrinos have small but nonzero masses. Neutrinos play an important role in elementary particle physics and in astrophysics, as you'll see in Chapter 26.

Gamma Decay

According to the nuclear shell model (Section 25.2), the nucleus has a ground state and excited states of higher energy. A radioactive decay, or collision with another particle, can leave the nucleus in an excited state. When it drops to the ground state, the excess energy emerges as a gamma-ray photon. For example, the alpha decay of ^{234}U leaves an excited state of ^{230}Th:

$$^{234}\text{U} \rightarrow {}^{230}\text{Th}^* + {}^4\text{He}$$

where the asterisk indicates an excited daughter nucleus. After a short time the thorium decays to its ground state, emitting a gamma-ray photon:

$$^{230}\text{Th}^* \rightarrow {}^{230}\text{Th} + \gamma$$

Like any photon, the gamma ray obeys the relationship $E = hc/\lambda$, with λ the photon's wavelength and E its energy. For example, this particular thorium decay involves an excited state $0.230\,\text{MeV} = 3.69 \times 10^{-14}\,\text{J}$ above the ground state, yielding a photon of wavelength

$$\lambda = \frac{hc}{E} = \frac{(6.626 \times 10^{-34}\,\text{J} \cdot \text{s})(3.00 \times 10^8\,\text{m/s})}{3.69 \times 10^{-14}\,\text{J}} = 5.39 \times 10^{-12}\,\text{m}$$

✓**TIP**

Alpha, beta, and gamma are the common forms of radioactive decay. A few nuclides decay by emission of a proton, neutron, or carbon nucleus or else undergo spontaneous fission.

Radioactive Decay Series

Radioactive decay occurs because a nucleus is unstable. Often the daughter nucleus is also unstable, and there may follow a whole series of decays before a stable product is reached. Figure 25.8 shows an example of such a **decay series**. A long series is unavoidable for heavy nuclides like uranium, because, as you know, there's no stable nucleus heavier than ^{209}Bi.

Uranium is often found in soil and rock, and among the nuclides in its decay series (Figure 25.8) is ^{222}Rn, a radioactive gas with half-life 3.82 days (more on half-life in the next section). Radon seeps into basements, and it's a carcinogen even in low concentrations. Commercial radon detectors let you check the radon concentration in your basement. For many people, indoor radon exposure is the dominant source of radiation they receive—often exceeding that from medical procedures, cosmic rays, nuclear power plants, and other sources.

Radiation and Life

The high-energy particles that comprise nuclear radiation can harm living organisms. Radiation damages cells directly by ionizing molecules, causing cell death. It can also damage DNA, inducing mutations or initiating cancer. On the other hand, radiation is used to treat cancer, because it selectively damages faster-growing cells. (That's also one reason radiation is especially dangerous to fetuses and growing children.) Beta emitters such as ^{89}Sr and ^{131}I are sometimes implanted near tumors to provide continuous radiation doses. Iodine is readily absorbed by the thyroid gland, so iodine isotopes are particularly useful in treating thyroid cancer.

Because they're uncharged, gamma rays are generally more penetrating and therefore more dangerous than alpha and beta radiation. However, gamma rays have important medical uses. So-called *gamma cameras* image the body's internal organs after a patient is administered a gamma-emitting radioisotope. Technetium-99 is particularly useful because cancer cells tend to trap it, enabling more precise identification of malignant tumors.

The biological effects of radiation depend on the radiation type and energy. Two SI units of **radiation dose** are relevant here. The **gray** (Gy) is equal to 1 joule of radiation energy absorbed per kilogram of absorbing material. The **sievert** (Sv) has the same units, but is weighted for the biological effects of different radiation types and energies. Alpha particles, if they make it into the body, are usually more damaging than gamma rays, making 1 Gy of alpha radiation more dangerous than 1 Gy of gammas. But 1 Sv of each has very nearly the same biological effect.

High radiation doses are lethal; 4.5 Sv, for example, kills about 50% of humans, with somewhat lower doses causing radiation sickness, hair loss, burns, and other generally survivable effects. Doses below about 0.1 Sv aren't as well understood, although it's generally assumed that the dangers—cancer and mutations—scale linearly with dose. Statistically, a one-time dose of 1 mSv (0.001 Sv) is believed to be associated with a 1-in-10,000 risk of developing cancer over your lifetime—compared with a 42% risk of cancer from all causes.

We're all exposed to radiation, from both natural sources and nuclear technologies (Figure 25.9). In the United States, some 80% of the average dose is natural; sources include cosmic rays, natural radioisotopes within your body, and especially indoor radon. Medical procedures contribute nearly all the rest, and effects from normal operation of nuclear power plants are, on average, negligible. Your so-called background radiation depends on where you live, with higher altitudes receiving more cosmic rays, and uranium-bearing rocks meaning more radon and other uranium decay isotopes. Occupation also affects your background dose, with nuclear industry workers, x-ray technicians, and airline crews subject to higher-than-average doses. (Airline crews receive high-altitude cosmic radiation.)

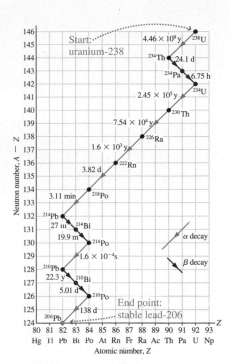

FIGURE 25.8 The most likely decay path for ^{238}U, ending in the stable isotope ^{206}Pb. Times are half-lives.

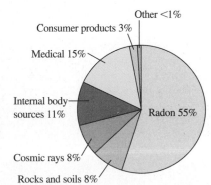

FIGURE 25.9 Natural (gray) and artificial (blue) sources of radiation for the average U.S. resident. The net annual background radiation dose is 3.6 mSv.

GOT IT? Section 25.3 A nuclide's mass number changes in (choose all that apply) (a) alpha decay; (b) β^- decay; (c) β^+ decay; (d) electron capture; (e) none of the above.

25.4 Activity and Half-Life

The decay of any particular radioactive nucleus is a random event. But there's a statistical pattern, much like tossing a pair of dice. Suppose you want to roll a total of 2 on the two dice. Then each die must come up 1—something that, with six-sided dice, has a probability of 1 in 36. Now you might get your 2 on the first toss, or it might take a while. But if you roll many times, you'll find that your 2 comes up just about 1/36 of the time. That's how it is with a radioactive sample: Individual decays are random, but when you have huge numbers of nuclei there's a clear pattern to the overall decay.

> ✓ **TIP**
>
> The analysis in this section is true for all forms of radioactive decay—alpha, beta, and gamma—so we won't distinguish between decay types.

Activity and the Decay Constant

In a given time, there's a certain probability that one given radioactive nucleus will decay. With two nuclei, there's twice the chance that the nucleus will decay. So the **decay rate**—the number of decays per unit time—is proportional to the number N of nuclei. Expressing this as an equation,

$$\frac{\Delta N}{\Delta t} = -\lambda N$$

where λ is the **decay constant** for the particular radioactive nuclide. The minus sign shows that the number of nuclei *decreases* with time, making $\Delta N/\Delta t$ negative. The **activity** (symbol R) is simply the absolute value of the decay rate:

$$R = \left| \frac{\Delta N}{\Delta t} \right| = \lambda N \quad \text{(Activity; SI unit: Bq)} \tag{25.4}$$

Activity is what you measure using a Geiger counter or similar device, counting decays per second from a radioactive sample. The SI unit of activity is the becquerel (Bq), with 1 Bq = 1 decay/s. An older unit often still used is the Curie (Ci), defined as the activity of 1 g of naturally occurring radium and equal to about 3.7×10^{10} Bq—that is, 37 billion decays/s.

A Geiger counter contains a thin wire electrode with a high potential difference (≈ 1 kV) between the wire and tube. Incoming alpha, beta, or gamma radiation ionizes gas in the tube, and the electric charge of the ion is detected. The resulting signal can be sent to a meter and a loudspeaker, resulting in the characteristic "clicks" of the Geiger counter.

Half-Life and the Radioactive Decay Law

Applying calculus to Equation 25.4 shows that the number N of nuclei should decay exponentially with time t, and experiment confirms this:

$$N = N_0 e^{-\lambda t} \quad \text{(Radioactive decay)} \tag{25.5}$$

where N_0 is the number of nuclei at $t = 0$.

Any quantity whose rate of change is proportional to the quantity itself undergoes exponential growth (positive proportionality) or decay (negative proportionality, as with radioactivity). Examples include money in a bank at fixed interest, growth of bacteria in a Petri dish, decay of a radioactive sample, and charge or potential difference on a discharging capacitor (Chapter 17).

It's convenient to describe radioactive decay in terms of **half-life**, $t_{1/2}$, defined as the time for half the nuclei in a radioactive sample to decay. After one half-life, half the original sample remains $N_0/2$; after two half-lives, there's half that, or $N_0/4$, and so on (Figure 25.10). There's a direct connection between the decay constant λ and half-life $t_{1/2}$. When $t = t_{1/2}$, Equation 25.5 becomes $(N_0/2) = N_0 e^{-\lambda t_{1/2}}$, or $e^{\lambda t_{1/2}} = 2$. Here we cancelled the factor N_0 and used the fact that $e^{-x} = 1/e^x$. We can further unravel our equation with the natural logarithm, which is the inverse of the exponential function: $\ln(e^x) = x$. Applying this rule,

$$\ln(e^{\lambda t_{1/2}}) = \lambda t_{1/2} = \ln 2$$

or

$$t_{1/2} = \frac{\ln 2}{\lambda} \quad \text{(Half-life and decay constant; SI unit: s)} \quad (25.6)$$

The half-life tells you how rapidly a particular nuclide will decay. Table 25.2 lists half-lives for the dominant decay modes of some radioactive nuclides, and they are shown graphically in Figure 25.4. Note the huge range from fractions of a second to billions of years. Appendix D lists more half-lives of many other nuclides.

If you're using a radioactive isotope in an instrument, the half-life can be important. Most smoke alarms contain ^{241}Am, an alpha emitter with half-life 433 years. The alpha particles ionize nitrogen and oxygen molecules in the air, and a low-level electric current results. When smoke is present, some alphas attach to smoke particles, and the ion current drops, triggering the alarm. With such a long half-life, you won't have to worry about getting a new radioactive source within the detector's lifetime—or yours!

EXAMPLE 25.6   **Medical Imaging with ^{18}F**

The isotope ^{18}F is widely used in medical imaging, because its 110-min half-life means it doesn't stay radioactive in the patient's body for too long. (a) If a patient is administered 10.0 μg of ^{18}F, what's the initial activity? (b) How much ^{18}F remains in the body after 24 h?

ORGANIZE AND PLAN Equation 25.4 gives the activity $R = \lambda N$, while Equation 25.6 gives the decay constant in terms of half-life: $\lambda = \ln 2/t_{1/2}$. You can find the initial number N_0 of radioactive nuclei using the dose and ^{18}F's molar mass, 18 g/mol. Then Equation 25.5 gives the number of nuclei remaining at any time: $N = N_0 e^{-\lambda t}$.

Known: $t_{1/2} = 110$ min $= 6600$ s; mass $m = 10.0$ μg; molar mass $= 18.0$ g/mol.

SOLVE (a) The initial number of nuclei is

$$N_0 = 1.0 \times 10^{-5}\,\text{g} \times \frac{1\,\text{mol}}{18.0\,\text{g}} \times \frac{6.022 \times 10^{23}\,\text{nuclei}}{1\,\text{mol}} = 3.35 \times 10^{17}\,\text{nuclei}$$

Then the initial activity is

$$R = \lambda N_0 = \frac{\ln 2}{t_{1/2}} N_0 = \frac{\ln 2}{6600\,\text{s}}(3.35 \times 10^{17}) = 3.52 \times 10^{13}\,\text{s}^{-1} = 3.52 \times 10^{13}\,\text{Bq}$$

where 1 Bq $= 1$ s^{-1}.

cont'd.

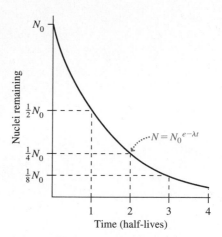

FIGURE 25.10 Exponential decay of a radioactive sample. Half the remaining sample decays in each half-life.

TABLE 25.2 Selected Half-Lives for Radioactive Nuclides

Nuclide	Decay mode(s)	Half-life
^{214}Rn	α	270 ns
^{217}Ra	α	1.7 s
^{12}N	β^+ or EC	11 ms
^{17}F	β^+ or EC	65 s
^{218}Po	α or β^-	3.1 min
^{239}U	β^-	23.5 min
^{239}Np	β^-	2.36 d
^{222}Rn	α	3.82 d
^{131}I	β^-	8.0 d
^{73}As	β^+ or EC	80 d
^{60}Co	β^-	5.27 y
^{90}Sr	β^-	28.8 y
^{14}C	β^-	5730 y
^{239}Pu	α or spontaneous fission	24,110 y
^{235}U	α or spontaneous fission	7.04×10^8 y
^{238}U	α or spontaneous fission	4.47×10^9 y
^{50}V	β^-, β^+, or EC	1.4×10^7 y

(b) After 24 h ($86,400$ s), the fraction of the original sample remaining is

$$\frac{N}{N_0} = e^{-\lambda t} = e^{-\ln(2)t/t_{1/2}} = e^{-\ln(2)(86,400\text{ s})/6600\text{ s}} = 1.15 \times 10^{-4}$$

Only about one ten-thousandth of the original sample remains, about a nanogram.

REFLECT The initial activity is on the order of 10^{13} Bq, high enough for diagnostic imaging. The 635-keV positrons from ^{18}F's decay travel only about 2 mm in tissue before they annihilate, which enables high-resolution imaging. Our calculation at 24 h may be an overestimate, because biological processes help flush the isotope out of the body; the so-called *biological half-life* depends on the specific chemical compound into which the ^{18}F is incorporated.

MAKING THE CONNECTION   What's the activity of the ^{18}F after 24 h?

ANSWER Activity is proportional to the number of nuclei, so the activity is also reduced by the factor 1.15×10^{-4}, to 4.05×10^9 Bq.

EXAMPLE 25.7   **Decay of ^{18}F**

For ^{18}F of the preceding example, how much time elapses before only 1% of the original sample remains?

ORGANIZE AND PLAN The fraction of nuclei remaining is related to time by the radioactive decay law (Equation 25.5 describes the radioactive decay: $N = N_0 e^{-\lambda t}$), whereas in the preceding example $\lambda = \ln 2/t_{1/2}$. We want 1% of the original sample to remain, so $N = N_0/100$.

Known: $t_{1/2} = 110$ min $= 6600$ s; $N/N_0 = 0.010$.

SOLVE With 1% of the nuclei remaining, $N_0/100 = N_0 e^{-\lambda t}$ or, using $e^{-x} = 1/e^x$, $e^{\lambda t} = 100$. To solve, take the natural logarithm of both sides: $\ln(e^{\lambda t}) = \lambda t = \ln 100$. Solving for t,

$$t = \frac{\ln 100}{\lambda} = \frac{\ln 100}{\ln 2}t_{1/2} = \frac{\ln 100}{\ln 2}(6600\text{ s}) = 4.38 \times 10^4\text{ s}$$

REFLECT That's just about 12 h—another indication that this short-half-life isotope doesn't stay around very long.

MAKING THE CONNECTION   How many half-lives does it take to reduce a radioactive sample's activity to $1/1000$ of its initial value?

ANSWER A calculation similar to this example gives $t/t_{1/2} = \ln 1000/\ln 2 = 9.97$—meaning it takes very nearly 10 half-lives for a thousand-fold activity reduction. But here's a quicker approach: Your calculator confirms that $2^{10} = 1024$; thus, after 10 half-lives, activity is reduced by a factor of about 1000. After another 10 half-lives, the activity is down by a factor of a million.

Radioisotope Dating

Radioisotope dating is a technique that provides reliable information about the age of a sample. Best known is **radiocarbon dating**, using ^{14}C. Cosmic rays impinging on the upper atmosphere liberate neutrons, which react with nuclei of ordinary atmospheric nitrogen (^{14}N), yielding ^{14}C and a proton: $^1n + {}^{14}\text{N} \rightarrow {}^1p + {}^{14}\text{C}$. Carbon-14 is a β^- emitter with half-life 5730 years. The ^{14}C mixes with common ^{12}C and joins the carbon cycle. Plants photosynthesize and animals eat, taking in ^{14}C. When an organism dies, carbon uptake ceases. The isotope ^{12}C is stable, but ^{14}C decays. Therefore, the ratio ^{14}C/^{12}C decreases with time, and by comparing this ratio in an ancient sample with its contemporary value (1.20×10^{-12}), scientists can determine the sample's age.

Carbon-14 dating is limited to formerly living things and can't be used on aquatic samples, which don't interact directly with the atmosphere. Radiocarbon dating is good to about 50,000 years, when ^{14}C activity becomes too low for reliable measurements. Also, atmospheric ^{14}C varies as a result of solar activity, but scientists can account for this using other date markers, such as tree rings or artifacts known to come from a certain period.

Longer-lived isotopes date geological events that occurred millions (My) to billions (Gy) of years ago. Some of the more useful include ^{40}K($t_{1/2} = 1.28$ Gy) and ^{238}U ($t_{1/2} = 4.47$ Gy).

Kennewick Man

In 1996, ancient human bones were discovered near Kennewick, Washington, and their owner was dubbed "Kennewick man." Analysis of the bones indicated a $^{14}C/^{12}C$ ratio of 4.34×10^{-13}. What's Kennewick man's age?

ORGANIZE AND PLAN The current $^{14}C/^{12}C$ ratio is 1.20×10^{-12}. Since ^{12}C is stable while ^{14}C decays, the $^{14}C/^{12}C$ ratio falls at the same rate as the ratio N/N_0 for ^{14}C. Thus,

$$\frac{^{14}C/^{12}C \text{ (old)}}{^{14}C/^{12}C \text{ (current)}} = \frac{N}{N_0} = e^{-\lambda t}$$

Knowing this ratio, we can find the time t.

Known: $t_{1/2} = 5730$ years for ^{14}C; $^{14}C/^{12}C$ ratio (old) = 4.34×10^{-13}; $^{14}C/^{12}C$ ratio (current) = 1.20×10^{-12}.

SOLVE Using the two $^{14}C/^{12}C$ ratios,

$$\frac{^{14}C/^{12}C \text{ (old)}}{^{14}C/^{12}C \text{ (current)}} = \frac{4.34 \times 10^{-13}}{1.20 \times 10^{-12}} = 0.362$$

Thus $0.362 = e^{-\lambda t}$. Taking the natural logarithm of both sides:

$$\ln(e^{-\lambda t}) = -\lambda t = \ln(0.362)$$

or

$$t = -\frac{\ln(0.362)}{\lambda} = -\frac{\ln(0.362)}{\ln(2)}t_{1/2} = -\frac{\ln(0.362)}{\ln(2)}(5730 \text{ y})$$

$$= 8400 \text{ y}$$

REFLECT Corrections for varying atmospheric ^{14}C concentration revised this age to 9200 years.

MAKING THE CONNECTION If the bones are really 9200 years old, what does this indicate about the ^{14}C content of the atmosphere at that time, relative to today?

ANSWER A sample that's only 8400 years old contains more ^{14}C than one 9200 years old, so there must have been more ^{14}C in the atmosphere 9200 years ago than today.

GOT IT? Section 25.4 You begin with 10,000 radioactive nuclei. How many half-lives have elapsed when 625 nuclei remain? (a) 2; (b) 3; (c) 4; (d) 5; (e) 6.

25.5 Nuclear Fission

Nuclear reactions realize the alchemists' dream of **transmuting** one element into another. Alpha decay creates an element two atomic numbers lower than the original nuclide, while $Z \rightarrow Z + 1$ in negative beta decay. The cosmic-ray interaction we introduced with radiocarbon dating is another example, turning nitrogen into carbon.

Artificial Transmutations

Rutherford produced the first artificial transmutation in 1919, bombarding nitrogen with alpha particles. The result was

$$^{14}N + {}^4He \rightarrow {}^{17}O + {}^1p$$

where, as you've seen before, ^1p designates the proton. As in every nuclear reaction, electric charge is conserved here. Mass number is usually conserved, but may not be in extremely high energy reactions where new particles are created (you'll see such reactions in Chapter 26). Alpha particles from radioactive materials provide a ready source of particles to bombard nuclei, but protons, electrons, and neutrons can also induce transmutations. For a particular reaction to occur, there has to be sufficient mass-energy in the reacting particles to produce the products. That includes the energy Δmc^2 associated with a mass difference between the initial and final particles, as well as any kinetic energy in the reacting particles. In Rutherford's reaction, for example,

$$\Delta mc^2 = [M(^{14}N) + M(^4He)]c^2 - [M(^{17}O) + M(^1H)]c^2$$

where we've used the mass of atomic hydrogen ^1H in place of the proton so electron masses will cancel. Using the atomic masses in Appendix D, you'll find $\Delta mc^2 = -1.2$ MeV. Thus the products (oxygen and hydrogen) have more mass than the reactants (nitrogen and helium). There would be insufficient energy for this reaction but for the fact that the polonium source Rutherford used produced alphas with 7.7 MeV each of kinetic energy. That's more than enough to make up the 1.2-MeV deficit, so this transmutation works just fine.

SOLVE The number of uranium nuclei is

$$\text{number of nuclei} = 1000 \text{ g} \times \frac{1 \text{ mol}}{235 \text{ g}} \times \frac{6.022 \times 10^{23} \text{ nuclei}}{1 \text{ mol}}$$

$$= 2.56 \times 10^{24} \text{ nuclei}$$

Therefore, the energy released is

$$E = 200 \text{ MeV} \, (2.56 \times 10^{24}) = 5.12 \times 10^{26} \text{ MeV}$$

That's a big number, so we convert to joules:

$$E = 5.12 \times 10^{26} \text{ MeV} \times \frac{1.60 \times 10^{-13} \text{ J}}{1 \text{ MeV}} = 8 \times 10^{13} \text{ J}$$

REFLECT That's still a big number—and it's about what we'd get from chemical reactions involving 20,000 tonnes (20 million kg) of chemical reactants. Release it all at once, and you've got a bomb. In fact, 1 kg is just about the amount of uranium that fissioned in the bomb that destroyed Hiroshima.

MAKING THE CONNECTION What mass of ^{235}U is needed to power your house for a day, assuming an average consumption of 1 kW?

ANSWER The total energy needed is $1000 \text{ W} \times 86{,}400 \text{ s} = 9 \times 10^7 \text{ J}$. Our example gave $8 \times 10^{13} \text{ J}$ for 1 kg of ^{235}U, so we'll need $(9 \times 10^7)/(8 \times 10^{13} \text{ J/kg}) \sim 10^{-6} \text{ kg} = 1 \text{ mg}$.

Chain Reactions

Neutrons induce fission in ^{235}U, and each fission event releases several neutrons. If, on average, more than one of those neutrons goes on to cause another fission event, the result is a **chain reaction**, with the fission rate growing exponentially (Figure 25.13). The idea of a chain reaction occurred to the visionary Hungarian physicist Leo Szilard (1898–1964) in 1933, several years before nuclear fission was confirmed. Szilard recognized that a chain reaction would liberate large amounts of energy, either in a useful and controlled way or all at once in a fearsome explosion. Both possibilities were soon realized.

Fission was discovered on the eve of World War II, and the military implications were obvious. During the war, physicists working at Los Alamos, New Mexico, developed the first nuclear weapons. One design used ^{235}U, the other ^{239}Pu. So confident were physicists in the uranium bomb that it was never tested; the first such bomb was used in warfare, and destroyed the Japanese city of Hiroshima on August 6, 1945. Three days later a plutonium weapon devastated Nagasaki; that design had been tested a month earlier in New Mexico. Because only ^{235}U (and not ^{238}U) fissions readily under normal circumstances, scientists built a huge plant at Oak Ridge, Tennessee, that enriched uranium by increasing significantly the proportion of the rare U-235 isotope. And because plutonium doesn't exist in nature, they also built a facility in Hanford, Washington, using nuclear reactors to produce plutonium-239.

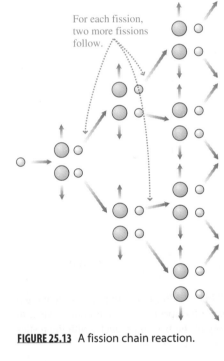

For each fission, two more fissions follow.

FIGURE 25.13 A fission chain reaction.

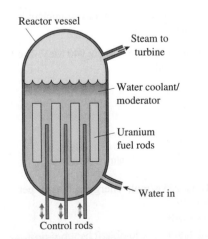

Reactor vessel
Steam to turbine
Water coolant/ moderator
Uranium fuel rods
Water in
Control rods

FIGURE 25.14 A boiling-water reactor, in which steam produced in the reactor vessel directly drives a turbine generator.

Nuclear Reactors

Szilard's other prediction—that a controlled nuclear chain reaction could provide useful energy—also came to pass. The first nuclear reactor achieved a chain reaction in 1942, helping physicists to understand the fission process on their way to nuclear weapons. The idea behind a nuclear reactor is to create a controlled chain reaction in which, on average, exactly one neutron from each fission event goes on to fission another nucleus. Then the reaction releases energy at a steady rate, and the reactor is said to be **critical**. Careful design and operation are needed to keep the reaction from going **supercritical**, meaning that the fission rate grows exponentially. The catastrophic 1986 Chernobyl nuclear accident involved just such a runaway chain reaction.

Figure 25.14 shows the essential elements of one common nuclear reactor design. The fuel rods shown contain uranium enriched to about 4% U-235. They're surrounded by a **moderator**, in this case plain water, which slows the fission-released neutrons and greatly increases the chances that they'll induce additional fission in U-235 rather than being absorbed in U-238. **Control rods**, made from neutron-absorbing material, can move in or out among the fuel rods to control the reaction rate. If the reactor threatens to become supercritical, extra control rods are inserted to shut down the chain reaction. In the simplest design for a complete nuclear power plant, the reactor in Figure 25.14 replaces the fossil-fueled boiler in a conventional power plant. A more complex design keeps the reactor water under pressure so it doesn't boil; the hot water then exchanges

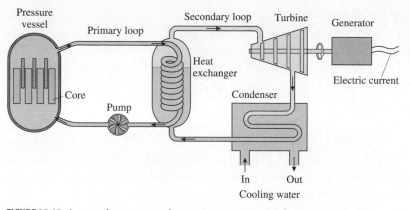

FIGURE 25.15 A complete power plant using a pressurized-water reactor, the most common type of nuclear power reactor in the United States. Some energy is lost through the condenser to the cooling water, as required by the second law of thermodynamics.

energy with a secondary system where water boils to drive a steam turbine and generator (Figure 25.15). Today the world gets some 15% of its electricity from nuclear reactors; in the United States the figure is 20% and in France it's over 75%.

One by-product of fission in nuclear reactors is the plutonium isotope ^{239}Pu, which also fissions readily. Despite enrichment, most uranium in the reactor is still ^{238}U, which absorbs neutrons:

$$^{238}U + {}^1n \rightarrow {}^{239}U$$

The ^{239}U is radioactive and undergoes beta decay:

$$^{239}U \rightarrow {}^{239}Np + \beta^- \quad (t_{1/2} = 23.35\,\text{min})$$

Neptunium-239 is also radioactive:

$$^{239}Np \rightarrow {}^{239}Pu + \beta^- \quad (t_{1/2} = 2.36\,\text{days})$$

Although ^{239}Pu also undergoes beta decay, its half-life is 24,110 years. Thus plutonium builds up within the fuel rods. Some of it fissions to provide additional energy, but much remains and can be separated chemically from leftover uranium and fission products. We've already noted that plutonium can fuel nuclear weapons, which makes for one geopolitically and scientifically significant link between nuclear power and weapons. Another link is uranium-235, whose enrichment for nuclear reactor fuel can be carried further to make weapons-grade material.

Nuclear reactors produce electric power and propel ships; they also create plutonium and a host of other dangerously radioactive isotopes. But by-products of fission are also useful; excess neutrons, for example, are used to produce radioisotopes for medical procedures. This brings us back to Fermi's original idea about inducing radioactivity through neutron activation, but now it's done not just for scientific research but also for human health.

GOT IT? Section 25.5   The most easily fissionable known nuclides are (choose all that apply) (a) ^{235}U; (b) ^{236}U; (c) ^{238}U; (d) ^{239}Pu; (e) ^{240}Pu.

25.6 Fusion

The curve of binding energy (Figure 25.5) shows that energy is released when two smaller nuclei combine. Fusion of light nuclei powers the Sun, and therefore it's responsible for nearly all life as well as most of the energy that runs civilization. Here we'll explore fusion energy, including our attempts to harness this nearly limitless energy source.

Energy from Fusion

As with fission and other nuclear reactions, there's a great enough mass difference between initial and final particles in fusion that we can calculate the energy release using $E = mc^2$. For example, here's a fusion reaction that occurs in stars:

$$^2\text{H} + {}^3\text{H} \rightarrow {}^4\text{He} + {}^1n$$

As usual, mass number and charge are conserved in this reaction. The energy released is

$$\Delta mc^2 = [M(^2\text{H}) + M(^3\text{H})]c^2 - [M(^4\text{He}) + M(^1n)]c^2$$

Using values from Appendix D, $\Delta mc^2 = 0.01888\,\text{u}\cdot c^2$; applying the conversion $\text{u}\cdot c^2 = 931.5\,\text{MeV}$ gives $\Delta mc^2 = 17.6\,\text{MeV}$. As with fission, this is vastly greater than energies released in chemical reactions.

CONCEPTUAL EXAMPLE 25.11 **The Nuclear Difference**

Compare the approximate energy released in this fusion reaction with (a) a typical fission reaction and (b) a chemical reaction like the burning of coal ($\text{C} + \text{O}_2 \rightarrow \text{CO}_2$), which releases 4.1 eV. Also compare these reactions on the basis of energy released per mass of reactants.

SOLVE You've seen that fission reactions release around 200 MeV. Our fusion reaction's energy is about 20 MeV, so the fission reaction releases 10 times the energy. But the fission reactants (^{235}U and a neutron) have a mass around 236 u, while the fusion reactants involve five nucleons, or about 5 u. Then the energy per unit mass is just under 1 MeV/u ($200\,\text{MeV}/236\,\text{u}$) for fission and about 4 MeV/u ($20\,\text{MeV}/5\,\text{u}$) for fusion. So fusion releases about four times as much energy on a per-mass basis. Both nuclear reactions' energy releases vastly exceed that from the chemical reaction: 200 MeV/4 eV, or a factor of 50 million for fission and 5 million for fusion, per reaction. One ^{12}C and $^{16}\text{O}_2$ have a total mass of 44 u, so burning carbon gives about 0.09 eV per u. Then fission's per-mass yield is about 10^7 times greater than that of carbon burning, while fusion's is about 4×10^7 times greater.

REFLECT The curve of binding energy (Figure 25.5) confirms our fission-fusion comparison; going up the steep left-hand side of the curve (fusion) clearly results in a greater energy release per nucleon. The comparison with the reaction $\text{C} + \text{O}_2 \rightarrow \text{CO}_2$ shows why nuclear fuels are so much more potent than chemical fuels. That's why a nuclear power plant is refueled once a year with a few truckloads of uranium, while a comparable coal plant receives many 110-car trainloads of coal each week. It's also the reason that a single nuclear bomb can destroy an entire city.

Fusion in Stars

You've seen that fusion is a potent energy source. But fusion isn't easy to initiate and doesn't happen spontaneously under normal conditions. Two nuclei repel by the electric force, so they need very high energies if they're to overcome the electrostatic repulsion and get close enough for the short-range nuclear force to grab and fuse them.

The interior of a star is one place where fusion does occur. There the high temperature (on the order of $10^7\,\text{K}$) gives nuclei enough kinetic energy to overcome electrostatic repulsion. That temperature is too high for electrons to remain bound to nuclei, so the material is a **plasma**, or ionized gas. Because of its massive gravity, a star's core is also at high density, enhancing encounters between nuclei that lead to fusion.

✓**TIP**

Because of the high temperatures needed for fusion, a fusion reaction is also called a *thermonuclear reaction*.

Ordinary stars like the Sun "burn" hydrogen to helium through a sequence of fusion reactions called the **proton-proton** cycle:

$$^1H + {}^1H \rightarrow {}^2H + \beta^+$$
$$^1H + {}^2H \rightarrow {}^3He$$
$$^3He + {}^3He \rightarrow {}^4He + 2\,{}^1H$$

The first two reactions occur twice for each occurrence of the third; the net result is to convert four ^1H nuclei (protons) into one helium-4 and two positrons:

$$4\,{}^1H \rightarrow {}^4He + 2\beta^+$$

There's also a neutrino from the first reaction, but we aren't showing it. Each fusion reaction releases energy, and the positrons annihilate with electrons, adding to the total energy: 24.7 MeV for the net reaction. Radiation and convection carry that energy out from the Sun's core to the solar surface, where it emerges as the sunlight that sustains life on Earth.

Origin of the Elements

In the first half hour after the Big Bang event that started our universe, hydrogen fused to form helium and trace amounts of deuterium and lithium. All the other chemical elements were formed in stars.

As stellar fusion continues, hydrogen declines and helium accumulates. Near the end of its life, a star like the Sun fuses helium to carbon in its central core. Several reactions are involved, but the net effect is

$$3\,{}^4He \rightarrow {}^{12}C$$

In our Sun fusion won't continue beyond this point, but in heavier stars the density and temperature are high enough that fusion continues to form heavier elements, up to iron.

EXAMPLE 25.12 **Helium Burning**

Find the energy released in the reaction $3\,{}^4He \rightarrow {}^{12}C$.

ORGANIZE AND PLAN As usual with nuclear reactions, the mass difference between initial and final particles is large enough that we can use $E = \Delta mc^2$ to find the energy release.

Known: $M({}^4He) = 4.0026$ u; $M({}^{12}C) = 12.0000$ u; $1\,u \cdot c^2 = 931.5$ MeV.

SOLVE For this process

$$E = \Delta mc^2 = [3M({}^4He)]c^2 - [M({}^{12}C)]c^2$$
$$= [3(4.0026\,u) - 12.0000\,u]c^2 = 0.0078\,u \cdot c^2$$

so

$$E = 0.0078\,u \cdot c^2 \times \frac{931.5\,\text{MeV}}{u \cdot c^2} = 7.27\,\text{MeV}$$

REFLECT Each helium-burning reaction produces less energy than a proton-proton reaction. However, helium burning occurs at a faster rate. When the Sun begins its helium-burning phase, the increased energy production will expand our star into a red giant.

MAKING THE CONNECTION One possible carbon-burning reaction that takes place in stars much heavier than the Sun is $2\,{}^{12}C \rightarrow {}^{24}Mg$. What is the energy released in this process?

ANSWER 13.9 MeV.

Once carbon is available, a range of fusion reactions produces all the elements up to iron. Probabilities of the various reactions determine the abundances of the chemical elements; following hydrogen and helium, the next most abundant are oxygen, carbon, neon, iron, nitrogen, and silicon. Normal fusion processes can't build elements heavier than iron because that's the most tightly bound nucleus (recall Figure 25.5). In very massive stars, neutron-capture reactions produce nuclei heavier than iron, using available energy to drive these reactions. And when the most massive stars explode as supernovae, a host of nuclear reactions during the explosion produces the whole range of heavier-than-iron elements.

Those stellar explosions, along with winds and other mass-loss events, disperse the newly formed elements into interstellar space. There they condense to form new stars and planetary systems—giving rise eventually to life and ourselves. So we're made, literally, of stardust!

Fusion Weapons

Fusion requires high temperature and density. Stars' immense gravity readily provides that combination, but on Earth it's difficult to achieve. Furthermore, the beginning reaction of the proton-proton chain occurs with low probability, making it especially difficult to mimic stellar fusion. Our only complete "success" with fusion is through **thermonuclear weapons**, also called "hydrogen bombs." These devices use a fission weapon to generate heat and pressure that trigger explosive fusion in a mixture of the hydrogen isotopes deuterium (^{2}H) and tritium (^{3}H). The tritium is produced by neutron capture, using neutrons from the fission trigger. Unlike fission, which occurs in stages (Figure 25.11), fusion of the entire fuel mass takes place almost instantaneously, resulting in a much larger explosion. The fission bombs built in World War II had the equivalent explosive power of 10^4 tons (10 kilotons) of chemical explosive, but fusion weapons can yield 10^7 tons (10 megatons) or more.

Fusion Reactors

For decades, scientists have sought to control fusion for energy production. Two main approaches are under development, each taking a different approach to the need for high temperature and sufficient density to facilitate fusion. Experimental devices of both types are operating, but none has yet reached the "breakeven" point where it produces as much fusion energy as it takes to run the device. The most immediately promising reaction in either scheme is deuterium-tritium (D-T) fusion: ^{2}H + ^{3}H → ^{4}He + ^1n, which yields 17.6 MeV.

Magnetic confinement uses the magnetic force on moving charges (Chapter 19) to confine the hot fusion plasma. Figure 25.16 shows the most promising magnetic-confinement device, the **tokamak**. Its toroidal shape keeps magnetic field lines within the machine's plasma chamber, helping prevent charged particles from escaping. A combination of resistive heating and energy injected via radio waves or beams on neutral particles raises the temperature to more than 10^8 K—hotter than the Sun's core. Tokamaks in the United States, England, and Japan have produced megawatt-level bursts of fusion energy for short times, approaching breakeven, but they're far from being practical fusion power sources. An international collaboration is building a tokamak called ITER in France; when it begins operation around 2016 it should produce 500 MW of fusion power, comparable to a medium-sized power plant.

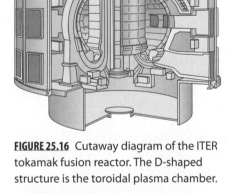

FIGURE 25.16 Cutaway diagram of the ITER tokamak fusion reactor. The D-shaped structure is the toroidal plasma chamber.

✓**TIP**

The word *tokamak* is a Russian acronym for "toroidal chamber in magnetic coils."

The second approach to controlled fusion is **inertial confinement**. Here a pea-sized pellet of deuterium and tritium is bombarded from all directions with powerful laser beams. That compresses and heats the pellet, creating pressure and temperature sufficient to initiate fusion. Fusion occurs so quickly that the fusing particles' inertia keeps them from leaving the reaction site, hence *inertial* confinement. The most advanced inertial confinement experiment is the National Ignition Facility at Livermore National Laboratory in California (Figure 25.17). Its 192 laser beams deliver 500 trillion watts—many times the output of all the world's power plants—for a mere 20 ns to "ignite" fusion in the fuel pellet.

FIGURE 25.17 Target chamber for the NIF inertial fusion facility undergoes installation. The chamber includes ports for 192 converging laser beams.

If we succeed with either fusion approach, fuel won't be a problem. There's enough natural deuterium (^{2}H) in the oceans that each gallon of seawater contains the energy equivalent of 300 gallons of gasoline!

Chapter 25 in Context

Here we've come to a fairly complete understanding of the atomic nucleus. We know it's composed of neutrons and protons, bound together by the strong force. The curve of binding energy quantifies the strength of that binding and shows that iron is the most tightly bound nucleus. The number of protons determines the chemical element, but different isotopes of the same element have different numbers of neutrons. Some isotopes are stable, but others undergo radioactive decay. The principal forms of decay are *alpha*, *beta*, and *gamma*, and the energy released in each nuclear decay is associated with a difference in masses of the parent nucleus and the decay products. The *half-life* determines the timescale for a particular nuclide's decay. More energetic nuclear reactions include *fission* and *fusion*. Fission supplies the world with some 15% of its electricity, while fusion powers the Sun and other stars and is responsible for the creation of most elements. Both fission and fusion provide the awesome destructive power of nuclear weapons, while controlled fusion remains a scientific and technical challenge.

Looking Ahead In Chapters 24 and 25, we've peered first into the atomic world, and then into the nucleus at the atom's core. In Chapter 26 we'll focus on even smaller entities: the individual subatomic particles. You'll meet the "zoo" of new particles discovered through the 20th century, and you'll see how the known particles can be combined into a simple scheme that describes the most fundamental interactions. We'll end Chapter 26, and the book itself, with the connection between particle physics and cosmology—the study of the large-scale universe and its evolution.

Nuclear Structure

(Section 25.1) The **nucleus** contains **protons** and **neutrons**, held together by the **strong force**. Each element has different **isotopes** with the same number of protons but different numbers of neutrons. The nucleus is roughly spherical and grows as the number of nucleons (protons and neutrons) increases.

Mass number: $A = Z + N$, where Z = **atomic number**, N = **neutron number**

Nuclear radius: $r = R_0 A^{1/3}$

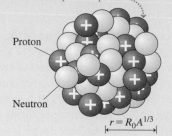

Protons and neutrons form roughly into the shape of a sphere.

Proton

Neutron

$r = R_0 A^{1/3}$

The Strong Force and Nuclear Stability

(Section 25.2) The **strong force** acts between proton-proton, proton-neutron, or neutron-neutron pairs. **Binding energy** measures how tightly the nucleus is bound. Nuclei with certain **"magic numbers"** of protons or neutrons are especially stable. The most stable nuclei have even numbers of protons, neutrons, or both.

Binding energy: $E_b = (\Delta m)c^2 = (Zm_p + Nm_n + Zm_e - M(^A X))c^2$

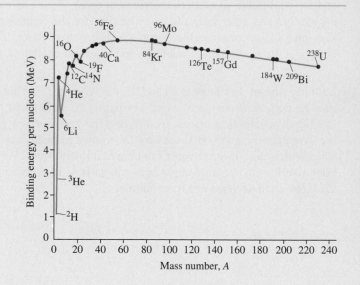

Radioactivity

(Section 25.3) The primary forms of **radioactive decay** are **alpha, beta,** and **gamma**. Alpha radiation consists of **alpha particles** (α), which are helium nuclei. Beta radiation consists of two types of **beta particles:** electrons (β^-) and positrons (β^+). Gamma radiation is electromagnetic radiation, consisting of high-energy photons called **gamma rays** (γ). Radioactive decay involves a difference in mass-energy, which shows up as kinetic energy of the decay products.

Alpha decay: $^A_Z X \rightarrow ^{A-4}_{Z-2} Y + ^4_2 He$

β^- decay: $^A_Z X \rightarrow _{Z+1}^{A} Y + e^-$ (neutrino is not shown)

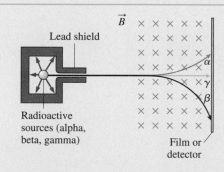

Lead shield

Radioactive sources (alpha, beta, gamma)

Film or detector

Activity and Half-Life

(Section 25.4) Each radioactive **nuclide** has a **half-life**, which tells how quickly the decay progresses.

Activity is measured using a **Geiger counter** or similar device, counting the number of decays per second from a radioactive sample. Radioactive isotopes such as ^{14}C are used in **radioisotope dating** to determine ages of ancient samples.

Activity: $R = \left| \dfrac{\Delta N}{\Delta t} \right| = \lambda N$

Radioactive decay law: $N = N_0 e^{-\lambda t}$

Half-life and decay constant: $t_{1/2} = \dfrac{\ln 2}{\lambda}$

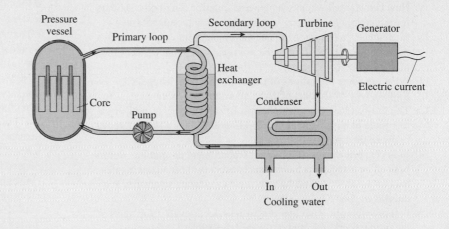

Nuclear Fission

(Section 25.5) Nuclear transmutations can be induced by bombarding nuclei with alphas or other subatomic particles. **Neutron bombardment** leads to **fission** in some nuclei, accompanied by an enormous release of energy. A **chain reaction** results when more than one neutron from each fission event goes on to split another nucleus.

Typical fission reaction:
$^{1}n + {}^{235}U \rightarrow {}^{102}Mo + {}^{131}Sn + 3{}^{1}n$

Fusion

(Section 25.6) **Fusion** of light elements also releases large amounts of energy. Extreme pressure and temperature are required for fusion; such conditions exist inside stars, but a net energy release from controlled fusion has yet to be achieved.

Typical fusion reaction: $^{2}H + {}^{3}H \rightarrow {}^{4}He + {}^{1}n$

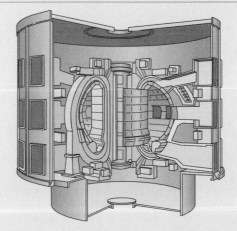

NOTE: Problem difficulty is labeled as ■ straightforward to ■■■ challenging. Problems labeled BIO are of biological or medical interest.

Conceptual Questions

1. How can a nuclide be unstable and yet still occur in nature?
2. Explain (without using a formula) why the nuclei ^{60}Fe and ^{60}Co should have the same radius.
3. Which do you expect to have a greater number of stable isotopes: Sn or Sb?
4. Why does the atomic number Z determine chemical properties, while the chemical properties are largely independent of the neutron number N?
5. The nuclide ^{79}Br is stable. Would you expect ^{75}Br to be stable or unstable? Why?
6. Explain why ^{31}P is stable, but ^{32}P and ^{37}P are not.
7. Why is a stable nucleus with $Z > N$ unlikely? Are there any such stable nuclei?
8. The isotopes ^{75}Br and ^{85}Br are both beta emitters. Which is more likely to be the β^- emitter and which the β^+ emitter?
9. How can you tell whether a particular nuclide is a good candidate for alpha decay? How can you prove it using atomic masses?
10. In β^+ decay, a proton in the nucleus changes into a neutron and a positron is emitted. How is this possible, given that the neutron's mass is larger than the proton's?
11. What are some limitations of radiocarbon dating using ^{14}C?
12. Explain why ^{222}Rn occurs in the uranium decay series.
13. The gray and sievert are both units of radiation dose, both equal to 1 joule of absorbed energy per kilogram. What's the difference between the two units?
14. In his experiments with neutron bombardment of uranium, Fermi noticed an increase in the measured activity if the incoming neutrons were first slowed by passing them through paraffin. Why did this happen?
15. Is it possible for a nuclear reaction to occur even if the mass of the products exceeds that of the initial particles?
16. In nuclear fission, which has more mass, reactants or products? Is the same true for fusion?

Multiple-Choice Problems

17. The number of neutrons in ^{90}Zr is (a) 50; (b) 51; (c) 52; (d) 53.
18. The approximate density of any nucleus is (a) 4×10^{15} kg/m^3; (b) 6×10^{16} kg/m^3; (c) 2×10^{17} kg/m^3; (d) 6×10^{18} kg/m^3.
19. The ratio of the diameter of the ^{104}Pd nucleus to that of ^{26}Mg is approximately (a) 1.4; (b) 1.6; (c) 1.8; (d) 2.0.
20. The total binding energy of a stable nucleus with mass number 60 is about (a) 100 MeV; (b) 300 MeV; (c) 400 MeV; (d) 500 MeV.
21. Of these four nuclides, the one with the largest total binding energy is (a) ^{55}Mn; (b) ^{66}Zn; (c) ^{72}Ge; (d) ^{84}Kr.
22. Of these four nuclides, the one with the largest binding energy per nucleon is (a) ^{60}Ni; (b) ^{96}Mo; (c) ^{113}In; (d) ^{146}Nd.
23. The β^- decay of ^{20}O produces (a) ^{21}O; (b) ^{20}F; (c) ^{20}N; (d) ^{21}F.
24. What's the approximate energy release in alpha decay of ^{226}Ac? (a) 2.5 MeV; (b) 4.0 MeV; (c) 5.5 MeV; (d) 7.0 MeV.
25. The isotope ^{17}F has a 65-s half-life. If you start with a 10-g sample of ^{17}F, after 32.5 s the amount remaining is (a) 7.5 g; (b) 7.1 g; (c) 6.2 g; (d) 5.0 g.

26. Iodine-131's half-life is 8.0 days. Ten percent of the original sample of this isotope remains after (a) 22.7 days; (b) 24.9 days; (c) 26.6 days; (d) 28.1 days.
27. Complete the following fission process: $^{1}n + \,^{235}$U $\rightarrow\,^{144}$Ba + _____ + $2^{1}n$. (a) ^{88}Kr; (b) ^{89}Kr; (c) ^{90}Kr; (d) ^{91}Kr.
28. Complete the following fusion process: ^{3}He + ^{4}He $\rightarrow\,^{6}$Li + _____. (a) ^{1}n; (b) ^{1}H; (c) ^{2}H; (d) ^{2}He.

Problems

Section 25.1 Nuclear Structure
29. ■ Write the isotope symbol ^{A}X for nuclei with (a) 24 protons and 25 neutrons; (b) 43 protons and 51 neutrons; (c) 82 protons and 108 neutrons.
30. ■ What are the neutron numbers of these nuclides: (a) ^{13}C; (b) ^{51}V; (c) ^{79}Br; (d) ^{136}Ba?
31. ■ Find the numbers of protons and neutrons in ^{80}Br and ^{80}Kr.
32. ■ Write out the full nuclide symbol $^{A}_{Z}X$ for the following isotopes: (a) oxygen-17; (b) xenon-139; (c) iridium-191; (d) lead-208.
33. ■ Find the nuclear radii for (a) ^{7}Li; (b) ^{20}Ne; (c) ^{133}Cs; (d) ^{239}Pu.
34. ■■ (a) What's the ratio of the diameters of the nuclei ^{77}Se and ^{14}N? (b) What's the ratio of the densities of these nuclei?
35. ■■ What's the nuclide whose nuclear diameter is closest to twice that of the ^{27}Al nucleus?
36. ■■ Find a stable nucleus with half the volume of the ^{46}Ti nucleus.
37. ■■■ Consider a large nucleus with diameter 10 fm. (a) Suppose an electron is confined within this nucleus. Use the Heisenberg uncertainty principle (Chapter 23) to estimate the electron's minimum kinetic energy. (b) Electrons emitted in beta decay normally have kinetic energies on the order of 1 MeV. Use this fact to argue that an electron cannot be confined to a nucleus.

Section 25.2 The Strong Force and Nuclear Stability
38. ■■ Compute the binding energy of (a) ^{14}N and (b) ^{28}Si.
39. ■■ For the two nuclides in the preceding problem, compute the binding energy per nucleon. Compare with the data shown in Figure 25.5.
40. ■■ (a) Which nuclide do you expect to have the larger binding energy per nucleon: ^{136}Ba or ^{144}Sm? (b) Calculate the binding energy per nucleon for both.
41. ■■ Suppose that the centers of the two protons in a ^{4}He nucleus are separated by approximately 2.0 fm. Find the electrostatic energy associated with the repulsive force between the protons. Compare with the binding energy of the ^{4}He nucleus.
42. ■■ How much energy is required to remove one proton from a ^{16}O nucleus?
43. ■■■ (a) Find the energy to remove (a) one proton and (b) one neutron from a ^{32}S nucleus. (c) Explain the difference between your answers to parts (a) and (b).
44. ■■ *Mirror nuclides* are pairs with the numbers Z and N reversed, such as ^{21}Ne and ^{21}Na. (a) Find the binding energies of ^{21}Ne and ^{21}Na. (b) What does the size of the difference between their binding energies suggest about the charge independence of the strong force?

45. ■■ (a) Use the astronomical data from Appendix E to find the binding energy of the Earth-Sun system. (b) How much energy would be required to remove the Earth completely from the solar system? (c) How would the mass of the Earth-Sun system change if Earth were removed from the solar system?

Section 25.3 Radioactivity

46. ■ Write the full reactions for the alpha decay of (a) ^{205}At; (b) ^{216}Rn; (c) ^{211}Ac.

47. ■ Write the full reactions for the β^- decay of (a) ^{67}Cu; (b) ^{85}Kr; (c) ^{112}Pd.

48. ■ Complete the following reactions: (a) ^{71}Zn $\rightarrow$ ^{71}Ga + _____ ; (b) ^{66}Ge $\rightarrow$ _____ + β^+; (c) ^{213}Ac $\rightarrow$ _____ + ^{4}He; (d) _____ $\rightarrow$ ^{210}Bi + γ.

49. ■ Complete the following reactions: (a) ^{158}Tm + β^- $\rightarrow$ _____ ; (b) ^{178}Hf $\rightarrow$ ^{178}Lu + _____ ; (c) ^{216}Fr $\rightarrow$ ^{212}At + _____ ; (d) ^{8}Be $\rightarrow$ _____ + ^{4}He.

50. ■■ Find the energy released in the alpha decay of ^{227}Ac.

51. ■■ Find the energy released in the β^- decay of ^{210}Pb.

52. ■■ Determine whether the following nuclides will undergo alpha decay: (a) ^{214}Po; (b) ^{199}Hg (mass = 198.968 262 u).

53. ■■ The EPA estimates that the limit for safe exposure to indoor radon-222 is 4 pCi per liter of air. (a) What is this activity in Bq? (b) At this rate, how many nuclei decay in 1 day?

54. ■■ Trace the steps in the uranium series that lead to ^{222}Rn, listing all the decays starting with ^{238}U and ending with ^{222}Rn.

55. ■■ An alpha particle and β^- particle, each with kinetic energy 40 keV, are sent through a 1.5-T magnetic field. The particles move perpendicular to the field, as in Figure 25.6. Find the curvature radius for each particle's trajectory. *Hint:* The kinetic energy is small enough to neglect relativity.

Section 25.4 Activity and Half-Life

56. ■■ You begin with 50,000 radioactive nuclei, and after 2.5 h only 12,500 of them remain. What's the half-life of this nuclide?

57. BIO ■■ **PET scans.** Oxygen-15, used in PET scans, has a 2.0 min half-life. A hospital cyclotron produces 2.60 mg of ^{15}O. (a) It's delivered to the diagnostic facility 6.0 min later. How much ^{15}O remains at this time? (b) After another 4.0 min, the ^{15}O is injected into a patient. How much is there at that point?

58. ■■ A radioactive sample containing 125×10^{15} nuclei has activity 2.57×10^{12} Bq. What's this nuclide's half-life?

59. ■■ The activity of a ^{60}Co sample measures at 3.90×10^{11} Bq. What's the mass of the cobalt-60?

60. BIO ■■ **Radioactivity in the body.** Potassium is an essential element that normally comprises about 0.30% of a person's body mass. 0.012% of potassium is the radioactive ^{40}K, with half-life 1.28×10^9 years. What's the potassium activity in a 60-kg person?

61. ■■ A sealed container has 25 μg of radon-222. (a) What's the sample's activity? (b) Find the activity and the amount of radon remaining after 30 days.

62. BIO ■■ **Analyzing ancient humans.** In 1991, hikers discovered the frozen remains of a human (the "Iceman") in a Swiss glacier. Measurements of ^{14}C beta emission from the Iceman revealed an activity of 0.121 Bq/g of carbon. How old was the Iceman?

63. ■■ The hydrogen isotope tritium (^{3}H) is used as a neutron source in nuclear weapons, so it's produced continually for military stockpiles. Tritium's half-life is 12.3 years. If we stopped producing tritium today with 2500 kg stockpiled, how much would remain after 100 years?

64. ■■ Smoke detectors use the isotope ^{241}Am, with half-life 433 years. (a) If you keep a smoke detector for 5 years, by what factor is ^{241}Am activity reduced relative to when it was new? (b) How many years pass before the activity falls to 99% of its initial value?

65. ■■■ The uranium isotopes ^{235}U and ^{238}U were present in roughly equal amounts when the solar system was formed. Today only 0.72% is the lighter isotope, with the remainder U-238. Use these data and half-lives from Table 25.2 to estimate the age of the solar system.

Section 25.5 Nuclear Fission

66. ■ Complete the following transmutation processes: (a) ^{2}H + ^{16}O $\rightarrow$ ^{14}N + _____ ; (b) ^{1}H + ^{7}Li $\rightarrow$ ^{1}n + _____ ; (c) ^{4}He + ^{13}C $\rightarrow$ _____ + ^{1}n.

67. ■■ Compute the energy released or required for each of the transmutations in the preceding problem.

68. ■ Complete the following transmutation processes: (a) ^{1}n + _____ $\rightarrow$ ^{4}He + ^{17}O; (b) ^{4}He + ^{88}Sr $\rightarrow$ ^{3}H + _____ ; (c) ^{2}H + ^{28}Si $\rightarrow$ + ^{27}Al + _____ .

69. ■■ Compute the energy released or required for the transmutations in the preceding problem.

70. ■■ A common neutron source uses alpha particles striking beryllium, causing the reaction ^{4}He + ^{9}Be $\rightarrow$ ^{12}C + ^{1}n. If the incoming alphas have kinetic energy 5.0 MeV, find the total kinetic energy of the reaction products (carbon and neutron).

71. ■ Complete the following fission reactions: (a) ^{1}n + ^{235}U $\rightarrow$ ^{144}Ba + _____ + 3^1n; (b) ^{1}n + ^{235}U $\rightarrow$ ^{91}Br + _____ + 2^1n; (c) ^{1}n + ^{239}Pu $\rightarrow$ ^{142}Xe + _____ + 2^1n.

72. ■■ Compute the energy released in the reaction in part (a) of the preceding problem.

73. ■ Complete the following fission reactions: (a) ^{1}n + _____ $\rightarrow$ ^{144}Ba + ^{92}Sr + 4^1n; (b) ^{1}n + ^{235}U $\rightarrow$ ^{97}Y + ^{137}I + _____ ; (c) ^{1}n + ^{239}Pu $\rightarrow$ ^{117}Ag + _____ + 3^1n.

74. ■■ Compute the energy released in the reaction in part (a) of the preceding problem. *Note:* The mass of ^{92}Sr is 91.911 030 u.

75. ■■ A fission reactor produces 1000 MW of electric power. Assume it operates at 30% efficiency, with an average of 200 MeV produced in each fission event. At what rate is the ^{235}U fuel consumed?

76. BIO ■■ **Fission dangers.** The isotope ^{90}Sr is a common fission product and is dangerous because it's easily absorbed by the body, especially bones, because it's similar chemically to calcium. Strontium-90's half-life is 28.8 years. What fraction of an absorbed dose of this isotope remains after 1 year? After 10 years?

77. ■■ The energy released in a nuclear explosion is stated as the equivalent mass of the chemical explosive TNT, usually in thousands of tons (kilotons; kt) or megatons (Mt). Exploding 1 g of TNT releases about 1000 calories = 4.184 kJ of energy. Early fission weapons yielded around 15 kilotons. How much uranium-235 had to fission to produce this explosive yield, assuming 200 MeV per fission event? Compare with the total mass of uranium in early bombs, about 50 kg.

Section 25.6 Fusion

78. ■ Complete the following fusion reactions: (a) ^{4}He + ^{4}He $\rightarrow$ ^{6}Li + _____ ; (b) ^{4}He + ^{3}He $\rightarrow$ ^{2}H + _____ ; (c) ^{2}H + ^{3}H $\rightarrow$ ^{1}n + _____ .

79. ■■ Find the energy released in each reaction in the preceding problem.

80. ■■ Find the energy released in each of these fusion reactions: (a) ^{4}He + ^{3}He $\rightarrow$ ^{7}Be; (b) ^{2}H + ^{2}H $\rightarrow$ ^{3}H + ^{1}H; (c) ^{12}C + ^{1}H $\rightarrow$ ^{13}N.

81. ■ ■ Find the energy released in each of these fusion reactions: (a) ^{3}He + ^{3}He → ^{4}He + 2^{1}H; (b) ^{1}H + ^{7}Li → 2^{4}He; (c) ^{3}He + ^{2}H → ^{4}He + ^{1}H.

82. ■ ■ ■ Approximately three-fourths of Earth's surface is covered by water, with an average ocean depth of about 3 km. (a) Deuterium makes up 0.015% of the hydrogen in water. How many deuterium nuclei are in all the oceans? (b) Suppose this deuterium were used in the fusion reaction ^{2}H + ^{2}H → ^{1}H + ^{3}H, which yields 4.0 MeV of energy. What's the total energy available from fusion of all that deuterium? Compare with the world's yearly energy use, about 4×10^{20} J. How many years' supply of deuterium would we have?

83. ■ ■ ■ Suppose your car uses 400 gallons of gasoline per year, with each gallon producing 1.3×10^8 J of energy. If you had a fusion-powered car using the reaction ^{2}H + ^{3}H → ^1n + ^{4}He, what mass of fusion fuel would be required instead of that 400 gal of gasoline?

84. ■ ■ ■ Show that the net energy released in the proton-proton cycle is about 26.7 MeV. Remember to include the annihilation of the positrons.

General Problems

85. ■ ■ Boron can absorb a slow neutron in the process ^1n + ^{10}B → ^{11}B + γ. Find the energy and wavelength of the emitted gamma ray.

86. ■ ■ (a) Find the radius of the ^{4}He nucleus. (b) Suppose that the helium atom, with a filled 1s electron shell, has the same ground-state radius as the Bohr hydrogen atom (Chapter 24). Find the density of the ^{4}He nucleus and the density of the ^{4}He atom. Compare the two densities.

87. BIO ■ ■ ■ **Medical imaging with ^{99}Tc*.** The isotope ^{99}Tc* is commonly used in several different types of medical imaging procedures. It emits a 140-keV gamma ray (comparable to the energies of diagnostic x rays) and has half-life 6.01 h. (a) What's the change in the nuclear mass after emission of the gamma ray? (b) What's the activity of 0.50 μg of ^{99}Tc* injected into a patient? (c) How much of ^{99}Tc* remains after one week? After 30 days?

88. ■ ■ ■ (a) Write a general formula for the energy released in β^+ decay of the nuclide ^{A}X. (b) Apply your formula to find the energy released in β^+ decay of ^{59}Ni.

89. BIO ■ ■ **Cancer risks.** A transatlantic airplane flight exposes you to a cosmic radiation dose of approximately 25 μSv. What's your lifetime risk of developing cancer from one such flight? Repeat for a PET scan, where your radiation dose is 5 mSv.

90. ■ ■ ■ (a) Show that the nucleus ^{8}Be will undergo alpha decay. (b) Write out the full reaction for this process. (c) Assuming that the parent nucleus was at rest before the decay, what are the speeds of the two daughter nuclei?

91. ■ ■ It's possible but difficult to realize the alchemists' dream of synthesizing gold. One reaction involves bombarding ^{198}Hg with neutrons, producing ^{197}Au and one other particle. Write the equation for the full reaction.

92. ■ ■ Suppose nuclei must be within a distance of 3 fm for the strong force to become effective. What temperature is required in order to initiate fusion of ^{2}H and ^{3}H? Assume a thermal energy of $\frac{3}{2}kT$ per nucleon.

93. BIO ■ ■ **Safe for human consumption?** Following the 1986 Chernobyl nuclear accident, a Swedish official claims that ^{131}I contamination in milk would be reduced to safe levels in 5 days. You're asked to verify this claim. The initial activity level was 2900 Bq/L of milk, and Sweden's limit is 2000 Bq/L. Is the 5-day figure accurate?

94. ■ ■ ■ Recall that the solar constant—the flux of solar energy reaching Earth's vicinity—is about 1400 W/m^2. If the Sun's energy originates in the proton-proton cycle, at what rate (kg/s) does the Sun lose mass? Compare the yearly mass loss with the Sun's total mass.

Answers to Chapter Questions

Answer to Chapter-Opening Question

Radiocarbon dating provides the age. Living things maintain a steady level of radioactive carbon-14, but at death ^{14}C decays at a predictable rate, while the stable ^{12}C does not. Measurement of the ^{14}C/^{12}C ratio therefore gives a precise age.

Answers to GOT IT? Questions

Section 25.1   (a) 1.26 r_A

Section 25.2   (e) ^{56}Fe > (d) ^{40}Ca > (c) ^{208}Pb > (a) ^{235}U > (b) ^{3}He

Section 25.3   (a) alpha decay

Section 25.4   (c) 4

Section 25.5   (a) ^{235}U and (d) ^{239}Pu

26 A Universe of Particles

■ What does this picture have to do with the origin of the universe?

To Learn

By the end of this chapter you should be able to

■ Describe the nature of antiparticles.
■ Explain the relation between particle exchange and the fundamental forces, and identify the exchange particles associated with each force.
■ Distinguish leptons from hadrons, and mesons from baryons.
■ Explain how conservation laws determine allowed particle reactions.
■ Describe the six types of quarks.
■ Explain how quarks combine to form mesons and baryons.
■ Describe the operation of linear accelerators and synchrotrons.
■ Tell how particle physics and cosmology together help explain the earliest instants of the universe.
■ Explain the role of the Hubble expansion and cosmic microwave background in establishing the Big Bang theory.
■ Describe the relative abundances of the three constituents of the universe: ordinary matter, dark matter, and dark energy.

We'll begin this chapter in the world of elementary particles; we'll end it with a look at the entire cosmos. In between you'll learn about the numerous subatomic particles, how they're classified, and how they interact. Particle interactions involve the four fundamental forces, and you'll see how those forces themselves are described in terms of particle exchanges. We'll look at the particle accelerators that give physicists their glimpse into the subatomic world. Then we'll take a brief foray into cosmology—the study of the structure and evolution of the entire universe. We'll come full circle to understand how modern cosmology and particle physics are intimately related. Thus we'll end by linking physics at the largest and smallest scales.

26.1 Particles and Antiparticles

In Chapter 24 we noted that the idea of fundamental particles goes back to ancient Greece. A key step in understanding particles was Thomson's 1887 identification of the electron (Chapter 18). That was followed by the identification of the proton as the hydrogen nucleus and Chadwick's 1932 discovery of the neutron (Chapter 25). The proton, neutron,

611

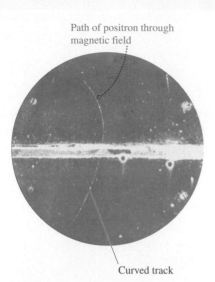

Path of positron through magnetic field

Curved track

FIGURE 26.1 In Carl Anderson's 1932 cloud chamber photo, the curved track is the path of a positron traveling through a uniform magnetic field.

Antihydrogen and Positronium

Physicists have recently succeeded in combining positrons and antiprotons to create antihydrogen atoms. In these atoms the light, positively charged positron orbits the massive, negatively charged antiproton, with the antiprotons created at Fermilab using the device shown here. The dynamics work out exactly as for ordinary hydrogen, so the Bohr orbits, transitions, and spectra of antihydrogen are identical to those of hydrogen. Another hydrogenlike "atom" is *positronium*, consisting of an electron and positron in what, classically, would be orbits around their common center of mass. Positronium is unstable, decaying when the positron and electron annihilate.

and electron would seem to form a tidy set of fundamental particles, the building blocks of all atoms.

The Positron

That simple picture of three fundamental particles didn't last long. Just months after Chadwick discovered the neutron, American physicist Carl Anderson identified the positron in a cloud chamber (Figure 26.1). The positron is an example of an **antiparticle**. Most antiparticles have the same mass and opposite charge as their corresponding particles. In Section 23.3 you saw that an electron-positron pair can be formed from the energy of a photon. In Chapter 25 you learned that positrons also arise in positive beta decay.

Creating an electron-positron pair requires energy $E = 2mc^2$, because the total mass of the pair is $2m$. Conversely, you've seen (Section 23.3) that an electron-positron pair can annihilate, forming two 511-keV gamma-ray photons:

$$e^- + e^+ \rightarrow 2\gamma$$

Pair production and annihilation is one of the more striking confirmations of Einstein's mass-energy relation.

✓ **TIP**

When discussing positrons and electrons outside the context of beta decay, the symbol e^+ is normally used for the positron and e^- for the electron.

Antiprotons and Antineutrinos

Since the positron discovery, physicists have identified many other antiparticles. In 1955, Emilio Segre (1905–1989) and Owen Chamberlain (1920–2006) discovered the **antiproton**, with mass $m = m_p = 1.67 \times 10^{-27}$ kg and charge $-e$. Antiparticles are generally designated with a bar over the ordinary particle's symbol; thus $\overline{p}$ is the antiproton. (The exception is the positron, designated e^+ or β^+.)

One reaction that creates antiprotons is the collision of two protons:

$$p + p \rightarrow p + p + p + \overline{p} \qquad (26.1)$$

This may seem to violate some conservation law, because you start with two particles of equal mass and end up with four particles, each of that same mass. Remember, though, that mass m is associated with rest energy mc^2. The reaction in Equation 26.1 works because the two initial protons have large kinetic energy, which gets transformed into rest energy. Note that the reaction produces a proton-antiproton *pair*. That's necessary to conserve electric charge—a fundamental quantity that's conserved in all particle reactions.

✓ **TIP**

Electric charge is always conserved, even if the number of particles is not.

Another antiparticle is the **antineutrino**. Recall from Section 25.3 that neutrinos are emitted as part of the beta decay process. Positron decay is accompanied by emission of a neutrino (symbol ν, the Greek letter "nu"), such as

$$^{55}\text{Fe} \rightarrow {}^{55}\text{Mn} + \beta^+ + \nu$$

Beta decay that produces an electron (β^-) also entails emission of an antineutrino (symbol $\overline{\nu}$); thus

$$^{14}\text{C} \rightarrow {}^{14}\text{N} + \beta^- + \overline{\nu}$$

Beta decay is *always* accompanied by the emission of a neutrino or antineutrino. In Chapter 25 we didn't show the neutrino, but it's always there.

You might wonder how a neutral particle such as the neutrino has an antiparticle. Here particle and antiparticle have the same charge (zero) and mass, so it's not obvious what distinguishes them. There are, in fact, more subtle distinctions, which you'll see in Section 26.3. Similarly, the neutron and antineutron are different particles, even though both have zero charge. However, some particles are their own antiparticles; a common example is the photon.

EXAMPLE 26.1 **Antiproton Creation**

Two protons moving in opposite directions collide, causing the reaction $p + p \rightarrow p + p + p + \bar{p}$. What's the minimum kinetic energy for each proton?

ORGANIZE AND PLAN Figure 26.2 shows the situation before and after the collision, where we've defined the x-axis to be along the colliding protons' line of motion. With the minimum initial energy, there's no energy left over as kinetic energy after the reaction, so the products are at rest. Momentum conservation then requires that the incoming protons have equal speed, and hence equal kinetic energy. The energy

Before collision:

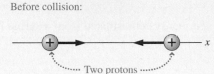

···· Two protons ····

After collision:

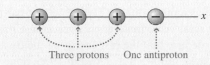

Three protons   One antiproton

FIGURE 26.2 Colliding protons produce a proton-antiproton pair.

needed to create a proton and antiproton, each with mass m_p, is $E = m_{\text{total}}c^2 = 2m_pc^2$.

Known: Proton mass $m_p = 1.67 \times 10^{-27}$ kg; $c = 3.00 \times 10^8$ m/s.

SOLVE Using the proton mass, the energy required is

$$E = 2m_pc^2 = 2(1.67 \times 10^{-27}\,\text{kg})(3.00 \times 10^8\,\text{m/s})^2$$
$$= 3.006 \times 10^{-10}\,\text{J}$$

The incident protons share this energy equally, so each has kinetic energy $K = 1.503 \times 10^{-10}$ J $= 938$ MeV.

REFLECT The kinetic energy of each incident proton is equal to the rest energy of a proton (or antiproton). You could skip this calculation and just look up the proton's rest energy—about 938 MeV (see Chapter 25), equivalent to the value computed here. That's substantial kinetic energy, requiring a particle accelerator capable of nearly 1 GeV

MAKING THE CONNECTION What happens if the colliding protons have larger kinetic energies than the minimum we found here?

ANSWER The excess appears as kinetic energy of the four particles (three protons and one antiproton). If there's enough, it might also create additional particle-antiparticle pairs.

Elementary Particles

The particles we've introduced in this section are just a few of the many known particles. We'll introduce others in Sections 26.2 and 26.3. By the mid-20th century, the number of different particles was getting out of hand, and the whole collection was dubbed the "particle zoo." Physicists tried to find the truly **elementary particles**—the smallest, simplest particles that can't be broken down further. Those particles composed of more than one elementary particle are **composite**. Through the 1960s and 1970s, physicists gained a better understanding of elementary and composite particles, and a comprehensive model of elementary particles emerged. We'll explore these developments in Sections 26.2 and 26.3.

GOT IT? Section 26.1 Identify all the statements that are true regarding antiparticles, relative to their corresponding particle. The antiparticle has (a) more mass; (b) the same mass; (c) more charge; (d) the opposite charge; (e) zero charge.

26.2 Particles and Fundamental Forces

Recall from Section 9.1 that Newton, who developed a mathematical theory of gravitation, was at a loss to explain how gravity works on a fundamental level. Physicists adopted the "action-at-a-distance" view to describe gravity acting across the vast reaches of empty space to attract two bodies that have no apparent physical contact. Early views of electric

TABLE 26.1 Relative Strengths of the Fundamental Forces

Force	Relative strength
Strong	1
Electromagnetic	10^{-2}
Weak	10^{-10}
Gravitational	10^{-38}

Students experience repulsive forces with each throw and catch.

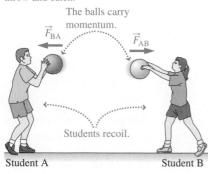

(a)

Students experience attractive forces when they grab each other's ball.

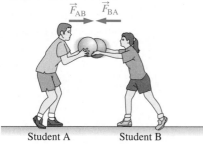

(b)

FIGURE 26.3 Action-at-a-distance forces explained by the exchange of particles: (a) repulsive forces; (b) attractive forces.

and magnetic forces were similar. In Chapter 15 you learned how the *field* concept gives an alternate view, in which one object creates a gravitational, electric, or magnetic field in the vicinity of another object, and the local field gives rise to the force. Although a useful tool, the field concept still doesn't answer Newton's original question about how these forces really work. One of the great advances in 20th-century physics was an understanding that particles play a central role in explaining "action-at-a-distance" forces. Recall that there are four fundamental forces, listed along with their relative strengths in Table 26.1.

The Strong Force and Yukawa's Meson

Chemistry students are familiar with the idea that exchanging particles helps explain forces. The covalent bonds in common molecules like O_2, for example, result from the two oxygen atoms sharing electrons. Figure 26.3 shows a physical model of a particle exchange leading to a force. Tossing a ball to your friend or catching a ball thrown toward you involves a transfer of momentum and hence a force, as you learned in Chapter 6: $\vec{F}_{net} = \Delta\vec{p}/\Delta t$. A force that acts continuously, such as gravity or electromagnetism, would require an ongoing exchange of particles. Figure 26.3 shows that either a repulsive force or an attractive force can be explained this way.

✓ **TIP**

Remember that momentum is conserved in a system isolated from external forces.

Remember from Chapter 25 that the strong force acts between any pair of nucleons: proton-proton, proton-neutron, or neutron-neutron. The Japanese physicist Hideki Yukawa (1907–1981) developed the idea of using particle exchange to explain the strong force. Physicists refer to an exchanged particle as a **mediator**, or carrier, of that particular force. Below we explain how Yukawa used the relativistic mass-energy relation (Chapter 20) and the uncertainty principle (Chapter 23) to deduce that the rest energy of the mediating particle should be about 130 MeV. When Yukawa proposed this idea in 1935, none of the then-known particles had rest energy anywhere near this value. The particle that mediates the strong force came to be called a **pi-meson**, or simply **pion** (symbol π). Note that the pion's rest energy is intermediate between that of electrons (about 0.5 MeV) and nucleons (about 940 MeV). That's why the new particle was called a **meson**, derived from the Greek μέσο (meso), meaning middle. In Section 26.3 you'll meet many other mesons.

In 1938, a group led by Carl Anderson discovered a particle in cosmic radiation that appeared to be a good candidate for Yukawa's meson. It had charge $-e$ and rest energy just over 100 MeV. However, it was soon found that this particle did not interact strongly with nucleons, so it was ruled out as the mediator of the strong force. We now call this particle the **muon**, and we'll describe it further in Section 26.3.

Yukawa's pi-meson was finally identified in 1947 by the English physicist Cecil Powell (1903–1969) and Italian physicist Giuseppe Occhialini (1907–1993). There are actually three different pions: a neutral pion π^0 with rest energy 135 MeV, and charged pions π^+ and π^-, each having a rest energy of 140 MeV and charges $\pm e$. The rest energies are all quite close to Yukawa's original estimate, consistent with pi mesons (pions) being the strong-force mediators.

Particle exchanges are illustrated using **Feynman diagrams**, named for the American physicist Richard Feynman (1918–1988), who pioneered their use. Figure 26.4 is a Feynman diagram showing the pion mediating the strong force between two nucleons.

Yukawa's Method

Here's how Yukawa estimated the pion's mass. The same method works for other force-mediating particles. Producing a particle with the pion's mass requires an equivalent energy $E = m_\pi c^2$. One form of Heisenberg's uncertainty principle relates uncertainties in energy and time:

$$\Delta E \Delta t \geq \frac{h}{2\pi}$$

As discussed in Chapter 23, the absolute minimum uncertainty is given by $\Delta E \Delta t \geq h/4\pi$. The less restrictive $h/2\pi$ is often used in approximations like this one. By associating the energy uncertainty with the energy needed to create the pion, you can determine the minimum time for the process as

$$\Delta t = \frac{h}{2\pi \Delta E} = \frac{h}{2\pi m_\pi c^2}$$

The greatest distance a particle could possibly go in this time is $R = c\Delta t$:

$$R = c\Delta t = \frac{h}{2\pi m_\pi c}$$

which we then solve for the pion's mass:

$$m_\pi = \frac{h}{2\pi R c}$$

We identify R with the average range of the strong force, about 1.5 fm. Using the known values then gives:

$$m_\pi = \frac{h}{2\pi R c} = \frac{6.626 \times 10^{-34}\,\text{J}\cdot\text{s}}{2\pi(1.5 \times 10^{-15}\,\text{m})(3.00 \times 10^8\,\text{m/s})} = 2.34 \times 10^{-28}\,\text{kg}$$

This is intermediate between the electron mass (9.1×10^{-31} kg) and nucleon mass (1.7×10^{-27} kg). Rather than mass, it's common to give the pion's rest energy:

$$E_{\text{rest}} = m_\pi c^2 = (2.34 \times 10^{-28}\,\text{kg})(3.00 \times 10^8\,\text{m/s})^2 = 2.11 \times 10^{-11}\,\text{J} = 130\,\text{MeV}$$

The process we've just outlined works for any exchange particle. In general, the particle's mass m is related to the interaction range R by

$$m = \frac{h}{2\pi R c} \qquad \text{(Mass of force-mediating particle; SI unit: kg)} \qquad (26.2)$$

Other Forces and Particles

With this understanding of the strong force and the pion, it makes sense to describe the other three fundamental forces (electromagnetic, weak, and gravitational) in terms of particle exchange. The electromagnetic force is mediated by photons. This is plausible, because you know from Chapter 20 that accelerated charges emit electromagnetic radiation. However, it's impossible to observe photons traveling between charged particles, because they're emitted and absorbed too quickly. For this reason, photons that mediate the electromagnetic force are called **virtual photons**. Figure 26.5a is a Feynman diagram corresponding to the repulsion of two electrons.

The weak force governs the process of beta decay (Section 25.3), including the decay of a neutron into a proton, electron, and antineutrino:

$$n \rightarrow p^+ + e^- + \bar{\nu}$$

In the 1960s, physicists suggested that the weak force was mediated by several heavy exchange particles. Those particles, first observed in 1983, are known as W^+, W^- (with charges $+e$ and $-e$), and the neutral particle Z^0. They're extremely massive, with rest energy 80.4 GeV for the W particles and 91.2 GeV for the Z^0. By comparison, the rest energies of the neutron and proton are just below 1 GeV. Figure 26.5b is a Feynman diagram for neutron decay.

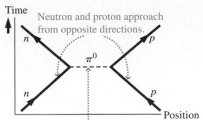

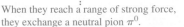

When they reach a range of strong force, they exchange a neutral pion π^0.

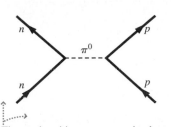

Time and position axes are omitted on a completed diagram.

FIGURE 26.4 A Feynman diagram, used to illustrate interactions between elementary particles. Here it's the strong force acting between a proton and neutron.

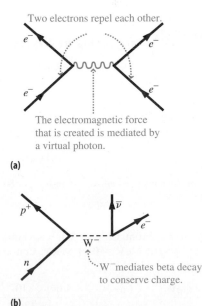

(a)

(b)

FIGURE 26.5 Feynman diagrams for (a) the electromagnetic force and (b) the weak force.

The final exchange particle to complete the picture is the **graviton**, predicted to mediate the gravitational force. The graviton has not yet been observed. That's because the gravitational force is many orders of magnitude smaller than the other fundamental forces, so the graviton must interact very weakly with matter. In Table 26.2, we summarize the main properties of the fundamental forces and the particles that mediate them.

TABLE 26.2 Fundamental Forces and Exchange Particles

Force	Relative strength	Mediating particle(s)	Rest energy of mediator	Range of force
Strong	1	π^0, π^+, and π^-	135 MeV (π^0) 140 MeV (π^+)	Up to 3 fm
Electromagnetic	10^{-2}	Photon	0	Infinite
Weak	10^{-10}	Z^0, W^+, and W^-	91.2 GeV (Z^0) 80.4 GeV (W)	$\ll 1$ fm
Gravitational	10^{-38}	Graviton	0	Infinite

CONCEPTUAL EXAMPLE 26.2 **Photon Mass**

Use the fact that the electromagnetic force acts over an infinite distance to predict the photon's mass.

SOLVE As shown in Equation 26.2, the mass of a mediating particle varies inversely with the range of the interaction:

$$m = \frac{h}{2\pi Rc}$$

where m is the mass of the mediating particle and R is the range. For the electromagnetic force, the range is infinite. Therefore, the mass of the mediating particle (the photon) should be zero.

REFLECT This prediction agrees with observation. You know from Chapter 20 that the photon's mass is zero.

EXAMPLE 26.3 **Range of the Weak Force**

Use the masses of the weak-force mediators to estimate the range of the weak force.

ORGANIZE AND PLAN Equation 26.2 relates particle mass m and interaction range R:

$$m = \frac{h}{2\pi Rc}$$

Known: Rest energies $m_{W^+}c^2 = m_{W^-}c^2 = 80.4$ GeV; $m_{Z^0}c^2 = 91.2$ GeV.

SOLVE We first convert rest energies into masses, so we can work in SI units. From Chapter 25, you know the conversion factor $1\,u \cdot c^2 = 931.5$ MeV $= 0.9315$ GeV. Converting the W^+ and W^- rest energy:

$$\frac{80.4 \text{ GeV}}{c^2} \times \frac{1\,u \cdot c^2}{0.9315 \text{ GeV}} \times \frac{1.661 \times 10^{-27} \text{ kg}}{1\,u} = 1.43 \times 10^{-25} \text{ kg}$$

Thus the range of the weak interaction mediated by a W particle is approximately

$$R = \frac{h}{2\pi mc} = \frac{6.626 \times 10^{-34} \text{ J} \cdot \text{s}}{2\pi (1.43 \times 10^{-25} \text{ kg})(3.00 \times 10^8 \text{ m/s})}$$
$$= 2.46 \times 10^{-18} \text{ m}$$

A similar calculation for the more massive Z^0 particle gives a slightly smaller range:

$$R = 2.16 \times 10^{-18} \text{ m}$$

REFLECT The weak force has an extremely short range. This makes sense physically, given that the beta decay takes place within a single nucleon.

MAKING THE CONNECTION Use the data from this example to determine the time interval over which the weak interaction occurs.

ANSWER From the discussion leading to Equation 26.2, based on the mediating particle traveling no faster than c, $R = c\Delta t$. Then $\Delta t = R/c = 8.2 \times 10^{-27}$ s for the W and 7.2×10^{-27} s for the Z. This is far shorter than the half-lives of known radioactive nuclides.

GOT IT? Section 26.2 Rank in decreasing order the masses of the exchange particles that mediate each of the four fundamental forces: (a) strong; (b) electromagnetic; (c) weak; (d) gravitational.

Reviewing New Concepts

- An antiparticle has the same mass and opposite charge as its corresponding particle. More subtle differences distinguish neutral particles from their antiparticles.
- An elementary particle cannot be broken into smaller particles, while a composite particle can.
- The four fundamental forces work via particle exchange.
- Pions mediate the strong force between nucleons; W and Z particles mediate the weak force; photons mediate the electromagnetic force; and gravitons are thought to mediate the gravitational force, although they have not been observed.

26.3 Classifying Particles

You've now been introduced to a substantial list of particles: electron and positron, proton, neutron, neutrino, muon, pion, photon, and the weak-force mediators W and Z. However, this barely scratches the surface! Through the 20th century, physicists greatly expanded this list, aided by accelerators with ever higher energies. They developed a classification scheme, with particles in groups having similar characteristics. This classification shows which particles are elementary and which are composite. It also leads to conservation laws that will help you understand reactions among particles.

Leptons

Leptons are the lightest particles, their name derived from the Greek $\lambda\epsilon\pi\tau o\zeta$ (leptos), meaning small or thin. Electrons and muons are leptons, and there's a third type of lepton called a **tau** particle (or **tauon**), which is the most massive. The electron, muon, and tau all carry charge $-e$. Each has an antiparticle with charge $+e$; they're also leptons. An important distinction between leptons and other particles is that leptons do not experience the strong force. They interact electromagnetically as well as through the weak force—as evidenced by their involvement in beta decay.

Neutrinos, too, are leptons. You saw in Section 26.1 that beta decay is accompanied by the emission of a neutrino or antineutrino, so there's a close relationship between the electron and its neutrino. There are actually three kinds of neutrinos, corresponding to the electron, muon, and tau. They're designated ν_e, the electron neutrino; ν_μ, the muon neutrino; and ν_τ, the tau neutrino. Each neutrino has an antineutrino with the same mass as the corresponding neutrino. Thus there are three different leptons and their three neutrinos, and each of these six particles has an antiparticle, making a total of 12 leptons (Table 26.3).

Table 26.3 lists the electron and all three neutrinos as stable, but the other two leptons are unstable, with mean lifetimes far less than 1 s. (The mean lifetime of a particle is the reciprocal of the decay constant λ introduced in Chapter 25, which makes it just a factor

TABLE 26.3 Leptons and Their Properties

Lepton name	Symbol	Antiparticle	Rest energy (MeV)	Mean lifetime (s)	Main decay modes
Electron	e^-	e^+	0.511	Stable	
Electron neutrino	ν_e	$\bar{\nu}_e$	$<3\times10^{-6}$	Stable	
Muon	μ^-	μ^+	106	2.2×10^{-6}	$e^-\nu_\mu\bar{\nu}_e$
Muon neutrino	ν_μ	$\bar{\nu}_\mu$	<0.19	Stable	
Tau	τ^-	τ^+	1780	2.9×10^{-13}	$\mu^-\nu_\tau\bar{\nu}_\mu, e^-\nu_\tau\bar{\nu}_e$
Tau neutrino	ν_τ	$\bar{\nu}_\tau$	<30	Stable	

of ln 2 less than the half-life.) When these particles decay, they create multiple leptons. For example, the decay mode for the muon is

$$\mu^- \rightarrow e^- + \nu_\mu + \bar{\nu}_e$$

Later we'll introduce conservation laws that explain why leptons and their neutrinos pair in this manner.

Hadrons

Hadrons are particles that interact via the strong force (as well as the electromagnetic force if the hadrons are charged hadrons). You've already encountered several hadrons: protons, neutrons, and pions. However, there are many more, some of which are listed in Table 26.4.

TABLE 26.4 Table of Hadrons

Particle name	Symbol	Anti-particle	Rest energy (MeV)	Mean lifetime (s)	Main decay modes	Spin	Baryon number B	Strangeness number S	Charm number C
Mesons									
Pion	π^-	π^+	140	2.6×10^{-8}	$\mu^+ \nu_\mu$	0	0	0	0
	π^0	Self	135	8.4×10^{-17}	2γ	0	0	0	0
Kaon	K^+	K^-	494	1.2×10^{-8}	$\mu^+ \nu_\mu, \pi^+ \pi^0$	0	0	1	0
	K_S^0	$\overline{K}_S^0$	498	9.0×10^{-11}	$\pi^+ \pi^-, 2\pi^0$	0	0	1	0
	K_L^0	$\overline{K}_L^0$	498	5.1×10^{-8}	$\pi^\pm e^\mp \nu_e, 3\pi^0,$ $\pi^\pm \mu^\mp \nu_\mu,$ $\pi^+ \pi^- \pi^0$	0	0	1	0
Eta	η^0	Self	548	5×10^{-19}	$2\gamma, 3\pi^0$ $\pi^+ \pi^- \pi^0$	0	0	0	0
Charmed D's	D^+	D^-	1870	1.0×10^{-12}	$e^+, K^\pm, K^0,$ $\overline{K}^0$ + anything	0	0	0	1
	D^0	$\overline{D}^0$	1865	4.1×10^{-13}	Same as D^+	0	0	0	1
	D_S^+	$\overline{D}_S^-$	1968	5.0×10^{-13}	Various	0	0	1	1
Bottom B's	B^+	B^-	5280	1.6×10^{-12}	Various	0	0	0	0
	B^0	$\overline{B}^0$	5280	1.5×10^{-12}	Various	0	0	0	0
J/Psi	J/ψ	Self	3097	10^{-20}	Various	0	0	0	0
Upsilon	$\Upsilon(1S)$	Self	9460	10^{-20}	Various	0	0	0	0
Baryons									
Proton	p	$\bar{p}$	938.3	Stable (?)		$\frac{1}{2}$	1	0	0
Neutron	n	$\bar{n}$	939.6	886	$pe^- \bar{\nu}_e$	$\frac{1}{2}$	1	0	0
Lambda	Λ	$\overline{\Lambda}$	1116	2.6×10^{-10}	$p\pi^-, n\pi^0$	$\frac{1}{2}$	1	-1	0
Sigmas	Σ^+	$\overline{\Sigma}^-$	1189	8.0×10^{-11}	$p\pi^0, n\pi^+$	$\frac{1}{2}$	1	-1	0
	Σ^0	$\overline{\Sigma}^0$	1193	7.4×10^{-20}	$\Lambda\gamma$	$\frac{1}{2}$	1	-1	0
	Σ^-	$\overline{\Sigma}^+$	1197	1.5×10^{-10}	$n\pi^-$	$\frac{1}{2}$	1	-1	0
Xi	Ξ^0	$\overline{\Xi}^0$	1315	2.9×10^{-10}	$\Lambda\pi^0$	$\frac{1}{2}$	1	-2	0
	Ξ^-	Ξ^+	1321	1.6×10^{-10}	$\Lambda\pi^-$	$\frac{1}{2}$	1	-2	0
Omega	Ω^-	Ω^+	1672	0.82×10^{-10}	$\Lambda K^-, \Xi^0 \pi^-$	$\frac{3}{2}$	1	-3	0
Charmed lambda	Λ_C^-	$\overline{\Lambda}_C$	2286	2.0×10^{-13}	Various	$\frac{1}{2}$	1	0	1

Review of particle physics, Particle Data Group, *Physics Letters B* **667**, 1 (2008).

Hadrons fall into two categories: **mesons** and **baryons**. Table 26.4 shows that mesons and baryons differ in important ways. One obvious difference is the spin (introduced in Chapter 24), which is zero for mesons and ½ or ³⁄₂ for baryons. Baryons also carry a **baryon number** of 1, which simply says they're baryons. We'll explain the role of baryon number when we discuss conservation laws for particle reactions. Note that every particle listed in Table 26.4 has an antiparticle with the same mass and opposite charge. Baryon number is also opposite; each antibaryon has baryon number -1.

Table 26.4 lists two other properties, called charm and strangeness, which some hadrons have and others do not. In the 1970s, physicists found that they needed these properties to distinguish all the observed mesons and baryons. Physicists have learned that hadrons and baryons aren't elementary but composite. They're composed of quarks, two of which are named charmed and strange, and that's the source of these unusually named quantum numbers. Baryons are composed of three quarks, and mesons are composed of a quark and antiquark. We'll explore quarks further in Section 26.4.

✓**TIP**

Baryon comes from the Greek $\beta\alpha\rho\acute{v}\zeta$ (baryos), meaning heavy; meson comes from the Greek $\mu\acute{\varepsilon}\sigma o$ (meso), meaning middle. Meson masses are generally between those of leptons and baryons.

Conservation Laws and Particle Reactions

Baryon number, charm, and strangeness help explain which particle reactions are allowed and which aren't. Generally, baryon number is conserved. Strangeness and charm are conserved in reactions involving the strong force, but not in those involving the weak force. Electric charge and energy are conserved in all reactions. Here, as always, energy includes rest energy.

For example, Section 26.1 considered the formation of antiprotons via the reaction

$$p + p \rightarrow p + p + p + \bar{p}$$

which can occur when two high-energy protons collide. There are just two baryons (protons) before the reaction occurs, with a net baryon number $1 + 1 = 2$. What's the baryon number after the reaction? The antiproton has a baryon number -1, so the net baryon number is $1 + 1 + 1 - 1 = 2$, and baryon number is conserved. Electric charge is also conserved, so this reaction is possible, provided the two colliding protons have enough energy to create the two new particles.

CONCEPTUAL EXAMPLE 26.4 **Decay of the Λ Particle**

The lambda (Λ) particle (Table 26.4) is an uncharged, heavy baryon that can decay via the reactions $\Lambda \rightarrow p^+ + \pi^-$ and $\Lambda \rightarrow n + \pi^0$. Examine both these reactions from the standpoint of conservation laws.

SOLVE It's best to consider the conservation laws one at a time. For energy conservation, note that a lambda particle's mass is larger than the total mass of a proton and pion or neutron and pion. Therefore, a lambda particle has enough rest energy for these decays. Charge is conserved in both decays, because the net charge of the proton and π^- is $e - e = 0$. Similarly, the neutron and neutral pion are both uncharged. Baryon number is conserved in each decay, because the lambda has baryon number 1, and so do the proton and neutron. The pions are mesons, with baryon number zero. Thus the net baryon number is 1 on each side of both reactions. None of the particles involved carry charm, so there's no issue with conservation of charm. The lambda particle has strangeness -1, but none of the decay products do. Thus, strangeness is not conserved. This indicates that the weak force is involved in this decay, because strangeness need not be conserved in weak-force reactions.

REFLECT Note that these are the only two likely decay modes for the lambda particle. Baryon conservation dictates that the lambda must decay into another baryon, which must have less mass than the lambda in order to conserve energy. This leaves only the proton and neutron. With the rest energy of the proton or neutron committed, the only mesons light enough to form from the leftover energy are pions.

Baryon number conservation explains why the proton is not observed to decay. To conserve baryon number, a proton would have to decay into another baryon. But there are no lighter baryons, so it would be impossible for a proton to decay and conserve energy. Some recent theories have suggested that baryon number conservation is not an absolute law, so that the proton might decay. Experiments indicate that the proton's mean lifetime cannot be less than 10^{35} years, making proton decay very infrequent if it ever occurs. (The age of the universe is on the order of 10^{10} years.)

Lepton number is also conserved. However, there's a different lepton number for each of the three families of leptons:

- Electrons and electron neutrinos have electron lepton number $L_e = 1$;
- Muons and muon neutrinos have a muon lepton number $L_\mu = 1$;
- Tauons and tau neutrinos have a tau lepton number $L_\tau = 1$.

For each type of particle the other two lepton numbers are zero. Each antiparticle has a lepton number -1. For example, the positron e^+ and antineutrino $\bar{\nu}_e$ each have a lepton number $L_e = -1$, while $L_\mu = L_\tau = 0$ for both.

Lepton conservation helps explain reactions involving leptons. For example, consider the beta decay

$$n \rightarrow p^+ + e^- + \bar{\nu}_e$$

The neutron and proton are hadrons, with lepton number zero. Beta decay produces an electron, with lepton number $L_e = 1$. For lepton number to be conserved, we need a particle with $L_e = -1$, making the total lepton number zero, as it was initially. The antineutrino serves this purpose. Another example of lepton number conservation is electron-positron pair production, introduced in Section 23.3. One way to create an electron-positron pair is with the energy of a gamma-ray photon. Creating a single electron would violate lepton number conservation, but an electron-positron pair has lepton numbers of $+1$ and -1, which add to zero. You might argue that you could create an electron with another positively charged particle to conserve charge, but you'd still need another particle with $L_e = -1$, to cancel the electron's lepton number. Only the positron and antineutrino have $L_e = -1$.

CONCEPTUAL EXAMPLE 26.5 **Muon Decay**

Examine the process of muon decay $\mu^- \rightarrow e^- + \bar{\nu}_e + \nu_\mu$ from the standpoint of conservation laws.

SOLVE The particles here are all leptons. With the muon and electron and their corresponding neutrinos involved, we need to consider both electron lepton number L_e and muon lepton number L_μ. The appropriate numbers are

> Muon and muon neutrino: $L_\mu = 1$
>
> Electron: $L_e = 1$
>
> Electron antineutrino: $L_e = -1$

Therefore, the muon lepton number is $L_\mu = 1$ both before and after the reaction. The electron lepton number is zero before the reaction, and after the reaction the two L_e numbers add to zero. Therefore, both lepton numbers are conserved. Electric charge is also conserved; neutrinos are uncharged, so the charge before and after the reaction is the same: $-e$.

REFLECT It wasn't necessary to consider baryon number in this reaction, which only involved leptons. And there's no problem with energy conservation, because the electron is lighter than the muon and the neutrinos have nearly zero rest energy.

Neutrinos were originally believed to be massless, and were that the case, conservation laws for the different lepton numbers would hold absolutely. However, it's recently become evident that neutrinos have small but nonzero mass. That means conservation of individual lepton numbers is, on occasion, violated. Indeed, neutrinos produced in the Sun appear to "oscillate," changing from one neutrino type to another, on their way to Earth.

GOT IT? Section 26.3 Which of the following reactions are forbidden, based on violation of a conservation law? (a) $\pi^- \rightarrow \mu^- + \nu_\mu$; (b) $\pi^- \rightarrow \mu^- + \bar{\nu}_\mu$; (c) $\pi^+ \rightarrow \mu^+ + \nu_\mu$; (d) $\pi^+ \rightarrow \mu^+ + \bar{\nu}_\mu$.

26.4 Quarks

At this point you've met the leptons, which are all elementary and fit into a nice scheme. The hadrons, on the other hand, might seem out of control! That's because hadrons aren't elementary, but are composed of smaller particles called **quarks**. The quark model was proposed independently by American physicists Murray Gell-Mann and George Zweig in 1963 and has proved extremely useful in understanding hadrons. Here you'll see how combinations of quarks describe all possible hadrons.

Quark Properties

Table 26.5 lists the six quarks and their most important properties. Note the quarks' fractional charge—a striking difference from particles we've discussed previously, which have charges that are multiples of the elementary charge e or are neutral. We can't detect quarks' fractional charges directly, however, because it appears impossible to isolate a single quark; more on this shortly. Each quark has a corresponding, oppositely charged **antiquark**. For example, the up quark (u) has charge $2e/3$ while the anti-up ($\bar{u}$) has charge $-2e/3$.

The following rules govern the composition of hadrons:

> Baryons are composed of three quarks, or three antiquarks for the antiparticle.
>
> Mesons are composed of a quark and an antiquark.

For example, a proton is a baryon composed of two up quarks and one down quark (Figure 26.6a). The notation for quark composition is simply to list the three quarks: *uud*. Note that this makes the proton's net charge $2e/3 + 2e/3 + (-e/3) = +e$, as expected. The neutron's quark composition is *udd*, giving zero net charge (Figure 26.6b). The antiparticle results from the corresponding antiquarks. Thus the antiproton is $\overline{uud}$, with net charge $-2e/3 + (-2e/3) + e/3 = -e$. The antineutron is $\overline{udd}$.

You can see that many combinations of three quarks, each with charge $\pm e/3$ or $\pm 2e/3$, will yield a baryon with net charge $+e$, 0, or $-e$. However, three quarks with charge $2e/3$ combine to yield a baryon with net charge $2e$. An example is the Δ^{++} particle, with quark composition *uuu*.

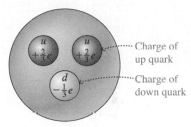

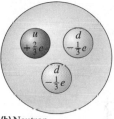

(a) Proton

(b) Neutron

FIGURE 26.6 Quark configuration for (a) a proton (*uud*) and (b) a neutron (*udd*).

TABLE 26.5 Quark Properties

Quark name	Symbol	Mass* (GeV/c^2)	Charge	Baryon number	Strangeness number S	Charm number C	Bottomness number B	Topness number T
Up	u	0.0015 to 0.0033	$2e/3$	$\frac{1}{3}$	0	0	0	0
Down	d	0.0035 to 0.0060	$-e/3$	$\frac{1}{3}$	0	0	0	0
Strange	s	0.07 to 0.130	$-e/3$	$\frac{1}{3}$	-1	0	0	0
Charmed	c	1.16 to 1.34	$2e/3$	$\frac{1}{3}$	0	1	0	0
Bottom	b	4.13 to 4.37	$-e/3$	$\frac{1}{3}$	0	0	-1	0
Top	t	169 to 174	$2e/3$	$\frac{1}{3}$	0	0	0	1

Antiquarks, $\bar{u}$, $\bar{d}$, $\bar{s}$, $\bar{c}$, $\bar{b}$, and $\bar{t}$, have opposite signs for charge, baryon number, S, C, B, and T.

*Review of particle physics, Particle Data Group, *Physics Letters B* **667**, 1 (2008).

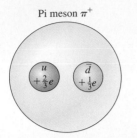

Pi meson π^+

FIGURE 26.7 Mesons consist of a quark and an antiquark. Shown here is the positive pi meson π^+, with quark configuration $u\bar{d}$.

TABLE 26.6 Quark Compositions of Selected Mesons and Baryons

Baryon Particle	Quark composition	Meson Particle	Quark composition
p^+	uud	π^+	$u\bar{d}$
n	udd	π^-	$\bar{u}d$
Λ	uds	K^+	$u\bar{s}$
Σ^+	uus	K^0	$d\bar{s}$
Σ^0	uds	D^+	$c\bar{d}$
Ξ^0	uss	D^0	$c\bar{u}$
Ξ^-	dss	B^+	$u\bar{b}$
Ω^-	sss	B^0	$d\bar{b}$
Λ_C^+	udc	Υ	$b\bar{b}$
Δ^{++}	uuu	J/Ψ	$c\bar{c}$
Δ^+	uud	B_S^0	$s\bar{b}$
Δ^0	udd	B_C^0	$c\bar{b}$
Δ^-	ddd	ϕ	$s\bar{s}$

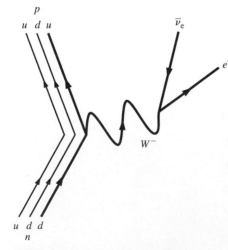

FIGURE 26.8 Feynman diagram for beta decay, showing the transformation of a down quark into an up quark in the process.

A meson is a quark-antiquark combination. The pion π^+ has composition $u\bar{d}$ (Figure 26.7). This combination gives the correct charge since $2e/3 + e/3 = +e$. The negative pion π^- has composition $\bar{u}d$, and is the antiparticle to the π^+. Table 26.6 lists quark compositions of a number of mesons and baryons.

You can understand reactions and decays involving hadrons in terms of transformations of the underlying quarks. Consider again the neutron's β^- decay:

$$n \rightarrow p^+ + e^- + \bar{\nu}_e$$

The neutron's quark composition is udd, and the proton's is uud. The electron and neutrino are leptons, so they don't contain quarks. Thus, β^- decay involves the transformation of one down quark into an up quark (Figure 26.8).

CONCEPTUAL EXAMPLE 26.6 **Positron Decay and Electron Capture**

Discuss the quark transformation that occurs during positron decay and electron capture.

SOLVE Recall from Chapter 25 that positron decay can be viewed as a proton changing into a neutron, accompanied by emission of a positron. Lepton conservation (Section 26.3) dictates that a neutrino is also emitted, so the process looks like this:

$$p^+ \rightarrow n + \beta^+ + \nu_e$$

The positron and neutrino don't involve quarks. The transformation of a proton into a neutron involves a change of quarks: $uud \rightarrow udd$. It appears that an up quark has transformed into a down quark. Similarly, electron capture can be viewed as

$$p^+ + e^- \rightarrow n + \nu_e$$

where again the neutrino is required by lepton conservation. The quark transformation is $uud \rightarrow udd$, which again indicates an up quark changing to a down quark.

REFLECT It shouldn't be surprising that positron decay and electron capture involve the same underlying process, because on a larger scale, both involve the transformation of a proton to a neutron. In Chapter 25 you learned that both these processes are likely for nuclei that are relatively proton-rich and need to change a proton to a neutron to approach stability.

CONCEPTUAL EXAMPLE 26.7 **Λ Decay**

In Conceptual Example 26.4, you saw one possible decay of the lambda particle (a baryon): $\Lambda \rightarrow p^+ + \pi^-$. Describe the quark transformation in this decay and in the one that follows it.

SOLVE From Table 26.6, the Λ's quark composition is uds. You already know that the proton's quark composition is uud, and the negative pion's quark composition (also in Table 26.6) is $\bar{u}d$. Therefore, the underlying transformation in the decay $\Lambda \rightarrow p^+ + \pi^-$ is from uds to uud and $\bar{u}d$. You can picture this as a strange quark changing into a down quark, and a pair $u\bar{u}$ being created. There's no problem with conservation laws when a quark and its antiquark are created, because all the conserved numbers (including charge) cancel.

The secondary decay is that of the pion, which you can see from Table 26.4 is

$$\pi^- \rightarrow \mu^- + \bar{\nu}_\mu$$

This is interesting in that the π^- (quark composition $\bar{u}d$) gets replaced by leptons, which contain no quarks. You can think of this as the down quark changing to an up (as in β^- decay), followed by annihilation of the resulting $u\bar{u}$ pair.

REFLECT Note that the annihilation of the pair $u\bar{u}$ in the last step is just the reverse of this pair's creation in the original lambda decay.

Color

Quarks have half-integer spin, so they obey the Pauli exclusion principle. Then how can two up quarks exist within the proton (uud), apparently having the same set of quantum numbers? How can there be two down quarks in the neutron (udd)? The answer lies in another property that distinguishes two otherwise identical quarks. According to **quantum chromodynamics** (QCD), each quark has a "color," designated red (R), green (G), or blue (B). We stress that these are just names, and have nothing to do with visual colors. Antiquarks come in three different colors, called antired ($\overline{R}$), antigreen ($\overline{G}$), and antiblue ($\overline{B}$). The color metaphor is useful, because the three colors together cancel, leaving a colorless baryon, analogous to white light being made from a mixture of all colors (Chapter 20). All baryons contain the three different colors of quarks, so each baryon is colorless. Similarly, the quark and antiquark in a meson must be of the same type (such as $R\overline{R}$), which also cancels to form a colorless particle.

Quantum chromodynamics describes the binding of quarks to form mesons and baryons. Analogous to electric charge, quarks of like color repel, and quarks of different colors attract. This explains the binding of quark and antiquark in a meson, and of three quarks in a baryon. The mediator for the force between quarks is a particle called a **gluon**. The force that binds nucleons, which we've called the strong or nuclear force, is actually a residual manifestation of the color force between quarks.

The understanding of quarks is a remarkable triumph of theoretical and experimental physics in the last 50 years. This is particularly so because individual free quarks cannot be observed. You might expect that a high-energy particle could dislodge a quark from a hadron. But that's impossible, because the color force between quarks actually *strengthens* with increasing quark separation. If a particle has enough energy, a new quark-antiquark pair can be created, as you saw in Conceptual Example 26.7, but a single quark can't be dislodged. This experimental fact is known as **quark confinement**. Despite quark confinement, high-energy scattering experiments have revealed the properties of quarks that we've outlined in this section.

The Standard Model

The description of matter in terms of leptons and quarks is called the **standard model**. The standard model classifies the six quarks and six leptons into **families** and **generations**, as shown in Figure 26.9. This 21st-century "periodic table" of sorts contains just 12 fundamental particles! Of course, there are also antiparticles for each quark and lepton. And there are the force-mediating particles: the photon for electromagnetism, W and Z particles for the weak force, and gluons for the color force. Although we've speculated about the graviton as the mediator of the gravitational force, gravity is not included in the standard model. In fact, incorporating gravity into the quantum-based descriptions applicable to the other forces is a formidable and ongoing challenge in physics.

To understand the significance of families and generations, consider the particle properties listed in Tables 26.3 and 26.5. For example, quarks in the first family (u, c, t) all have charge $+2e/3$, while quarks in the other family (d, s, b) all have charge $-e/3$. The two lepton families have similar charge characteristics, with members of the family e, μ, and τ carrying charge $-e$, while the neutrinos are neutral. Each succeeding generation of quarks and leptons becomes more massive. Normal matter is made entirely from particles in the first generation. The second- and third-generation quarks and leptons decay into first-generation particles. The quark and lepton groups are distinguished by the fact that quarks are always found in combinations that form baryons or mesons, while leptons are always found individually and don't bind to form other subatomic particles.

Quark families		Lepton families		Generation
u	d	e^-	ν_e	I
c	s	μ	ν_μ	II
t	b	τ^-	ν_τ	III

FIGURE 26.9 Classification of quarks and leptons by family and generation. Each particle in the table has an antiparticle, not shown.

GOT IT? Section 26.4 Which one or more of the following reactions results in the transformation of quarks from one color to another? (a) $\mu^- \rightarrow e^- + \overline{\nu}_e + \nu_\mu$; (b) $n \rightarrow p^+ + \beta^- + \overline{\nu}_e$; (c) $\Xi^+ \rightarrow \Lambda^0 + \pi^+$; (d) $K^- \rightarrow \mu^- + \pi^0$.

26.5 Particle Accelerators

Most of the particles we've described in this chapter have short decay times (Table 26.4). Therefore, they don't generally occur in nature—although some are created briefly by high-energy cosmic rays, and these give physicists one way to study particles and their interactions. But the primary experimental tools for particle physics are accelerators that boost subatomic particles to highly relativistic energies. Here we'll describe different particle accelerators and their capabilities. We've already introduced major types of particle accelerators in other chapters, and we'll build on those discussions.

Linear Accelerators

The simplest design is a **linear accelerator** (also called a **linac**). Electric fields accelerate charged particles along a straight evacuated tube. Recall that a particle carrying one elementary charge e gains 1 eV of kinetic energy for every volt of potential difference (Section 16.2). Thus, a 100-MV accelerator can produce 100-MeV electrons or protons; that is, the particles each have kinetic energy of 100 MeV. The highest energy linac, at the Stanford Linear Accelerator Center (SLAC), is about 3 km long and produces 50-GeV electrons. Although this is enormous compared with the electrons' rest energy, it's not nearly as high as the energies produced in other types of accelerators.

EXAMPLE 26.8 **Speed of a High-Energy Electron**

What's the speed of the electrons in the 50-GeV linac at SLAC? Express the answer as a fraction of the speed of light.

ORGANIZE AND PLAN This kinetic energy is much higher than the electron's rest energy (0.511 MeV), so relativity is essential here. Useful relativistic relationships from Chapter 20 are $E = K + E_0$ and $E = \gamma mc^2$, where the relativistic factor is $\gamma = 1/\sqrt{1 - v^2/c^2}$.

Known: Electron rest energy $mc^2 = 0.511$ MeV; $K = 50$ GeV.

SOLVE Because $K \gg E_0$ in this case, a good approximation is $E = K + E_0 \approx K$. Solving $E = \gamma mc^2$ for the relativistic factor,

$$\gamma = \frac{E}{mc^2} = \frac{5.0 \times 10^{10} \text{ eV}}{5.11 \times 10^5 \text{ eV}} = 9.78 \times 10^4$$

Now solving $\gamma = 1/\sqrt{1 - v^2/c^2}$ for v,

$$\gamma^2 = 1/(1 - v^2/c^2)$$

$$v^2/c^2 = 1 - 1/\gamma^2 = 1 - \frac{1}{9.56 \times 10^9} = 0.999\,999\,999\,895$$

Thus

$$v = (0.999\,999\,999\,948)c$$

REFLECT That's amazingly close to the speed of light!

MAKING THE CONNECTION Would protons in this accelerator be traveling faster than, slower than, or the same speed as the electrons?

ANSWER Protons have more mass (and therefore more rest energy) than electrons, so they travel slower if given the same energy. But only a little slower! Rework the example with the proton rest energy, 938 MeV, and you'll find $v = 0.99982c$.

Cyclotrons and Synchrotrons

We described **cyclotrons** and **synchrotrons** in Section 18.3. Both use magnetic fields to "steer" charged particles around circular paths in evacuated tubes, while electric fields increase the particles' energies. In a cyclotron (Figure 18.15), the magnetic field fills the entire machine, and this makes high-energy machines impractical and expensive. More fundamentally, cyclotrons require that the frequency of the particles' circular motion (Equation 18.5) be independent of particle energy. But that holds only for nonrelativistic energies. Although cyclotrons are useful in such applications as radioisotope production for medical diagnostics, they're not viable tools for high-energy particle physics.

In a synchrotron, a large ring of fixed radius carries the charged particles. The ring can be made much larger than a cyclotron, because the magnetic field is needed only within the thin ring itself, not in the area it surrounds. And it's easy to vary the field in synchronism with the increasing particle energy—hence "synchrotron." The Fermi National Lab (Fermilab) in Illinois has a synchrotron with 1.0-km radius that accelerates protons to 1 TeV (10^{12} eV). Upon achievement of the 1-TeV energy, the device was dubbed the "tevatron." The tevatron was responsible for identifying the top quark in 1995.

The world's largest accelerator is the Large Hadron Collider (LHC), a synchrotron at the CERN facility on the Swiss-French border (see chapter-opening photo). With a radius of 4.3 km, the LHC accelerates protons to 7 TeV. Oppositely directed proton beams collide, producing effective collision energy of 14 TeV. Gigantic detectors measure the energies and trajectories of the multiple particles produced in these collisions. Computers analyze nearly a billion collisions that occur each second, searching for new or unusual events.

Stationary Targets and Colliding Beams

Early accelerators slammed their high-energy particle beams into stationary targets. But colliding beams have a greater effective energy. If each of two colliding particles has kinetic energy K, then the total energy available to create new particles is $2K$ (Figure 26.10a). With a stationary target, momentum conservation requires the center of mass of the system of interacting particles to continue in motion, and that reduces the energy available to make new particles or initiate reactions (Figure 26.10b). Using relativistic dynamics, it can be shown that the energy available when an incident particle of mass m_1 and kinetic energy K strikes a stationary-target particle of mass m_2 is

$$E = \sqrt{(m_1c^2 + m_2c^2)^2 + 2m_2c^2K} \tag{26.3}$$

✓**TIP**

Oppositely charged particles, such as electrons and positrons, curve in opposite directions in a magnetic field. Thus oppositely charged particles traverse a synchrotron ring in opposite directions.

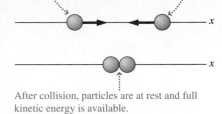
Two fast-moving particles approach collision.

After collision, particles are at rest and full kinetic energy is available.

(a)

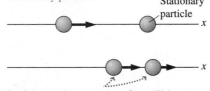
A fast-moving particle approaches collision with a stationary particle.

Stationary particle

Particles continue to move after collision, so only part of the original kinetic energy is available.

(b)

FIGURE 26.10 Comparison of (a) colliding-beam and (b) stationary-target strategies.

EXAMPLE 26.9 | **Stationary Targets versus Colliding Beams**

Compare the energies available from a 1.0-TeV proton accelerator when (a) two 1.0-TeV protons collide head-on and (b) a 1.0-TeV proton strikes another proton in a stationary target.

ORGANIZE AND PLAN For colliding particles each with kinetic energy K, the total available energy is $2K$. For the stationary-target experiment, the energy is given by Equation 26.3:

$$E = \sqrt{(m_1c^2 + m_2c^2)^2 + 2m_2c^2K}$$

Known: Proton rest energy $mc^2 = 938$ MeV $= 0.938$ GeV; $K = 1.0$ TeV $= 1000$ GeV.

SOLVE Here it's convenient to express all energies in GeV. The total energy of the colliding beams is $2K = 2000$ GeV. For the stationary-target experiment,

$$E = \sqrt{(m_1c^2 + m_2c^2)^2 + 2m_2c^2K}$$
$$E = \sqrt{(0.938 \text{ GeV} + 0.938 \text{ GeV})^2 + 2(0.938 \text{ GeV})(1000 \text{ GeV})}$$
$$= 43.4 \text{ GeV}$$

REFLECT The colliding beams provide almost 50 times the energy of the stationary-target experiment!

MAKING THE CONNECTION How much energy is available when a 50-GeV electron from SLAC strikes a stationary proton?

ANSWER Using Equation 26.3, the answer is 9.7 GeV.

The preceding example shows that colliding beams make more energy available. On the other hand, making particle beams collide is quite a technical feat. Imagine firing bullets toward each other and hoping they hit! The synchrotron itself is an amazing device, as oppositely directed beams of charged particles are kept tightly bunched while they circle repeatedly around a thin tube many kilometers in circumference. Mostly the beams just pass through each other, but there are enough collisions in the detector regions to yield statistically meaningful results.

A significant difficulty with synchrotrons is that the particles continually lose energy by electromagnetic radiation (Chapter 20). This so-called **synchrotron radiation** is due to the fact that circular motion entails acceleration (Chapter 3), and accelerated charges radiate electromagnetic waves (Chapter 20). At some point, the circling charges radiate as much energy as they gain from the accelerator, so their kinetic energy can't increase further. Linear accelerators don't experience such great losses, so physicists are hoping some day to construct a 30-km-long linac that produces 1-TeV particles.

26.6 Particles and the Universe

Accelerators probe subatomic particles and their interactions. But they're also "time machines," creating conditions that last existed a fraction of a second after the Big Bang event that began our universe. The past few decades have been an exciting time for physics and **cosmology**—the study of the universe and its evolution—as physicists established connections between the very smallest entities—subatomic particles—and the entire universe.

Strong evidence points to the universe beginning in a hot, dense condition characterized by rapid expansion—the **Big Bang**. In broad strokes, the evolution of the universe is the story of that expansion and associated cooling. At first, it's so hot that thermal energy is too high for even the simplest composite particles to form. But as the universe cools, ever more complex structures occur: protons and neutrons, nuclei, atoms, stars, galaxies, planets, living things, conscious brains, and everything else in our everyday world. Particle accelerators, built by those brains on at least one of those planets, re-create the high-energy conditions that prevailed in the early universe and thus afford a glimpse of the universe in its earliest moments. Here we'll review current understandings of the origin and evolution of the universe.

Cosmic Expansion

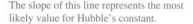
The slope of this line represents the most likely value for Hubble's constant.

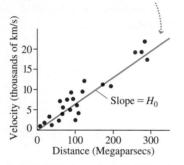

FIGURE 26.11 A modern Hubble plot, showing recession velocity versus distance for many galaxies.

In Chapter 20 we noted how astronomer Edwin Hubble used the Doppler shift for light to conclude that the universe is expanding. Hubble measured Doppler shifts and distances to a number of galaxies and found that distant galaxies are receding from us at speeds proportional to their distances. Figure 26.11 is a modern plot of this **Hubble relation**, which is expressed as a simple equation:

$$v = HR \qquad \text{(Hubble relation)} \qquad (26.4)$$

Here H is the **Hubble parameter**, which changes gradually on the billion-year timescale of the expanding universe. Its current value is the **Hubble constant**, H_0, about 72 km/s/Mpc. The distance unit here is the megaparsec (Mpc), equal to 3.26 million light years or 3.09×10^{22} m.

The Hubble expansion might suggest that we're at the center of the universe, since distant galaxies are all receding from us. But an observer on any other galaxy would see the same thing. If the universe extends forever—the current view of most cosmologists—then there's no issue of a "center." And if the universe is finite, then general relativity tells us it's the four-dimensional analog of a closed, two-dimensional surface like a balloon. Figure 26.12 shows that as the balloon expands, each dot—representing a galaxy—sees all the others move away at speeds proportional to their distances.

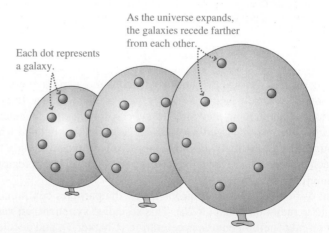
As the universe expands, the galaxies recede farther from each other.

Each dot represents a galaxy.

FIGURE 26.12 Analogy for an expanding universe of finite extent. The balloon's two-dimensional surface represents the four dimensions (3 space, 1 time) of the actual universe.

EXAMPLE 26.10 **Age of the Universe**

Use the current value of the Hubble constant, $H_0 = 72$ km/s/Mpc, to estimate the age of the universe in years. Assume that the galaxies have always been moving apart with their current speeds.

ORGANIZE AND PLAN In the Big Bang theory, the universe begins with today's observable universe at a single point. We can use the Hubble relation to find out when that was. If their speeds haven't changed, two galaxies that today are a distance R apart have been moving at speed $v = H_0 R$ (Hubble relation, Equation 26.4, with $H = H_0$) since the Big Bang. So they've been moving for time $T = R/v$. Substituting v from the Hubble relation gives $T = 1/H_0$ for the time since the galaxies were together. Since R doesn't enter this equation, it holds for any pair of galaxies. Therefore, the entire observed universe would have been at the same point at this time, so T is our estimate for the age of the universe. H_0 is in mixed units, so we'll need to convert to all SI.

Known: $H_0 = 72$ km/s/Mpc, 1 Mpc $= 3.09 \times 10^{22}$ m (from text).

SOLVE

$$T_{universe} = \frac{1}{H_0} = \frac{1 \text{ s} \cdot \text{Mpc}}{72 \text{ km}} \times \frac{1 \text{ km}}{1000 \text{ m}} \times \frac{3.09 \times 10^{22} \text{ m}}{1 \text{ Mpc}}$$

$$= 4.29 \times 10^{17} \text{ s}$$

With 1 year $(y) = 3.16 \times 10^7$ s, this is

$$T_{universe} = 4.29 \times 10^{17} \text{ s} \times \frac{1 \text{ y}}{3.16 \times 10^7 \text{ s}} = 1.36 \times 10^{10} \text{ y}$$

or 13.6 billion years.

REFLECT That's a good estimate, but don't consider it exact. The precise relation between the Hubble constant and age of the universe depends on how cosmic expansion varies with time.

MAKING THE CONNECTION How far is the most distant galaxy, assuming it's moving at nearly the speed of light c relative to us?

ANSWER An object moving at the speed of light travels 1 light year each year. According to our calculation, it's been moving for 13.6 billion years, so it's now 13.6 billion light years away. That distance defines the edge of our visible universe.

The Cosmic Microwave Background

Through the mid-20th century the Big Bang competed with alternative theories that incorporated universal expansion yet claimed the universe had no beginning. But a discovery in the 1960s made those alternatives untenable and laid a solid foundation for modern cosmology. During 1963–1965, radio astronomers Arno Penzias and Robert Wilson at Bell Laboratories discovered a background signal with wavelength around 1 mm, which seemed to come from all directions. After unsuccessful attempts to eliminate the signal as radio "noise," Penzias and Wilson concluded that it was coming from throughout the cosmos. At nearby Princeton University, physicist Robert Dicke had earlier predicted that the high-energy radiation from a hot Big Bang would cool as the universe expanded, to a current temperature around 3 K. Dicke's prediction provided the explanation for the observed **cosmic microwave background** (CMB). You can think of this as blackbody radiation that was originally at a high temperature, with a corresponding short wavelength. That radiation has suffered an extreme redshift as the universe expanded and now appears as longer wavelength microwave radiation.

Quantitative evidence supports these ideas. Recall from Chapter 23 that the blackbody radiation per unit wavelength from a source at a temperature T peaks at wavelength λ_{max} given by Wien's law: $\lambda_{max} T = 2.898 \times 10^{-3}$ m $\cdot$ K. Careful measurements show that the cosmic microwave background has $\lambda_{max} = 1.06$ mm, corresponding to a temperature of 2.73 K. The fit between data and the blackbody radiation curve for that temperature is remarkable (Figure 26.13). The cosmic microwave background originated some 380,000 years after the Big Bang, when the universe had cooled to about 3000 K. At that point electrons could join protons to make hydrogen atoms, without being disrupted through collisions at the high thermal energies that had prevailed earlier. The formation of atoms meant the universe was, for the first time, composed largely of neutral particles. The charged particles that dominated earlier interact strongly with electromagnetic waves, and so the universe had been opaque. But when neutral atoms formed, it became transparent. Photons present at that time could then travel throughout the universe with only a small chance of subsequent interactions. It's those photons—now shifted to longer wavelengths and lower energies—that constitute the cosmic microwave background.

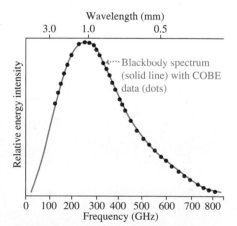

FIGURE 26.13 Spectrum of the cosmic microwave background radiation is a near-perfect fit to a 2.73-K blackbody (solid line). Data (dots on graph) are from COBE, the Cosmic Background Explorer satellite.

The CMB comes from all directions, and it's remarkably uniform. Once we subtract for the motion of our Milky Way relative to nearby galaxy clusters, the apparent temperature as deduced from the CMB blackbody curve (Figure 26.13) varies with direction in the sky by only about 1 part in 100,000. But those tiny variations contain a wealth of information about the early universe, going back even farther than the 380,000-year origin of the CMB. Primordial density fluctuations, which led to the large-scale structure of galaxies, galaxy clusters, and superclusters that we observe today, are encoded in the CMB variations. So is the imprint of sound waves that resonated through the early universe, and this helps us deduce and confirm details of Big Bang evolution.

Among those details is a period of extraordinarily rapid expansion—so called **inflation**—whose occurrence some 10^{-35} s after the beginning accounts for several essential features of today's universe. Study of the cosmic microwave background has taken cosmology from a vague science to an exact one, where we can pin down quantities like the age of the universe to within a few percent.

Particle Physics and Cosmology

In a chapter on particle physics, why are we talking about the Big Bang and the large-scale universe? Because particle physics is the "third leg" of modern cosmology, joining the Hubble expansion and the cosmic microwave background in helping us understand the evolution of the universe. You already saw (Chapter 25) how nuclear physics explains the origin and abundances of the chemical elements, including the helium-to-hydrogen ratio that was established in the universe's first half hour. Particle physics pushes us back even farther, to a time when high thermal energies made reactions like those in our accelerator experiments commonplace. Conditions in the first microseconds and earlier already contained the "seeds" of the structure in today's universe, so understanding particle interactions in the earliest instants is crucial in charting the universe's subsequent evolution. For example, the Relativistic Heavy Ion Collider (RHIC) at Brookhaven National Laboratory collides 100-GeV gold ions to create a "quark-gluon plasma"—a soup of quarks and the force-mediating gluons that bind them—characteristic of the universe just a few microseconds old, when the temperature was some 10^{12} K. These and many other particle physics experiments confirm the details of Big Bang evolution; conversely, modern cosmology sheds light on particle physics.

Particle physics is as much about the forces between particles as it is about the particles themselves. Although today we count four fundamental forces, physicists believe that at least three of them—the electromagnetic, weak, and strong forces—are manifestations of a single force and that at high enough energies they become one. Indeed, such **unification** of the electromagnetic and weak force was demonstrated both theoretically and experimentally in the late 20th century, in part through the identification of the W and Z particles. Direct evidence for further unification of the "electroweak" and strong forces requires energies orders of magnitude beyond our present capabilities, but 21st-century particle accelerators may provide indirect evidence for such unification. Figure 26.14 summarizes the evolution of the universe, including times and temperatures when the fundamental forces were unified.

Gravity is another story. Einstein's general relativity describes gravity as the curvature of a continuous spacetime. Reconciling this picture with quantum physics has proved extraordinarily difficult. One attempt at merging gravity with the other forces is **string theory**, which treats the fundamental entities in nature not as particles but as tiny stringlike loops. Vibrations of the loops should correspond to the known particles, making string theory potentially capable of explaining another quandary: why the particles have the masses they do. However, there's not a single string theory but many, and it's not clear that any of them describes our universe. Furthermore, none appear at this point capable of experimental verification. Time will tell whether string theory is or is not a fruitful approach to fundamental physics. In developing the ultimate merging of general relativity and quantum physics, physicists hope they'll discover a **Theory of Everything**—a description of all happenings in the physical world in terms of a single universal interaction.

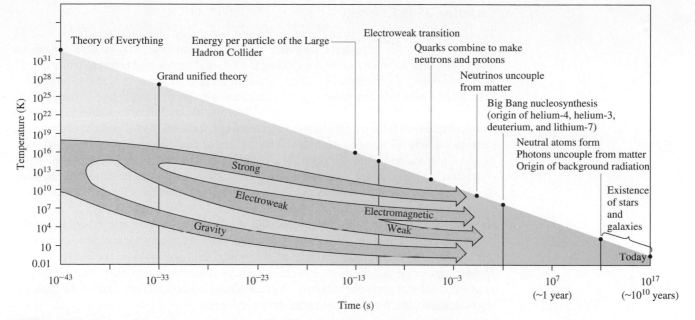

FIGURE 26.14 Evolution of the universe, showing unification of fundamental forces at the earliest time. Note the logarithmic scales.

Dark Stuff

Optical telescopes show astronomers stars and other visible matter in distant galaxies; infrared and radio telescopes reveal clouds of interstellar gas. But the motions of stars—inferred from Doppler measurements—suggest the presence in galaxies of much more gravitating matter than our telescopes can see. Furthermore, cosmological models suggest that the overall density of the universe is greater than we infer from the matter we observe. So there must be invisible **dark matter**.

Particle physics and cosmology combine to tell us something remarkable: Dark matter cannot be the ordinary stuff that comprises everything we see, from ourselves and our planet to the most distant stars. It's something different, something we know of only through its gravitational influence. In particular, dark matter consists not of quarks or leptons, but of some new, hitherto undetected particles. Physicists have several candidates and are actively seeking to detect dark matter particles. The task is challenging because dark matter interacts extremely weakly with ordinary matter.

What we do know about dark matter is that it comprises some 23% of the universe's mass. Ordinary matter, in contrast, accounts for a mere 4%. So what's the rest? It's an even more mysterious **dark energy**. Einstein included a concept similar to dark energy when he first developed his general theory of relativity, but quickly abandoned the idea. Through the 20th century, physicists and cosmologists believed the expansion of the universe was slowing, as the separating galaxies did work against their mutual gravitation. But a 1998 discovery turned this idea on its head: Studies of distant supernovas revealed that the expansion was not slowing but accelerating! That implies a kind of "antigravity" operating at the largest scales. Dark energy is the source of that "antigravity." What is dark energy? We don't know, any more than we know what dark matter is. But we do know that dark energy is the dominant "stuff," comprising approximately 73% of the universe. Here then, is a humbling thought: At the end of a book that purports to survey most all of physics—the science that provides our most basic understanding of physical reality—we find that we're ignorant about the stuff that makes up 96% of the universe!

Chapter 26 in Context

This chapter has taken us from the very smallest—the truly *elementary particles, quarks* and *leptons*—to the large-scale structure of the entire universe. You've seen how three quarks combine to make *baryons*, including the familiar proton and neutron, while quark-antiquark pairs constitute *mesons*. The "glue" in these combinations is the strong force. Like other forces, it's described quantum mechanically by an interchange of particles, in this case, *gluons*. Leptons, in contrast, don't feel the strong force; they interact through electromagnetism (photon interchange) and the weak force (W and Z particle exchanges). Leptons include the familiar electron and its associated neutrino, and two families of more massive leptons and their neutrinos. The elementary particles carry physical properties characterized by quantum numbers; some, like electric charge, are familiar. Others, such as *strangeness*, have no everyday manifestation. Conservation laws for the various quantum numbers govern what particle reactions can occur. Today we study high-energy particle reactions using particle accelerators, but those reactions were commonplace in the universe's earliest instants. Particle physics is therefore intimately connected with cosmology in the search for understanding of the origin and evolution of the universe. The *Hubble expansion* and the *cosmic microwave background* join particle physics in giving us our current picture of the universe—a place where mysterious dark matter and dark energy dominate, leaving ordinary matter only a 4% role.

Looking Ahead You're finished with this book and with your introduction to physics. But physics awaits you in the world beyond your classroom—it governs the natural phenomena all around us, and it's at the heart of the technologies that enrich your life. You'll continue to learn of new discoveries in fundamental physics, and you'll benefit from yet-to-be developed physics-based technologies. We hope we've given you both an understanding and an appreciation for the role physics plays and will continue to play in your life.

CHAPTER 26 SUMMARY

Particles and Antiparticles

(Section 26.1) **Elementary particles** can't be broken into smaller pieces; **composite particles** are made of combinations of elementary particles.

Antiparticles have the same mass and opposite charge as their corresponding particle.

Antiproton creation: $p + p \rightarrow p + p + p + \bar{p}$

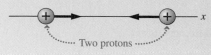
Before collision:

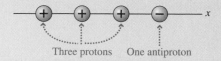
Two protons

After collision:

Three protons   One antiproton

Particles and Fundamental Forces

(Section 26.2) The **fundamental forces** work via **exchange particles** that **mediate** the interactions.

The mass of an exchange particle is inversely related to the range of the force.

Mass of mediating particle, as a function of interaction range:

$$m = \frac{h}{2\pi Rc}$$

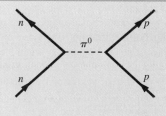

Classifying Particles

(Section 26.3) **Leptons** are the lightest particles and are elementary. **Hadrons** come in two varieties: **mesons** (composed of a **quark** and **antiquark**) and **baryons** (composed of three quarks).

Each different kind of particle is subject to a set of conservation laws, but **charge conservation** and **energy conservation** (including rest energy) hold in all reactions.

Muon decay: $\mu^- \rightarrow e^- + \bar{\nu}_e + \nu_\mu$

Quarks

(Section 26.4) There are six quarks, and six corresponding antiquarks. Baryons are composed of three quarks, or three antiquarks for the antiparticle. Mesons are composed of a quark and an antiquark.

According to **quantum chromodynamics** (QCD), each quark has a "color": red (R), green (G), or blue (B). The gluon mediates the force between different "colors" of quarks. The force between quarks increases with distance, making quarks **confined**. Thus it's impossible to have an isolated, free quark.

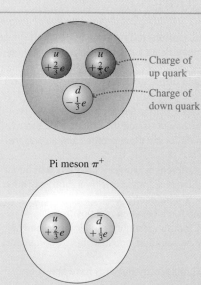
Pi meson π^+

Particle Accelerators

(Section 26.5) Linear accelerators and **synchrotrons** accelerate charged subatomic particles to extremely high energies, into the TeV range.

Colliding-beam interactions provide more useful energy than when the particle beam strikes a **stationary target**.

Energy available in stationary-target experiment:
$$E = \sqrt{(m_1c^2 + m_2c^2)^2 + 2m_2c^2K}$$

Energy available in colliding-beam experiment: $E = 2K$

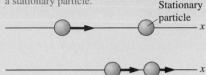

A fast-moving particle approaches collision with a stationary particle.

Stationary particle

Particles continue to move after collision, so only partial kinetic energy is available.

Particles and the Universe

(Section 26.6) Three lines of evidence point to the universe beginning in a hot, dense **Big Bang**.

(1) Hubble's discovery of galactic redshifts shows that the galaxies are receding from each other with speeds proportional to their distances.

(2) The **cosmic microwave background radiation** (CMB) is a remnant of the time 380,000 years after the beginning, when electrons and protons combined to make hydrogen atoms. Subtle nonuniformities in the CMB provide a wealth of information about the early universe.

(3) Particle physics combines with astrophysical evidence to yield a universe containing only 4% ordinary matter made from quarks and leptons. Another 23% is unknown **dark matter**, and the remaining 73% is **dark energy** whose "antigravity" effect is accelerating the universe's expansion.

Hubble relation: $v = HR$

In the present epoch, $H = H_0 = 72$ km/s/Mpc

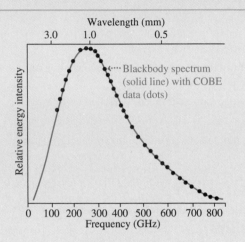

Wavelength (mm)

Blackbody spectrum (solid line) with COBE data (dots)

NOTE: Problem difficulty is labeled as ■ straightforward to ■ ■ ■ challenging. Problems labeled BIO are of biological or medical interest.

Conceptual Questions

1. Why is it easier to create proton-antiproton pairs by colliding beams of protons, rather than firing a proton beam into a stationary target?

2. Discuss the similarities and differences between the photon and neutrino.

3. For a particle that mediates a particular force, what is the relationship between the particle's mass and the range of the force?

4. Two protons collide, and out come three protons and an antiproton. How is this possible?

5. Discuss similarities and differences between the pion and muon. When the muon was first detected, physicists thought it may have been a pion. Why was there confusion?

6. What particle mediates the force between two positrons?

7. Of the 12 fundamental matter particles (six leptons and six quarks), which can only be found as constituents of composite particles?

8. All quarks have spin $\frac{1}{2}$. Explain why baryons have spin $\frac{1}{2}$ or 3/2 but mesons all have spin 0.

9. How does the photon differ from the fundamental matter particles (leptons and quarks)? In what ways is the photon similar to these particles?

10. Why is it necessary to attach the property of "color" to quarks?

11. Why is synchrotron radiation a feature of circular particle accelerators but not linear accelerators?

12. Explain the origin of the cosmic microwave background.

Multiple-Choice Problems

13. In reactions that create new particles, a conserved quantity is (a) the number of particles; (b) kinetic energy; (c) electric charge.

14. In β^- decay, the particles produced are an electron and (a) a neutrino; (b) an antineutrino; (c) a photon; (d) a proton.

15. The rest energy of the negative pion (π^-, mediator of the nuclear force) is 140 MeV. Based on the pion's rest energy, the range of the nuclear force should be about (a) 0.35 fm; (b) 1.0 fm; (c) 1.4 fm; (d) 1.8 fm.

16. Which one of the following reactions is allowed? (a) $p + p \rightarrow p + p + n$; (b) $p + p \rightarrow p + \pi^+ + \gamma$; (c) $p \rightarrow n + \pi^+$; (d) $p + \bar{p} \rightarrow \Lambda + \bar{\Lambda}$.

17. Which one of the following reactions is allowed? (a) $\mu^- + p \rightarrow n + \bar{\nu}_\mu$; (b) $\mu^- + p \rightarrow n + \nu_\mu$; (c) $e^- + p \rightarrow n + \bar{\nu}_e$; (d) $\mu^- \rightarrow n + e^- + \bar{\nu}_\mu + \nu_e$.

18. Complete the following reaction: $\tau^- + p \rightarrow n +$ _____. (a) e^-; (b) e^+; (c) ν_τ; (d) $\bar{\nu}_\tau$.

19. The quark compositions of four mesons are given. Which has strangeness -1? (a) $u\bar{s}$; (b) $s\bar{s}$; (c) $s\bar{u}$; (d) $c\bar{c}$.

20. The quark compositions of four mesons are given. Which meson has a charm number of -1? (a) $c\bar{d}$; (b) $c\bar{u}$; (c) $s\bar{c}$; (d) $c\bar{c}$.
21. The predominant constituent of the universe is (a) ordinary matter; (b) dark matter; (c) dark energy.

Problems

Section 26.1 Particles and Antiparticles

22. ■ How much energy is required to produce a neutron-antineutron pair?
23. ■ ■ An electron and positron at rest annihilate and form two photons, which have the same energy and emerge in opposite directions. Find the energy and wavelength of each photon.
24. ■ ■ Repeat the preceding problem if the electron and positron were moving toward each other at $0.5c$ in the lab frame.
25. ■ ■ If you want to produce a neutron-antineutron pair using colliding neutron beams, what should be the speed of the colliding neutrons?
26. ■ ■ Consider the beta decay $^{10}\text{Be} \rightarrow {}^{10}\text{B} + \beta^-$, with the beryllium nucleus initially at rest. (a) How much energy is released? (b) What other particle is produced? (c) What are the possible energies of that other particle?
27. ■ ■ The starship *Enterprise* on the television program *Star Trek* was powered by a matter-antimatter reactor. Suppose you have a 25,000-metric-ton starship and wish to accelerate it to $0.01c$. What is the total mass of matter and antimatter required, assuming a 100% efficient engine?

Section 26.2 Particles and Fundamental Forces

28. ■ The positive pion π^+ has rest energy 140 MeV. Use this value to estimate the range of the strong force.
29. ■ Physicists have proposed a particle called the *Higgs particle*, which would explain why other particles have the masses they do. If the Higgs' rest energy is 150 GeV, what's the range of the force it mediates?
30. ■ ■ What mass do you expect the graviton to have, if it is detected? What difficulties might there be in detecting a particle with this mass?
31. ■ Physicists believe that the omega particle (ω), with rest energy of 782 MeV, might be the mediator of a short-range repulsive force. What would be the range of that force?
32. ■ The neutral pion is unstable, with lifetime around 8.4×10^{-17} s. (a) Use the uncertainty principle to estimate the interaction time when the pion mediates the nuclear force, and compare with the pion's lifetime. Repeat for the charged pions, with lifetimes of 2.6×10^{-8} s.
33. ■ ■ To use high-energy particles as probes of other subatomic particles, the probe particles' de Broglie wavelength must be comparable to the distance being probed. Suppose you want to probe an elementary particle on a distance scale of about 1.0×10^{-18} m. Find the kinetic energy of (a) a proton and (b) an electron used to probe this distance scale.
34. ■ ■ ■ Draw a Feynman diagram for the β^+ decay of ^{22}Na. (Remember that β^+ decay produces a positron and a neutrino.)

Section 26.3 Classifying Particles

35. ■ ■ Fill in the following reactions with the missing particle or particles: (a) $\mu^- \rightarrow e^- + \underline{\quad}$; (b) $\mu^+ \rightarrow e^+ + \underline{\quad}$.
36. ■ ■ Fill in the following reactions with the missing particle or particles: (a) $\tau^- \rightarrow e + \underline{\quad}$; (b) $\tau^+ \rightarrow e^+ + \underline{\quad}$; (c) $\tau^- \rightarrow \mu^- + \underline{\quad}$; (d) $\tau^+ \rightarrow \mu^+ + \underline{\quad}$.

37. ■ ■ A μ^- and μ^+ at rest annihilate to form two gamma rays, which emerge in opposite directions. Find the energy and wavelength of each gamma ray.
38. ■ ■ Fill in the following reactions with the missing particle or particles: (a) $\text{K}^- \rightarrow \mu^- + \underline{\quad}$; (b) $\Lambda \rightarrow n + \underline{\quad}$.
39. ■ ■ Fill in the following reactions with the missing particle or particles: (a) $\underline{\quad} + p \rightarrow n + e^+$; (b) $\text{K}^+ \rightarrow \mu^+ + \underline{\quad}$.
40. ■ The neutral pion decays by emitting two gamma rays. If the pion is initially at rest, what are the energy and wavelength of the gamma rays?
41. ■ ■ Which of the following decays or reactions is allowed? For those that aren't allowed, explain why not. (a) $\Lambda \rightarrow p + \pi^-$; (b) $\pi^+ \rightarrow \mu^+ + n + \bar{\nu}_\mu$; (c) $p + \pi^- \rightarrow p + \Sigma^+$.
42. ■ ■ Which of the following decays or reactions is allowed? For those that aren't, explain why not. (a) $\Xi^- \rightarrow \Lambda + \pi^-$; (b) $e^+ + \pi^0 \rightarrow p + \nu_e$; (c) $\Xi^0 \rightarrow \bar{p} + \pi^-$.
43. ■ ■ The Σ^+ particle decays via the reaction $\Sigma^+ \rightarrow p + \pi^0$. For this reaction, discuss how the following conservation laws apply: electric charge, mass-energy, baryon number, lepton number, spin, strangeness.
44. ■ ■ ■ Trace all possible decay sequences for the Ω^- baryon down to stable particles.
45. ■ ■ Consider the decay of the Ω^- particle via the reaction $\Omega^- \rightarrow \Lambda + \text{K}^-$. (a) If the omega was originally at rest, how much energy is available for the kinetic energy of the two products? (b) Which of the products gets more kinetic energy?
46. ■ ■ ■ For the decay in the preceding problem, find the kinetic energy of each product.

Section 26.4 Quarks

47. ■ The Ω^- particle has a quark configuration sss. What do you expect for the quark configuration of the Ω^+?
48. ■ The Ξ^- particle is a baryon with strangeness -2 that contains a down quark and a strange quark. (a) What must be the third quark? (b) What do you expect for the quark configuration and electric charge of the Ξ^-'s antiparticle?
49. ■ ■ The D^0 meson has quark configuration cu. (a) What's the quark configuration of the $\overline{\text{D}}^0$? (b) Which properties are the same for the D^0 and its antiparticle, and which are different?
50. ■ ■ The D_S^- meson has $+1$ for both strangeness and charm. What quark-antiquark combination produces this meson?
51. ■ ■ For the proton, Σ^0, and Δ^+, show that the electric charge, baryon number, strangeness, and charm are all equal to the sums of the corresponding numbers for the constituent quarks.
52. ■ ■ Repeat the preceding problem for these particles: K^+, D^+, and B^+.
53. ■ ■ Trace the evolution and transformation of quarks in the reaction $\Sigma^+ \rightarrow p + \pi^0$ and the subsequent decay of the neutral pion.
54. ■ ■ Trace the evolution and transformation of quarks in the reaction $\Sigma^- \rightarrow n + \pi^-$ and the subsequent decay of the negative pion.
55. ■ ■ What are all possible quark configurations for baryons with charge (a) $2e$ and (b) $-2e$?

Section 26.5 Particle Accelerators

56. ■ ■ Find the speed of 7-TeV protons at the LHC accelerator at CERN.
57. ■ Find and compare the speeds of (a) 1-GeV electrons and (b) 1-GeV protons.
58. ■ ■ Consider the formation of antiprotons via the reaction $p + p \rightarrow p + p + p + \bar{p}$. Find the accelerator energies needed to initiate this reaction in (a) a colliding-beam experiment and (b) a stationary-target experiment.

59. ■■ The Alternating Gradient Synchrotron at Brookhaven National Laboratory in New York accelerates protons to a kinetic energy of 33 GeV. Find the time required for a proton to traverse the accelerator's 800-m circumference.

60. ■■ When a 50-GeV electron in the SLAC linear accelerator strikes a proton in a stationary target, how much energy is available to create new particles?

61. ■■ Compare the energy available between colliding-beam and stationary-target experiments with proton-proton collisions for each of these kinetic energies: (a) 1.0 GeV; (b) 10 GeV; (c) 100 GeV.

62. ■■ The cyclotron frequency is given by Equation 18.5 as $f_c = |q|B/2\pi m$, but this expression has to be corrected for relativistic particles to $f_c = \dfrac{|q|B}{2\pi m}\sqrt{1 - v^2/c^2}$. Assuming a uniform 2.5-T magnetic field throughout a cyclotron, find the cyclotron frequency required for electrons (a) moving at $0.99c$ and (b) with kinetic energy 1.0 GeV.

Section 26.6 Particles and the Universe

63. ■ Find the temperature corresponding to a thermal energy of 7 TeV, the particle energy in the Large Hadron Collider.

64. ■■ Hydrogen atoms first appeared when the temperature of the universe had dropped to about 3000 K. To what thermal energy does this correspond? Compare with hydrogen's ionization energy.

65. ■ The median wavelength of the cosmic microwave background is 1.504 mm. Use Equation 23.2 to find the corresponding blackbody temperature.

66. ■■ The parsec, used in expressing the Hubble constant, is defined as the distance to an object such that the Earth-Sun distance would subtend an angle of 1 s (1/3600 of a degree) when seen from the object. Use this definition to express the parsec in (a) light years and (b) meters.

67. ■ The most distant objects observed are some 13 billion light years from Earth. How fast are they receding from us?

68. ■ Find the distances to galaxies whose redshifts imply that they're moving away from us at (a) 2.5 Mm/s; (b) 15 Mm/s; (c) $0.20c$; (d) $0.80c$.

69. ■■■ In the early universe, there was enough thermal energy to create neutrons via the reaction $e^- + p^+ \rightarrow n + \nu_e$. (a) Compute the minimum energy required to initiate this reaction. (b) Find the temperature corresponding to the energy you found in part (a), and use Figure 26.14 to estimate when in the history of the universe this process was most likely to occur.

General Problems

70. ■■ Are either or both of these decay schemes possible for the tau particle: (a) $\tau^- \rightarrow e^- + \bar{\nu}_e + \nu_\tau$ and (b) $\tau^- \rightarrow \pi^- + \pi_0 + \nu_\tau$?

71. ■■ Is the interaction $p + p \rightarrow p + \pi^+$ allowed? If not, what conservation law does it violate?

72. ■■ The Σ^+ and Σ^- particles have quark compositions *uus* and *dds*, respectively. Are the Σ^+ and Σ^- each other's antiparticles? If not, give the quark compositions of their antiparticles.

73. ■■ The mass of the photon is assumed to be zero, but experiments put an upper limit of only 5×10^{-63} kg on the photon's mass. What would be the range of the electromagnetic force if the photon mass were actually at this upper limit?

74. ■■ The Tevatron at Fermilab accelerates protons to 1 TeV kinetic energy. (a) What's this in joules? (b) How far would a 1-g mass have to fall near Earth's surface to gain this much energy?

75. ■■■ (a) Find the relativistic factor γ for a 7-TeV proton in the Large Hadron Collider. (b) Find an accurate value for the proton's speed, in terms of c.

76. ■■■ The neutral lambda particle has mass 1116 MeV/c². (a) Find the minimum energy needed for colliding protons and antiprotons to produce a lambda-antilambda pair. (b) To what proton speed does this correspond?

77. ■■■ A so-called *muonic atom* is a hydrogen atom with the electron replaced by a muon, with mass 207 times the electron's. Find (a) the size and (b) the ground-state energy of the muonic atom.

78. ■■■ Astronomers describe redshifts in the spectra of distant objects using the quantity Z, defined as the shift in wavelength divided by the original wavelength λ_0: $Z = (\lambda - \lambda_0)/\lambda_0$. Use the relativistic Doppler formula (Equation 20.9), along with the Hubble relation, to find distances to objects with redshifts Z of (a) 0.50; (b) 1.5; (c) 7.0, roughly the redshift of the most distant galaxies observed.

Answers to Chapter Questions

Answer to Chapter-Opening Question

The circle marks the buried tunnel of the Large Hadron Collider, the world's largest particle accelerator. For physicists and cosmologists, it's a "time machine" that recreates for a fleeting instant conditions that prevailed in the first microsecond of the universe.

Answers to GOT IT? Questions

Section 26.1 (b) the same mass and (d) the opposite charge
Section 26.2 (c) weak $>$ (a) strong $>$ (b) electromagnetic $=$ (d) gravitational
Section 26.3 (a) $\pi^- \rightarrow \mu^- + \nu_\mu$ and (d) $\pi^+ \rightarrow \mu^+ + \bar{\nu}_\mu$
Section 26.4 (b) $n \rightarrow p^+ + \beta^- + \bar{\nu}_e$, (c) $\Xi^+ \rightarrow \Lambda^0 + \pi^+$, and (d) $K^- \rightarrow \mu^- + \pi^0$

Mathematics

Quadratic Formula

If $ax^2 + bx + c = 0$, then $x = \dfrac{-b \pm \sqrt{b^2 - 4ac}}{2a}$.

Circumference, Area, Volume

Where $\pi \simeq 3.14159\ldots$

circumference of circle	$2\pi r$
area of circle	πr^2
surface area of sphere	$4\pi r^2$
volume of sphere	$\frac{4}{3}\pi r^3$
area of triangle	$\frac{1}{2}bh$
volume of cylinder	$\pi r^2 l$

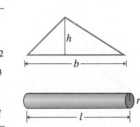

Trigonometry

definition of angle (in radians); $\theta = \dfrac{s}{r}$

2π radians in complete circle

1 radian $\simeq 57.3°$

Trigonometric Functions

$\sin\theta = \dfrac{y}{r}$

$\cos\theta = \dfrac{x}{r}$

$\tan\theta = \dfrac{\sin\theta}{\cos\theta} = \dfrac{y}{x}$

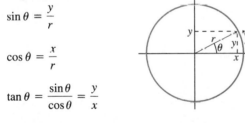

Values at Selected Angles

$\theta \rightarrow$	0	$\dfrac{\pi}{6}$ (30°)	$\dfrac{\pi}{4}$ (45°)	$\dfrac{\pi}{3}$ (60°)	$\dfrac{\pi}{2}$ (90°)
$\sin\theta$	0	$\dfrac{1}{2}$	$\dfrac{\sqrt{2}}{2}$	$\dfrac{\sqrt{3}}{2}$	1
$\cos\theta$	1	$\dfrac{\sqrt{3}}{2}$	$\dfrac{\sqrt{2}}{2}$	$\dfrac{1}{2}$	0
$\tan\theta$	0	$\dfrac{\sqrt{3}}{3}$	1	$\sqrt{3}$	∞

Graphs of Trigonometric Functions

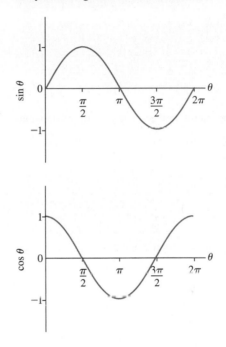

Trigonometric Identities

$\sin(-\theta) = -\sin\theta$

$\cos(-\theta) = \cos\theta$

$\sin\left(\theta \pm \dfrac{\pi}{2}\right) = \pm\cos\theta$

$\cos\left(\theta \pm \dfrac{\pi}{2}\right) = \mp\sin\theta$

$\sin^2\theta + \cos^2\theta = 1$

$\sin 2\theta = 2\sin\theta\cos\theta$

Laws of Cosines and Sines

Where A, B, C are the sides of an arbitrary triangle and α, β, γ are the angles opposite those sides:

Law of cosines

$C^2 = A^2 + B^2 - 2AB\cos\gamma$

Law of sines

$\dfrac{\sin\alpha}{A} = \dfrac{\sin\beta}{B} = \dfrac{\sin\gamma}{C}$

Exponentials and Logarithms

$e^{\ln x} = x$ $\ln e^x = x$ $e = 2.71828\ldots$

$a^x = e^{x \ln a}$ $\ln(xy) = \ln x + \ln y$

$a^x a^y = a^{x+y}$ $\ln\left(\dfrac{x}{y}\right) = \ln x - \ln y$

$(a^x)^y = a^{xy}$ $\ln\left(\dfrac{1}{x}\right) = -\ln x$

$\qquad\qquad\qquad 10^{\log x} = x$

Approximations

For $|x| \ll 1$, the following expressions provide good approximations to common functions:

$e^x \simeq 1 + x$

$\sin x \simeq x$

$\cos x \simeq 1 - \tfrac{1}{2}x^2$

$\ln(1 + x) \simeq x$

$(1 + x)^p \simeq 1 + px$ (binomial approximation)

Expressions that don't have the forms shown may often be put in the appropriate form. For example:

$$\frac{1}{\sqrt{a^2 + y^2}} = \frac{1}{a\sqrt{1 + \dfrac{y^2}{a^2}}} = \frac{1}{a}\left(1 + \frac{y^2}{a^2}\right)^{-1/2} \approx \frac{1}{a}\left(1 - \frac{y^2}{2a}\right)$$

for $y^2/a^2 \ll 1$, or $y^2 \ll a^2$.

Unit Vector Notation

An arbitrary vector $\vec{A}$ may be written in terms of its components A_x, A_y, A_z and the unit vectors $\hat{\imath}, \hat{\jmath}, \hat{k}$ that have length 1 and lie along the x-, y-, z-axes:

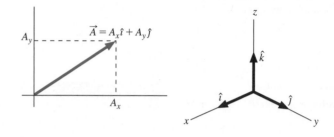

The International System of Units (SI)

This material is from the U.S. edition of the English translation of the seventh edition of "Le Système International d'Unités (SI)," the definitive publication in the French language issued in 1991 by the International Bureau of Weights and Measures (BIPM). The year the definition was adopted is given in parentheses.

length (meter): The meter is the length of the path traveled by light in vacuum during a time interval of 1/299 792 458 of a second. (1983)

mass (kilogram): The kilogram is equal to the mass of the international prototype of the kilogram. (1889)

time (second): The second is the duration of 9 192 631 770 periods of the radiation corresponding to the transition between the two hyperfine levels of the ground state of the cesium-133 atom. (1967)

electric current (ampere): The ampere is that constant current which, if maintained in two straight parallel conductors of infinite length, of negligible circular cross section, and placed 1 meter apart in vacuum, would produce between these conductors a force equal to 2×10^{-7} newton per meter of length. (1948)

temperature (kelvin): The kelvin, unit of thermodynamic temperature, is the fraction 1/273.16 of the thermodynamic temperature of the triple point of water. (1967)

amount of substance (mole): The mole is the amount of substance of a system that contains as many elementary entities as there are atoms in 0.012 kilogram of carbon-12. (1971)

luminous intensity (candela): The candela is the luminous intensity, in a given direction, of a source that emits monochromatic radiation of frequency 540×10^{12} hertz and that has a radiant intensity in that direction of (1/683) watt per steradian. (1979)

SI Base and Supplementary Units

Quantity	SI Unit Name	Symbol
Base Unit		
Length	meter	m
Mass	kilogram	kg
Time	second	s
Electric current	ampere	A
Thermodynamic temperature	kelvin	K
Amount of substance	mole	mol
Luminous intensity	candela	cd
Supplementary Units		
Plane angle	radian	rad
Solid angle	steardian	sr

SI Prefixes

Factor	Prefix	Symbol
10^{24}	yotta	Y
10^{21}	zetta	Z
10^{18}	exa	E
10^{15}	peta	P
10^{12}	tera	T
10^{9}	giga	G
10^{6}	mega	M
10^{3}	kilo	k
10^{2}	hecto	h
10^{1}	deka	da
10^{0}	—	—
10^{-1}	deei	d
10^{-2}	centi	c
10^{-3}	milli	m
10^{-6}	micro	μ
10^{-9}	nano	n
10^{-12}	pico	p
10^{-15}	femto	f
10^{-18}	atto	a
10^{-21}	zepto	z
10^{-24}	yocto	y

Some SI Derived Units with Special Names

Quantity	SI Unit			
	Name	Symbol	Expression in Terms of Other Units	Expression in Term of SI Base Units
Frequency	hertz	Hz		s^{-1}
Force	newton	N		$m \cdot kg \cdot s^{-2}$
Pressure, stress	pascal	Pa	N/m^2	$m^{-1} \cdot kg \cdot s^{-2}$
Energy, work, heat	joule	J	$N \cdot m$	$m^2 \cdot kg \cdot s^{-2}$
Power	watt	W	J/s	$m^2 \cdot kg \cdot s^{-3}$
Electric charge	coulomb	C		$s \cdot A$
Electric potential, potential difference, electromotive force	volt	V	J/C	$m^2 \cdot kg \cdot s^{-3} \cdot A^{-1}$
Capacitance	farad	F	C/V	$m^{-2} \cdot kg^{-1} \cdot s^4 \cdot A^2$
Electric resistance	ohm	Ω	V/A	$m^2 \cdot kg \cdot s^{-3} \cdot A^{-2}$
Magnetic flux	weber	Wb	$V \cdot s$	$m^2 \cdot kg \cdot s^{-2} \cdot A^{-1}$
Magnetic field	tesla	T	Wb/m^2	$kg \cdot s^{-2} \cdot A^{-1}$
Inductance	henry	H	Wb/A	$m^2 \cdot kg \cdot s^{-2} \cdot A^{-2}$
Radioactivity	becquerel	Bq	1 decay/s	s^{-1}
Absorbed radiation dose	gray	Gy	J/Kg, 100 rad	$m^2 \cdot s^{-2}$
Radiation dose equivalent	sievert	Sv	J/Kg, 100 rem	$m^2 \cdot s^{-2}$

Conversion Factors

The listings below give the SI equivalents of non-SI units. To convert from the units shown to SI, multiply by the factor given; to convert the other way, divide. For conversions within the SI system, see the table of SI prefixes in Appendix B, Chapter 1, or the inside front cover. Conversions that are not exact by definition are given to, at most, four significant figures.

Length

1 inch (in) = 0.0254 m

1 foot (ft) = 0.3048 m

1 yard (yd) = 0.9144 m

1 mile (mi) = 1609 m

1 nautical mile = 1852 m

1 angstrom (Å) = 10^{-10} m

1 light year (ly) = 9.46×10^{15} m

1 astronomical unit (AU) = 1.5×10^{11} m

1 parsec = 3.09×10^{16} m

1 fermi = 10^{-15} m = 1 fm

Mass

1 slug = 14.59 kg

1 metric ton (tonne; t) = 1000 kg

1 unified mass unit (u) = 1.661×10^{-27} kg

Force units in the English system are sometimes used (incorrectly) for mass. The units given below are actually equal to the number of kilograms multiplied by g, the acceleration of gravity.

1 pound (lb) = weight of 0.454 kg

1 ton = 2000 lb = weight of 908 kg

1 ounce (oz) = weight of 0.02835 kg

Time

1 minute (min) = 60 s

1 hour (h) = 60 min = 3600 s

1 day (d) = 24h = 86,400 s

1 year (y) = 365.2422 d = 3.156×10^7 s

Area

1 hectare (ha) = 10^4 m^2

1 square inch (in^2) = 6.452×10^{-4} m^2

1 square foot (ft^2) = 9.290×10^{-2} m^2

1 acre = 4047 m^2

1 barn = 10^{-28} m^2

1 shed = 10^{-30} m^2

Volume

1 liter (L) = 1000 cm^3 = 10^{-3} m^3

1 cubic foot (ft^3) = 2.832×10^{-2} m^3

1 cubic inch (in^3) = 1.639×10^{-5} m^3

1 fluid ounce = 1/128 gal = 2.957×10^{-5} m^3

1 barrel = 42 gal = 0.1590 m^3

1 gallon (U.S.; gal) = 3.785×10^{-3} m^3

1 gallon (British) = 4.546×10^{-3} m^3

Angle, Phase

1 degree (°) = $\pi/180$ rad = 1.745×10^{-2} rad

1 revolution (rev) = 360° = 2π rad

1 cycle = 360° = 2π rad

*The length of the year changes very slowly with changes in Earth's orbital period.

Speed, Velocity

$1 \text{ km/h} = (1/3.6) \text{ m/s} = 0.2778 \text{ m/s}$ $\qquad$ $1 \text{ ft/s} = 0.3048 \text{ m/s}$

$1 \text{ mi/h (mph)} = 0.4470 \text{ m/s}$ $\qquad$ $1 \text{ ly/y} = 3.00 \times 10^8 \text{ m/s}$

Angular Speed, Angular Velocity, Frequency, and Angular Frequency

$1 \text{ rev/s} = 2\pi \text{ rad/s} = 6.283 \text{ rad/s (s}^{-1})$ $\qquad$ $1 \text{ rev/min (rpm)} = 0.1047 \text{ rad/s (s}^{-1})$

$1 \text{ Hz} = 1 \text{ cycle/s} = 2\pi \text{ s}^{-1}$

Force

$1 \text{ dyne} = 10^{-5} \text{ N}$ $\qquad$ $1 \text{ pound (lb)} = 4.448 \text{ N}$

Pressure

$1 \text{ dyne/cm}^2 = 0.10 \text{ Pa}$ $\qquad$ $1 \text{ lb/in}^2 \text{ (psi)} = 6.895 \times 10^3 \text{ Pa}$

$1 \text{ atmosphere (atm)} = 1.013 \times 10^5 \text{ Pa}$ $\qquad$ $1 \text{ in H}_2\text{O (60°F)} = 248.8 \text{ Pa}$

$1 \text{ torr} = 1 \text{ mm Hg at } 0°\text{C} = 133.3 \text{ Pa}$ $\qquad$ $1 \text{ in Hg (60°F)} = 3.377 \times 10^3 \text{ Pa}$

$1 \text{ bar} = 10^5 \text{ Pa} = 0.987 \text{ atm}$

Energy, Work, Heat

$1 \text{ erg} = 10^{-7} \text{ J}$ $\qquad$ $1 \text{ Btu}^* = 1.054 \times 10^3 \text{ J}$

$1 \text{ calorie}^* \text{ (cal)} = 4.184 \text{ J}$ $\qquad$ $1 \text{ kWh} = 3.6 \times 10^6 \text{ J}$

$1 \text{ electronvolt (eV)} = 1.602 \times 10^{-19} \text{ J}$ $\qquad$ $1 \text{ megaton (explosive yield: Mt)}$

$1 \text{ foot-pound (ft} \cdot \text{lb)} = 1.356 \text{ J}$ $\qquad$ $= 4.18 \times 10^{15} \text{ J}$

Power

$1 \text{ erg/s} = 10^{-7} \text{ W}$ $\qquad$ $1 \text{ Btu/h (Btuh)} = 0.293 \text{ W}$

$1 \text{ horsepower (hp)} = 746 \text{ W}$ $\qquad$ $1 \text{ ft} \cdot \text{lb/s} = 1.356 \text{ W}$

Magnetic Field

$1 \text{ gauss (G)} = 10^{-4} \text{ T}$

Radiation

$1 \text{ curie (ci)} = 3.7 \times 10^{10} \text{ Bq}$ $\qquad$ $1 \text{ rad} = 10^{-2} \text{ Gy}$

$\qquad\qquad\qquad\qquad\qquad\qquad\qquad\qquad$ $1 \text{ rem} = 10^{-2} \text{ Sv}$

Energy Content of Fuels

Energy Source	Energy Content
Coal	$29 \text{ MJ/kg} = 7300 \text{ kWh/ton} = 25 \times 10^6 \text{ Btu/ton}$
Oil	$43 \text{ MJ/kg} = 39 \text{ kWh/gal} = 1.3 \times 10^5 \text{ Btu/gal}$
Gasoline	$44 \text{ MJ/kg} = 36 \text{ kWh/gal} = 1.2 \times 10^5 \text{ Btu/gal}$
Natural gas	$55 \text{ MJ/kg} = 30 \text{ kWh/100 ft}^3 = 1000 \text{ Btu/ft}^3$
Uranium (fission)	
$\quad$ Normal abundance	$5.8 \times 10^{11} \text{ J/kg} = 1.6 \times 10^5 \text{ kWh/kg}$
$\quad$ Pure U-235	$8.2 \times 10^{13} \text{ J/kg} = 2.3 \times 10^7 \text{ kWh/kg}$
Hydrogen (fusion)	
$\quad$ Normal abundance	$7 \times 10^{11} \text{ J/kg} = 3.0 \times 10^4 \text{ kWh/kg}$
$\quad$ Pure deuterium	$3.3 \times 10^{14} \text{ J/kg} = 9.2 \times 10^7 \text{ kWh/kg}$
$\quad$ Water	$1.2 \times 10^{10} \text{ J/kg} = 1.3 \times 10^4 \text{ kWh/gal} = 340 \text{ gal gasoline/gal}$
H_2O	
$\quad$ 100% conversion, matter	
$\quad\quad$ to energy	$9.0 \times 10^{16} \text{ J/kg} = 931 \text{ MeV/u} = 2.5 \times 10^{10} \text{ kWh/kg}$

*Values based on the thermochemical calorie; other definitions vary slightly.

Properties of Selected Isotopes

Subatomic Particle Masses

Particle	Mass (u)	Mass (kg)
Electron	$5.48\,580 \times 10^{-4}$	9.1094×10^{-31}
Proton	1.007 276	1.6726×10^{-27}
Neutron	1.008 665	1.6749×10^{-27}
Alpha particle	4.001 506	6.6447×10^{-27}

Atomic Number (Z)	Element	Symbol	Mass Number (A)	Atomic Mass* (u)	Abundance (%) or Decay Mode [†] (if Radioactive)	Half-Life (if Radioactive)
0	(Neutron)	n	1	1.008 665	β^-	10.6 min
1	Hydrogen	H	1	1.007 825	99.985	
	Deuterium	D	2	2.014 102	0.015	
	Tritium	T	3	3.016 049	β^-	12.33 y
2	Helium	He	3	3.016 029	0.00014	
			4	4.002 603	≈ 100	
3	Lithium	Li	6	6.015 123	7.5	
			7	7.016 005	92.5	
4	Beryllium	Be	7	7.016 930	EC, γ	53.3 d
			8	8.005 305	2α	6.7×10^{-17} s
			9	9.012 183	100	
5	Boron	B	10	10.012 938	19.8	
			11	11.009 305	80.2	
			12	12.014 353	β^-	20.4 ms
6	Carbon	C	11	11.011 433	β^+, EC	20.4 ms
			12	12.000 000	98.89	
			13	13.003 355	1.11	
			14	14.003 242	β^-	5730 y
7	Nitrogen	N	13	13.005 739	β^-	9.96 min
			14	14.003 074	99.63	
			15	15.000 109	0.37	
8	Oxygen	O	15	15.003 065	β^+, EC	122 s
			16	15.994 915	99.76	
			18	17.999 159	0.204	
9	Fluorine	F	19	18.998 403	100	
10	Neon	Ne	20	19.992 439	90.51	
			22	21.991 384	9.22	
11	Sodium	Na	22	21.994 435	β^+, EC, γ	2.602 y
			23	22.989 770	100	
			24	23.990 964	β^-, γ	15.0 h
12	Magnesium	Mg	24	23.985 045	78.99	
13	Aluminum	Al	27	26.981 541	100	
14	Silicon	Si	28	27.976 928	92.23	
			31	30.975 364	β^-, γ	2.62 h
15	Phosphorus	P	31	30.973 763	100	
			32	31.973 908	β^-	14.28 d
16	Sulfur	S	32	31.972 072	95.0	
			35	34.969 033	β^-	87.4 d

cont'd.

Atomic Number (Z)	Element	Symbol	Mass Number (A)	Atomic Mass* (u)	Abundance (%) or Decay Mode † (if Radioactive)	Half-Life (if Radioactive)
17	Chlorine	Cl	35	34.968 853	75.77	
			37	36.965 903	24.23	
18	Argon	Ar	40	39.962 383	99.60	
19	Potassium	K	39	38.963 708	93.26	
			40	39.964 000	β^-, EC, γ, β^+	1.28×10^9 y
20	Calcium	Ca	40	39.962 591	96.94	
24	Chromium	Cr	52	51.940 510	83.79	
25	Manganese	Mn	55	54.938 046	100	
26	Iron	Fe	56	55.934 939	91.8	
27	Cobalt	Co	59	58.933 198	100	
			60	59.933 820	β^-, γ	5.271 y
28	Nickel	Ni	58	57.935 347	68.3	
			59	58.934 352	β^+, EC	7.6×10^4 y
			60	59.930 789	26.1	
			64	63.927 968	0.91	
29	Copper	Cu	63	62.929 599	69.2	
			64	63.929 766	β^-, β^+	12.7 h
			65	64.927 792	30.8	
30	Zinc	Zn	64	63.929 145	48.6	
			66	65.926 035	27.9	
33	Arsenic	As	75	74.921 596	100	
35	Bromine	Br	79	78.918 336	50.69	
36	Krypton	Kr	84	83.911 506	57.0	
			89	88.917 563	β^-	3.2 min
38	Strontium	Sr	86	85.909 273	9.8	
			88	87.905 625	82.6	
			90	89.907 746	β^-	28.8 y
39	Yttrium	Y	89	89.905 856	100	
43	Technetium	Tc	98	97.907 210	β^-, γ	4.2×10^6 y
47	Silver	Ag	107	106.905 095	51.83	
			109	108.904 754	48.17	
48	Cadmium	Cd	114	113.903 361	28.7	
49	Indium	In	115	114.903 88	95.7; β^-	5.1×10^{14} y
50	Tin	Sn	120	119.902 199	32.4	
53	Iodine	I	127	126.904 477	100	
			131	130.906 118	β^-, γ	8.04 d
54	Xenon	Xe	132	131.904 15	26.9	
			136	135.907 22	8.9	
55	Cesium	Cs	133	132.905 43	100	
56	Barium	Ba	137	136.905 82	11.2	
			138	137.905 24	71.7	
			144	143.922 73	β^-	11.9 s
61	Promethium	Pm	145	144.912 75	EC, α, γ	17.7 y
74	Tungsten	W	184	183.950 95	30.7	
76	Osmium	Os	191	190.960 94	β^-, γ	15.4 d
			192	191.961 49	41.0	
78	Platinum	Pt	195	194.964 79	33.8	
79	Gold	Au	197	196.966 56	100	
80	Mercury	Hg	202	201.970 63	29.8	
81	Thallium	Tl	205	204.974 41	70.5	
			210	209.990 069	β^-	1.3 min
82	Lead	Pb	204	203.974 044	β^-, 1.48	1.4×10^{17} y
			206	205.974 46	24.1	
			207	206.975 89	22.1	
			208	207.976 64	52.3	
			210	209.984 18	α, β^-, γ	22.3 y
			211	210.988 74	β^-, γ	36.1 min
			212	211.991 88	β^-, γ	10.64 h
			214	213.999 80	β^-, γ	26.8 min

cont'd.

Atomic Number (Z)	Element	Symbol	Mass Number (A)	Atomic Mass* (u)	Abundance (%) or Decay Mode† (if Radioactive)	Half-Life (if Radioactive)
83	Bismuth	Bi	209	208.980 39	100	
			211	210.987 26	α, β^-, γ	2.15 min
84	Polonium	Po	210	209.982 86	α, γ	138.38 d
			214	213.995 19	α, γ	164 μs
86	Radon	Rn	222	222.017 574	α, β	3.8235 d
87	Francium	Fr	223	223.019 734	α, β^-, γ	21.8 min
88	Radium	Ra	226	226.025 406	α, γ	1.60×10^3 y
			228	228.031 069	β^-	5.76 y
89	Actinium	Ac	227	227.027 751	α, β^-, γ	21.773 y
90	Thorium	Th	228	228.028 73	α, γ	1.9131 y
			232	232.038 054	100; α, γ	1.41×10^{10} y
92	Uranium	U	232	232.037 14	α, γ	72 y
			233	233.039 629	α, γ	1.592×10^5 y
			235	235.043 923	0.72; α, γ	7.038×10^8 y
			236	236.045 563	α, γ	2.342×10^7 y
			238	238.050 786	99.275; α, γ	4.468×10^9 y
			239	239.054 291	β^-, γ	23.5 min
93	Neptunium	Np	239	239.052 932	β^-, γ	2.35 d
94	Plutonium	Pu	239	239.052 158	α, γ	2.41×10^4 y
95	Americium	Am	243	243.061 374	α, γ	7.37×10^3 y
96	Curium	Cm	245	245.065 487	α, γ	8.5×10^3 y
97	Berkelium	Bk	247	247.070 03	α, γ	1.4×10^3 y
98	Californium	Cf	249	249.074 849	α, γ	351 y
99	Einsteinium	Es	254	254.088 02	α, γ, β^-	276 d
100	Fermium	Fm	253	253.085 18	EC, α, γ	3.0 d

*The masses given throughout this table are those for the neutral atom, including the Z electrons.
†"EC" stands for electron capture.

Astrophysical Data

Sun, Planets, Principal Satellites

Body	Mass (10^24 kg)	Mean Radius (10^6 m Except as Noted)	Surface Gravity (m/s²)	Escape Speed (km/s)	Sidereal Rotation Period* (days)	Mean Distance from Central Body† (10^6 km)	Orbital Period	Orbital Speed (km/s)	Eccentricity	Semimajor Axis (10^9 m)
Sun	1.99×10^6	696	274	618	36 at poles 27 at equator	2.6×10^{11}	200 My	250		
Mercury	0.330	2.44	3.70	4.25	58.6	57.6	88.0 d	48	0.2056	57.6
Venus	4.87	6.05	8.87	10.4	−243	108	225 d	35	0.0068	108
Earth	5.97	6.37	9.81	11.2	0.997	150	365.3 d	30	0.0167	149.6
Moon	0.0735	1.74	1.62	2.38	27.3	0.385	27.3 d	1.0		
Mars	0.642	3.38	3.74	5.03	1.03	228	1.88 y	24.1	0.0934	228
Phobos	9.6×10^{-9}	9–13 km	0.001	0.008	0.32	9.4×10^{-3}	0.32 d	2.1		
Dcimos	2×10^{-9}	5–8 km	0.001	0.005	1.3	23×10^{-3}	1.3 d	1.3		
Jupiter	1.90×10^3	69.1	26.5	60.6	0.414	778	11.9 y	13.0	0.0483	778
Io	0.0888	1.82	1.8	2.6	1.77	0.422	1.77 d	17		
Europa	0.479	1.57	1.3	2.0	3.55	0.671	3.55 d	14		
Ganymede	0.148	2.63	1.4	2.7	7.15	1.07	7.15 d	11		
Callisto	0.107	2.40	1.2	2.4	16.7	1.88	16.7 d	8.2		
and 13 smaller satellites										
Saturn	569	56.8	11.8	36.6	0.438	1.43×10^3	29.5 y	9.65	0.0560	1430
Tethys	0.0007	0.53	0.2	0.4	1.89	0.294	1.89 d	11.3		
Dione	0.00015	0.56	0.3	0.6	2.74	0.377	2.74 d	10.0		
Rhea	0.0025	0.77	0.3	0.5	4.52	0.527	4.52 d	8.5		
Titan	0.135	2.58	1.4	2.6	15.9	1.22	15.9 d	5.6		
and 12 smaller satellites										
Uranus	86.6	25.0	9.23	21.5	−0.65	2.87×10^3	84.1 y	6.79	0.0461	2870
Ariel	0.0013	0.58	0.3	0.4	2.52	0.19	2.52 d	5.5		
Umbriel	0.0013	0.59	0.3	0.4	4.14	0.27	4.14 d	4.7		
Titania	0.0018	0.81	0.2	0.5	8.70	0.44	8.70 d	3.7		
Oberon	0.0017	0.78	0.2	0.5	13.5	0.58	13.5 d	3.1		
and 11 smaller satellites										
Neptune	103	24.0	11.9	23.9	0.768	4.50×10^3	165 y	5.43	0.0100	4500
Triton	0.134	1.9	2.5	3.1	5.88	0.354	5.88 d	4.4		
and 7 smaller satellites										

*Negative rotation period indicates retrograde motion, in opposite sense from orbital motion. Periods are sidereal, meaning the time for the body to return to the same orientation relative to the distant stars rather than the Sun.

†Central body is galactic center for Sun, Sun for planets, and planet for satellites.

Answers to Odd-Numbered Problems

Chapter 1
Answers to odd-numbered multiple-choice problems
9. (b)
11. (c)
13. (a)
15. (c)
17. (a)

Answers to odd-numbered problems
19. (a) 1.395×10^4 m;
    (b) 2.46×10^{-5} kg;
    (c) 3.49×10^{-8} s;
    (d) 1.28×10^9 s
21. 10^9 kg
23. (a) 1.083×10^{21} m³;
    (b) 5.51×10^3 kg/m³, about 5.5 times the density of water
25. 1.389×10^{-3}
27. 31.29 m/s
29. 2.29 m
31. 5.08 m
33. (a) 30.14; (b) 11740
35. 1.993×10^{-26} kg
37. (a) 1 mi = 1.609 km;
    (b) 1 kg = $10^9 \mu$g;
    (c) 1 km/h = 0.278 m/s;
    (d) 1 ft³ = 0.0283 m³
39. 7.842×10^3 m/s
41. (a) 0.385 AU; (b) 1.52 AU;
    (c) 5.20 AU; (d) 30.1 AU
43. (a) 4; (b) 8
45. (a) 1.283 s; (b) 499 s; (c) 1.50×10^4 s
47. $T: \sqrt{m/k}$
49. $v: \sqrt{gh}$
51. (a) 1; (b) 3; (c) 3; (d) 5
53. 1.50×10^2 cm²
55. 2.700 g/cm³
57. 3×10^9
59. 0.1 mm
61. (a) Earth; (b) $\rho_E/\rho_V = 1.063$
63. (a) 619.54 kg/m³;
    (b) The average density is only about 60% of the density of water
65. (a) 16.76 m/s; (b) 1.64
67. about 5×10^{25} atoms
69. (a) 2.563 kg; (b) 4.826×10^{25}

Chapter 2
Answers to odd-numbered multiple-choice problems
13. (c)
15. (a)

17. (b)
19. (c)
21. (c)
23. (c)
25. (b)

Answers to odd-numbered problems
27. $\Delta x = 0$, and s (my notation for total distance) = 400 m
29. (a) $\Delta x = 0$, $s = 960$ km;
    (b) $\Delta x = 160$ km, $s = 1120$ km;
    (c) $\Delta x = 80$ km, $s = 1200$ km.
31. 500 s or 8 min 20 s
33. 4.44 m/s
35. (a) 204 min = 3.40 h;
    (b) 485.3 km/h
37. $\vec{v} = 58$ km/h, $+x$; average speed is 194 km/h
39. 3.96 m/s
41. (a) -3.16 mi/h;
    (b) 55.38 s for each mile
43. 2.2 s
45. 0.755c
47. $v(t) = -(55/12) + (5/3)t$
49. (i) 0 − 7.5 s: 2.33 m/s²;
    (ii) 7.5 − 12.5 s: 0 m/s²;
    (iii) 12.5 − 20 s: -2.67 m/s²

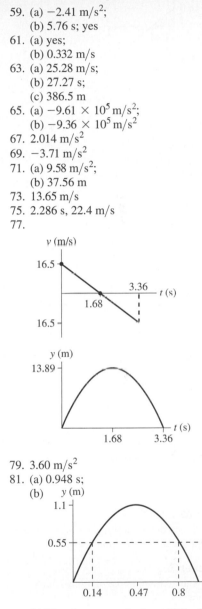

51. (a) a is greatest in the interval 0 − 7.5 s;
    (b) a is smallest in the interval 12.5 − 20 s;
    (c) a is zero in the interval 7.5 − 12.5 s;
    (d) maximum: 2.33 m/s², min: -2.67 m/s²
53. (a) 26.8 m/s; (b) 6.7 m/s²
55. 7.8 m/s
57.

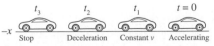

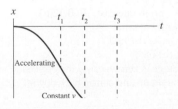

59. (a) -2.41 m/s²;
    (b) 5.76 s; yes
61. (a) yes;
    (b) 0.332 m/s
63. (a) 25.28 m/s;
    (b) 27.27 s;
    (c) 386.5 m
65. (a) -9.61×10^5 m/s²;
    (b) -9.36×10^5 m/s²
67. 2.014 m/s²
69. -3.71 m/s²
71. (a) 9.58 m/s²;
    (b) 37.56 m
73. 13.65 m/s
75. 2.286 s, 22.4 m/s
77.

79. 3.60 m/s²
81. (a) 0.948 s;
    (b)
    (c) The player spends about 70% of the time at or above half of the maximum height. This gives the appearance that the player is "hanging" in the air.
83. (a) 138.6 m/s; (b) 1742.4 m;
    (c) 184.8 m/s; (d) 44 s
85. -4.95 m/s²

87. $y = 20 + 4t - 4.9t^2$; $v_y = 4 - 9.8t$

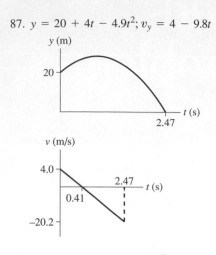

89. (a) 11.2 m/s; (b) -448 m/s^2
91. (a) -5.99 m/s^2; (b) 2.24 s; (c) -12.9 m/s^2
93. $x = 25t - 1.25t^2$; $v_x = 25 - 2.5t$
    $x(0) = 0, x(2) = 45, x(4) = 80,$
    $x(6) = 105, x(8) = 118.75,$
    $x(10) = 125$ (in meters)

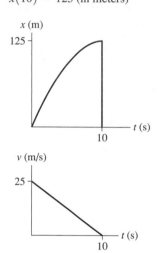

95. (a) 0.554 s, 1.915 s;
    (b) 6.67 m/s, -6.67 m/s
97. 31.25 m from the left, assuming the moving sidewalk moves to the right.
99. 90 s
101. 14.67 s, 161.3 m
103. 3.57 s; yes, flight time is 3.57 s so the opponent still has time.

Chapter 3
Answers to odd-numbered multiple-choice problems
15. (b)
17. (c)
19. (a)
21. (c)
23. (a)
25. (b)

Answers to odd-numbered problems
27. 53.1°, 36.9°, 90°
29. 1.60 m

31. (a) 15.3 cm;
    (b) 52.3°, 37.6°, 90°
33. 0.575 km
35. (a) 117.1°; (b) 199°; (c) $-73.3°$
37. (i) for $\vec{r}_1$, 5.61 m, $-64.76°$;
    (ii) for $\vec{r}_2$, 3.69 m, 164.6°
39. $(5.95\text{ m})\hat{\imath} - (6.05\text{ m})\hat{\jmath}$
41. 130.6°
43. (a) (1.11 m, 6.3 m); (b) (11.26 m, 6.5 m);
    (c) $(0, -10\text{ m})$
45. $\vec{T} = \vec{R} + \vec{S} = (10.5\text{ m})\hat{\imath} + (7.8\text{ m})\hat{\jmath}$;
47. $(58.52\text{ m})\hat{\imath} - (2.93\text{ m})\hat{\jmath}$
49. (a) 28.56 m/s; (b) $-(18.18\text{ m/s})\hat{\imath}$;
    (c) $(28.56\text{ m/s})\hat{\jmath}$
51. $(9.33 \times 10^{-4}\text{ m/s})\hat{\imath} +$
    $(7.17 \times 10^{-4}\text{ m/s})\hat{\jmath}, 1.18 \times 10^{-3}\text{ m/s}$
53. $(3.54\text{ m/s}^2)\hat{\imath} - (8.54\text{ m/s}^2)\hat{\jmath}$
55. (a) (245.9 m/s, 917.6 m/s);
    (b) $(4.47\text{ m/s}^2)\hat{\imath} + (16.68\text{ m/s}^2)\hat{\jmath}$
57. (a) $-(68\text{ m/s})\hat{\imath}$;
    (b) $-(9.07 \times 10^4\text{ m/s}^2)\hat{\imath}$,
    magnitude: 9.07×10^4 m/s^2, direction: $-\hat{\imath}$, opposite of the initial direction of the ball's velocity
59. (a) (1.14 m/s^2, -0.15 m/s^2);
    (b) $(11.4\text{ m/s})\hat{\imath} - (1.5\text{ m/s})\hat{\jmath}$, speed:
    11.5 m/s
61. (a) $\Delta\vec{v} = +(2.546\text{ m/s})\hat{\jmath}$
    (b) $\Delta\vec{v}' = +(0.142\text{ m/s})\hat{\imath} +$
    $(2.403\text{ m/s})\hat{\jmath}$
63. (a) 74.4 m; (b) 3.896 s; (c) 18.6 m
65. (a) 23 m; (b) 1.3 m
67. (a) 18.66 m/s; (b) 1.392 s;
    (c) $+(13.4\text{ m/s})\hat{\imath} - (13.64\text{ m/s})\hat{\jmath}$
69. 24.5 m
71. (a) 383.4 m/s; (b) 55.3 s
73. 720 m
75. 9.073 m/s
77. Between 7.98° and 20.76°
79. Between 20.4° and 26.57°
81. 0.0265 m/s^2; the ratio is given by
    $\cos 38°/\cos 0° = 0.788$
83. (a) At top of the loop, gravity provides the source of centripetal force,
    $mg = mv^2/r$, so $a_r = g$;
    (b) 8.46 m/s
85. (a) 2.51×10^{-5} s; (b) 7.5×10^{13} m/s^2
87. 630.6 m/s^2
89. 4.74×10^5 m/s^2
91. (d)
93. (a) 400 m;
    (b) With A $= (0, 0)$, we have
    $\Delta\vec{v} = (80\text{ m})\hat{\imath} - (76.4\text{ m})\hat{\jmath}$
95. (a) $(21.2\text{ m/s})\hat{\imath}$;
    (b) $(21.2\text{ m/s})\hat{\imath} - (17.8\text{ m/s})\hat{\jmath}$;
    (c) 67.2 m
97. (a) 38.7°, north due east; (b) 0.32 h;
    (c) (10 km, 8.0 km)
99. (a) 11.2 m/s; (b) 1.62 s
101. (a) 45°; (b) 68°;
    (c) the range is zero because the projectile goes straight up and comes back down

105. 2.72×10^{-3} m/s^2, about 2.78×10^{-4} g
107. 0.125 m

Chapter 4
Answers to odd-numbered multiple-choice problems
17. (a)
19. (c)
21. (c)
23. (a)
25. (c)
27. (c) under the assumption of 25 m/s for the speed of the car.

Answers to odd-numbered problems
29. 2.2 N to the right
31. 17.43 m/s^2, upward
33. 0.184 N, to the left
35. 0.0698 N, 111.4°
37.

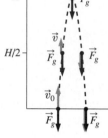

39. 1.64×10^{-21} m/s^2
41. 0.361 m/s
43. 7.9×10^5 N
45. 75 m/s
47. 15.77 kg
49. $(-156.1\text{ N})\hat{\imath} + (108\text{ N})\hat{\jmath}$
51. 31.98 m/s
53. (a) 477.8 N on hand;
    (b) 19.11 N on head;
    (c) 54.9 N on each leg
55. 2.97 m/s^2
57. (a) 291.2 N; (b) 2184 N
59. 7.82°
61. 113 N
63. (a) 12 m/s^2;
    (b) 18 N, 12 N, 6.0 N;
    (c) $F_{12} = 18$ N, $F_{23} = 6$ N
65. 1.1×10^4 N
67. (a) 286.7 N; (b) 296.8 N
69. 2.2 m/s^2
71. (a) -0.441 m/s^2; (b) 6.8 m
73. (a) -0.0396 m/s^2; (b) 37.87 s;
    (c) 0.0040
75. (a)

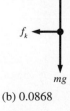

(b) 0.0868

77. (a)

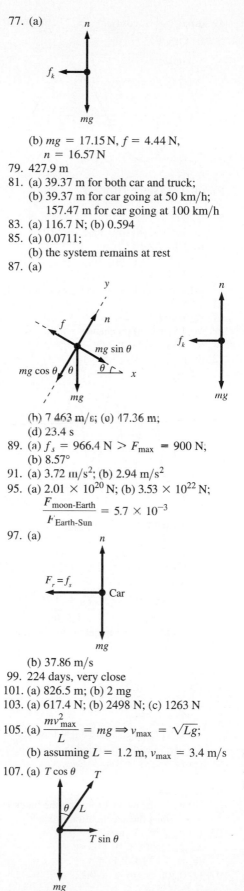

(b) $mg = 17.15$ N, $f = 4.44$ N,
    $n = 16.57$ N
79. 427.9 m
81. (a) 39.37 m for both car and truck;
    (b) 39.37 m for car going at 50 km/h;
        157.47 m for car going at 100 km/h
83. (a) 116.7 N; (b) 0.594
85. (a) 0.0711;
    (b) the system remains at rest
87. (a)

(h) 7.463 m/s; (c) 17.36 m;
(d) 23.4 s
89. (a) $f_s = 966.4$ N $> F_{\max} = 900$ N;
    (b) 8.57°
91. (a) 3.72 m/s²; (b) 2.94 m/s²
95. (a) 2.01×10^{20} N; (b) 3.53×10^{22} N;
    $\dfrac{F_{\text{moon-Earth}}}{F_{\text{Earth-Sun}}} = 5.7 \times 10^{-3}$
97. (a)

(b) 37.86 m/s
99. 224 days, very close
101. (a) 826.5 m; (b) 2 mg
103. (a) 617.4 N; (b) 2498 N; (c) 1263 N
105. (a) $\dfrac{mv_{\max}^2}{L} = mg \Rightarrow v_{\max} = \sqrt{Lg}$;
    (b) assuming $L = 1.2$ m, $v_{\max} = 3.4$ m/s
107. (a) $T\cos\theta$

(b) $2\pi\sqrt{\dfrac{L\cos\theta}{g}}$;

(c) as $\theta \to 0$, $T \to 2\pi\sqrt{\dfrac{L}{g}}$

109. 0.069
111. (a)

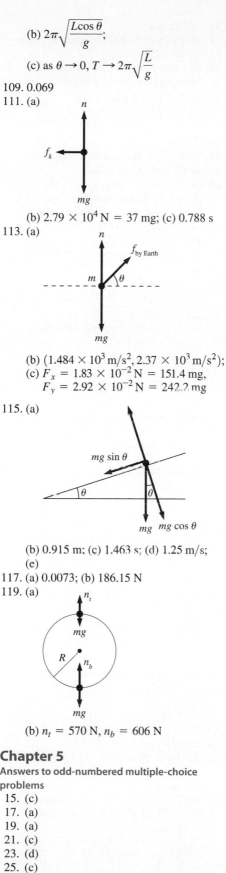

(b) 2.79×10^4 N = 37 mg; (c) 0.788 s
113. (a)

(b) $(1.484 \times 10^3 \text{ m/s}^2, 2.37 \times 10^3 \text{ m/s}^2)$;
(c) $F_x = 1.83 \times 10^{-2}$ N = 151.4 mg,
    $F_y = 2.92 \times 10^{-2}$ N = 242.2 mg
115. (a)

(b) 0.915 m; (c) 1.463 s; (d) 1.25 m/s;
(e)
117. (a) 0.0073; (b) 186.15 N
119. (a)

(b) $n_t = 570$ N, $n_b = 606$ N

Chapter 5
Answers to odd-numbered multiple-choice problems
15. (c)
17. (a)
19. (a)
21. (c)
23. (d)
25. (c)
27. (c)

Answers to odd-numbered problems
29. 1890 J
31. (a) −2104 N; (b) −3.05 × 10⁵ J

33. −2.59 J
35. (a) 5.85 J; (b) −5.85 J; (c) 8.5 J
37. (a) 4620 J; (b) −1813 J; (c) 2807 J
39. (a) 3.17 N; (b) 1.9 J; (c) −1.9 J;
    (d) 0
41. (a) 3.92 m/s²;
    (b) 0.294 J on mass 1, and 0.196 J on
        mass 2;
43. (a) 11.76 N/m; (b) 0.833 m
45. 4.77 × 10⁴ N/m
47. 6.0 J
49. (a) 1.75 J; (b) 1.3125 J;
    (c) 3.5 J; (d) −1.75 J
51. (a) −20 J; (b) 20 J; (c) 80 J;
    (d) 80 J; (e) 20 J
53. 5.39 cm
55. (a) 0.0392 m; (b) 0.0784 m;
    (c) 0.136 m
57. (a) 3.97 kg; (b) 1220 J; (c) 76.25 J
59. (a) 1.64 × 10⁸ J; (b) 1.366 × 10⁵ N;
    (c) yes
61. (a) −15.3 J; (b) −0.83 N
63. (a) 34.45 J; (b) −34.45 J; (c) 25.45 m;
    (d) 1.705 J; (e) 22.33 m/s
65. (a) 14 m/s; (b) 7.5 m
67. 8008 N
69. (a) 6.5 × 10⁴ J; (b) 7.30 × 10⁴ J
71. (a) 1.219 m/s; (b) 1.716 m/s;
    (c) 2.211 m/s
73. 2.58 × 10⁶ J
75. 0
77. (a) 5.02 × 10⁵ J;
    (b) 0.14 glass
79. about 9300 times; no
81. (a) 32.87 J; (b) 37.84 m/s
83. (a) 5.25 m/s; (b) 69 J
85. 11.38 m/s
87. 0.486 m
89. (a) 6.86 m/s; (b) 1.8 m
91. (a) −1.0 × 10⁴ J; (b) 2.1 × 10⁴ J;
    (c) 4.747 m/s
93. 1.014 m
95. 4.915 m/s
97. 2990 N/m
99. 2.55 × 10⁵ J
101. 8.983 × 10¹² W
103. (a) 2058 J on person, 603.68 on chair;
     (b) 171.5 W
105. 22.5 m/s
107. 0.617 MJ
109. (a) 3.822 J;
     (b) a straight line that starts at (0,0) and
         ends at (0.728, 10.5);
     (c) 3.82 J
111. (a) 95.2 J; (b) 1.586 W;
     (c) Blood is viscous and the passage
         through the blood vessels is much longer
         than the height of the person; some energy
         turns into thermal energy and is not
         recoverable.
113. 1.34 miles
115. (a) 24 J; (b) 12 J
117. (a) 0.288 m; (b) 11.66 N/m
119. (a) $x = 1$ m; (b) $x = 2.917$ m
121. (a) 17.96 m; (b) 20.02 m/s

123. -18.29 J
125. (a) 2.217 m/s; (b) 1.097 m;
    (c) 5.33 m/s

Chapter 6
Answers to odd-numbered multiple-choice problems
19. (a)
21. (b)
23. (d)
25. (a)
27. (d)
29. (c)
31. (b)

Answers to odd-numbered problems
33. 196.875 N, 1575 kg·m/s
35. 467.2 kg·m/s
37. 10.687 kg·m/s
39. (a) $1\,N\cdot s = 1\,(kg\cdot m/s^2)\cdot s = 1\,kg\cdot m/s$;
    (b) 53.57 m/s
41. (a) 0.2156 N;
    (b) 2.59×10^{-4} kg·m/s;
    (c) 2.59×10^4 kg·m/s
43. (a) $(1.53\,kg\cdot m/s)\hat{\imath} + (1.08\,kg\cdot m/s)\hat{\jmath}$;
    (b) $(2.02\,kg\cdot m/s)\hat{\imath} + (1.31\,kg\cdot m/s)\hat{\jmath}$
45. (a) 28.95 kg·m/s, 86.84 kg·m/s;
    (b) 279.3 J, 837.9 J
47. (a) 678.4 kg·m/s; (b) 479.7 kg·m/s;
    (c) 0
49. (a) Let $\vec{v}_i = v_i(\cos\theta\hat{\imath} - \sin\theta\hat{\jmath})$,
    $\vec{v}_f = v_f(\cos\theta\,\hat{\imath} + \sin\theta\,\hat{\jmath})$;
    $\Delta\vec{p} = (0.267\,kg\cdot m/s)\hat{\jmath}$;
    (b) $(-0.0346\,kg\cdot m/s)\hat{\imath} + (0.247\,kg\cdot m/s)\hat{\jmath}$;
    (c) $(10.68\,N)\hat{\jmath}$ for (a) and $(-1.384\,N)\hat{\imath} + (9.88\,N)\,\hat{\jmath}$ for (b)
51. (a) 1.5 kg·m/s for 0 to 0.5 s, and 3.0 kg·m/s for 0.5 to 1.0 s;
    (b) 37.5 m/s
53. (a) 1.46 kg·m/s; (b) 919.3 kg
55. 0.477 m/s, in the $-x$-direction
57. 17.33 m/s
59. 1.61 m/s
61. $-0.186\,v_0$ for 60-kg person, and $0.814\,v_0$ for the 87.5-kg person (need to know v_0 to have numerical answers)
65. $v_f = v_i/2$; $K_i = \dfrac{1}{2}mv_i^2$, and
    $K_f = \dfrac{1}{2}(2m)v_f^2 = \dfrac{1}{4}mv_i^2 = \dfrac{K_i}{2} \neq K_i$,
    energy is not conserved
67. When the bullet hits the block; as the bullet + block system swings to height h.
69. (a) $mv/(m + M)$; (b) $\dfrac{M + m}{m}\sqrt{2gh}$
71. (a) 48.79 n/s;
    (b) 45.45%;
    (c) 47.88 m/s
73. Before: 9.70 m/s; after: -1.70 m/s
75. 0.551
77. 0.05 m/s, and 0.90 m/s

79. $(0.725\,m/s)(\hat{\imath} + \hat{\jmath})$
81. 15.4 m/s or 34.5 mi/h, which exceeds 30-mi/h speed limit
83. One ball 1.167 m/s, 45° above x-axis; other ball 1.167 m/s, 45° below x-axis
85. $(8.2 \times 10^{-22}\,kg\cdot m/s)\hat{\imath} + (3.1 \times 10^{-28}\,kg\cdot m/s)\hat{\jmath}$
87. $(11.5\,m/s)\hat{\imath} - (23\,m/s)\hat{\jmath}$
89. 0.40 m
91. 7.42×10^8 m, outside Sun's radius
93. 2.063 m from the pivot
95. (a) 0.30 m/s;
    (b) velocities are exchanged;
    (c) the CM velocity remains the same
97. $(x_{cm}, y_{cm}) = (54.16\,m, 28.0\,m)$
99. (a) 36.39 cm; (b) 12.55 cm;
    (c) your muscles contract, pulling the tendons
103. $(-3.01 \times 10^{-24}\,N)\hat{\imath} - (4.62 \times 10^{-24}\,N)\hat{\jmath}$
105. (a) $(101.8\,N)\hat{\imath}$; (b) 0.042 m/s
107. 62.2 cm/s, in the opposite direction
109. (a) 3.27×10^{-4} s; (b) 2.30×10^4 N;
    (c) 1.875 m/s; (d) 0.221
111. 9.6 km/h
113. (a) 4.276 m/s; (b) 5.55 m/s
115. (a) 323 s $\approx$ 5.4 min;
    (b) 838 s $\approx$ 14 min
117. (a) $(x_{cm}, y_{cm}) = (0, 100.8\,cm)$;
    (b) $(0, 108.9\,cm)$
119. (a) before: $(0, 98.4\,cm)$, after: $(0, 100.8\,cm)$;
    (b) 2.4 cm
123. (a) $-v_i/3$; (b) 5

Chapter 7
Answers to odd-numbered multiple-choice problems
17. (c)
19. (b)
21. (b)
23. (c)
25. (d)
27. (a)
29. (d)

Answers to odd-numbered problems
31. (a) periodic; (b) 3.17×10^{-8} Hz
33. 0.375 s
35. (a) 6.25×10^{-5} s; (b) 1.01×10^5 rad/s
37. $T \sim \sqrt{m/k}$
39. 0.067 N/m
41. (a) 4.9 m; (b) 4.62 m; (c) 3.54 m
43. 154 N/m
45. 0.088 kg
47. $x(t) = (0.5\,m)\cos(8.71t)$, with $T = 0.721$ s

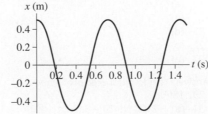

49. (a) $x(t) = A\cos(\omega t + \pi/2)$ (plot with $A = 1$, $\omega = 1$)

51. 0.268 m
53. (a) 0.733 s; (b) 0.0187 m
55. T/8
57. 0.58 m
59. (a) 14.74 J; (b) 4.89 m/s; (c) 0.203
61. (a) 10 s; (b) 10.13 kg;
    (c) $v_{max} = 0.47$ m/s, $a_{max} = 0.296\,m/s^2$
63. (a) 0.295 m; (b) 0.609 J; (c) 1.32 m/s
65. 5.83 Hz
67. (a) 10 Hz; (b) 62.83 rad/s
69. (a) 4.17 m; (b) no change
71. (a) 1.28 s; (b) 0.407 m
73. $\theta(t) = (5°)\cos(2.556t)$

75. 2.16 s
77. 0.007 s
79. lightly damped since ω_{damped} is real and positive
81. (a) $\omega_{damped} = \sqrt{\dfrac{k}{m} - \left(\dfrac{b}{2m}\right)^2} = 6.778$ rad/s > 0;
    (b) $T_{damped} = 0.927$ s, $T_0 = 0.91$ s;
    (c) Since $e^{-bT_{damped}/2m} = 0.3$, the amplitude will damp below $A/2$ in one oscillation
83. 0.045 kg/s
85. (a) 196.5 N/m; (b) 21.6 s
87. (a) 6.67×10^{-5} N/m;
    (b) 3.446×10^{-10} kg
89. (a) $T = 0.555$ s;
    (b) $T = T_1/\sqrt{2}$, where $T_1 = 0.785$ is the period with one spring
91. (a) $f_1 = 2f_2 \Rightarrow \omega_1 = 2\omega_2$;
    (b) $a_{1,max} = 4a_{2,max}$
93. (a) 0.144 m;
    (b) when the normal force on the block is zero (when the acceleration exceeds g)
95. (a) $v_{max} = 754$ m/s, $a_{max} = 5.68 \times 10^{16}\,m/s^2$;
    (b) 5.70×10^{-21} J
97. (a) 0.306 m; (b) A = 0.222 m; $T = 0.583$ s

Chapter 8
Answers to odd-numbered multiple-choice problems
15. (d)
17. (a)
19. (b)
21. (a)
23. (d)
25. (c)
27. (b)
29. (a)

Answers to odd-numbered problems
31. 1.267×10^4 m
33. 314 rad/s, about half that of the *E. Coli's* flagellum.
35. 245 rad/s = 2340 rpm
37. -4.12×10^{-22} rad/s^2
39. 1757.4 turns
41. (a) -119.7 rad/s^2; (b) 408 rev; (c) 241 m
43. -6.13×10^{-22} rad/s^2
45. 5.02 rad/s
47. (a) 97.1 rad/s^2; (b) 84.9 rad/s
49. (a) 16.82 m/s; (b) 0.128 rad/s^2
51. (a) 1.09×10^4 m; (b) 4.74 GB
53. (a) 1.047 rad/s; (b) 1.1 m/s
55. 2.57×10^{29} J
57. 11 J
59. (a) 7.94×10^{-5} kg m^2;
    (b) trans: 35.1 J; rotation: 0.627 J;
    $K_{trans}/K_{rot} \approx 56$
61. 40.26 rad/s
65. 102.1 rad/s
67. $\dfrac{K_{trans}}{K_{tot}} = \dfrac{2}{3}, \dfrac{K_{rot}}{K_{tot}} = \dfrac{1}{3}$
69. 5.42 rad/s
71. $\dfrac{5}{7} g \sin\theta < g \sin\theta$
73. (a) 0.933 m; (b) the θ dependence cancels out (i.e. the ratio of the times taken to reach the bottom by the two objects does not depend on θ)
75. 22.75 N · m
77. 33.33 pN
79. (a) 83.3 N · m; (b) 0.346 rad/s^2
81. 61.67 cm
83. (a) 130.95 N; (b) 194.2 N
85. (a) 0.702 kg; (b) 10.9 N
87. 2.66×10^{40} kg · m^2/s, 10^6 times greater than that due to rotation
89. 2.22 rad/s
91. (a) 45.5 kg · m^2/s, into the page as viewed from above; (b) 4.55 N · m, out of the page, as viewed from above
93. $-z$-direction.
95. -5.96×10^{16} N · m, down along the axis of rotation
97. (a) 2.94 Nm; (b) 752.6 rad/s^2
99. (a) 52.5°; (b) -260.66 J; (c) -284.5 N · m
101. (a) solid sphere wins; (b) solid cylinder: 1.4 m; hollow ball: 1.26 m; hollow cylindrical shell: 1.05 m
103. 1663 N · m

Chapter 9
Answers to odd-numbered multiple-choice problems
17. (a)
19. (d)
21. (b)
23. (b)
25. (b)
27. (b)

Answers to odd-numbered problems
29. 1.41×10^{-9} N
31. (a) 1.347×10^{26} N;
    (b) $F_{Earth-Sun} = 3.548 \times 10^{22}$ N, $F/F_{Earth-Sun} = 3.8 \times 10^4$
33. 3.63×10^{-47} N
35. Saturn: 11.1 m/s^2; Jupiter: 25.94 m/s^2
37. 6.48×10^{23} kg; very close to the given value of 6.42×10^{23} kg
39. (a) $F_g = 1.27 \times 10^{-10}$ N;
    (b) 6.17×10^{-10} N · m
41. (a) 3187 m; (b) 32090 m;
    (c) 3.446×10^5 m
43. 0.866
45. 2.12×10^{10} m
47. approximately 3.36×10^{18} m^3/s^2 for Mercury, Venus and Jupiter
49. $T_A/T_B = 2.828$
51. (a) 1.43×10^4 s; (b) $h = 2.175 R_E$.
53. 2.732 years
55. (a) -3.24×10^{35} J;
    (b) -5.27×10^{34} J; (a)/(b) = 6.15
57. (a) 443 m/s; (b) 4120 m/s; (c) 8747 m/s
59. 1.775×10^{32} J; 996 m/s
61. -1.267×10^{10} J
63. 2375 m/s
65. 6.71×10^7 m
67. 6160 s = 1.7 h
69. (a) 2439 m/s; (b) 2.1×10^8 m/s
71. 8.84×10^7 m = $51 R_{moon}$
75. (a) 3.8×10^{28} J; (b) -7.6×10^{28} J;
    (c) -3.8×10^{28} J
77. (a) 4.0×10^{10} J; (b) 1478 kg;
    (c) 7360 m/s
79. 7060 s
81. 1.65×10^{12} m/s^2
83. (a) 3.44×10^{-5} m/s^2;
    (b) 3.22×10^{-5} m/s^2;
    (c) 2.2×10^{-6} m/s^2
85. 0.1796 m/s^2
87. b, d, e, a, c
89. 6.2 m; 22.8 m
91. (a) 4.597×10^{10} m/s; (b) 0.2065
93. (a) 2068 km; (b) 804 km
95. 3.78 m/century
97. 2200 m
99. (a) 2950 m; (b) 2.95×10^{14} m

Chapter 10
Answers to odd-numbered multiple-choice problems
15. (d)
17. (b)
19. (c)
21. (d)
23. (c)

Answers to odd-numbered problems
25. 1.24 m^3
27. 1.69
29. 0.657 L
31. 1.1×10^{-7} m
33. 7.69 kg
35. 0.31 m
37. 10.34 m, not practical
39. (a) 5.81×10^7 Pa; (b) -3.63×10^{-4}
41. 20.1 m
43. 23.8 km
45. 0.898 m
47. 2.29×10^3 kg
49. 1.81×10^6 N
51. $F_b = 0.836$ N; $W = mg = 686$ N, $F_b \ll W$
53. (a) 63700 N; (b) 6.31 m^3; (c) 0.904
55. 500 kg/m^3
57. 1033.4 kg/m^3
59. (a) 3.31 N; (b) 2.08 N
61. 17.2%
63. 0.11 m/s^2
65. 2.67×10^{-4} m^3/s
67. (a) 2.67×10^{-4} m^3/s; (b) 18 m/s
69. 6.0×10^{-5} m^3/s
71. 113.6 kPa
73. 3.83 m/s
75. 1800 N, outward
77. decreases by 2.6%
79. 111.4 Pa
81. 0.27 m^2
83. (a) 143 kg; (b) 0.50 m; (c) 7170 Pa
85. 2.28×10^4 N; not likely
87. (a) 1.96 N; (b) 1.33×10^{-2} m/s;
    (c) 5.89 m/s
89. (a) 309 m^3; (b) 8.4 m
91. 1.5×10^4 Pa
93. (a) 23.7 N; (b) 23.6 N
95. 7.196×10^{-6} m^3/s

Chapter 11
Answers to odd-numbered multiple-choice problems
15. (d)
17. (b)
19. (b)
21. (b)
23. (b)
25. (d)

Answers to odd-numbered problems
27. (a) 0.566 m/s; (b) 1.13 m/s
29. 288 km
31. (d) $v = \omega/k$
37. 21.5 cm
39. (a) 235.2 m/s; (b) 8.86×10^{-4} kg/m
41. G: 392 Hz, 588 Hz; D: 588 Hz, 882 Hz; A: 880 Hz, 1320 Hz; E: 1318 Hz, 1977 Hz
43. (a) 192.96 m/s
45. 493.8 Hz 4 times every 10 seconds
47. 701.4 m
49. (a) 12005 m; (b) 36.27 s.
51. 2.48×10^{-8} W
53. $\frac{1}{2}$
55. (a) 89.18 dB; (b) 3.16%

57. (a) −6.02 dB; (b) −20 dB; (c) −40 dB
59. (a) 89.2 km; (b) 892 m.
61. 5.01
63. 238.15 Hz
65. 19.94 Hz, 59.83 Hz, 99.71 Hz, 139.6 Hz
67. (a) 1.53 m; (b) 0.327 m; (c) 0.164 m;
 (d) 0.071 m
69. 2048 Hz
71. 6.82 m/s
73. (a) 247.3 Hz; (b) 238 Hz
75. 381.4 Hz
77. (a) 1.2×10^5 Hz, 2.86×10^{-3} m;
 (b) 603 Hz, 0.569 m
79. 343 m/s
83. (a) a factor of 4; (b) −6.0 dB
85. 31.7 dB, louder than a quiet whisper
 at 20 dB
87. 54.3 kHz
89. 0.138 m
91. 0.80 W
93. 0.0148 m/s

Chapter 12
Answers to odd-numbered multiple-choice problems
15. (d)
17. (b)
19. (b)
21. (c)
23. (b)

Answers to odd-numbered problems
25. −15°C
27. (a) −320.8°F, 77.15 K;
 (b) 620.6°F, 600.15 K
29. 5.4°F
31. (a) 59.4°F; (b) −18°C, − 0.4°F
33. 100.76°F
35. (a) −40°C = −40°F; (b) 233.15 K
37. (a) 50.0072 m; (b) 49.982 m
39. (a) 1.26×10^{-4} m; 4.87×10^{-6} m
41. -3.12×10^{-9} m
43. 5.2 s
47. 0.2532 m²
49. (a) 39.95 g; (b) 11 g; (c) 52.47 g; (d) 528 g
51. 3.90×10^{27}
53. (a) Decrease; (b) 8.43 cm
55. (a) in order He, Ne, Ar, Kr, Xe, Rn:
 0.164 kg/m³, 0.818 kg/m³, 1.635 kg/m³,
 3.426 kg/m³, 5.37 kg/m³, 9.08 kg/m³;
 (b) only He and Ne
57. (a) 264.36 kPa; (b) 253.7 kPa
59. (a) 1.31 g; (b) 1.253 g
61. 56.14 cm³
63. (a) less; (b) 1.9×10^4 kg
65. (a) 1845 m/s; (b) 764.3 m/s
67. 1172 K
69. 1.098
71. (a) 1.2×10^{-19} J; (b) $E/E_{ion} = 0.055$
75. −0.138
77. (a) 5.52×10^{-3} m; (b) 1.392 cm
79. (a) 19.3633 m; (b) 19.3721 m
81. 532 kPa
83. 162.3 atm
85. v238/v235 = 0.996
87. 0.266 m

Chapter 13
Answers to odd-numbered multiple-choice problems
23. (a)
25. (b)
27. (a)
29. (d)
31. (a)

Answers to odd-numbered problems
33. 1.17×10^6 J
35. 1.22×10^7 J
37. 57.5 Cal
39. 27930 J, or 6.67 Cal
41. 3.74 kW
43. (a) 99.2 J/°C; (b) 19.84 kJ
45. 19.25 °C
47. 46 g (46 mL)
49. 427 kg
51. 268 s(=4 min 28 s)
53. 8841 J/kg
55. 438 J
57. 34.5 °C, at constant volume
59. (a) 0.147 kWh; (b) 2.35 cents
61. 4.62×10^4 J
63. −5900 J
65. 266 s
67. 1.67×10^5 J
71. 3.0 g
73. 0.131 kg
75. 15 W
77. $4.6
79. (a) 0.004 m²°C/W;
 (b) 0.0267 m²°C/W;
 (c) 0.133 m²°C/W;
 RStyrofoam > R_wood > R_glass
81. 0.085 m²°C/W; smaller than R-19 which
 has a value of 3.346 m²°C/W
83. (a) 1387 W/m²; (b) 3.6×10^6 m²
85. 0°C, 80 g of ice
87. (a) ice-water mixture;
 (b) 328.6 g of water and 171.4 g of ice
89. (a) 1.68×10^5 J; (b) 84 s
91. 172 s
93. (a) 4.8×10^4 W (using $k = 40$ W/°C m
 for steel, helpful if given in Table 13.4);
 (b) 1.06 °C/s at constant volume
95. 4.27×10^3 W
97. 76.8 K
99. 255 K
101. 3.14×10^6 s
103. −25 °C
105. 90 minutes
107. 0.425 °C

Chapter 14
Answers to odd-numbered multiple-choice problems
17. (d)
19. (a)
21. (b)
23. (b)
25. (c)

Answers to odd-numbered problems
27. 64.86 J
29. 210 W

31. (a) 415 J; (b) volume decreases;
 (c) −1081.65 J, volume increases
33. 6810 J; 20430 J
35. −258 J
37. (a) 11.7 °C, assuming at room temperature initially;
 (b) 25.3 J
39. (a) 4/3; (b) 215.3 J
41. (a) 32.42 atm; (b) 258.6 J; (c) 791.7 K
43. 1565.5 J
45. (a) 11.21 L; (b) 0.379 atm; 684.5 J
47. (a) $W_{A \to B} = -9.0 \times 10^5$ J;
 $W_{B \to C} = 0$; $W_{C \to D} = +3.0 \times 10^5$ J;
 $W_{D \to A} = 0$;
 (b) $W_{net} = -6.0 \times 10^5$ J;
49. (a) decreases; (b) 1519.5 J
51. (a) 25.12 atm; (b) 728.4 K
53. (a) 288 kJ; (b) 46.9 kJ
55. (a) 3.36 J/K; (b) −3.36 J/K
57. Melting: 121.98 J/K; boiling: 605.9 J/K
59. 2.72 J/K
61. 254.5 J/K
63. 0.34
65. (a) 0.353; (b) 600 MJ
67. 0.268
69. 1.23×1015 J
71. 2280 W
73. 0.353
75. (a) $52.7; (b) $17.94 saved
81. $C(52, 5) = 2.6 \times 10^6$
83. (a) 3.3 kJ; (b) 0.392 mol
85. 1.35
87. (a) 571.1 kPa; (b) 437.74 J
89. (a) 0.532; (b) 570 MW
91. (a) 399.4 J; (b) 2.65×10^5 Pa
93. (a) 288.2 K; (b) 15.5 kJ
95. (a) 4.29; (b) $(COP)_{max} = 11.42$
97. (a) 0.635; (b) 0.762
99. (a) 40 kPa; (b) 83.33 kPa; (c) about 80 kJ

Chapter 15
Answers to odd-numbered multiple-choice problems
15. (b)
17. (b)
19. (c)
21. (c)
23. (d)
25. (d)

Answers to odd-numbered problems
27. (a) 1.76×10^{11} C; (b) 9.58×10^7 C
29. (a) 3×10^{26} electrons
31. 4.8 mC
33. 0.38 m
35. (a) 10^{10} N directed straight up; (b) The
 weight is 1.64×10^{-26} N, insignificant
 compared to the electrical force.
37. 36.4 cm
39. 5 m. This means that gravity can be
 ignored in atomic or quantum calculations.
41. $\tan \theta \sin^2 \theta = \dfrac{kQ^2}{4L^2 mg}$

43. (a) Between the charges, 36.6 cm from the smaller charge. (b) Both magnitude and sign of the third charge cancel out of the equations for the Coloumb force when those forces are set equal to each other.
45. 2.77 N directed away from the center of triangle.
47. 8.6×10^{-6} N directed away from the center of the square.
49. (1.78 m, 2.29 m)
51. $(9.92 \times 10^{-9}$ N, -3.29×10^{-8} N)
53. 0.0014 N/C
55. a, d, b, c
57. (a) 8600 N/C in the $+x$-direction; (b) 3200 N/C in the $-x$-direction
59. 11°
61. 940,000
63. 8.85×10^{-8} C/m^2
65. (a) 5.15×10^{11} N/C; (b) 8.24×10^{-8} N
67. $x = 0.53$ m
69. (a) 23.6 kN/C in the $-x$-direction; (b) 38.0 kN/C at an angle 71.0° below the $+x$ axis
71. 5500 N/C
73. 185 N/C, 30° below the $-x$-axis
75. (a) Down; (b) $r = 523$ nm $(d = 1.05 \ \mu m)$
77. 4.13×10^6 m/s, 19.1° below horizontal
79. (a) 9.22×10^6 m/s; (b) 2.54×10^{-15} m. This is on the order of the diameter of a small nucleus.
81. (a) 136 N/C; (b) The same direction as the electron's original motion.
83. 4.6×10^{13}
85. (a) Place the third charge at $y = 2.72$ m; (b) 0
87. (a) 3.33×10^{-12} kg; (b) (i) Up; (ii) Down
89. (a) 3.4×10^8; (b) 6.6×10^{-9} N, directed vertically upward; (c) The bee's weight is 1.2×10^{-3} N, roughly 180,000 times the electrical force.
91. $\{+40.3 \ \mu C, -6.91 \ \mu C\}$ or $\{-40.3 \ \mu C, +6.91 \ \mu C\}$
93. (a) 2.64 mm; (b) 1.32 mm
95. 4.3 μm compression

Chapter 16
Answers to odd-numbered multiple-choice problems
19. (b)
21. (d)
23. (c)
25. (b)
27. (d)
29. (c)
31. (c)

Answers to odd-numbered problems
33. 3.15×10^{14} J
35. (a) earth: 4.4×10^4 C, moon: 546 C; (b) earth: 8.7×10^{11} C/m^2, moon: 1.4×10^{11} C/m^2
37. -1.99×10^7 J

39. 15.82 J
41. (b) 5.25 m/s
43. (a) 140 V; (b) 3.11×10^{-8} C, positive
45. (a) 3.56×10^{-10} C; (b) 2.2×10^9
47. 15.49 m/s
49. 1.15×10^5 V
51. (a) 3.2×10^{-19} C; (b) 3.2×10^{-19} C
53. 2.03×10^3 V/m
55. 1.496×10^6 m/s
57. (a) 3.85×10^4 V/m; (b) -6.15×10^{-15} N
59. -12.65 nC
61. (a) negative; (b) 1.8×10^4 V/m; (c) 4.8×10^{-19} C; (d) 3
63. 2.80 cm
65. (a) 0.70 F; (b) let $C_1 = 0.25$ F, $C_2 = 0.45$ F total Q: 16.8 C, $Q_1 = 6.0$ C, $Q_2 = 10.8$ C
67. (a) 17.14 μc; (b) 8.57 V; (c) $V_1 = V_2 = 3.43$ V; (d) $Q_1 = 3.43 \ \mu$C, $Q_2 = 13.72 \ \mu$C;
69. 9.0 μF
71. 1.0 μF
73. 2.0 μF
77. (a) 648 J; (b) 0.259 MW
79. (a) 0.443 nF; (b) 22.125 nC; 553 nJ
81. 1250 nF, in parallel
83. (a) 8.87 pF; (b) 106.42 pC; (c) 1920 V
85. 0.0127 mm
87. 2.475 mC
93. (a) 10^5 J; (b) 2.127×10^4 V
97. (a)

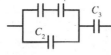

(b) 4.27 μF

Chapter 17
Answers to odd-numbered multiple-choice problems
19. (b)
21. (c)
23. (c)
25. (c)
27. (d)
29. (b)
31. (a)

Answers to odd-numbered problems
33. 1.29×10^{-5} m/s
35. $8.75 \times 10^{-5} \ \Omega$
37. 7.78 kA
39. (a) 8.1 Ω; (b) 0.081 Ω; (c) $8.1 \times 10^{-4} \ \Omega$
41. Within an uncertainty of $\Delta R = 0.0228 \ \Omega$
43. 4
45. (a) 125 s; (b) 12 J
47. 4.48 Ω
49. (a) 3.6 C; (b) 5040 C; (c) 6048 J
51. 3.6 V, 1.94 Ω
53. (a) 149 Ω; (b) 0.048 A (in 250-Ω resistor) and 0.032 A (in 370-Ω resistor)

59. 40 Ω
61. 75 V
63. (a) 2.5 V; (b) let $(R_1, R_2, R_3) = (20 \ k\Omega, 30 \ k\Omega, 75 \ k\Omega)$, $(I_1, I_2, I_3) = (0.125$ mA, 0.0833 mA, 0.0333 mA)
65. 230.4 J
67. 0.042 A
69. (a) 14.58 A; (b) 1.05 MJ = 0.292 kWh; (c) $1.31
71. 12
73. (a) 870 W; (b) 1.81°C
75. (a) 0.034 nC; (b) 0.316 nC; (c) 1.69 nC
77. (a) 0.693; (b) 1.228
79. (a) 2.4 mC; (b) 0.635 mA
81. (a) 0.48 C; (b) 0.0229 C more
83. (a) 5.735 mC, 5.479 mA; (b) 39.4 mC, 2.417 mA
87. 20.8 Ω, 1.84 V
89. 3.367 μF
91. (a) 6800 Ω; (b) 1.632 s
95. (a) 4800; (b) 1.0 A

Chapter 18
Answers to odd-numbered multiple-choice problems
23. (b)
25. (d)
27. (d)
29. (c)
31. (a)
33. (d)

Answers to odd-numbered problems
35. (a) 0.1995 μN, down; (b) 0.1995 μN, up
37. (a) 5.30 N, vertically downward; (b) 5.30 N, vertically upward; (c) 5.30 N, vertically upward; (d) 5.30 N, vertically downward; (e) 0
39. (a) 1.50×10^{-18} N, west; (b) 9.0×10^8 m/s^2, west
41. (a) 0.1088 pN, $-z$-direction; (b) 6.8×10^5 N/C, $-z$-direction
43. (2025 N/C), $-x$-direction
45. 1.87×10^{-28} kg; $m_\mu/m_p = 0.112$, $m_\mu/m_e = 205$
47. (a) 1.885×10^5 m/s; (b) 0.052 T
49. 0.69 mT
51. (a) 85.37 μm; (b) 3.655 mm; (c) 3.644 mm
53. 0.0876 T
55. 38.59 mA
57. (a) 54208 T; (b) 54.208 T
59. assuming current flows eastward, B = 73.3 μT, north
61. (a) 0; (b) $-(0.09$ N $\cdot$ m)$\hat{j}$; (c) $(0.0636$ N $\cdot$ m)$(\hat{i} - \hat{j})$
63. $(0.01875$ N$)\hat{j}$ on segment along x, $(0.01875$ N$)\hat{i}$ on segment along y, and $-(0.01875$ N$)(\hat{i} + \hat{j})$ along diagonal; the net force is zero

65. (a) 70.5 μN, west;
    (b) 75 μN, 70° north of up;
67. 11.04 A $\cdot$ m^2
69. R = 1.99 cm, 12 turns
71. 0.0768 T
73. 2.92 μT
75. (a) 133.3 μT; (b) 16.67 μT;
    (c) 83.33 N/m, repulsive
77. 3.35 $\times$ 10^{-3} N/m, toward each other, force attractive
79. (a) 0;
    (b) 1.326 $\times$ 10^{-4} N/m, toward the center of square
81. (a) 2.356 $\times$ 10^{-4} T; (b) 0
83. (a) (8.6 $\times$ 10^{-7} N), toward the wire;
    (b) (8.6 $\times$ 10^{-7} N), toward the loop;
    (c) (8.6 $\times$ 10^{-7} N), away from the wire;
    (d) (8.6 $\times$ 10^{-7} N), away from the loop
85. (a) 0.102 T; (b) 0.02356 T;
    (c) 0.0628 T in both cases
87. $-(147.6 \text{ m/s})\hat{\imath}$
89. $(0.06 \text{ N})\hat{\imath}$
91. 0.726 T
93. 22.62°, west of magnetic north
95. 1000 m

Chapter 19
Answers to odd-numbered multiple-choice problems
17. (a)
19. (c)
21. (a)
23. (a)
25. (b)
26. (c)

Answers to odd-numbered problems
27. 0.0736 Wb
29. 0.0117 Wb
31. 0.2873 A
33. (a) Clockwise;
    (b) 0.6136 mA; 0.3375 mV
35. 35
37. (a) 0.0751, clockwise; (b) 12.98 mW
39. (a) 1.764 V; (b) 3.528 V
41. (b) 1.225 V
45. (a) 0.01178 V; (b) 1508 V
47. 50
49. (a) 17.5 Ω;
    (b) All power would be lost.
51. 0.06 s
53. (a) 11.5 mH; (b) 9.325 A;
55. (a) 4.8 A; (b) 4.15 A; (c) 0.03475 s;
    (d) 4.15 A
57. Between 2516 and 3559 Hz
59. Increase C fourfold
61. (a) 0.0393 s; (b) 0.0731 J;
    (c) 8.55 mC
63. (a) 0.0687 A; (b) 16.51 W
65. (a) 280 V; (b) 560 W
67. (a) 169.7 V; (b) 49 Hz;
    (c) $\mathcal{E}(t)$ = (169.7 V) sin(308t)
69. (a) 31.83 Hz; (b) 79.58 Hz, 150 Ω
71. 129.5 Ω
73. (a) 0.33929 A; (b) 0.33938 A

75. (a) Capacitive reactance: 1061 Ω, inductive reactance: 84.82 Ω;
    (b) 1666 Ω; (c) 31.27 μF
77. (a) $-27.7°$; (b) 0.885; (c) 75.3 W
79. (a) 516.4 rad/s; (b) 10.87 W
81. 42
83. 1102.9 V
85. (a) 3.573 A/s; (b) Opposite
87. (a) 5.07 $\times$ 10^{16} J;
    (b) 3380 s, or 56.3 minutes

Chapter 20
Answers to odd-numbered multiple-choice problems
13. (d)
15. (d)
17. (b)
19. (d)
21. (d)
23. (a)

Answers to odd-numbered problems
25. 4.74 $\times$ 10^{14} Hz
27. 273 nm
29. 468 nm
31. (a) 0.50 μT; (b) 7.5 N/C
33. (a) 1.18 $\times$ 10^{-4} Pa; (b) 5.9 $\times$ 10^{-5} Pa
35. (a) 0.15 MHz, radio waves;
    (b) 150 MHz, radio waves;
    (c) 150 GHz, microwaves;
    (d) 1.5 $\times$ 10^{14} Hz, IR;
    (e) 1.5 $\times$ 10^{17} Hz, UV
37. 10^9 $-$ 10^{12} Hz
39. 1.50 m
41. 3 $\times$ 10^8 m/s
43. t_{perp} = 35.36 s; $t_{\text{perp}} < t_{\text{parallel}}$
45. (a) Earth: 133.3 s, ship: 133.3267 s;
    (b) Earth: 4.44 s, ship: 4.24 s;
    (c) Earth: 1.778 s, ship: 1.176 s;
    (d) Earth: 1.3468 s, ship: 0.190 s;
47. (a) 10 ns; (b) 11.547 ns;
    (c) 3.464 m; (d) 3 $\times$ 10^8 m/s
49. 0.141 c
51. 59.945 s
53. 0.9999995 c
55. 709 m
57. (a) Yes, 771.442 m $<$ 75 m; (b) 0.033 s
59. 0.9803 c
61. 0.9396 c
63. (a) 534.6 nm; (b) 488.45 nm;
    (c) 311.77 nm; (d) 123.88 nm;
    (e) 38.28 nm
65. 0.47 c, to the right
67. $f_0\sqrt{\dfrac{1 + v/c}{1 - v/c}}$; 2.29 $\times$ 10^{15} Hz
69. 34.4/min
71. 0.999949 c
75. 1.022 Mev
77. 0.866 c
79. (a) 0.14 c; (b) 0.910527 c
81. (a) 2.733 $\times$ 10^{-20} kg $\cdot$ m/s,
    E = 51.1 Me V; (c) 100
83. 0.995 c
85. 7.94 MW
87. 1.067 $\times$ 10^{-16} kg

89. 0.96 c
91. 0.93237 c

Chapter 21
Answers to odd-numbered multiple-choice problems
17. (b)
19. (c)
21. (a)
23. (c)
25. (b)

Answers to odd-numbered problems
27. 6 m
29. (a) 4.0 m; (b) 0.50 m/s
31. (a)

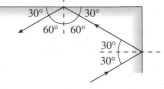

    (b) 24 cm
33. (a) 116.7 cm; (b) 53.85 cm
39. (a) 187.5 cm; (b) real; (c) -1.5
41. di = 52.2 cm; M = -1.5, image inverted
45. (a) Upright;
    (b) 2.25 cm from center of the ball;
    (c) 1.5 cm
47. (a) 40 cm;
    (b) image located at di = -15 cm,
    upright, reduced with M = 0.25
49. Diamond with n = 2.42, v = 0.413 c = 1.28 $\times$ 10^8 m/s
51. (a) 2.29 $\times$ 10^8 m/s; (b) 1.948 $\times$ 10^8 m/s;
    (c) 1.27 $\times$ 10^8 m/s; (d) 1.24 $\times$ 10^8 m/s
53. (a) 14.9° (b) 12.92°
55. 0.55 m
57. 0.93 m
59. θ_c = 24.4°
61. (a) 22.48°; (b) 0.469 mm
67. (a) d_o = 60 cm, di = 30 cm;
    (b) d_o = 30 cm, di = 60 cm
59. (a) Convex; (b) Inverted; (c) do = 15 cm
71. (a) 4.91 mm; (b) 9.0 mm; (c) 54 mm
73. 4.89 cm
75. (a) 6.56; (b) 17.67
77. (a) 0.172°; (b) 1.72°
79. 1.87 cm
85. (a) 23.0 cm; (b) Switching the two values with R_1 = 20 cm, and R_2 = -40 cm gives the same result.
87. -2.67 diopters
89. 2.03 diopters, assuming glasses 2 cm from eyes
91. (a) $f = \dfrac{R}{n - 1}$; (b) 13 cm
93. $n_2 = 1 + \dfrac{1}{3}(n_1 - 1)$
95. (a) 19.95 cm;
    (b) M = -0.5536, image size reduced
97. (a) 72 cm;
    (b) di = -4.5 cm, or 4.5 cm inside the surface

99. (i) for $M = -1/2$: (a) do $= 56$ cm,
    di $= 28$ cm; (b) image real;
    (c) $f = 18.67$ cm;
    (ii) for $M = -2$: (a) do $= 28$ cm,
    di $= 56$ cm; (b) image real;
    (c) $f = 18.67$ cm;
101. converging lens with $P = +3.33$ diopters
103. 7.56 mm

Chapter 22

Answers to odd-numbered multiple-choice problems

15. (d)
17. (d)
19. (b)
21. (b)
23. (c)

Answers to odd-numbered problems

25. (a) 76.16 nm; (b) 96.03 nm;
    (c) 105.96 nm
27. (a) No violet light;
    (b) Violet light
29. 542.6⁴

6.
65.
67.
69.
71.
73. 1
75. 0.
77. 1.
79. 0.0
81. 210

Chapte

Answers to   numbered multiple-choice problems

17. (b)
19. (c)
21. (d)
23. (d)
25. (d)
27. (d)

Answers to odd-numbered problems

29. 6.25×10^{18}
31. (a) 139.6 g;
    (b) 2.4×10^{-7} C, too small to be noticed
33. 2282 K
35. 7431 K
37. (a) 1369 nm, infrared;
    (b) 2.18×10^{-5} m^2
39. 2.73 K
41. 951 K
43. 1.028×10^{15} Hz, 292 nm
45. (a) 1.12×10^{15} Hz, 267.5 nm; (b) 4.56 V
47. (a) 0.448 V; (b) 3.97×10^5 m/s
49. (a) 2.32 V; (b) 3.926×10^{-19} J
51. (a) 6.272×10^{-34} J · s, −5.34%;
    (b) 7.2×10^{14} Hz, 4.5×10^{-19} J
53. (a) 6.626×10^{-25} J;
    (b) 1.988×10^{-22} J
55. (a) 8.28×10^{-20} J; (b) 1.21×10^{20}
57. (a) UVB: 1.56×10^{18}, visible:
    2.77×10^{18}
59. (a) 4.53×10^{-10}; (b) 1387 W/m^2
61. 90°
63. 108.5°
65. 97.2 pm
67. (a) 1876.5 MeV;
    (b) Each photon has an energy of
    938.27 MeV
59. (a) 7.27×10^5 m/s; (b) 546 fm
1. 2.91×10^6 m/s
3. (a) 16.73 eV; (b) 1.46×10^{-21} J;
    (c) 3.65×10^{-22} J
    27.82 pm, very unlikely
    (a) 1.506×10^{-24} J;
    (b) 4.96×10^{-19} J
) 5.273×10^{-25} kg · m/s;
) 1.526×10^{-19} J; (c) 1.3 μm
    24.1 pm; (b) 4.2 pm; (c) 0.345 pm
    2478 fm, much smaller than the diam-
    of a proton
    32 fm; (b) 9.3827×10^8 eV
    $\times 10^{-26}$ m
    V
    7×10^{20}; (b) 6.1×10^9

4

-numbered multiple-choice

rs to odd-numbered problems

27. 1.92×10^7 m/s
29. 4.86 fm with Al target, much shorter than
    29.6 fm found in Ex. 24.1 for Au target
31. (a) Atom: 6.236×10^{-31} m^3, nucleus:
    7.238×10^{-45} m^3;
    (b) 2.68×10^3 kg/m^3, about 2.7 times
    the density of water;
    (c) 2.31×10^{17} kg/m^3, much greater than
    that found in atom for part (b)
33. (a) 13; (b) 28; (c) 41;

35. $2279 - 7458$ nm
37. (a) 3.4 eV; (b) 365 nm
39. (a) 2.19×10^6 m/s; (b) 1.09×10^6 m/s
41. (a) 486 nm; (b) 0.82 m/s
43. (a) 1.89 eV; (b) absorption; (c) 656 nm
45. 69
49. (a) −54.4 eV; (b) −122.4 eV
51. (a) $2s$; (b) $4p$; (c) $4f$; (d) $3d$
53. (a) Allowed, 656 nm;
    (b) Forbidden;
    (c) Allowed, 1875 nm;
    (d) Allowed, 102.6 nm
55. (a) $l = 1$;
    (b) $m_l = -1, 0, +1$;
    (c) $m_s = \pm\frac{1}{2}$
57. (a) $\dfrac{L_f}{L_i} = \dfrac{1}{\sqrt{3}}$; (b) $\dfrac{L_f}{L_i} = \sqrt{3}$;
    (c) $\dfrac{L_f}{L_i} = \dfrac{1}{\sqrt{2}}$
63. (a) Holes; (b) Electrons
65. 40 kV
67. 0.0248 fm
69. 6895 V
71. 2.33 eV
73. (a) 532 nm; (b) 1.68 W
75. $n - 5 \rightarrow 3$
79. (a) 35.3°, 65.9°, 90°, 114.1°, 144.7°;
    (b) $L_z = m_l \dfrac{h}{2\pi}$, $m_l = 2, 1, 0, -1, -2$
81. 3.02×10^{17} per second

Chapter 25

Answers to odd-numbered multiple-choice problems

17. (a)
19. (b)
12. (d)
23. (b)
25. (b)
27. (c)

Answers to odd-numbered problems

29. (a) ^{49}Cr; (b) ^{94}Tc; (c) ^{190}Pb;
31. For ^{80}Br: $Z = 35, N = 45$;
    ^{80}Kr: $Z = 36, N = 44$;
33. (a) 2.3 fm; (b) 3.26 fm;
    (c) 6.13 fm; (d) 7.45 fm
35. Any nuclide with A $= 216$ (e.g., ^{216}At)
45. (a) 2.66×10^{33} J; (b) 2.66×10^{33} J;
    (c) 2.93×10^{16} kg
47. (a) ^{67}Cu $\rightarrow \beta^- + ^{67}$Zn;
    (b) ^{85}Kr $\rightarrow \beta^- + ^{85}$Rb;
    (c) ^{112}Pd $\rightarrow \beta^- + ^{112}$Ag
49. (a) ^{158}Er; (b) β^+; (c) ^{4}He; (d) ^{4}He
51. ^{210}Bi $\rightarrow \beta^- + ^{210}$Po, 1.16 MeV
53. (a) 0.148 Bq/L;
    (b) 1.28×10^4/L, assuming constant rate
55. Electron: 0.45 mm; alpha: 19.3 mm
57. (a) 0.325 mg; (b) 0.081 mg
59. 9.32 mg
61. (a) 1.42×10^{11} Bq;
    (b) 6.16×10^8 Bq, with 0.108 μg
    remained

63. 8.92 kg
65. About 6.3×10^9 years
67. (a) 3.11 MeV released;
    (b) $Q = -10.4$ MeV, $K_{min} = 11.9$ MeV;
    (c) 2.21 MeV released
69. (a) $Q = -4.3$ MeV;
    (b) $Q = -2.92$ MeV;
    (c) $Q = -6.09$ MeV (I simply give the
    Q value, and not K_{min})
71. (a) ^{89}Kr; (b) ^{143}La;
    (c) ^{96}Zr
73. (a) ^{239}Np; (b) 2n; (c) ^{120}Ag
75. 3.5 kg/day
77. 0.776 kg, much smaller than the 50 kg
    used in early bomb
79. (a) $Q = -22.37$ MeV;
    (b) $Q = -7.46$ MeV;
    (c) 17.6 MeV
81. (a) 12.86 MeV; (b) 17.35 MeV;
    (c) 18.35 MeV
83. 0.0307 mol each (0.09 g of ^{3}H and 0.06g
    of ^{2}H)
85. 11.45 MeV, 0.11 pm
87. (a) 2.5×10^{-31} kg; (b) 9.74×10^{10} Bq;
    (c) After 1 week: 1.92×10^{-15} g; after
    30 days: 4.32×10^{-43} g
89. Flight: 2.5×10^{-6}, or 1 in 400,000; PET
    scan: 5×10^{-4}, or 1 in 2,000
91. ^{198}Hg $+ \, ^1n \rightarrow \, ^{197}$Ag $+ \, ^2$H
93. Yes; takes only 4.29 days to drop to
    2000 Bq/L

Chapter 26

Answers to odd-numbered multiple-choice problems
13. (c)
15. (c)
17. (b)
19. (c)
21. (c)

Answers to odd-numbered problems
23. 0.511 MeV, 2.43 pm
25. 0.866c
27. 1250 kg total (625 kg of matter and
    625 kg of anti-matter)
29. 1.32×10^{-18} m
31. 0.252 fm
33. (a) 1.24 TeV;
    (b) 1.24 TeV
35. (a) $\mu^- \rightarrow e^- + \bar{\nu}_e + \nu_\mu$;
    (b) $\mu^+ \rightarrow e^+ + \nu_e + \bar{\nu}_\mu$
37. 105.66 MeV, 11.7 fm
39. (a) $\bar{\nu}_e + p \rightarrow n + e^+$;
    (b) $K^+ \rightarrow \mu^+ + \nu_\mu$
41. (a) Allowed;
    (b) Not allowed, violation of baryon
        number conservation and muon
        lepton number conservation;
    (c) Not allowed, electric charge not
        conserved

45. (a) 62 MeV;
    (b) K^{-1} gets more
47. $\overline{sss}$
49. (a) $\bar{c}u$;
    (b) Same mass and lifetime, different
        charm number and decay products
55. (a) uuu, ccc, ttt; (b) $\overline{uuu}$, $\overline{ccc}$, $\overline{ttt}$
57. (a) ve = 0.999 999 87 c;
    (b) vp = 0.875 c, ve/vp = 1.143
59. (a) $2.668\,\mu s$
61. (a) Colliding beam: 2.0 GeV, with
        stationary target: 2.32 GeV;
    (b) Colliding beam: 20 GeV, with
        stationary target: 4.72 GeV;
    (c) Colliding beam: 200 GeV, with
        stationary target: 13.82 GeV
63. 5.42×10^{16} K
65. 2.73 K
67. 2.87×10^8 m/s
69. (a) 1.29 MeV;
    (b) 9.98×10^9 K; $< 10^{-3}$ s after Big
        Bang
71. No, baryon number not conserved
73. 7.0×10^{19} m
75. (a) 7462;
    (b) $v = c(1 - 8.98 \times 10^{-9})$
77. (a) 2.556 fm; (b) -2.815 keV

Credits

Chapter 1
Page 1: NASA/Goddard Space Flight Center. Page 4, Figure 1.2:(a) NASA (b) Andrew Syred/Photo Researchers. Page 5 (left): University of Texas at Austin News and Information Service. Page 5 (right): NASA. Page 9: Charles Schug/istockphoto.

Chapter 2
Page 18: Koichi Kamoshida/Getty Images. Page 30: Trip/Alamy. Page 33, Figure 2.20: Jim Sugar/CORBIS. Page 33 (bottom): Scott Brinegar/Disneyland/CORBIS.

Chapter 3
Page 41: Alfredo Estrella/AFP/Getty Images. Page 51, Figure 3.16: Richard Megna/Fundamental Photographs. Page 58: NASA.

Chapter 4
Page 65: Steve Fitchett/Getty Images. Page 71: NASA. Page 82: Kelly Wiard/Tire Rack.

Chapter 5
Page 94: Paolo Cocco/AFP/Getty Images. Page 109: Gilbert Iundt/CORBIS. Page 112: Richard McDowell/Alamy.

Chapter 6
Page 122: Jennifer M. Williams. Page 125: Harold & Esther Edgerton Foundation, 2009, courtesy of Palm Press, Inc. Page 139: Tetra Images/Alamy.

Chapter 7
Page 147: Karl Johaentges/Photolibrary. Page 152: NASA. Page 157 (left): Richard Chung. Page 157 (right): Robert Holmes/Alamy. Page 163, Figure 7.20a: University of Washington Libraries. Page 163 (right), Figure 13.25: AP Wide World.

Chapter 8
Page 169: NASA/Goddard Space Flight Center. Page 173, Figure 8.4: NOAA. Page 189: Anthony J Causi/CORBIS.

Chapter 9
Page 199: NASA/JPL. Page 210: NASA/JPL-Caltech. Page 212, Figure 9.15: AIP/Photo Researchers, Inc.

Chapter 10
Page 221: Chris Harvey/Getty Images. Page 227: Jayme Pastoric/U.S. Navy. Page 233, Figure 10.19: Don Farrall/Getty Images. Page 234: Courtesy Pagani Automobili s.p.a. Page 235, Figure 10.22: Courtesy Columbia University Physics Dept. Page 236, Figure 10.24: Herman Eisenbeiss/Photo Researchers, Inc.

Chapter 11
Page 242: Pelamis Wave Power. Page 244, Figure 11.4: Uri Haber-Schaim/Kendall/Hunt Publishing. Page 250: Getty Images.

Chapter 12
Page 263: Bryn Colton/CORBIS. Page 264: Samuel Ashfield/Photo Researchers, Inc. Page 268: Matt Meadows/Peter Arnold Inc. Page 277, Figure 12.14: Stefan Klein/istockphoto.

Chapter 13
Page 283: NOAA. Page 291: Scott Markewtiz/Getty Images. Page 294, Figure 13.15b: Richard Megna/Fundamental Photographs. Page 297, Figure 13.19b: Manuel G. Velarde.

Chapter 14
Page 304: Yann Guichaoua/Photolibrary. Page 317, Figure 14.13: Jim West/Alamy. Page 317 (bottom): Courtesy Richard Wolfson.

Chapter 15
Page 329: Uselmann Foto Design/Photolibrary. Page 333, Figure 15.5: Kristin Piljay. Page 345: Dante Fenolio/Photo Researchers, Inc. Page 349: Orchid Cellmark Inc., Farmers Branch, TX.

Chapter 16
Page 355: Yoav Levy/Phototake. Page 360: Science Photo Library/Photo Researchers, Inc. Page 362, Figure 16.8: USGS. Page 363, Figure 16.12a: David J. Green- electrical/Alamy. Page 368: Cindy Charles/PhotoEdit.

Chapter 17
Page 377: David Young-Wolff/Photo Edit, Inc. Page 385 (top): U.S. Department of Energy. Page 385 (bottom): Toyota Motor Sales, U.S.A., Inc. Page 397, Figure 17.26: Yoav Levy/Phototake.

Chapter 18
Page 403: Steve Bloom Images/Alamy. Page 404, Figure 18.1a: Stuart Field. Page 405: Dr. Terrence Beveridge/Visuals Unlimited/ Getty Images. Page 408, Figure 18.10: Brookhaven National Laboratory/Photo Researchers, Inc. Page 413: Ivan Massar.

Chapter 19
Page 427: NOAA. Page 431: Jack Hollingsworth/Getty Images.

Chapter 20
Page 453: David Nunuk/Photo Researchers, Inc. Page 460: NASA.

Chapter 21
Page 480: Yoav Levy/Phototake. Page 482, Figure 21.3: Martin Bough/Fundamental Photographs. Page 487 (top): Colin Underhill/ Alamy. Page 487, Figure 21.13: Drew Von Maluski. Page 490: Living Art Enterprises/Photo Researchers, Inc. Page 500, Figure 21.38: Giant Magellan Telescope, Carnegie Observatories. Page 500, Figure 21.39b: Omikron/Photo Researchers.

Chapter 22
Page 510: Larry Lilac/Alamy. Page 511, Figure 22.2: Erik Borg/Addison Wesley. Page 512, Figure 22.4a: OVIA IMAGES/Alamy. Page 514, Figure 22.8b: Courtesy of Jay M. Pasachoff. Page 515, Figure 22.9b: M. Cagnet et al. Atlas of Optical Phenomena Springer-Cerlag, 1962. Page 520, Figure 22.18: Reproduced by permission from "Atlas of Optical Phenomena", 1962, Michel Cagnet, Maurice Franco and Jean Claude Thrierr; Plate 18. Copyright (c) Springer-Verlag GmbH & Co KG. With kind permission of Springer Science and Business Media. Page 521: B.A.E. Inc./Alamy. Page 521, Figure 22.21: GIPhotoStock/Photo Researchers, Inc. Page 521, Figure 22.22: Chris Jones. Page 523, Figure 22.26: Omikron/Photo Researchers, Inc. Page 523, Figure 22.28: Richard Megna/Fundamental Photographs.

Chapter 23
Page 531: IBM Corporation. Page 540: Jean-Pierre Clatot/AFP/Getty Images. Page 543:

Index

Bernoulli's equation, 233–234, 235, 238
Bernoulli's principle, 235, *235*
buoyancy, 229–231, *230*, 238, 240
fluid motion, 231–235, 238
fluid pressure, 225–229, 238
measuring density in, 231, *231*
Pascal's principle, 227, 228
Poiseuille's law, 236
rotational motion, 172–173
surface tension, 236, *236*, 238
viscosity, 232, 236, *236*, 237t, 238
fluorine, electron configuration of, 571
flute, 254
flux. *See* magnetic flux
FM. *See* frequency modulation
focal length
corrective lenses, 502
lenses, 494, 495
magnification and, 497
mirror, 483, 485, 486
focal point
compound microscope, 498
lenses, 491
mirror, 483, 484, 486
food calories, 284, 300
food energy, 119
football, 189
force
centripetal force, 84
conservative forces, 105, 108, 112, 115
defined, 87
drag forces, 83, *83*, 98, 111, 112
force diagrams, 67, *67*
fundamental forces, 86, 613–616, 631
momentum, 123–127
motion and, 68, *68*
nature of, 65–66, *66*
net force, 67, *67*, 84, 87, 123–127
Newton's third law of motion, 71
nonconservative forces, 105, 111, 112, 115, 116
normal force, 75, *75*, 76, 78, 85
tension, 77–78, *77*
torque, 182–185, *182-185*
as vector quantity, 66
velocity and, 69
work done by constant force, 95–99, *100*, 116
See also electromagnetic force; gravitational force; strong force; weak force
force diagrams, 67, *67*, *159*
force-vs.-position graph, 100
frame of reference, 18–19
Franklin, Benjamin, 330
Franklin, Rosalind, 522
free-body diagram, 67
free fall
acceleration and, 32–35, 36, 202–203
apparent weightlessness, 216
kinematic equations for, 33, 36
Newton's third law of motion, 71
freezing
entropy of freezing water, 314
heat of transformation, 292

freezing point
calibration of thermometer and, 265
temperature scales, 264
of water, 265
frequency
of electromagnetic waves, 455, 460, 475
in periodic motion, 148, 164
sound waves, 248, 250
waves, 243
frequency modulation (FM), 458, 518
friction, 79–83, 87
causes of, 79
drag forces, 83, *83*, 87
kinetic friction, 79, *79*, 82, 83, 87, 96, 98, 105–106, *106*
rolling friction, 79, 81, *81*, 83, 87
static friction, 79, 81–82, *82*, 83, *83*, 85, 87
viscosity and, 236
frictional forces, 111
fuel cell cars, 281
fulcrum, 185
fundamental constants, 8, 14
fundamental forces, 86, 613–616, 631
fundamental wavelength, 247, 248
fuses, 392, 393
fusion. *See* nuclear fusion
fusion weapons, 604, 605

G

Galilean telescope, 499
Galileo, 33, 34, 51, 68, 70, 158, 181, 200, 203, 491
Galvani, Luigi, 360
gamma cameras, 593
gamma decay, 590, 592, 605, 606
gamma radiation, 590, *590*, 606
gamma rays, *457*, 460, 475, 590, 606
gases
Boyle's law, 271
characteristics of, 221–222
Charles's law, 271
compression of, 305–306, *306*, 324
diffusion in, 277–278, *277*, 279
equipartition theorem, 290, *290*, 300
expansion of, 306
Gay-Lussac's law, 271
ideal-gas law, 270–271, 273, 279
ideal gases, 269–274
kinetic theory of, 274–278, 279
phase changes, 291–295, 300
phase diagrams, 294, *294*
specific heat of, 288–289, 288t, 290, 300
thermal energy of, 276, 300
gasoline, thermal expansion coefficient of, 281
Gay-Lussac's law, 271
Geiger, Hans, 555
Geiger counter, 593, 607
gel electrophoresis, 349, *349*
Gell-Mann, Murray, 621
general relativity, 461, 475, 492

general theory of relativity, 216, 217
generators, 434, *434*, 448
geometrical optics, 480–505
compound microscope, 498, *498*
dispersion, 490–491, *491*, 503, 504
eye and vision, 500–503
law of reflection, 487
magnification, 481, 496–497, *496*
plane mirrors, 481–482, *481*, *482*, 487
reflecting telescopes, 499–500, *499*, *500*, 505
reflection, 481–487
refracting telescopes, 499, *499*, 500, 505
refraction, 480, 487–489, *487*, 503, 504
Snell's law, 488, 504
spherical mirrors, 483–487, *483*, *484*, *485*, 504, *504*
thin-lens equation, 494, 500, 502, 504
thin lenses, 491–496, *491–496*, 504
total internal reflection, 490, *490*, 504
geosynchronous satellites, 212, *212*, 213, 217, 220
germanium, 425
Germer, Lester, 544
Giant Magellan Telescope, *500*
giga- (prefix), 4t
Gilbert, William, 404
glasses. *See* eyeglasses
Global Positioning System (GPS), *214*, 216
global warming, 282
gluons, 623
gold
density of, 7
electron configuration of, 572
golf, 41, 53, 55, 88, 111, *111*, 127
GPS. *See* Global Positioning System
gram (unit), 2
graphite, Compton effect in, 542
grating spectrometer, 519, *519*
gravimeter, 203
gravitation, 199–217
acceleration in free fall, 32–35, 36, 202–203
as "action-at-a-distance" force, 205–206, 614
apparent weightlessness, 58, 215–216, 217
binary stars and galaxies, 215, *215*
Cavendish balance, 204, *204*, 205
gravitational acceleration, 33, 202, 203, 204, 217
gravitational potential energy, 9, 106, *106*, 107, 210–212, 217
Kepler's laws, 207–209, *207–209*, 212, 217
Newton's law of gravitation, 199–206, 208–209, 215, 217
satellites and, 212–215
tides and, 215, *215*, 217, 220
See also gravity
gravitation constant, 201, 205
gravitational acceleration, 33, 202, 203, 204, 217, 338
gravitational field
defined, 338
of Earth, 338–339, *338*
gravitational force, 201, 205t, 614t, 615, 616t